Multidisciplinary Computational Anatomy

Makoto Hashizume

Editor

Multidisciplinary Computational Anatomy

Toward Integration of Artificial Intelligence
with MCA-based Medicine

 Springer

Editor
Makoto Hashizume
Professor Emeritus
Kyushu University
Fukuoka
Japan

ISBN 978-981-16-4327-9   ISBN 978-981-16-4325-5   (eBook)
https://doi.org/10.1007/978-981-16-4325-5

This Springer imprint is published by the registered company Springer Nature Singapore Pte Ltd.
The registered company address is: 152 Beach Road, #21-01/04 Gateway East, Singapore 189721, Singapore

Foreword

Multidisciplinary Computational Anatomy (MCA) is a recent and rapidly evolving field of research which brings new perspectives to medicine in general and surgery in particular. It provides a personalized digital representation of the patient based on his/her medical images and other available information (clinical, biological, lifestyle, environment, etc.) as well as statistical information computed on populations.

This digital representation of the patient's anatomy and physiology provides a powerful basis to assist diagnosis and prognosis with quantitative markers, and to assist the therapy with computational tools for its planning, simulation, and rehearsal. This paves the way of a new era of digital medicine and surgery where the computer is used to augment the capabilities of the physician, not to replace him/her.

I had the privilege to follow the advances of the MCA project leaded by Professor Makoto Hashizume at an international meeting each year during the period 2014–2018. This book presents the cutting-edge and state-of-the-art results obtained by the major actors of this project in Japan. It provides a large and thorough description of the scientific principles of multidisciplinary computational anatomy along with major clinical applications.

The scientific principles include the mathematical, computational, and biophysical models required to describe and analyze the spatial and temporal evolutions of the anatomy through time and populations, measured with various modalities of structural or functional medical images. They also include some of the most recent artificial intelligence methods based on statistical learning from large databases of patients. The book provides the methods to model human and tumor growth, abdominal organs as well as the cranial nervous, cardio-respiratory, and musculoskeletal systems.

The clinical applications cover most of the medical disciplines, including oncology, digestive, and brain surgery, neurology, cardiology, pulmonology, hepatology, radiology, endoscopy, histopathology, embryology, etc. The clinical applications based on MCA models are also well illustrated with a number of emerging innovative imaging technologies.

This book carefully prepared by Prof. Makoto Hashizume and his eminent colleagues is very timely. It is an important contribution to science and medicine and it will be very useful to all the persons interested in digital medicine and surgery, including students, teachers, and researchers in many disciplines including medicine, biology, computer science, and applied mathematics.

The advances presented in this book, through the development of more faithful digital representations of the patients also called "digital twins" or "virtual patients," will definitely contribute to further improve the quality and precision of medical practice. This is for the benefit of *real* patients all over the world. This is the reason why I want to congratulate wholeheartedly all the contributors of this outstanding MCA project.

Nicholas Ayache, Ph.D.
Inria and MICCAI Society
Nice, France
December 2020

Preface

The project "Multidisciplinary Computational Anatomy and its Application to Highly Intelligent Diagnosis and Therapy (MCA: multidisciplinary computational anatomy in short)" was funded by the Japan Ministry of Education, Culture, Sports, Science and Technology (MEXT) Grant-in-Aid for Scientific Research on Innovative Areas in 2014. I acknowledge the MEXT for financial support during the period of 2014–2018 fiscal year. A grant for the international activity was also accepted by MEXT in 2017 and it has accelerated the global initiatives in collaboration with 31 foreign leading centers of excellence (COE) over the world. It greatly contributed to strengthen the scientific levels in mathematical statistics as well as computing and data sciences. Young investigators who were financially supported by this grant had a great opportunity to go abroad to accomplish collaborative research works internationally with those COE and contributed to establish the fundamentals of MCA together with them.

The purpose of this book is to introduce the basics and the state of the art of this technology and its clinical application, and to propose the "MCA-based Medicine." It contains not only the cutting-edge technologies produced by the MCA project, but also the basic mathematics and fundamentals. This book will be helpful and informative for both basic and clinical researchers wishing to systematically survey the state of the art on MCA for challenging to change or to improve the current medicine. MCA is a new frontier of science that establishes a mathematical analysis base for a comprehensive and useful understanding of the "dynamic, living human anatomy," and defines a new mathematical method for early detection and a highly intelligent diagnosis and treatment for the incurable or intractable diseases.

The "Personalized Digital Patient" is a digital representation of the patient's anatomy and physiology based on models whose parameters can be learnt automatically from real or simulated medical images and additional clinical, biological behavioral, and environmental data. Personalized digital patient's modeling pays the most important role in "Surgical Data Science," which allows us to design the management for patients in OR and to support the medical doctors making a final decision during surgery. What is called "Multidisciplinary" is defined as one point of space in digital human body, which composes multiple axes such as scale, function, time, and pathology axes. The clinical significance in each axis contains image resolution, modality of image, temporary phase such as second, minute, hour to age, or pathology of the normal to diseased stage, respectively. It shows multiple attributes or a variety of modality of data on each part of the digital human body. Each part is a representative space of data so that the project aims at establishment of unified notation, technology, or principle and development of clinical application of the human model, such as spatiotemporal model, function-anatomy model, or pathology-anatomy model. The project on MCA comprises scientific research on innovative areas based on medical images integrated with those information related to (1) the spatial axis, from a cell size to an organ size level, (2) the time series axis, from an embryo to postmortem body, (3) the functional axis, on physiology or metabolism which is reflected in a variety of medical image modality, and (4) the pathological axis, from a healthy physical condition to a diseased condition. In order to establish MCA model, all those data must be registered so that the data with multiple attributes are projected on the same virtual space on the 3D axis and a standard spatiotemporal model of the digital human body is formed.

The research group was consisted of three main "planned research groups" (A01, A02, and A03. Each consisted of three study groups and nine study groups in total), and sixty-two "public offered research groups" (thirty-one each during the periods 2015–2016 and 2017–2018, respectively). A01 is a group for foundations of Multidisciplinary Computational Anatomy, A02 is a group for Computer-aided Diagnosis and Surgery Systems, and A03 is a group for Clinical and Scientific Applications. The latter public offered groups which joined us during the 2 year each, were adopted to strengthen the contents of the project. The principle has been technically established by making such MCA scenarios as mathematical methods for MCA modeling; (1) brain MCA modeling, (2) musculoskeletal MCA modeling, (3) lifetime MCA modeling, (4) chest MCA modeling, and (5) abdominal MCA modeling. Owing to their great contribution to development of MCA, the results were introduced in the world-class level international journals and presented at the many international meetings. The total number of pure reviewed papers is more than 900 during the last 5 years. International collaborative research works were so much actively performed that many young researchers were grown up under direction of the world-class leaders at the center of excellence.

A novel registration method was introduced by developing an automatic detection of over 100 anatomical landmarks in medical CT. It showed a fair sensitivity without any abnormal training datasets. Because only images from normal cases were used in the training process, the proposed method requires no manual input of abnormal lesions. Their proposed method is expected to be beneficial in routine CT examination for patients with a cancer. A spatiotemporal statistical model often suffers from a problem caused by sparsely distributed data along a time axis, which makes the statistical modeling difficult. We overcome the difficulty in those statistical modeling by introducing a two-stage modeling algorithm. The first stage maps all data into a feature space with reduced dimension and the second stage performs statistical modeling with q-Gaussian based parameter estimation followed by interpolation of statistics along a time axis. The effectiveness of the proposed modeling approach was demonstrated in the modeling of anatomical landmarks and surfaces of human embryos of Kyoto collection. A spatiotemporal shape model was also developed with a nested constraint and a neighboring constraint. Eventually combination of the both constraints with topological changes was developed and applied into modeling spatiotemporal statistical variations of surfaces of brain, ventricles, and choroid plexuses of human embryos. A super resolution algorithm was developed; one is a dictionary based algorithm and the other is a deep learning based algorithm. A deep learning based super resolution technique and Generative adversarial network for super resolution (SRGAN) were extended to be applicable to three-dimensional (3D) low-resolution (LR) image. It was confirmed that the quality of the super resolution image was 2.42 dB higher than bi-cubic interpolation. Deep learning is now widely available, and redevelopment of the automated segmentation tools using deep learning will greatly improve the accuracy and facilitate their application to the MCA modeling. Therefore, deep learning based automated segmentation tools are developed so that functional and pathological modeling is investigated using automatically segmented anatomical structures. Segmentation accuracy of muscles significantly improved using U-net compared with the previous hierarchical multi-atlas method. Average symmetric surface distance was reduced from 1.75 mm by the previous method to 0.99 mm by U-net. Volume measurement error was reduced from 13.3% to 0.3% in average. A MR-CT image synthesis tool was developed using CycleGAN. Paired CT and MR dataset is not necessary, but just a bunch of unpaired CT and MR data having similar FOVs are used. The MCA modeling for the musculoskeletal system was developed, in which physiological units for functioning (muscle fibers) and transmission systems including connective tissues and physical connections are modeled. The new discipline provides a framework for personalized functional musculoskeletal anatomy modeling, which has not been available before.

Integration of those comprehensive set of medical imaging is required and clinically helpful. A computer needs to seamlessly understand human anatomy from macro- and micro-levels. The images of many different scales must be integrated so that multiscale image registration and navigation are the main topics in establishing the CAD and CAS system based on MCA. Tumor structure and growth analysis over time are good examples of MCA as well as human growth pattern. An automated real-time pathological prediction method was developed for endocytoscopic images. It is expected to contribute to early detection of colonic neoplastic lesion and to reduction of cancer incidence and death rate. They developed extraction method of the 3D microstructure of vascular network from high-resolution dual-energy CT images using high luminance synchrotron radiation micro CT and successfully extracted capillary beds throughout the pulmonary acinus by combination of the dual-energy SRμCT images. Modern medical imaging devices have granted the visualization of metabolic and functional information of a body which have raised the enormous interest in quantitative image analysis. A computational model has been established for functional imaging and a new analysis method was developed by using FDG-PET/CT images. A novel approach was developed using a deep learning technique for 3D CT images. The system enabled to measure metabolic volume, total lesion glycolysis, and effective dose based on the automated segmented results.

Marked development was also found in such clinical application system of the MCA model as surgical navigation, evaluation of the surgeon's skill, semi-automatic surgical robotic system with AI, surgical simulation system, and prediction of therapeutic effect as well as disease specific treatment model. A 4-dimensional human body model enabled deforming the skin and internal structure (organs, skeletal muscles, vasculature, etc.) associated with the whole body movement. In the surgical field, the changes in the position and shape of the target organ are expected to be predicted from the changes in the posture of the subject on the operation bed. A surgical navigation method and a surgical assist-robot have been proposed to guide the operation and relieve the surgeon's workload. A marker-base/marker-less AR navigation system and a compact OMS robot for precise positioning have been developed. They seamlessly integrated these two systems and developed an autonomous surgical system, which could exchange the roles between surgeons and surgical systems by making the robot be the primary operator and the surgeon be the surveillant. They proposed an intelligent autonomous surgical robot to approach the affected surgical area based on MCA model and limited intraoperative biological information.

MCA model is the integrated multiple prediction model with such as multiscale prediction model, temporal prediction model, anatomy-function prediction model, or anatomy-pathology prediction model. Thus, it was planned in a new international network group on MCA to establish the principle of modeling theories of the spatiotemporal anatomy, multiscale anatomy, functional anatomy, and multiscale pathology anatomy in addition to the multidisciplinary registration theory and correlation theory. Emphasis was set to a theory establishment and base construction of the mathematical study foundation, and to expand the range of the new system. The outcome might lead to one of the best solutions to overcome the difficulties in the current medicine. Overall, this topic is a scientific domain connected to the universal problems in people's health. The future perspective is toward the development of human resources as well as a new scientific field of mathematical statistics, information sciences, computing data science, and mechanical- and bio-engineering to medical applications based on MCA.

Finally, I would like to thank all the researchers who participated in the project on MCA. I would also like to express my sincere gratitude to the advisory committee members, Professors Nicholas Ayache (Inria Sophia Antipolis-Mediterranee), David Hawkes (University College London), Junichi Hasegawa (Chukyo University), Hiroshi Iseki (Waseda University), and Yoshihiro Kakeji (Kobe University). I sincerely thank Professor Emeritus Hidefumi Kobatake (Tokyo University of Agriculture and Technology) for his continuous support and heartful advice.

Our goal is to contribute to the promotion of people's health by proposing the "MCA-based Medicine." I hope the concept of MCA would be distributed and developed, expanding the new scientific field of frontier of science over the world.

Fukuoka, Japan   Makoto Hashizume
June 10, 2020

Contents

Part I

Introduction: Perspectives Toward MCA-Based Medicine

From Geometric Models to AI in Computer-Assisted Interventions

David J. Hawkes

Abstract

This chapter describes a programme of work undertaken at UCL and most recently at the Wellcome/EPSRC Centre for Interventional and Surgical Sciences (WEISS) to develop and use a hierarchy of computational models of deformable anatomy that includes pathology to guide biopsy and surgical interventions. The development of appropriate models and the concept of a multiscale hierarchy of complexity and function is described. Recent advances in machine learning that significantly improve usability are presented, leading to the generation of systems to meet unmet clinical needs in the treatment of prostate and liver cancer.

Keywords

Image-guided interventions · Prostate cancer · Liver cancer · Artificial intelligence · Machine Learning · Computer Modelling

1.1 Introduction

This chapter illustrates and contrasts our approach and progress for the minimally invasive MR guided biopsy and treatment of prostate cancer and for minimally invasive laparoscopic liver surgery. Figure 1.1 provides a schematic view of the hierarchical approach to model complexity. The challenges of clinical translation and system validation are described and the chapter finishes with some comments on clinical adoption.

The motivation for this work aligns very well with that of the Multidisciplinary Computational Anatomy (MCA)

D. J. Hawkes (✉)
Wellcome/EPSRC Centre for Surgical and Interventional Sciences, University College London, London, UK
e-mail: d.hawkes@ucl.ac.uk

Project. Both applications described below integrate information across spatial scales including information from pathology in the case of the prostate and dynamic information such as tissue motion and deformation over the time course of the intervention. In particular, this programme of work links to the gastric surgery work in Nagoya and a collaboration was established during a short sabbatical to Professor Mori's laboratory by the author in 2015 [1].

1.2 Prostate Cancer and the Introduction of Multiparametric MRI

In the UK in 2019 deaths from prostate cancer were about 11,900 making it the second most lethal cancer for males. The incidence is rising due to improved detection rates and greater awareness reaching 48,500 in 2019 in the UK, making it the most commonly diagnosed cancer in men [2]. These figures show that the majority of men diagnosed will not succumb to the disease. Overdiagnosis and unnecessary treatment of clinically non-significant disease is a concern.

The work with Professor Mark Emberton and his team at UCLH started in 2005, when it became apparent that an image-guided approach to biopsy and targeted focal therapy was feasible. At that time cancer risk was predominantly determined by clinical signs such as nocturia, poor urinary flow, digital rectal examination and raised prostate-specific antigen (PSA). Diagnosis was by transrectal ultrasound (TRUS)-guided biopsy, an essentially random sampling of the prostate gland. Depending on the results, treatment choices were either do nothing except monitor the patient, radical surgical prostatectomy or whole gland radiotherapy. The first risks escape of metastatic disease and impact on long-term survival, and the latter two lead to a significant risk of damage to critical nearby structures. Until recently the prostate was the only solid organ in which cancer was assessed and treated without imaging to locate disease.

At that time, advances in multiparametric MRI (mpMRI) were showing excellent results in detecting prostate cancer.

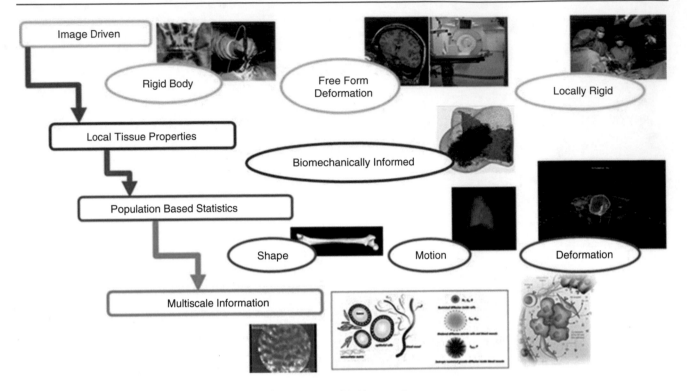

Fig. 1.1 A hierarchical approach in scale and complexity for image-guided interventions

In a significant study of 576 men, mpMRI was subsequently shown to be more sensitive than conventional transrectal ultrasound-guided biopsy [3]. Twenty-seven percent of men might avoid unnecessary biopsy and 18% might be diagnosed with the significant disease who might otherwise have been missed. Importantly mpMRI provides the motivation for our work as it provides a target for image-guided biopsy and, in appropriate cases, image directed focal therapies. While the display of the MR images during interventions can provide very useful guidance to the skilled and experienced operator it is a difficult task and has a steep learning curve. Compensation for the very different orientation and gland distortion due to the insertion of the rectal ultrasound probe is challenging. Below we present our solution to address this.

1.2.1 A Statistical Shape Model of the Prostate for Registration of MR-Derived Information to Transrectal Ultrasound

Models of the outer boundary of the prostate are segmented from T2-weighted MR and meshed to create a 3D model of the prostate, pelvis and rectum. This in turn is registered to the transrectal ultrasound images. This provides a mapping of the delineated pathology seen on the MRI to ultrasound coordinates to enable targeting of the biopsy needle in transperineal biopsy or targeting of focal ablation [4].

Segmentation was originally done manually by expert observers. Finite element (FE) methods were used to generate a model of this anatomy and of the transrectal ultrasound probe. FE methods were then used to generate a large number of deformations simulating a range of positions of the probe within the rectum. Typically, 500 deformations are computed per patient. A statistical motion model (SMM) was obtained by resampling the prostate volume and performing PCA on the resulting deformation coordinates. The SMM was registered to the transrectal ultrasound slices by finding the pose and deformation that maximised directional alignment of surface vectors of the model and the ultrasound detected prostate capsule. Registration was initialised interactively.

This system underwent trial in 8 patients and a Target Registration Error (TRE) of 2.42 mm was achieved using MR and US visible landmarks on 100 trial registrations.

Subsequently, the system was further developed using a generative SMM on a leave-one-out basis. A very similar TRE of 2.40 mm was obtained [5]. This process is illustrated in Fig. 1.2. This innovation meant that the SMM did not need to be regenerated for each subject.

These technologies were incorporated into a commercial system, the Smart Target System (now owned by Intuitive Fusion LLC, Miami, USA). 129 patients were selected with MRI visible lesions. A clinical trial was conducted comparing biopsy guidance using the Smart Target System and conventional visual alignment by the urologist. Each patient had

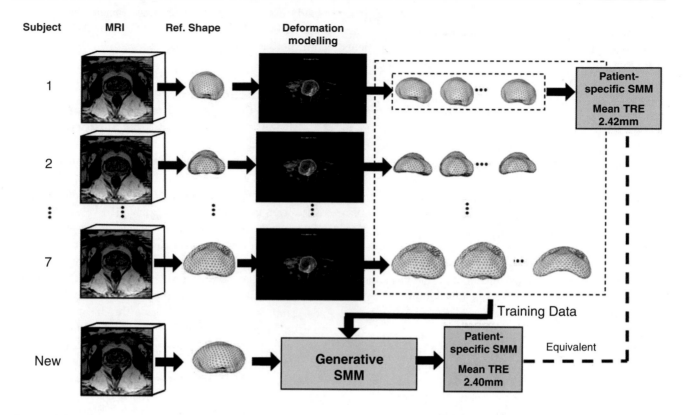

Fig. 1.2 Schematic showing how a generative SMM was used to register TRUS of the prostate with a model derived from mpMRI

both procedures, randomised in order. In total using both procedures 93 clinically significant cancers were detected and each procedure alone detected 80. Each procedure, therefore, missed 13 clinically significant cancers. This trial concluded that the computer-assisted guidance system is equivalent to an experienced urologist performing this task.

1.2.2 A Multi-Scale Model of Pathology

The MR diffusion signal is an imaging probe modulated by cellular scale changes and potentially is sensitive to structural changes associated with the growth of cancer. Although the resulting reconstructed image is at a millimetre scale the information corresponds to micron level changes. A new method developed in our lab, Vascular, Extracellular and Restricted Diffusion for Cytometry in Tumours (VERDICT), models the diffusion in intracellular water, extracellular water and water in the micro-vasculature. Fitting a simple geometric model comprising spheres and cylinders to the simulation of the diffusion signal enables characterisation of the shape and size of these components. This enables derivation of a number of parameters of cellular structure such as the Intracellular Volume Fraction (IVF). This in turn enables inference of changes associated with cancer development and progression [6]. This has recently been applied to cancer

grading in prostate cancer [7]. The signal has been spatially correlated with histopathology [8]. Results of a preliminary clinical trial on 42 men who had VERDICT scans followed by targeted biopsy showed that the estimated IVF was significantly higher in patients with Gleason Grade 3 + 4 and above, compared with Gleason Grade 3 + 3 and lower [9]. This is an important finding as this is the current threshold for determining whether a cancer is clinically significant and hence requiring treatment. The measure was also shown to be sufficiently reproducible to be used as a marker of significant cancer. More work is ongoing to qualify this as a cancer progression biomarker.

1.2.3 Integration of AI Methods

Examination of the clinical workflow reveals three areas where more automation would reduce user interaction, making delivery of the procedure more efficient and potentially more accurate. These comprise the initial interpretation of the mpMRI with identification and delineation of any lesion, delineation of the outer capsule of the prostate to generate the prostate model and finally fully automating the process of registration of that model to TRUS.

One example for automated lesion identification from our laboratory is described in [10] in which 148 men with lesions reported from mpMRI, and graded histologically as Gleason $> = 3 + 3$, were used to train and test a classifier with fivefold cross-validation. The images of 30 patients were then used to test the model. At a specificity threshold of 50%, the sensitivity was 0.93 and 0.88, respectively, which exceeded the performance of three experienced radiologists.

For automated registration of an MR-derived model to TRUS a system using correspondence between identified anatomical structures and ad-hoc landmarks was used to train a convolution neural network [11]. One hundred and eight image pairs from 76 patients were used to train with 12-fold cross-validation. Six or seven patients whose data was held back from training were used for validation. Overall registration accuracy, expressed as the TRE, was 3.6 mm. This is a promising result although not as good as the system described above using the statistical motion model. The advantage is a completely automated system which, once trained, delivered a registration in under one second.

Clearly, further development is needed and improved performance is expected but significant challenges remain. The results described above are from one centre and results must be generalisable to multiple centres, different MRI and TRUS devices and different patient populations, whose cancer risks may evolve over time.

1.3 Unmet Clinical Need: Laparoscopic Liver Surgery

Bowel cancer, leading to liver metastases, has the second-highest cause of cancer death in the UK with an incidence of over 42,000 and a death rate of over 16,000 in the UK. Primary liver cancer has an incidence of over 6000 with a death rate only slightly less and the incidence is rising rapidly [2]. Surgery to remove the cancerous lesion is a potentially curative procedure for many patients. Conventional surgery involves a very large abdominal incision with significant post-operative pain, long-term scarring and long recovery times. Minimally invasive laparoscopic surgery involves insufflation of the abdominal cavity with carbon dioxide followed by insertion through trocars of a laparoscope to provide a viewing port and specialised surgical instruments. Minimally invasive laparoscopy has significant patient and economic benefits in terms of less traumatic surgery and shorter recovery times, but the restricted surgical field of view increases the risk of accidental damage to critical structures such as blood vessels and the bile duct. Intra-operative complications can lead to reversion to conventional abdominal surgery with increased risk of morbidity and mortality. At present relatively few resections are undertaken laparoscopically in the UK due to the higher risk of damage to major blood vessels and the biliary tree. Image guidance with improved visualisation should lead to an increase in the proportion of laparoscopic procedures.

1.3.1 Methodological Steps to a Comprehensive Image Guidance System for Laparoscopic Surgery

CT imaging has been used for many years to guide and plan liver resection with protocols to optimise the hepatic arteries, portal veins and lesion visibility. From these data, the lobes of the liver, the vascular systems and CT visible lesions can be segmented and a 3D model visualised to plan surgical resections, both open and laparoscopic. Commercial systems are available for this task and we used the service provided by Visible Patient SAS, Strasbourg. Navigation is limited as the pose visualised is often far from the surgeon's view and no spatial correspondence between the 3D model and the surgeon's viewpoint is provided. There is no information as to the proximity of surgical tools and instruments to critical structures such as the vasculature, biliary tree and lesion.

Our goal is to establish spatial correspondence between the surgeon's view, the surgical tools he/she is using and the 3D pre-operative model. The challenge is that the liver is a soft deforming structure that moves and changes shape during breathing and whose shape significantly distorts during the procedure.

We have implemented and tested a clinical system based on an interactive alignment of a rigid 3D model with external optical tracking of the laparoscope and other surgical instruments. The system has been tested on a liver shaped phantom, in-vivo on a porcine model that provides an anatomy of similar size and shape to the human, and on nine patients [12]. The registration combines an intuitive manual alignment stage, surface reconstruction from an optically tracked stereo laparoscope and a rigid "iterative closest point" registration to register the intra-operative liver surface to the liver surface derived from CT. As part of this work, we proposed a simple and clinically feasible method for "hand–eye" calibration, which determines the position and orientation of the laparoscope camera relative to the markers attached to the external end of the laparoscope. It is based on a single invariant point viewed on the laparoscopic images [13]. This reduced the contribution of projection errors from 4.1 mm down to 2.0 mm for a laparoscope tracked with external optical markers. An example display is provided in Fig. 1.3.

We showed that surfaces could be reconstructed using video images from an optically tracked and calibrated stereo laparoscope at a rate of 370 ms/reconstruction using GPU technology with an accuracy ranging from 2.4 mm to 5.7 mm on a visually realistic rigid plastic liver phantom [14]. Such a system showed the potential for generating 3D reconstructions

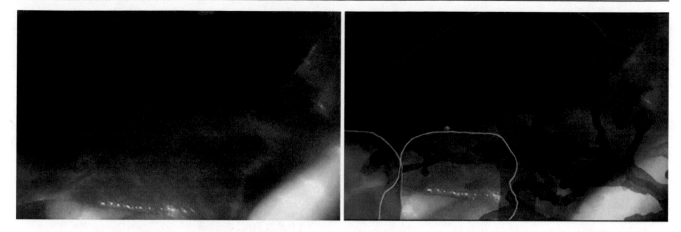

Fig. 1.3 Left image: the right liver lobe as seen through the laparoscopic camera. Right image: augmented view with liver outline (orange), veins (blue and purple), arteries (red), gall bladder (yellow) and a tumour (green) [12].

of liver surfaces without using structured light or other illumination which might be difficult to achieve during laparoscopy. It does, however, require a stereo laparoscope which is not readily available in clinical laparoscopy.

The nine clinical studies enabled us to collect representative laparoscopic data. We define the system accuracy via the re-projection error (RPE), the error of registered pre-operative CT derived data projected onto the video image. The system was capable of accuracies around 12 mm but there was significant variability.

We have also incorporated 2D laparoscopic ultrasound (LUS), taken with the probe in acoustic contact with the liver surface, which provides information on the location of vascular and biliary structures to guide resection avoiding damage to these critical structures. Using a porcine model in-vivo we obtained a mean TRE of 6.6 mm [15]. The landmarks were vessel cross-sections identified in both LUS images acquired with abdominal insufflation and vessel centrelines identified in pre-procedure CT acquired without insufflation. The TRE was reduced to 4.2 mm using a CT scan obtained during insufflation. This proof of concept showed the feasibility of a locally (~10 mm) rigid transformation model although significant practical problems remained, in particular in identifying corresponding vessels.

As in the prostate application above the methods based on deep learning promise significant reductions in user interaction and improved workflow. We have described a method using a CNN architecture comprising fully-convolutional deep residual networks with multi-resolution loss functions to delineate the liver outline from laparoscopic images [16]. The system was trained using 2050 video frames from six liver resections and seven laparoscopic staging procedures. Although this proof of concept was promising with a DICE score of >0.95 there were significant failures and further work is required to fully automate the process.

We recently demonstrated a method based on multi-labelled content-based image retrieval to align a temporal sequence of untracked LUS images of the liver surface and vasculature, with a segmented pre-operative CT scan [17]. Results on data collected on five patients showed that, by including a series of five untracked images in time, a single LUS image can be registered with accuracies ranging from 5.7 to 16.4 mm with a success rate of 78%. This method has the potential to change the paradigm of image-guided laparoscopic liver surgery by removing the need for an external tracker on the laparoscope and laparoscopic ultrasound probe.

These innovations, taken together, could dramatically change how laparoscopic surgery is undertaken, ultimately removing the need for external optical trackers. As with the prostate work, AI promises the means to overcome time-consuming and error-prone user interaction, and speed up certain of the more complex computations with learnt solutions.

1.4   Discussion and Conclusion

Unusually for a University-based Centre the Wellcome/ EPSRC Centre for Interventional and Surgical Sciences (WEISS) at UCL supports a fully ISO 13485:2016 compliant Quality Management System (QMS). This was set up originally for the Smart Target prostate image-guided biopsy system but has now been adopted by the Centre to support the development of regulatory compliant medical software and software-based systems. It is now used in a number of projects. Together with activity to facilitate best practice in technology design, development and evaluation, we can ensure that the safety of prototype devices can be demonstrated, supporting documentation exists and clinical evaluation can take place. This work has lowered the barriers to commercialisation, significantly de-risked development within an academic environment and enabled clinical translation. For the Smart Target system CE marking was obtained in 2016 and FDA Section 510(k) approval in 2017.

The liver work described above was developed with the SciKit-Surgery libraries which form a collection of tools for surgical navigation that can be bound together with Python, making use of standard graphical user interface components for rapid development of novel applications [18]. The liver surgery project also complies with the QMS.

In the last 2 years, two other clinical devices, initially developed within our Centre, have been commercialised. One by Odin Vision Ltd., provides AI-enabled endoscopy, and the other, Echo Point Medical Ltd., provides precision diagnosis and treatment of cardiovascular disease using novel fibre-optic sensor technology.

The rapid progress in machine learning methods has opened up significant opportunities in image-guided interventions and surgical navigation. Many applications have failed to achieve widespread clinical adoption as they are too time-consuming to use and require too much scarce expert time in doing mundane tasks such as image segmentation, lesion identification and assisting in image registration. Recent advances in machine learning are demonstrating that most, if not all, of these tasks can be automated. This will have a major impact on the cost-effectiveness of these procedures and progress in effective clinical implementations are expected.

The caveat is that systems must be reliable in the real world of the clinic. Sufficient training data is needed to ensure system robustness and systems must be validated for local environments. Subtle changes in clinical protocol, small differences in performance of different imaging and sensing technologies, changes in patient presentation and disease profiles can all reduce performance.

Effective training for the surgeons of tomorrow in these new and emerging technologies is paramount and a requirement for their future adoption. Patient-specific computational models and personalised treatment plans are becoming more widespread and will be integrated into training. This could also be used for external consultations and multidisciplinary team meetings.

The work described in this short paper has taken between one and two decades to reach its current level of performance. For the prostate work a commercial, regulatory-approved system is available, but improvements are still being made incorporating the latest AI methods. For the liver work, on the other hand, significant progress in different technologies has been made but a clinically effective system is still to be realised.

The research described in this chapter can be seen to complement well the work undertaken within the MCA project. We have demonstrated a strong multidisciplinary research environment, integrated structural and functional informa-

tion over different spatial scales including pathology and have shown examples of how the dynamics of tissue motion can be integrated.

What is clear in both applications described above is that close working between biomedical engineers, computer scientists and the clinical teams, comprising surgeons, interventionists, theatre nurses, radiologists and pathologists is absolutely vital. We have involved patient groups at all stages of development. This work only succeeds with the determination and passion of the developers to want to make a difference in patients' lives.

Acknowledgements This work was supported by the UK NIHR BRC at UCLH/UCL, HIC Fund (grants HICF-T4-310 and HICF-T4-317), a parallel funding partnership between the UK Department of Health and the Wellcome Trust, the UK NIHR i4i (grant II-LA-1116-20005), EPSRC and the Wellcome/EPSRC Centre for Interventional and Surgical Sciences (grant 203145Z/16/Z). The views expressed in this publication are those of the author and not necessarily those of the Wellcome Trust, NHS, NIHR or Department of Health. The author is grateful for the input from all the authors and co-authors of the cited work, but in particular Dan Stoyanov, Dean Barratt, Matt Clarkson, Yipeng Hu, Laura Panagiotaki, Steve Thompson, Seb Ourselin, Mark Emberton and Brian Davidson.

References

1. Hayashi Y, Misawa K, Oda M, Hawkes DJ, Kensaku MK. Clinical application of a surgical navigation system based on virtual laparoscopy in laparoscopic gastrectomy for gastric cancer. Int J Comp Assis Radiol Surg. 2016;11:827–36. https://doi.org/10.1007/s11548-015-1293-z.
2. Cancer Research UK. https://www.cancerresearchuk.org/health-professional/cancer-statistics/statistics-by-cancer-type/. Accessed 19 March 2021.
3. Ahmed HU, El-Shater Bosaily A, Brown LC, Gabe R, Kaplan R, Parmar MK, et al. Diagnostic accuracy of multi-parametric MRI and TRUS biopsy in prostate cancer (PROMIS): a paired validating confirmatory study. Lancet. 2017;389:815–2. https://doi.org/10.1016/S0140-6736(16)32401-1.
4. Hu Y, Ahmed HU, Taylor Z, Allen C, Emberton M, Hawkes DJ, Barratt D. MR to ultrasound registration for image-guided prostate interventions. Med Image Anal. 2012;16:687–703. https://doi.org/10.1016/j.media.2010.11.003.
5. Hu Y, Gibson E, Ahmed HU, Moore CM, Emberton M, Barratt DC. Population-based prediction of subject specific prostate deformation for MR-to-ultrasound image registration. Med Image Anal. 2015;26:332–44. https://doi.org/10.1016/j.media.2015.10.006.
6. Panagiotaki E, Walker-Samuel S, Siow B, Johnson SP, Rajkumar V, Pedley RB, et al. Noninvasive quantification of solid tumor microstructure using VERDICT MRI. Cancer Res. 2014;74(7):1902–12. https://doi.org/10.1158/0008-5472.CAN-13-2511.
7. Panagiotaki E, Chan RW, Dikaios N, Ahmed HU, O'Callaghan J, Freeman A, et al. Microstructural characterization of Normal and malignant human prostate tissue with vascular, extracellular, and restricted diffusion for cytometry in Tumours magnetic resonance imaging. Investig Radiol. 2015;50:218–27. https://doi.org/10.1097/RLI.0000000000000115.
8. Bailey C, Bourne RM, Siow B, Johnston EW, Appayya MB, Pye H, et al. VERDICT MRI validation in fresh and fixed prostate specimens using patient-specific moulds for histological and MR

alignment. NMR Biomed. 2019;32:e4073. https://doi.org/10.1002/nbm.4073.

9. Johnston EW, Bonet-Carne E, Ferizi U, Yvernault B, Pye H, Patel D, et al. VERDICT MRI for prostate cancer: intracellular volume fraction versus apparent diffusion coefficient. Radiology. 2019;291:391–7. https://doi.org/10.1148/radiol.2019181749.

10. Antonelli M, Johnston EW, Dikaios N, Cheung KK, Sidhu HS, Appayya MB, et al. Machine learning classifiers can predict Gleason pattern 4 prostate cancer with greater accuracy than experienced radiologists. Eur Radiol. 2019;29:4754–64. https://doi.org/10.1007/s00330-019-06244-2.

11. Hu Y, Modat M, Gibson E, Li W, Ghavami N, Bonmati E, et al. Weakly-supervised convolutional neural networks for multimodal image registration. Med Image Anal. 2018;49:1–13. https://doi.org/10.1016/j.media.2018.07.002.

12. Thompson S, Schneider C, Bosi M, Gurusamy K, Ourselin S, Davidson B, et al. In vivo estimation of target registration errors during augmented reality laparoscopic surgery. Int J CARS. 2018;13:865–74. https://doi.org/10.1007/s11548-018-1761-3.

13. Thompson S, Stoyanov D, Schneider C, Gurusamy K, Ourselin S, Davidson B, et al. Hand–eye calibration for rigid laparoscopes using an invariant point. Int J CARS. 2016;11:1071–80. https://doi.org/10.1007/s11548-016-1364-9.

14. Totz J, Thompson S, Stoyanov D, Gurusamy K, Davidson BR, Hawkes DJ, Clarkson MJ, Totz, et al. Fast semi-dense surface reconstruction from stereoscopic video in laparoscopic surgery. In: Stoyanov D, Collins DL, Sakuma I, Abolmaesumi P, Jannin P, editors. Information processing in computer-assisted interventions. IPCAI 2014. Lecture notes in computer science, vol. 8498. Cham: Springer. https://doi.org/10.1007/978-3-319-07521-1_22.

15. Song Y, Totz J, Thompson S, Johnsen S, Barratt D, Schneider C, et al. Locally rigid, vessel-based registration for laparoscopic liver surgery. Int J CARS. 2015;10:1951–61. https://doi.org/10.1007/s11548-015-1236-8.

16. Gibson E, Robu MR, Thompson S, Edward P, Schneider C, Gurusamy K, et al. Deep residual networks for automatic segmentation of laparoscopic videos of the liver. In: Webster III RJ, Fei B, editors. Medical imaging 2017: image-guided procedures, robotic interventions, and modeling. Proc. of SPIE, vol. 10135; 2017.

17. Ramalhinho J, Tregidgo HFJ, Gurusamy K, Hawkes DJ, Davidson B, Clarkson MJ. Registration of untracked 2D laparoscopic ultrasound to CT images of the liver using multi-labelled content-based image retrieval. IEEE Trans Med Imaging. 2021;40:1042–54. https://doi.org/10.1109/TMI.2020.3045348.

18. Thompson S, Dowrick T, Ahmad M, Xiao G, Koo B, Bonmati E, et al. SciKit-surgery: compact libraries for surgical navigation. Int J CARS. 2020;15:1075–84. https://doi.org/10.1007/s11548-020-02180-5.

Part II

Basic Principles of MCA: Fundamental Theories and Techniques

A Concept of Multidisciplinary Computational Anatomy (MCA)

2

Yoshitaka Masutani

Abstract

Multidisciplinary Computational Anatomy (MCA) is a multidisciplinary domain that literally extends Computed Anatomy (CA). The CA itself is relatively new, and MCA, which is its further extension, is a new academic field that is expected to develop in the future. It is important to have a bird's eye view and understanding of these academic fields in order to understand the contents of this book. Therefore, this section outlines the background from CA to MCA and the ideas in these disciplines.

Keywords

Computed anatomy · Multidisciplinary computational anatomy · Data representation space · Multi-attribute data space

2.1 Computational Anatomy

Computed Anatomy (CA) can be regarded as a radical enhancement of the quantitativeness and reliability of traditional anatomy through advanced data processing based on computer science, mathematics, and statistical methods. Similar to recent science and technology including "computational physics," "computational astronomy," and "computational biology," CA treats hypothesis-based analytical model construction. In addition, it approaches the essence of the human body based on large amounts of data processing and numerical calculations. The target data of CA is mainly the morphology of structures in the human body of various scales such as organs and tissues.

This is largely due to the dramatic development of computer technology, and medical imaging devices such as X-ray CT, MRI, and microscope.

The former enabled the processing of large amounts of data at high speed while the latter realizes accurate and precise acquisition of morphological information of structures in various scales. One of the major achievements of CA is the statistical shape model (SSM) of various organs and its application, which is based on the shape representation by using statistical methods such as principal component analysis (PCA) [1]. Although there are individual differences in the shape of organs, their general shapes are highly similar and it is often easy to determine the approximate correspondence between the shapes of each individual. Based on the correspondence among the sample shapes, SSM decomposes a group of shapes that have a certain tendency even though there are individual differences into common components and expresses the shape by synthesizing them. The generated instance of SSM is a compact representation of the "probable shape" of a specific organ and can enhance the reliability of organ recognition, which is a segmentation of the organ in new image data. That is, SSM plays a role as an expression of prior knowledge for understanding medical images by a computer. SSM-based segmentation of organs in medical images results in an optimization problem centered on determining weights in the synthesis of each shape component. Furthermore, the weights for shape components obtained in the process of patient-specific organ segmentation represent the shape characteristics peculiar to the target individual and can be used for diagnostic support for diseases associated with organ morphological changes such as liver fibrosis [2]. On the other hand, there are cases where it is not simple to determine the correspondence among the sample shapes for the construction of SSM. In particular, when the structural topology differs between individuals called "anomaly," for example, when the number of vertebrae differs, it is impossible to determine the correspondence. Computational anatomy also tackles these kinds of problems.

Y. Masutani (✉)
Graduate School of Information Sciences, Hiroshima City University, Hiroshima, Japan
e-mail: masutani@hiroshima-cu.ac.jp

The above analysis methods and concepts are an extension of traditional anatomy and biological morphology measurement called "morphometry." However, the CA approaches enable high-definition and large-scale data processing such as X-ray CT and MRI. It was realized for the first time and is the essence of CA. The MEXT grant project "Computational Anatomy[1]" in Japan [3] is aimed at building a CA model of organs based on a large amount of X-ray CT data of healthy adults, and also developing mathematical modeling methods such as SSM. Furthermore, the purpose was to develop these into medical image understanding and consequently support for diagnosis and treatment. Overseas research on the analysis of anatomical structures based on large-scale databases, such as the ADNI (Alzheimer's Disease Neuroimaging Initiative) project [4], has been widely conducted, and the field of CA continues to develop globally.

2.2 Multiplicity of Computational Anatomy

Here, we will describe what the multiplicity of CA, that is, Multidisciplinary Computational Anatomy (MCA) means, and how the subject of MCA differs from that of CA.

In general, "multiplicity" means increasing the elements or ideas and perspectives of things to two or more. In this case, the "element" corresponds to the conceptual or abstract thing of the element, way of thinking, or viewpoint of things, but in MCA, the "element" clearly refers to the type or attribute of data. Unlike previous CA approaches, we handle data groups of a wide variety of images such as multiple scales (resolution), multiple modalities (image types), multiple time phases (time-age, etc.), and multiple pathologies including healthy at the same time. This means that multiple data located at different coordinates on the "axis" often represented by concepts such as "scale axis," "modality axis," "time axis," and "pathology axis" are handled in a unified manner.[2] In general, conventional anatomy deals only with the morphology and structure of healthy organs and tissues, but one of the major features of MCA could be that it handles pathological tissues and structures due to diseases.

[1]"Computational Anatomy (CA)" and "Multidimensional Computational Anatomy (MCA)" refer to academic fields, but both are also the names of grant-in-aid research projects of the Ministry of Education, Culture, Sports, Science and Technology. Therefore, when the content of the project is the target, it is written as "~ project."

[2]It is often confused with "multi-axis" because it sets many axes such as scale, modality, age, and pathology. However, according to this idea, the previous CA is a "single-axis" CA, and it is necessary to always target data having multiple coordinates on any single axis. This is not always the case in CA research, and it is more natural to base it on points in the data representation space defined in this section.

2.3 Data Representation Space for MCA

The concept difference between CA and MCA can be understood when considered in the space composed of the above-mentioned multiple axes (Fig. 2.1). This space is defined as "data representation space." That is, it is based on a point in the data representation space, and one point in the space indicates data of a specific scale, modality, time phase, pathological condition, and other types and attributes. However, even if the data is from multiple patients, if the data has the same type and attributes, it is only one point in the space. Previous CA uses a large amount of data represented by a single element in the data representation space. That is, it deals with a limited subject of the disease such as CT abdominal images of healthy adults or brain MRI of Alzheimer's disease patients, to identify specific organ shapes and structures. On the other hand, in MCA aiming for a comprehensive understanding of the human body, we handle the wide variety of data as multiple points in the data representation space. In the MCA project up to now, the data representation space is constructed assuming the four axes of scale, function, time, and pathology mentioned above. However, it is also possible to expand the data representation space by adding new axes, for example, race, place of residence, and era of survival, etc. In addition, various interpretations and definitions are established for each axis, such as the case where it represents a continuous quantity such as time and scale, and the case where it contains discrete labels such as function and pathology. The pioneering part is also large, and future development is expected.

2.4 From Data Representation Space to Multi-Attribute Data Space

The data representation space is important for conceptually understanding multidimensional data, but further data processing is required for quantitative analysis and modeling that leads to an understanding of the human body. That is, registration (alignment) between a large amount of image data. This is also essential for the data handling in previous CA, and a model like SSM is constructed by associating the data of many individuals with anatomical landmarks [5]. In MCA, it is necessary to register all data such as different scales, different time phases, and different modalities in the same individual, not only between individuals. As a result, multiple and many individual data groups are mapped to the human body standardized in space-time, and multi-attribute data obtained from multiple individuals is distributed at certain spatiotemporal coordinates (x, y, z, t). A space can be newly constructed (Fig. 2.2). This conceptual human body is defined as a "spatiotemporal standard human body," and the space of data composed of any points thereof is called a "multi-attribute data space."

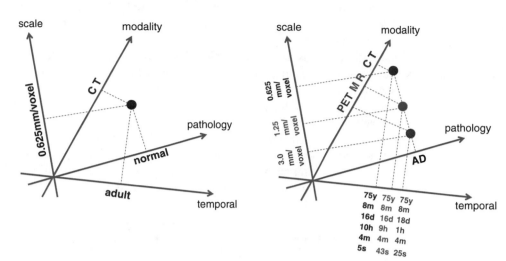

Fig. 2.1 Multiplicity in the data representation space

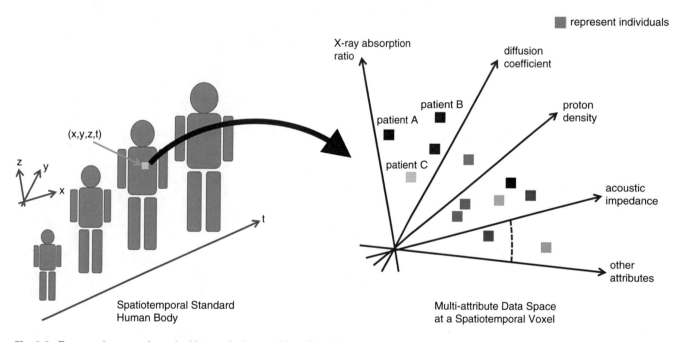

Fig. 2.2 From spatiotemporal standard human body to multi-attribute data space

The ultimate purpose of MCA is to find knowledge to understand the human body itself from the image data of many individuals with multiple attributes and to contribute to the advancement of diagnosis and treatment. For that sake, the following three steps are needed. The first step is to construct a spatiotemporal standard human body through registration between a large amount of data. The next is to find a certain distribution and law from a large number of data groups distributed in the multi-attribute data space at each position. Finally, it is necessary to perform appropriate modeling based on them. If we call the result an "MCA model," it can be said that this is a model that represents a kind of virtual human body and virtual patient.

2.5 Goal of MCA

As the final part of this section, the differences between CA and MCA and the goals of the latter are clarified with concrete examples.

In the previous CA approach, so-called "intergroup comparison" research is widely conducted, in which clinical image data of a subject consisting of two groups, such as patients and normal controls, is analyzed and the significant difference in features obtained from each group is compared. It is a typical example of research that makes comparisons using the ADNI database mentioned above. In the studies, VBM (Voxel-based morphometry) [6] was used to determine the volume of each part of the brain between the patient data group of Alzheimer's disease (AD) and the healthy person (control) data group. One of the purposes of this approach is to identify sites with significant differences between the two groups based on volume statistics (mean and standard deviation) of each site, that is, sites with large volume changes due to AD. Therefore, it can be regarded as an MCA approach because it targets binary data along the pathological axis of disease state and healthy state, but it is slightly insufficient in a sense for the following reasons. The goal of MCA is integrated modeling that includes the mechanism of disease development and progression that leads to an understanding of the human body itself. The above approach merely models and compares each group independently although it can be applied to the discrimination of diseases. However, it does not cover the modeling such as continuous changes of the image features due to the progression of the disease and also the mechanism of its development. Beyond the binary state description of healthy and disease, further modeling makes it possible to discuss the mechanism of progression, which are expression and regression analysis of the severity and progression of the disease by grade and stage, based on the features obtained from multi-modal images. That is, it turns out that a comparison of the two groups is too simple and not enough.

2.6 Challenge to "Comprehensive Understanding of the Human Body"

It is clear that the goal of MCA, which aims to comprehensively understand the human body, is in line with the goal of medical science. However, a major feature of MCA is that it is based on rich information supported by a large amount of multidimensional image data. On the other hand, as an emerging approach for analyzing medical images based on a large amount of similar data, a group of deep learning methods such as Deep Convolutional Neural Network (DCNN) have been attracting attention and being widely used in recent years due to their high performance [7]. It can be said that the training result implicitly expresses some knowledge about the human body captured in the image, but the correspondence with the systematic medical knowledge is not clear at present. That is, we just examine the output to use it. On the other hand, as mentioned above, the approach of CA

and MCA is an attempt to express knowledge explicitly by various models and to lead to an understanding of the image data of the human body and the human body itself. From the above, it seems that the methods of image analysis by deep learning and the approach of CA are contradictory at present, but it will be clarified in the future by detailed analysis of the training process and results of deep learning, which is being clarified. It is considered that it can be fused as a subset that supports the knowledge of CA and MCA.

What has been clarified so far in the MCA Project is that there are many technical problems to be solved, such as registration between images of different scales, which have not been dealt with in the past. There is also the problem that the amount of data required to construct a spatio-temporal standard human body and build a highly reliable model from a multi-attribute data space is insufficient. Although it may be difficult to complete a multidimensional anatomical model within the project period, some attempts have been made to build partial models. The approach of constructing a partial model based on multiple data, which is limited to a specific region or around the disease, will be integrated and developed to cover multiple diseases and the whole body in the future for the aim of completion of the MCA model.

This section outlines the background and concepts of a new discipline, MCA. In the following sections, some of the results of the "MCA Project", that is, the above attempts, will be shown more concretely, and the goals and current issues of the MCA will be further clarified.

References

1. Heimann T, Meinzer HP. Statistical shape models for 3D medical image segmentation: a review. Med Image Anal. 2009 Aug;13(4):543–63.
2. Hori M, Okada T, Higashiura K, Sato Y, Chen YW, Kim T, Onishi H, Eguchi H, Nagano H, Umeshita K, Wakasa K, Tomiyama N. Quantitative imaging: quantification of liver shape on CT using the statistical shape model to evaluate hepatic fibrosis. Acad Radiol. 2015;22(3):303–9.
3. Kobatake H, Masutani Y. Computational anatomy based on whole body imaging: basic principles of computer-assisted diagnosis and therapy. Springer; 2017.
4. Mueller SG, Weiner MW, Thal LJ, Petersen RC, Jack CR, Jagust W, Trojanowski JQ, Toga AW, Becket L. Ways toward an early diagnosis in Alzheimer's disease: the Alzheimer's disease neuroimaging initiative (ADNI). Alzheimers Dement. 2005;1:55–66.
5. Hanaoka S, Shimizu A, Nemoto M, Nomura Y, Miki S, Yoshikawa T, Hayashi N, Ohtomo K, Masutani Y. Automatic detection of over 100 anatomical landmarks in medical CT images: a framework with independent detectors and combinatorial optimization. Med Image Anal. 2017;35:192–214.
6. Ashburner J, Friston KJ. Voxel-based morphometry–the methods. Neuro Image. 2000;11(6):805–21.
7. Greenspan H, van Ginneken B, Summers RM. Guest editorial deep learning in medical imaging: overview and future promise of an exciting new technique, IEEE trans. Med Img. 2016;35(5):1153–9.

Construction of Multi-Resolution Model of Pancreas Tumor

3

Hidekata Hontani, Tomoshige Shimomura,
Tatsuya Yokota, Mauricio Kugler, Tomonari Sei,
Chika Iwamoto, Kenoki Ohuchida,
and Makoto Hashizume

Abstract

In this section, a method for constructing a multi-resolution model of pancreas tumor is described. The model is constructed from MR images of a KPC mouse that has a pancreatic tumor and a set of microscopic pathological images of the tumor. The multi-resolution model (explicitly or implicitly) represents the simultaneous probability density distribution of MR images and corresponding microscopic pathological images and would help to predict the distribution of pathological images that would be corresponded with each voxel in a pancreatic tumor region of a given MR image. Registration of the MR image and the pathological images is needed for the model construction, and we at first reconstruct a 3D microscopic pathology image of the pancreatic tumor from the set of pathology images. These images are obtained by slicing the extracted tumor into spatially continuous thin sections and by capturing each section under a microscope, and non-rigid registration is required for the 3D reconstruction from these microscopic images. A newly developed method for the non-rigid registration and a GAN-based method for the construction of the multi-resolution model is briefly described. In addition to the construction of the multi-resolution model, a non-rigid image registration method would be useful for constructing a temporal model that represents the growth of the tumor inside the body. Given a temporal series of MR images, one would be able to construct the temporal model by non-rigidly registering the MR images captured at different times for describing the temporal change of the tumor. For the description of the temporal change, one needs to non-rigidly register the organs around the tumor in the MR images, and the description of the temporal change is not easy because the non-rigid registration deforms not only the organs but also the tumor: The deformation of the tumor makes it difficult to describe its temporal change accurately. In this section, one non-rigid registration method that handles this problem is also briefly described.

3.1 Introduction

In this section, we describe a multi-resolution model of pancreatic cancer tumors. One of the objectives of the multidisciplinary computational anatomy project was to develop methods for constructing statistical models of the human body by integrating medical images along spatial, temporal, functional, and pathological axes. The multi-resolution model of pancreatic cancer tumors described in this section is constructed by integrating low-resolution MR images and high-resolution pathological microscopy images. This is an example of medical image integration along the spatial axis. MR images are low resolution but can be captured non-invasively. The microscopy images, on the other hand, can be captured only invasively but can be used for the definitive diagnosis of disease because of their high resolution. If the MR images and the corresponding pathological ones are statistically not independent, it may be possible to predict one from the other. If the pathological images can be predicted from MR images, it may be useful in clinical applications. In the following, we describe the construction of a multi-resolution model that can be used to predict candidate pathological microscopy images corresponding to each voxel of the tumor region in the MR image. The distribution of the candidate pathological images would change with respect to the voxel location in the tumor. One key technique for the model construction is the registration between the

H. Hontani (✉) · T. Shimomura · T. Yokota · M. Kugler
Nagoya Institute of Technology, Nagoya, Japan
e-mail: hontani@nitech.ac.jp

T. Sei
The University of Tokyo, Tokyo, Japan

C. Iwamoto · K. Ohuchida · M. Hashizume
Kyushu University, Fukuoka, Japan

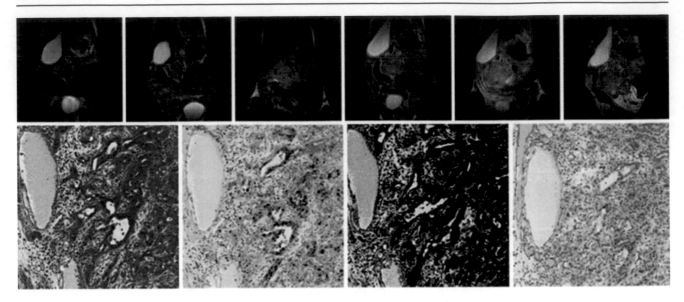

Fig. 3.1 Top: Temporal series of MR images of the KPC mouse. Bottom: Spatial series of pathological microscopic images of the extracted pancreatic cancer tumor (From left to right: H&E, Ki67, MT, and CD31)

MR image and pathological microscopy images. A deep generative model is used for the prediction of the candidate pathological images.

3.1.1 Data Description

In this study, we use a set of images of KPC mice, which are genetically engineered and are widely used as models for human pancreatic cancer research [1]. A temporal series of eight T1-enhanced MR images of the whole body of a KPC mouse was captured once a week. These images will be used for modeling of the growth of pancreatic cancer tumor and will be integrated with the multi-resolution model described in this section. The construction of the tumor growth model and the integration with the multi-resolution model are included in future works. The pancreatic tumor was extracted just after the last MR image was captured, and whole of the tumor was sliced into thin sections. The slices were then stained with four different stains (H&E, Ki67, MT, and CD31), and each of the stained slices was captured by a microscope. The spatial resolution of the MR images was 1.536 mm × 1.536 mm × 0.5 mm, and that of the microscopic images was 0.15 μm × 0.15 μm. The slice thickness was about 4. The microscopic images of the thin sections were used for reconstructing a 3D microscopic image of the extracted pancreas tumor: The microscopic images of the thin sections of the tumor were non-rigidly registered and were piled in order of the spatial series of the sections. Fusing the tumor region in the MR image captured at last and the reconstructed 3D microscopic image of the tumor, we finally

construct a multi-resolution model of the pancreatic tumor. Examples of images are shown in Fig. 3.1.

3.1.2 Image Registration for the Description of Tumor Growth

As described above, one key technique for the construction of the multi-resolution model is image registration. The image registration is important not only for the construction of the multi-resolution model but also for the construction of the tumor growth model. The criteria used for image registration will vary depending on its purpose. In this subsection, we describe an image registration method used to describe the temporal changes in position and shape of a pancreatic cancer tumor in the body by using a temporal series of MR images to show the difference from the criteria for the image registration methods used to construct the multi-resolution model, which will be described later.

Before the image registration of the MR images, we manually annotated the regions of the abdominal organs and the pancreatic tumor in the MR images. Figure 3.2 shows some of the annotated regions. As shown, one can observe the tumor region (***) is increased in size over time. The objective here is to describe the change in position and shape of the tumor in the body. The tumor growth would be spatially anisotropic, and the position of the tumor in the body changes over time. This change is caused by factors due to the non-rigidity of the body and factors associated with the anisotropic growth of the tumor. For the description of the tumor growth, it would be useful to explicitly distinguish between

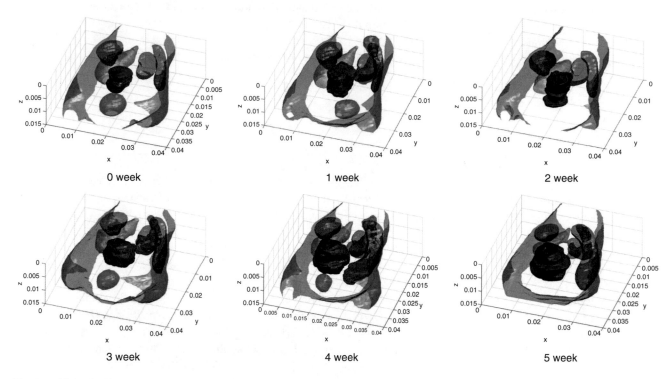

Fig. 3.2 Abdominal organ regions and tumor region manually labeled in the temporal series of the MR images. The regions of pancreatic cancer tumor (blue), kidneys (red), liver (orange), spleen (purple), and bladder (green) are indicated [2]

these two factors: non-rigid deformation of the body and the temporal change of the shape and size of the tumor. In order to compensate only the deformations due to the non-rigidity of the body and to explicitly describe only the temporal change caused by the tumor growth, we proposed an image registration method that can non-rigidly register the abdominal organ regions in the two given images while keeping the shape of the tumor in each image unchanged.

A large deformation diffeomorphic metric mapping (LDDMM) is one of the most widely employed methods for the non-rigid registration of images. Given two images, $I_0(\boldsymbol{x})$ and $I_1(\boldsymbol{x})$, the method can obtain a diffeomorphic mapping, ϕ_1, between them by minimizing the following cost function:

$$E(\boldsymbol{v}) = \int_0^1 \| \boldsymbol{v} \|_V^2 \, dt + \lambda \| I_0 \circ \phi_1^{-1} - I_1 \|^2, \tag{3.1}$$

where

$$\phi_t(\boldsymbol{x}) = \int_0^T \boldsymbol{v} dt + \phi_0 \tag{3.2}$$

for $t \in [0, 1]$ and ϕ_0 is an identity mapping. Let L denote a differentiation operator, then

$$\boldsymbol{f}.\boldsymbol{g}_V = L\boldsymbol{f}, L\boldsymbol{g}_{L^2}, \tag{3.3}$$

and $\| \boldsymbol{v} \|_V^2 = \boldsymbol{v}, \boldsymbol{v}_V$. The first term in the right-hand side is a regularization term and makes the deformation field, $\boldsymbol{v}$,

smooth enough so that the resultant mapping, ϕ_1, is diffeomorphic. The second term measures the distance between the deformed image and the other given one.

Applying the LDDMM method to the two MR images captured consecutively at different times, one can obtain the diffeomorphic mapping, ϕ_1, that maps from the older MR image to the other new one. Let I_0 and I_1 denote MR images captured earlier and later, respectively and let the tumor region in I_0 and I_1 be denoted by Ω_0 and Ω_1, respectively. As mentioned above, the abdominal organ regions located around the pancreatic tumor in I_0 and I_1 are manually annotated. The obtained diffeomorphic mapping, ϕ_1, would map each point, $\boldsymbol{y}_0^i$, in Ω_0 to a corresponding point, $\boldsymbol{y}_1^i$ in Ω_1, where $i = 1, 2, ..., N$ denotes the ID numbers of the points, and the deformation vector field, $\delta_{01}^i = \boldsymbol{y}_1^i - \boldsymbol{y}_0^i$ ($i = 1, 2, ...$), describes the temporal change of the tumor region. The mapping of the tumor region that is represented by the vector field, $\boldsymbol{v}_{01}^i$, though, would not be an identity map even when the size and shape of the tumor is unchanged between the two MR images were captured because the pose and position of the tumor would be changed. For describing the tumor growth, we need to explicitly compensate only the components of the deformation field caused by the non-rigid deformation of the body.

We modified an LDDMM method to develop a new method that rigidly registers pre-specified regions in given two images and non-rigidly registers the other regions. The new method can obtain a defeomorphic mapping from the older MR image, I_0, to the newer one, I_1, while keeping the shape and size of the tumor region, Ω_0, in I_0. Let $I_0' = I_0(\boldsymbol{x}')$ and $I_1' = I_1(\boldsymbol{u}')$, where $\boldsymbol{x}' \notin \Omega_0$ and $\boldsymbol{u}' \notin \Omega_1$, respectively. Let U denote a rotation matrix and $\boldsymbol{t}$ denote a translation vector. Let $\boldsymbol{q}_0^j$ ($j = 1, 2, \ldots, M$) denote points randomly and densely located inside the tumor region, Ω_0, in I_0. The newly developed method minimizes the following cost function [2]:

$$E_{\text{new}}(\boldsymbol{v}, U, \boldsymbol{t}) = \int_0^1 \| \boldsymbol{v} \|_V^2 \, dt + \lambda \| I_0' \circ \phi_1^{-1} - I_1' \|^2 + \gamma \sum_{j=1}^{M} \| \boldsymbol{t} + U\boldsymbol{q}_0^j - \phi_1 \circ \boldsymbol{q}_0^j \|_{L^2}^2, \tag{3.4}$$

where the third term in the right-hand side denotes the difference between the points rigidly mapped by U and $\boldsymbol{t}$ and those non-rigidly mapped by ϕ_1. The tumor region, Ω_0 is mapped by U and $\boldsymbol{t}$. Once the two MR images are registered by this method, the non-rigid deformation inside the body is compensated while the size and shape of the tumor region, Ω_0 in the older MR image, is maintained. Let Ω_0' denote the region obtained by transforming Ω_0 with the rotation, U, and the translation, $\boldsymbol{t}$. Then the residual difference between Ω_0' and Ω_1 can be interpreted as being due to the tumor growth. It should be noted that if the tumor keeps its size and shape between the two MR images are captured, and no residuals would be observed between Ω_0' and Ω_1.

3.1.3 3D Reconstruction of Microscopic Tumor Image

As described, the tumor extracted from the KPC mouse was divided into a spatial series of thin sections, each of which was stained and imaged under a microscope. Microscopic images are 2D, and MR images are 3D. We need to reconstruct a 3D microscopic image of the tumor in order to register each of the tumor regions in the microscopic images to the corresponding portion of the tumor region in the MR image. One can obtain a 3D microscopic image by simply stacking the 2D microscopic images, but the resultant spatial pattern of the tumor in the 3D microscopic image would be discontinuous and not smooth because of the translation, rotation, and non-rigid deformation of the thin sections. Non-rigid registration between the images of the neighboring thin sections is required to obtain an appropriate 3D microscopic image.

The smoothness of the spatial pattern of the reconstructed image is used as the criterion for the non-rigid registration. The similarity of pixel values, which is often used as the criterion for the non-rigid registration between images, should not be employed for the reconstruction of 3D microscopic images. Let the vertical direction along which the 2D microscopic images are stacked be represented by the z-axis. When the similarity of voxel values is employed as the criterion, voxels having similar values are aligned along the z-axis in the reconstructed 3D images and, as a result, each of the anatomical structures, such as the blood vessels, would have a linear, unnatural structure along the z-axis direction.

Landmarks corresponding between images are used for determining the deformation of images in the registration. Let the series of the 2D microscopic images be denoted by $J_j(\boldsymbol{x})$ ($j = 1, 2, \ldots, K$) and let the corresponding landmarks detected in J_j be denoted by $\boldsymbol{y}_j^l$. Let $\mathcal{T}^l$ denote the l-th trajectory of corresponding landmarks found through images. For the detection of the corresponding landmarks, we employ a simple template matching, in which the similarity is evaluated by the normalized cross-correlation (NCC). A set of the landmarks are randomly generated at portions in which the spatial gradient of the image is higher than the threshold T (i.e., $\| \nabla J_j \| > T$), and the corresponding landmark is detected by locally searching the point that has the maximum NCC value in the next image J_{j+1}. Let $\mathcal{T}^l = \left\{ \boldsymbol{y}_{j_l}^l, \boldsymbol{y}_{j_l+1}^l, \ldots, \boldsymbol{y}_{\dot{j}_l}^l \right\}$ denote a trajectory of the l-th landmark found through the j_l-th image, J_{j_l}, to the $\dot{j}_l$ image, $J_{\dot{j}_l}$, where $j_l < \dot{j}_l$ (see Fig. 3.3). The shape of each trajectory is not smooth because of the transformation of the thin sections. It should be reminded that each microscopic image, $J_j(\boldsymbol{x})$, is an image of the cross-section of the tumor and the landmarks corresponded in different images are located on the cross-sections of an identical anatomical structure. The proposed method non-rigidly deforms every given image so that all of the tra-

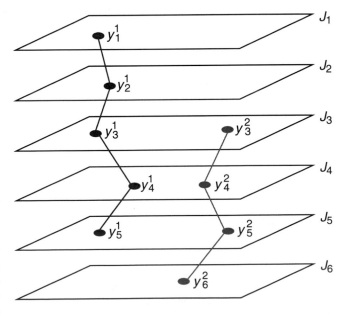

Fig. 3.3 Trajectories, $\mathcal{T}^1$ and $\mathcal{T}^2$, detected through images

jectories become smooth, and hence anatomical structures in the resultant 3D image would be smooth.

The trajectory-based method mentioned above works stably and accurately only if the detection of the corresponding landmark can be given up when the corresponding landmark is not present in the 2D microscopic image because of artifacts. Actually, one can find many artifacts, such as folds and wrinkles of the specimen and image blurs. Figure 3.4 shows an example of the image of a folded specimen. The artifacts obscure the true corresponding landmarks that would be detected if the artifacts did not exist, and corresponding false landmarks would be detected when the true corresponding landmarks are obscured. In order to determine that no corresponding landmark is found, the proposed method applies backward template matching. Given a source landmark y_j^l in the j-th image, J_j, the forward template matching detects the corresponding landmark, y_{j+1}^l, that has the maximum NCC value in the next image, J_{j+1}. Then a local image patch centered on the detected point, y_{j+1} in J_{j+1}, is set as a template, and the backward template matching detects a corresponding landmark from the j-th image, J_j. Let the detected point be denoted by $\hat{y}_j^l$. If the corresponding point detected by the forward template matching is correct, the point, $\hat{y}_j^l$, detected by the backward template matching should be located near the source point, y_j^l. The proposed method rejects the corresponding landmark, y_{j+1}^l, detected by the forward template matching if the backward template matching detects the corresponding point far from the source point:

$$\delta = \| \hat{y}_j^l - y_j^l \| < D, \qquad (3.5)$$

where D is a threshold of which value is experimentally determined in advance. Figure 3.4 shows an example of the rejection of the corresponding landmarks detected by the forward template matching.

Let the map non-rigidly deforms $J_j(x)$ be denoted by ϕ_j. Let $\tilde{y}_j^l = \phi_j \circ y_j^l$ denote the point to which y_j^l should be mapped. Given a set of landmarks, $\{y_j^{\ell_j}\}$, detected in J_j and given a set of the points to which the landmarks should be mapped, $\{\tilde{y}_j^{\ell_j}\}$, where $\ell_j \in \mathcal{L}_j = \{l_j^1, l_j^2, \ldots, l_j^{L_j}\}$ denotes the ID number of the landmarks detected in the j-th image, one can straightforwardly obtain ϕ_j that satisfies $\tilde{y}_j^{\ell_j} = \phi_j \circ y_j^{\ell_j}$ for any ℓ_j, e.g., by using a B-spline deformation method [3]. For smoothing the l-th trajectory, $\mathcal{T}^l = \{y_{j_l}^l, y_{j_{l+1}}^l, \ldots, y_{j'}^l\}$, the points to be mapped, y_j^{-l}, are determined by minimizing the total variance of each trajectory as follows:

$$\{\tilde{y}_{j_l}^l, \ldots, \tilde{y}_{j_l}^l\} = \arg \min_{x_{j_l}, \ldots, x_{j_l}} \left(\sum_{j=j_l}^{j_l-1} \| x_{j+1} - x_j \|^2 + \lambda \sum_{j=j_l}^{j_l} \| x_j - y_j^l \|^2 \right),$$

(3.6)

where x_j is a point in the j-th image, J_j ($j \in \{j_l, j_{l+1}, \ldots, j'_l\}$). The first term of the cost function to be minimized denotes the length between two points, x_j and x_{j+1}, in the neighboring two images and the second term denotes the distance from the l-th landmark in the j-th image, y_j^l to x_j. The first term smooths the trajectory by shortening it, and the second term prevents the resultant trajectory from being too far from the original trajectory, $\mathcal{T}^l$. Figure 3.5 shows an example of the change of the trajectories. Every trajectory is smoothed, as shown in the figure, and the given 2D images are non-rigidly deformed so that all detected landmarks are located on the smoothed trajectories. It should be noted that different from many conventional methods for the construction of 3D microscopic images, and the proposed method determines the non-rigid mapping of each image using not only its neighboring images but also give all images.

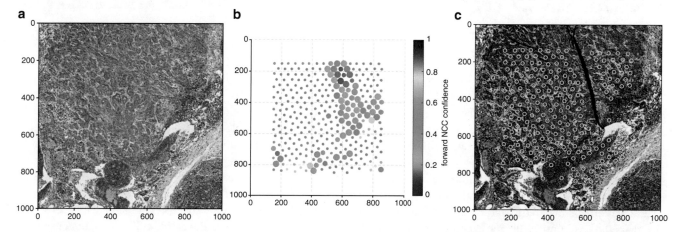

Fig. 3.4 Example of landmarks detected around an artifact. (**a**) A source image. (**b**): Each radius indicates δ in (Eq. 3.5). Near the artifacts, the backward template matching fails to find the corresponding landmarks close to the source points. (**c**) A target image with an artifact. The circles show the detected corresponding landmarks. The method successfully rejects false corresponding landmarks near the artifact

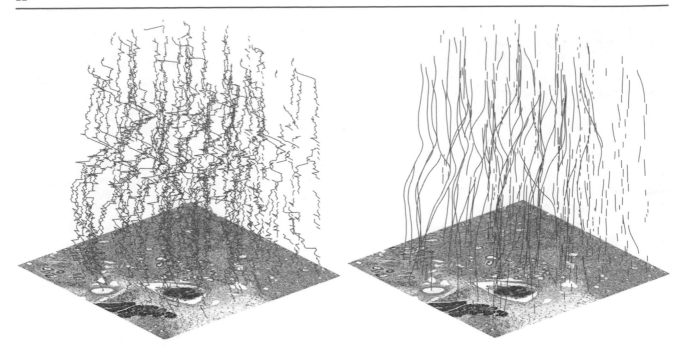

Fig. 3.5 Example of trajectories of corresponding landmarks detected through images. Left: Not-smooth trajectories before the images are not deformed. Right: Smoothed trajectories after the deformation

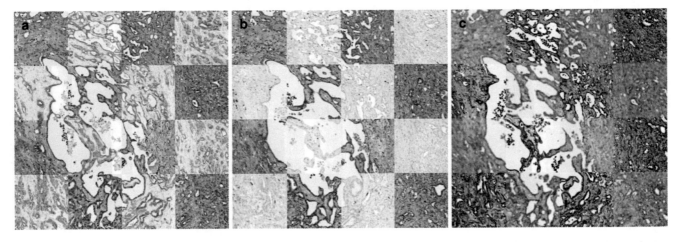

Fig. 3.6 Examples of registered two images of differently stained. The two images are stained with (**a**) H&E and CK19, (**b**) H&E and Ki67, and (**c**) H&E and MT, respectively [4]

Figure 3.6 shows some examples of the non-rigidly registered two neighboring images stained with different stains. As can be seen, the microstructures are smoothly reconstructed by the method. Figure 3.7 shows the cross-section of the reconstructed 3D image. The left panel shows the cross-section of a 3D image reconstructed without the non-rigid registration. As shown, the spatial pattern of the cross-section is not smooth if the non-rigid registration is not applied. The spatial pattern of the 3D image reconstructed by the proposed method is continuous and smooth. Because the thin sections are stained with different stains, alternatively, one

can see the horizontal stripes in the cross-section. The thin sections were stained with different stains for describing anatomically important structures in the reconstructed 3D image. For example, the Ki67 stain is used to enhance nuclei of tumor active cells. Extracting the stained nuclei from the Ki67-stained images and estimating the spatial density of the active nuclei in the reconstructed 3D tumor image, one can visualize the spatial distribution of the active cells in the tumor. Figure 3.8 shows the distribution of active cells in the tumor. As can be seen, the active cells are more densely distributed near the outer surface of the tumor.

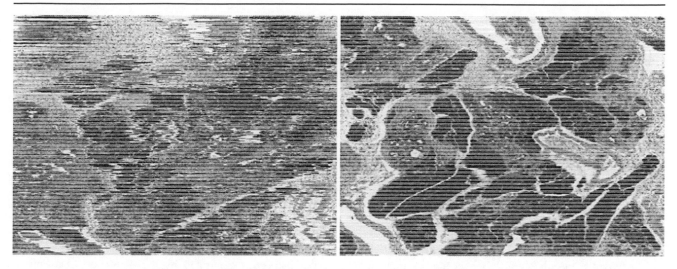

Fig. 3.7 Examples of the cross-sections of the reconstructed 3D pathology image. Horizontal stripes are presented because each slice image is stained saLeft: The cross-section of the 3D pathology image is constructed by only rigid registration. Right: The cross-section after the non-rigid registration

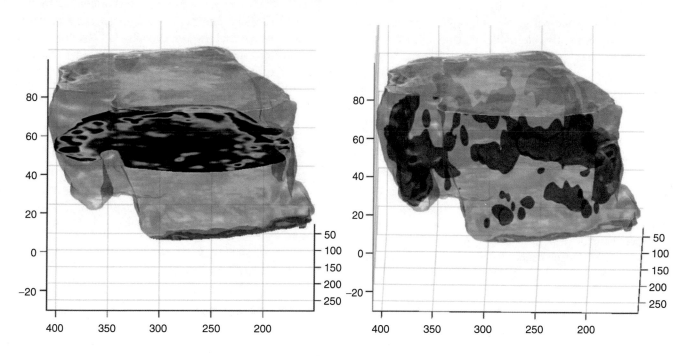

Fig. 3.8 3D distribution of active cells. The nuclei stained with Ki67 are extracted, and the density is estimated. Left: A heat map of the density in one cross-section. Right: 3D density distribution of the nuclei stained with Ki67

3.2 Construction of Multi-Resolution Model of Pancreatic Tumor Image

The multi-scale image model of pancreatic cancer tumors is constructed from the MR image and the corresponding pathological images that are registered to the MR image. Once the 3D microscopic image is reconstructed, one can register the 3D microscopic image with the tumor region in the last captured MR image. The tumor region in the MR image is at first manually labeled for the registration, and the tumor region in the reconstructed 3D pathology image is non-rigidly registered to the manually labeled region in the MR image. The outer surface of the tumor is significantly useful, especially for determining the global transformation of the tumor region from the 3D pathology image to the MR image. For determining the non-rigid deformation of the 3D pathology image inside the tumor, the spatial image pattern inside the tumor region in each 3D image should be referred to. A conventional mutual-information-based method for registering images from different modalities [***] can be employed for the non-rigid registration of the tumor regions. One problem here is the large difference in the spatial resolution. One voxel in the MR image corresponds to about 100×100 rectangular region in each microscopic image. In this study, the mutual information between the voxel values of the MR image and the pixel values averaged over the rectangular region corresponding to each voxel of the MR image is maximized for the registration.

After the 3D images are registered, we have a set of pairs of the voxel value of the MR image and a corresponding patch image in the pathology images. Let the m-th voxel value of the be denoted by v_m, and the corresponding microscopic image patch be denoted by $\boldsymbol{u}_m$, where $m = 1, 2, \ldots, M$. Given the set of the pairs, $\mathcal{S} = \{(v_m, \boldsymbol{u}_m) | m = 1, 2, \ldots, M\}$, we construct a multi-scale model that represents the relationship between v_m and $\boldsymbol{u}_m$ and that can predict the conditional probability distribution, $p(\boldsymbol{u}|v)$, where v denotes the voxel value inside the tumor region in a given MR image and $\boldsymbol{u}$ denotes the pathology image patch. Once the model is constructed, one can predict pathology image patches that would correspond to each voxel in a pancreatic tumor region in a given MR image by drawing image patches from the conditional probability distribution, $p(\boldsymbol{u}|v)$. The image patches predicted from $p(\boldsymbol{u}|v)$ would change with respect to the conditioning voxel value, v, in the tumor region of the MR image [5].

Deep generative models are useful for the drawing of image patches. A deep generative model trained with a training set of images independently drawn from an identical probability distribution can generate fake images by drawing samples from the identical probability distribution. Generative Adversarial Networks (GANs) are one of the most popular deep generative models among various types of ones. We modified an α-GAN [6]. α-GAN is a fusion of Variational Auto Encoder (VAE) [7] and GAN. VAE is one of the most fundamental image generative models: It consists of an encoder that maps an input image to a latent variable and a decoder that maps the latent variable to the input image. The latent variable is regularized so that it obeys the Gaussian distribution. One drawback of a conventional VAE is the image quality of the decoded images. They are often blurred. GAN is one of the most widely employed generative models. A GAN improves the performance of the image generator by training a discriminator that classifies true images from fake ones generated by the generator. Conventional GANs can generate high-quality images, but the mode collapse often happens and fails to generate sufficiently diverse images. α-GAN compensates for each shortcoming by fusing these two. Let $\boldsymbol{u}$ and z denote the images and the corresponding latent variables, respectively. Let $\boldsymbol{\theta}$ and $\boldsymbol{\eta}$ denote the parameters of the decoder and of the encoder that estimates the parameter values of the probability distribution function of the latent variable, respectively. The ELBO that is maximized by VAE is given as follows:

$$\mathcal{L}(\theta, \eta) = \mathbb{E}_{q_\eta(z|u)}\Big[\log p_\theta(\boldsymbol{u}|z)\Big] - \mathrm{KL}\Big[q_\eta(z|\boldsymbol{u}) \| p^*(z)\Big].$$

(3.7)

α-GAN represents $p_\theta(\boldsymbol{u}|z)$ using the output of the discriminator. Let the class of true images and of fake ones be denoted by $\mathcal{C}_{\mathrm{true}}$ and $\mathcal{C}_{\mathrm{fake}}$, respectively. The discriminator is trained to estimate the posterior probability distribution:

$$p(\mathcal{C}_{\mathrm{fake}}|\boldsymbol{u}) \propto p(\boldsymbol{u}|\mathcal{C}_{\mathrm{fake}}) p(\mathcal{C}_{\mathrm{fake}}).$$

(3.8)

Assuming $p(\mathcal{C}_{\mathrm{true}}) = p(\mathcal{C}_{\mathrm{fake}})$, we have

$$\mathbb{E}_{q_\eta(z|u)}\Big[\log p_\theta(\boldsymbol{u}|z)\Big] = \mathbb{E}_{q_\eta(z|u)}\left[\log \frac{p_\theta(\boldsymbol{u}|z)}{p^*(\boldsymbol{u})}\right] + \mathbb{E}_{q_\eta(z|u)}\Big[\log p^*(\boldsymbol{u})\Big],$$

(3.9)

where $p^*(\boldsymbol{u})$ denotes the true (unknown) probability distribution of true images. The second term in the right-hand side of (Eq. 3.9) is constant, and the first term can be represented by using the output of the discriminator as the right-hand side of the following equation:

$$\mathbb{E}_{q_\eta(z|u)}\Big[\log p_\theta(\boldsymbol{u}|z)\Big] = \mathbb{E}_{q_\eta}\left[\log \frac{\mathcal{D}_\phi(\mathcal{G}_\theta(z))}{1 - \mathcal{D}_\phi(\boldsymbol{u}^*)}\right],$$

(3.10)

where $\mathcal{G}_\theta(z)$ denotes the fake image generated by the generator (decoder) from the latent variable, z, $\mathcal{D}_\phi(\mathcal{G}_\theta(z))$ denotes the output of the discriminator when the generated fake image is input and $\mathcal{D}_\phi(\boldsymbol{u}^*)$ denotes the output of the discriminator against the true image, $\boldsymbol{u}^*$.

The second term of the right-hand side in (Eq. 3.7) is the distance between the probability density distribution of the latent variable encoded from the given image, $q_\eta(z|\boldsymbol{u})$, and the predetermined distribution of the latent variable, $p^*(z)$, for the regularization. We employ a standard Gaussian distribution for $p^*(z)$. Let z^* denote the latent variable drawn from $p^*(z)$. Similar to (Eq. 3.10), by introducing another discriminator, $\mathcal{D}_\psi$, that distinguishes the latent variables generated by the encoder from those sampled from the predetermined probability distribution, $p^*(z)$, we have the following representation:

$$-\mathrm{KL}\Big[q_\eta\left(z|u\right)\|\,p^*\left(z\right)\Big]=\mathbb{E}_{q_\eta(z|u)}\left[\frac{p^*\left(z\right)}{q_\eta\left(z|u\right)}\right]\simeq\mathbb{E}_{q_\eta(z|u)}\left[\log\frac{\mathcal{D}_\psi\left(z^*\right)}{1-\mathcal{D}_\psi\left(z\right)}\right],\tag{3.11}$$

where ψ denotes the parameters of the discriminator for the latent variables. Combining (Eq. 3.10) and (Eq. 3.11), we have

$$\mathcal{L}\left(\theta,\eta,\phi,\psi\right)=\mathbb{E}_{q_\eta(z|u)}\left[\log\frac{\mathcal{D}_\phi\left(\mathcal{G}_\theta\left(z\right)\right)}{1-\mathcal{D}_\phi\left(u^*\right)}+\log\frac{\mathcal{D}_\psi\left(z^*\right)}{1-\mathcal{D}_\psi\left(z\right)}\right].\tag{3.12}$$

In the above objective function, the discriminators are used to evaluate the realness of the generated images. As a result, the generative models obtained by optimizing the objective function shown in (Eq. 3.12) often suffer from mode collapse. α-GAN hence incorporates the directly measured distance between the generated image and its original one. Modeling $p_\theta(u|z)$ with Laplace distribution, i.e. $p_\theta\left(u|z\right)\propto\exp\left(-\lambda\,\|\,u-\mathcal{G}_\theta\left(z\right)\|\right)$, one can evaluate the realness as follows:

$$\mathcal{L}\left(\theta,\phi\right)=\mathbb{E}_{q_\eta(z|u)}\Big[-\lambda\,\|\,u-\mathcal{G}_\theta\left(z\right)\|\Big].\tag{3.13}$$

α-GAN combines (Eq. 3.13) with (Eq. 3.12) and solves the following problem:

$$\min_{\phi,\psi}\max_{\theta,\eta}\mathcal{L}\left(\theta,\eta,\phi,\psi\right)=\min_{\phi,\psi}\max_{\theta,\eta}\left\{\mathbb{E}_{q_\eta(z|u)}\left[-\lambda\,\|\,u-\mathcal{G}_\theta\left(z\right)\|+\log\frac{\mathcal{D}_\phi\left(\mathcal{G}_\theta\left(z\right)\right)}{1-\mathcal{D}_\phi\left(u^*\right)}+\log\frac{\mathcal{D}_\psi\left(z^*\right)}{1-\mathcal{D}_\psi\left(z\right)}\right]\right\}.\tag{3.14}$$

In this study, not only the voxel value of the MR image but also the distance between the voxel and the outer surface of the tumor, d, is also used for the condition. Let $c = (v, d)^\mathsf{T}$ denotes a two-tuple of the voxel value obtained for each voxel in the tumor region of the MR image that represents the voxel value and the distance from the outer surface. Then, we have $\mathcal{G}_\theta(z,c)$ and $q_\eta(z,c|u)$ instead of $\mathcal{G}_\theta(z)$ and $q_\eta(z|u)$ in (Eq. 3.14). The conditioned generator, $\mathcal{G}_\theta(z,c)$, is expected to sample images from the conditioned probability distribution, $p_\theta(u|c,z)$, and $q_\eta(z,c|u)$ maps a given image, u, not only to the latent variable, z, but to the conditions, c, that correspond to the given image, u.

Assume that for each voxel in the tumor region in a given MR image one can measure the voxel value, v, and the distance, d, from the outer boundary. Then, given the measured values, $c = (v, d)$, for some specific voxel in the given MR image, the trained deep generative model can generate fake microscopic images of H&E-stained specimen that would correspond to the voxel by the drawing of samples, z, from $p^*(z)$ followed by the computation of $\mathcal{G}_\theta(z,c)$. Figure 3.8 shows examples of the fake pathology images generated by the deep generative model. From an identical latent variable sampled from $p(\mathbf{z}^*)$, one can generate a variety of microscopic images by changing the conditional values of c, as shown in Fig. 3.9.

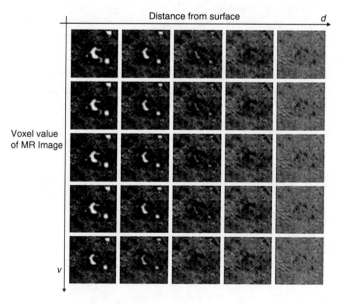

Fig. 3.9 An example of the change of the fake image generated from a same latent variable due to the change of the conditions

3.2.1   Summary

In this section, we described an outline of a method for constructing a multi-scale model of pancreatic cancer tumors from an MR image and microscopic pathology images. One of the objectives of the construction of the multi-scale model was to enable to prediction the corresponding pathology images from a given MR image of a pancreatic tumor. The prediction of the pathology image from the MR one is useful for clinical applications because MR images can be obtained non-invasively, and pathological images, on the other hand, can only be obtained invasively. One important process in the construction of the multi-scale model is the registration between given images. The multi-scale model constructed in this study represents the probability distribution of microscopic pathological images conditioned by the corresponding MR images. For constructing such a model, we need to register MR images and corresponding MR ones. For this registration, we reconstruct a 3D microscopic image from a spatial series of 2D microscopic images of the tumor and registration between these images is needed for the 3D reconstruction. The objective of the former registration between MR image and 3D microscopic one is to correspond each voxel in the tumor region of the MR image to a region of the microscopic pathology images for making a training set of pairs of a voxel in MR image and a region in pathology image. The objective of the latter registration is to make the spatial patterns of the resultant 3D microscopic image continuous and smooth. Once the training set of the pairs of a voxel in the MR image and the corresponding region in the pathology images, we can construct a deep generative model that can predict the pathology images that correspond to each voxel in the tumor region in the given MR image.

In the registration between the MR image and the reconstructed 3D pathology image, we employed a conventional registration method that maximizes the mutual information. In this study, we evaluated the mutual information between the voxel values of the MR image and the mean voxel values of the corresponding regions in the pathology images. We decided to use the mean pixel values because necrosis regions in H&E-stained pathology images would have colors different from those in the non-necrosis regions. In other words, we selected the image features used for the evaluation of the mutual information manually. Future works include to develop a method for selecting image features appropriate for registering between images from different modalities that have largely different spatial resolutions.

References

1. Lee JW, Komar CA, Bengsch F, Graham K, Beatty GL. Genetically engineered mouse models of pancreatic cancer: the kpc model (lsl-krasg12d/+; lsl-trp53r172h/+; pdx-1-cre), its variants, and their application in immuno-oncology drug discovery. Curr Protoc Pharmacol. 2016;73(1):14–39.
2. Tamura Y, Yokota T, Kugler M, Triquet V, Sei T, Iwamoto C, Ohuchida K, Hashizume H, Hontani H. Partial rigid diffeomorphism for measuring temporal change of pancreatic cancer tumor. In: International workshop on advanced image technology (IWAIT) 2019, vol. 11049. International Society for Optics and Photonics; 2019. p. 110491H.
3. Rueckert D, Aljabar P, Heckemann RA, Hajnal JV, Hammers A. Diffeomorphic registration using b-splines. In: International conference on medical image computing and computer-assisted intervention. Springer; 2006. p. 702–9.
4. Kugler M, Goto Y, Tamura Y, Kawamura N, Kobayashi H, Yokota T, Iwamoto C, Ohuchida K, Hashizume M, Shimizu A, et al. Robust 3d image reconstruction of pancreatic cancer tumors from histopathological images with different stains and its quantitative performance evaluation. Int J Comput Assist Radiol Surg. 2019;14(12):2047–55.
5. Shimomura T, Mauricio K, Yokota T, Iwamoto C, Ohuchida K, Hashizume M, Hontani H. Construction of a generative model of h&e stained pathology images of pancreas tumors conditioned by a voxel value of mri image. In: Computational pathology and ophthalmic medical image analysis. Springer; 2018. p. 27–34.
6. Rosca M, Lakshminarayanan B, Warde-Farley D, Mohamed S. Variational approaches for auto-encoding generative adversarial networks. arXiv preprint arXiv:1706.04987; 2017.
7. Kingma DP, Welling M. Auto-encoding variational bayes. arXiv preprint arXiv:1312.6114; 2013.

Fundamental Technologies for Integration of Multiscale Spatiotemporal Morphology in MCA

4

Akinobu Shimizu, Naoki Kobayashi, and Hayaru Shouno

Abstract

This chapter presents the achievements of multiscale spatiotemporal statistical models during the whole period of the multidisciplinary computational anatomy project. Shimizu et al. built spatiotemporal statistical models along a time axis of embryos and children, in which a modeling method to deal with a small sample of data was developed and smoothness constraints along a time axis were introduced into the spatiotemporal statistical model. Modelling of organs with nested and neighbouring constraints was also studied by their group. Multiscale models were constructed, in which Shimizu et al. presented super-resolution techniques by dictionary learning and deep learning, and Shouno et al. studied super-resolution under a noisy environment in order to solve mapping problems among multimodal images. Kobayashi et al. developed algorithms for image understanding of microscopic images of a KPC mouse, where 3D tissues structures were recognised from pancreatic serial section images and from hyperspectral images.

Keywords

Spatiotemporal statistical model · Embryo · Children · KPC mouse · Super-resolution

A. Shimizu (✉)
Institute of Engineering, Tokyo University of Agriculture and Technology, Tokyo, Japan
e-mail: simiz@cc.tuat.ac.jp

N. Kobayashi
Faculty of Health and Medical Care, Saitama Medical University, Saitama, Japan

H. Shouno
Graduate School of Information Engineering, The University of Electro-Communications, Tokyo, Japan

4.1 Introduction

The goal of this study was to develop spatiotemporal statistical models that describe statistical variations of anatomical features (e.g. points and surfaces) and grey values (e.g. CT) along a temporal axis of a human. Furthermore, super-resolution algorithms for a multiscale model were developed. Reconstruction and understanding of a microscopic volume and application of the developed statistical models, such as segmentation with the models were also challenged to develop the multiscale model.

This chapter presents the achievements of multiscale spatiotemporal statistical models during the whole period of the multidisciplinary computational anatomy project from fiscal year (FY) 2014 to FY 2018. Section 4.2 presents spatiotemporal statistical models of time series data along time or scale axes of embryos and children. Section 4.3 shows super-resolution techniques for multiscale model. Section 4.4 describes developed algorithms of image understanding of microscopic images of a KPC mouse.

4.2 Spatiotemporal Statistical Model of Time Series Data

Shimizu et al. focused on spatiotemporal statistical variations of embryo, and children, which included lots of problems to be solved in the modelling, such as rapid growth or topological changes of anatomical features (e.g. surfaces and landmarks). This section shows the achievements of the research group. Note that terminology on spatiotemporal data and models refers to the paper [1].

4.2.1 Modelling with a Small Sample of Data

A spatiotemporal statistical model often suffers from a problem caused by sparsely distributed data along a time axis, which makes the statistical modelling difficult. Shimizu et al. proposed a two-stage modelling algorithm, in which first stage maps all data into a feature space with reduced dimension and second stage performs statistical modelling with q-Gaussian based parameter estimation followed by interpolation of statistics along a time axis [2, 3]. The combination of the dimensionality reduction and the q-Gaussian based parameter estimation makes the statistical modelling robust under the condition of a small sample of data. The effectiveness of the proposed modelling approach was demonstrated in the modelling of anatomical landmarks and surfaces of human embryos of Kyoto collection [4].

4.2.2 Modelling with Smoothness Constraint along a Time Axis

Smoothness of statistics along a time axis is a key technology in the spatiotemporal statistical modelling. The contributions are twofolds as shown below.

First, the best method was explored for the interpolation of statistics in the second stage of the spatiotemporal modelling [2, 3] presented in Sect. 4.2.1. The best combination was exhaustively searched among all possible pairs of three interpolation methods for average vector and six for covariance matrix as shown below.

- Interpolation methods for neighbouring average vectors: Linear, B-Spline, Information geometry.

- Interpolation methods for neighbouring covariance matrices: Linear (rotation), tensor B-Spline, Affine-invariant, Log-Euclidean, Wasserstein geometry, Information geometry.

Figure 4.1 presents a spatiotemporal statistical model of landmark points on the face of human embryos of Kyoto collection [4], where *information geometry* was selected as optimal for both average vector and covariance matrix.

Second contribution is that another smoothness constraint was introduced in the spatiotemporal statistical modelling of children's liver scanned with inconsistent time intervals. Temporal regularisation was introduced to realise smooth changes in directions of neighbouring principal axes [5], which was inspired by the paper [1]. In addition, adaptive kernel regression was introduced to deal with inconsistent time intervals. The effectiveness was demonstrated using CT data of Children's National Health System in Washington DC.

4.2.3 Modelling with Nested and Neighbouring Constraints

Topological constraints are important for modelling of statistical variation of nested structures and/or neighbouring structures so as to prevent unnatural leakage of inner structures and overlap between neighbouring structures. Shimizu

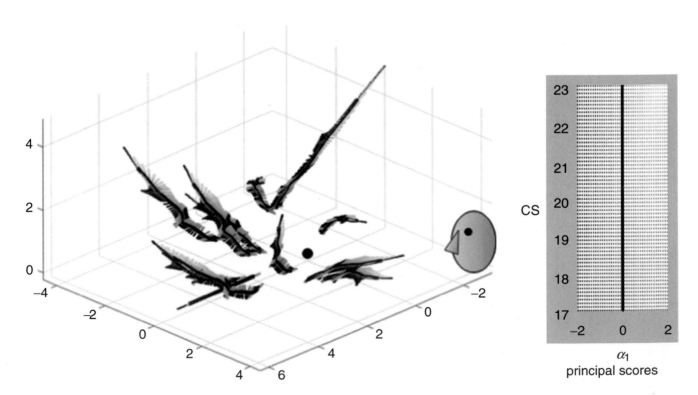

Fig. 4.1 A Spatio-temporal model of twelve landmark points on face using information geometry based interpolation. Average trajectories with variation along first principal modes are displayed

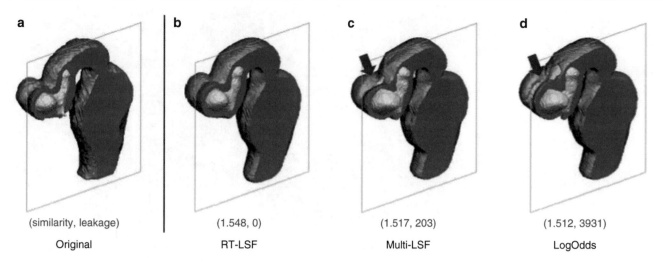

a	b	c	d
(similarity, leakage)	(1.548, 0)	(1.517, 203)	(1.512, 3931)
Original	RT-LSF	Multi-LSF	LogOdds

Fig. 4.2 Original nested shapes and the reconstructed shapes from the three SSMs (Fig. 4.4 in [6])

et al. proposed a spatiotemporal statistical shape model (SSM) with a nested constraint [6] and a neighbouring constraint.

Figure 4.2 shows an original nested shape of the Kyoto collection [4]. Remains are the reconstructed shapes by the proposed method and two conventional methods. It was confirmed from the figure that the proposed RT-LSF-based SSM has no leakage, while conventional SSMs have leakages as denoted by red arrows.

4.2.4   Modelling of Topological Changes along a Time Axis

Some anatomical structures have topological changes in shape along a time axis. For example, choroid plexus typically appears after Carnegie stage 19. Shimizu et al. developed a level set distribution model that describes the appearance and disappearance of surfaces seamlessly. Eventually, the abovementioned nested and neighbouring constraints were combined with topological changes, which were used in modelling spatiotemporal statistical variations of surfaces of the brain, ventricles and choroid plexuses of human embryos of Kyoto collection [4].

4.3   Multiscale Model

Shimizu et al. and Shouno et al. developed super-resolution (SR) methods for multiscale models, one of which is a dictionary-based method and the others are deep learning-based methods.

4.3.1   Dictionary-Based Super-Resolution [7, 8]

Dictionary learning is a popular approach for SR of a single frame. Shimizu et al. employed the anchored neighbourhood regression (ANR) approach of the paper [9] to bridges between micro-CT volumes with different resolutions. It used ridge regression to learn exemplar neighbourhoods offline and uses these neighbourhoods to precompute projections to map low resolution (LR) patch volumes onto the high resolution (HR) domain. The advantage of the approach was the higher computational efficiency against other dictionary-based methods while keeping high performance.

The proposed method was applied to two micro-CT volumes resected from a KPC mouse of pancreatic cancer, whose resolutions are 9 [μm] and 20 [μm], respectively. After registration between the two volumes using a landmark-based algorithm, the dictionary of patch volumes was constructed from a large number of pairs between LR and HR patch volumes. It was confirmed that the PSNR by the proposed method was 2.42 [dB] higher than tri-cubic interpolation.

4.3.2   Deep Learning-Based Super-Resolution [10]

Shimizu et al. developed a deep learning-based SR technique. Generative adversarial network for SR (SRGAN) [11] was extended to be applicable to three-dimensional (3D) LR images. 3D SRGAN iteratively optimises loss functions of a discriminator and a generator, in which 3D ResNet was used

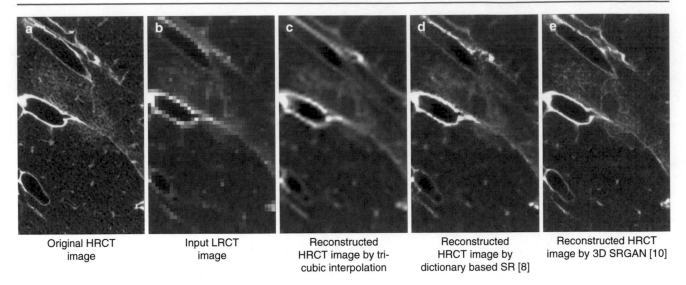

| Original HRCT image | Input LRCT image | Reconstructed HRCT image by tri-cubic interpolation | Reconstructed HRCT image by dictionary based SR [8] | Reconstructed HRCT image by 3D SRGAN [10] |

Fig. 4.3 Original HRCT, input LRCT and reconstructed HRCT images from conventional methods and proposed method (3D SRGAN [10])

as the generator and 3D CNN was employed as the discriminator. Finally, 3D ResNet was applied to an LR CT volume of lung and HR CT volume whose resolution is eight times higher than that of LR CT was reconstructed.

Figure 4.3 shows slice images of original HRCT volume and input LR one of the test data. Reconstruction of HRCT volume was carried out by tri-cubic interpolation, 3D SRGAN and dictionary-based SR, respectively. The figure demonstrated that 3D SRGAN successfully reconstructed the HRCT image from the input LR one. Peak signal-to-noise ratio of the HRCT volume was 25.41 [dB], which was higher than those of tri-cubic interpolation and dictionary-based SR.

4.3.3 Super-Resolution Problem under the Noisy Environment

In this study, Shouno et al. treat an SR problem under a noisy environment in order to solve the mapping problem among multimodal images. The SR problem is to generate an HR image from its corresponding LR image. In this field, there exists a lot of SR systems that work well for the natural images obtained by several digital cameras, which have very high contrast and low noise components. In several medical imaging devices, however, there exists some systematic noise intrinsically. Thus, in this study, a noisy observation is assumed because it is difficult to omit the noises which are involved in the observation. For this purpose, two deep learning-based systems are prepared. One is called "Very-Deep Super-Resolution (VDSR)" for the SR part; the other is called "Denoising Convolution Neural Network (DnCNN)" for denoising [12, 13]. The VDSR and the DnCNN are one of

state-of-the-art models in each field. Both of them are based on the ResNet system, and the network architecture is very similar except for the batch normalisation layers.

To solve the SR problem, Shouno et al. focus on the combination of deep learning systems. In the manner of using a deep learning approach, it is not so unnatural to use a single deep learning architecture to solve both SR and denoising problems simultaneously. The other approach is to prepare expert deep learning systems to solve SR and denoising problems independently. The left of Fig. 4.4 shows the concept of these approaches. The first model is to apply a single deep learning model to solve both SR and denoising problems. The second model is to apply denoising problem solver at first and then apply SR problem solver. The third is to apply SR problem solver at first and then apply denoising problem solver.

At first, these three approaches are adopted to solve the natural images dataset for evaluation. To evaluate the abilities of these methods, downscaled with AWGN images are prepared. In the assessment, controlling the noise strength, which is appeared in the standard deviation of the Gaussian noise, image restoration ability is measured by PSNR and SSIM. And then, these three approaches are also adopted for Micro-CT SR problems. The original Micro-CT images, which show a part of the lung, are provided from Kumamoto University Hospital.

In the training of each deep learning system, transfer-style learning is adopted; that is, a natural image dataset is applied for training SR and denoising deep learning techniques at first. Two hundred and ninety-one natural images for transfer-style learning are prepared. After that, the deep learning systems are trained with a small number of Micro-CT images.

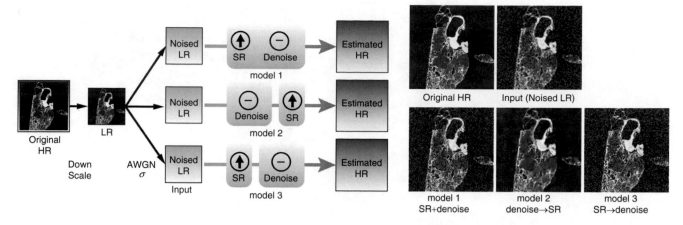

Fig. 4.4 Left: Schematic diagram of SR systems approach (HR: high-resolution, LR: low-resolution), Right: Bottom line pictures show the result of each model

The right of Fig. 4.4 shows a typical result of SR images. The top row pictures show the original HR image and the input of the system, which is the corresponding LR image with AWGN corruption. In the assessment process, the additive white Gaussian noise with the standard deviation as $\sigma = 0.03$ is added for each input. The bottom row shows the estimated HR image results. The left one shows the result of model 1, that is solving SR and denoising problem in a single deep learning system. The middle and the right ones show the results of model 2 and model 3, respectively. These two models are using separated deep learning systems for SR and denoising. From these SR images results, model 2 looks just better than the other two models, since it can eliminate noisy components in the background. In the PSNR evaluation, 18.33 [dB] is obtained for the model 1, 21.58 [dB] for the model 2, and 18.13 [dB] for the model 3, respectively.

Under the several corruption rate environments, that is $\sigma = 0.01{\sim}0.1$, the result performance of model 2 is better than those of model 1 and model 3 in the meaning of both PSNR and SSIM value.

The DL system is applied to the noisy SR problems. VDSR and DnCNN are adopted for the SR and denoising problem solvers, respectively. From the quantitative point of view, model 2 looks better than the other models. Seeing Fig. 4.4, the result of model 2 shows the clearer meaning of background noise rather than the other. However, some region of high spatial frequency looks terrible. This result comes from the spatial signal frequency is similar to the spatial noise frequency. In model 2, the first system is denoising DL, which might eliminate fine-resolution information. Thus, taking out the information from the estimated corruption information should be considered.

4.4 Image Processing of Pathological Images

This section shows the achievements on image understanding of pancreatic serial section images and from hyperspectral images.

4.4.1 3D Tissue Structure from Pathological Images

Kobayashi et al. target the discovery of small lesions which Computed Tomography (CT) or Magnetic Resonance Imaging (MRI) missed, and have studied registration of the three-dimensional (3D) structure obtained by pathological images and the 3D structure obtained by CT or MRI. In this study, the goals are image segmentation and 3D reconstruction of various structures from pancreatic serial sections of KPC mouse.

Materials are about 1000 pancreatic serial sections stained with Haematoxylin and Eosin (HE), 65 sections stained with Cytokeratin 19 immunostaining (CK), and 65 sections stained with Masson trichrome (MT). These slices were taken with a virtual slide scanner at about 7 times magnification. The image size of each slice is about 15,000 × 10,000 pixels. Though image segmentation of slides are necessary for 3D reconstruction, manual segmentations are impractical because of the large number of slides and pixels. Therefore, only 25 region of interests (ROIs) were manually segmented, and remains were automatically segmented using texture feature (co-occurrence [14] and run-length matrix feature [15]) and support vector machine (SVM) [16]. In the SVM, a linear kernel function was used for high-speed processing

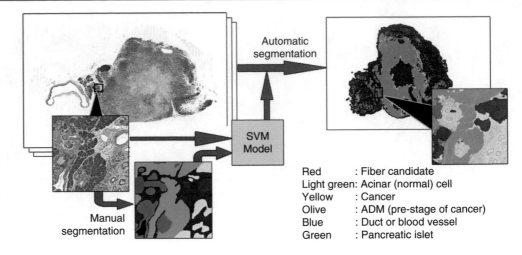

Fig. 4.5 HE stained images, ROI, manually segmentation image, and automatically segmentation result

and stepwise feature selection was adopted to improve accuracy.

The left side of Fig. 4.5 shows the image taken of the HE-stained specimen, part of the ROI, and manually segmented image. The right side of Fig. 4.5 shows the 3D rendered automatic segmentation result and ROI. Comparing the manual segmentation in the lower left of Fig. 4.5 and the ROI in the right of Fig. 4.5, acinar cells (light green) and the pancreatic islet (green) could be accurately segmented; however, cancer (yellow), ADM (olive), and fibre candidate regions (red) were difficult to identify. Furthermore, it has been difficult to identify fibres accurately, which are fine structures, on a pixel-by-pixel basis. Overall, as a result of performing fivefold cross-validation on a pixel-by-pixel basis, the accuracy was 96% [17, 18].

Future plan is to evaluate the results in 2D and 3D images from a pathologist view and to make clear the inadequate point for precise segmentation.

4.4.2 Hyperspectral Image Processing of Pathological Images

With the development of a virtual microscope technology called whole slide imaging (WSI) in 2006, people have been raising expectations for computer diagnosis supports. A WSI system mostly takes RGB images. However, it is a known fact that it is very difficult to accurately extract tissues and give a cancer diagnosis with RGB images. In this context, this research proposes a method to more accurately extract tissues and give a cancer diagnosis using Hyperspectral Imaging (HSI) than using RGB images [19].

The method proposed by Kobayashi et al. involves pixel-by-pixel cancer nuclei detection and tissue detection by employing techniques such as spectral selection and bag of features (BoFs) for analysing the pathological images. A mouse's HE-stained pancreatic cancer tissue specimen was

Table 4.1 Accuracy by tenfold cross-validation

	HSI + BoF	HSI	RGB
Average of accuracy	0.96	0.91	0.84
Average of sensitivity	0.96	0.91	0.83
Average of specificity	0.96	0.92	0.86

used for the experiment. The spectral features were extracted from an image captured by a hyperspectral camera (EBA Japan CO., LTD., NH-3). An image of size 752 × 480 pixels was obtained, with the wavelength ranging in value from 420 nm to 720 nm, and having 61 bands at 5 nm intervals.

1. Tissue classification by HSI.

Five tissues (nucleus, sinusoid, lymphocytes, fibres and cytoplasm) are classified and accuracies are evaluated by using subject-based three-hold cross-validation in the experiment. The codebook number of BoF is set to 80.

The classification accuracy is compared between the proposed method using HSIs and a random forest algorithm using RGB images. The overall rate of the proposed method is 68% and is improved by 11% compared to the RGB-based method. In particular, the classification accuracies of fibres and cytoplasm are improved by 5–24% than the RGB-based method.

2. Cancer extraction by HSI.

The method of tenfold cross-validation was thereby applied for randomly dividing the set of samples into ten approximately equal-sized parts. The classification accuracy of three methods, namely BoF + his (Hyperspectral Imaging), HSI, and RGB was then compared. The codebook number for BoF was set to a value of 150.

The classification results obtained from the tenfold cross-validation method are shown in Table 4.1. The obtained results show that the accuracy rate of the HSIs is 91%, which in turn has an accuracy rate greater by 7% compared to the RGB images. Furthermore, the accuracy rate can be increased to 96% by adding the BoF feature to HSI.

From the result, consequently, the accuracy of tissue classifications and cancer diagnoses was enhanced with HIS images rather than with RGB images. By adding BoF to HIS images, their accuracy has been further enhanced. BoF includes spatial information around respective picture elements. While many traditional researches on HIS images used the differences in transmissions by wavelength and waveforms, few of them paid attention to spatial information. It is possible to apply to HIS Images a texture analysis method such as a density co-occurrence matrix; however, this method is not feasible because it takes excessive calculation efforts. Meanwhile, the proposal method enables us to figure out with simple calculations the characteristics of HIS images including their spatial information. It is necessary from now on to discuss more effective objects and parameters of BoF [20].

Kobayashi et al. also proposed a capturing system with multispectral filter array (MSFA) technology [21–23]. Therein, a mosaicked image captured using an MSFA is demosaicked to reconstruct multispectral images (MSIs) [24, 25]. Joint optimisation of the spectral sensitivity of the MSFAs and demosaicking is considered, and pathology-specific multispectral imaging is proposed [26, 27]. This optimises the MSFA and the demosaicking matrix by minimising the reconstruction error between the training data of haematoxylin and eosin-stained pathological tissue and a demosaicked MSI using a cost function. The effectiveness of the proposed MSFA and demosaicking is demonstrated by comparing the recovered spectrum and RGB image with those obtained using a conventional method.

4.5 Conclusion

This chapter presented the achievements of multiscale spatiotemporal statistical models during the whole period of the multidisciplinary computational anatomy project from fiscal year (FY) 2014 to FY 2018. Shimizu et al. studied on spatiotemporal statistical model along a time axis of embryos and children as well as modelling of organs with nested and neighbouring constraints. Super-resolution techniques by dictionary learning and deep learning were developed by Shimizu et al. and Shouno et al. Kobayashi et al. developed algorithms for image understanding of microscopic images of a KPC mouse.

Acknowledgements Studies of Subsection 4.2 were approved by the ethics committee of the Graduate School and Faculty of Medicine of Kyoto University (R0316, R0989) and the ethics committee of Tokyo University of Agriculture and Technology (No. 30-28). Study of a part of Subsection 4.2.2 was approved by the ethics committee at Children's National Medical Center (approval no. 00003792) and the ethics committee of Tokyo University of Agriculture and Technology (No. 30-31). Study of a part of Subsection 4.3 was approved by the ethics committee at Fukui University (approval no. 20100064) and the ethics committee of Tokyo University of Agriculture and Technology (No. 30-27).

The studies in Sect. 4.2 were collaboration researches with Prof. Hontani (Nagoya Institute of Technology), Prof. Matsuzoe (Nagoya Institute of Technology), Prof. Yamada (Kyoto University), Prof. Takakuwa (Kyoto University) and Dr. Linguraru (Children's National Health System). The micro-CT volumes resected from a KPC mouse of pancreatic cancer in Sections 4.3 and 4.4 was provided by Prof. Hashizume (Kyushu University), Prof. Ouchida, Dr. Iwamoto (Kyushu University) and Prof. Mori (Nagoya University). The studies in Sect. 4.3.2 were collaboration researches with Prof. Kido (Yamaguchi University) and Prof. Inai (Fukui University). The authors would like to thank all these people for valuable data, discussions and comments.

References

1. Durrleman S, Pennec X, Trouvé A, Braga J, Gerig G, Ayache N. Toward a comprehensive framework for the spatiotemporal statistical analysis of longitudinal shape data. Int J Comput Vis. 2013;103(1):22–59.
2. Kishimoto M, Saito A, Takakuwa T, Yamada S, Matsuzoe H, Hontani H, Shimizu A. A spatiotemporal statistical model for eyeballs of human embryos. IEICE Trans Inf Syst. E100-D, 7:1505–15.
3. Kasahara K, Saito A, Takakuwa T, Yamada S, Matsuzoe H, Hontani H, Shimizu A. A spatiotemporal statistical shape mode of the brain surface during human embryonic development. Adv Biomed Eng. 2018;7:146–55.
4. Kameda T, Yamada S, Uwabe C, Suganuma N. Digitization of clinical and epidemiological data from the Kyoto collection of human embryos: maternal risk factors and embryonic malformations. Congenit Anom. 2012;52(1):48–54.
5. Saito A, Nakayama K, Porras AR, Mansoor A, Biggs E, Linguraru MG, Shimizu A. Construction of a spatiotemporal statistical shape model of pediatric liver from cross-sectional data. Proc of Medical Image Computing and Computer-Assisted Intervention., Part I, LNCS. 2018;11071:676–83.
6. Saito A, Tsujikawa M, Takakuwa T, Yamada S, Shimizu A. Level set distribution model of nested structures using logarithmic transformation. Med Image Anal. 2019;56:1–10. https://doi.org/10.1016/j.media.2019.05.003.
7. Shimizu A, Hontani H, Kobayashi N, Shono H, Mori K, Iwamoto C, Ouchida K, Oda Y, Hashizume M. A multi-scale and multi-modality statistical model of pancreas. Int J Comput Assist Radiol Surg. 2016;10(suppl 1):S165–6.
8. Ishida J, Saito A, Shimizu A. Multi-modality statistical model between a thoracic CT volume and photographs of lung cryosections of a cadaver. Int J Comput Assist Radiol Surg. 2017;12(suppl 1):S192.
9. Timofte R, Smet VD, Gool LV. Anchored neighborhood regression for fast example-based super-resolution. Proc of IEEE International Conference on Computer Vision. 2013;
10. Tozawa K, Saito A, Kido S, Inai K, Kimura H, Shimizu A. Super resolution of a lung CT volume using a generative adversarial network. Int J Comput Assist Radiol Surg. 2018;13(suppl 1):S170.
11. Ledig C, Theis L, Huszar F, Caballero J, Cunningham A, Acosta A, Aitken A, Tejani A, Totz J, Wang Z, Shi W. Photo-realistic single image super-resolution using a generative adversarial network. Proc of CVPR. 2017:4681–90.

12. Jiwon K, Lee JK, Lee KM. Accurate image super resolution using very deep Convoutional networks. In: IEEE Conf. On CVPR; 2016. p. 1646–54.

13. Zhang K, Zuo W, Chen Y, Meng D, Zhang L. Beyond a Gaussian Denoiser: residual learning of deep CNN for image Denoising. IEEE Trans Image Processing. 2017;26(7):3142–55.

14. Haralick RM, Shanmugam K, Dinstein I. Texture feature for image classification. IEEE Trans Syst Man Cybernet. 1973;SMC-3:6.

15. Galloway MM. Texture analysis using gray level run lengths. Computer Graphics and Image Processing. 1975;4(2):172–9.

16. Cortes C, Vapnik V. Support-vector networks. Mach Learn. 1995;20:273–97.

17. Komagata H, Ishikawa M, Shinoda K, Kobayashi N, Iwamoto C, Ohuchida K, Hashizume M. 3-Dimensional Anatomical Reconstruction from Pancreas Serial Sections Using Texture Features and Machine Learning. In: Proc. of the 6th IIEEJ International Conference on Image Electronics and Visual Computing (IEVC 2019); 2019.

18. Komagata H, Ishikawa M, Shinoda K, Kobayashi N, Iwamoto C, Ohuchida K, Hashizume M. 3D reconstruction of pancreatic ducts and collagen fibers from pathological images of pancreas serial sections. In: Image Electronics and Visual Computing Workshop; 2017.

19. Ishikawa M, Okamoto C, Shinoda K, Komagata H, Iwamoto C, Ohuchida K, Hashizume M, Shimizu A, Kobayashi N. Detection of pancreatic tumor cell nuclei via a hyperspectral analysis of pathological slides based on stain spectra. OSA Biomedical Optics Express. 2019;10(9):4568–88.

20. Hashimoto E, Ishikawa M, Shinoda K, Hasegawa M, Komagata H, Kobayashi N, Mochidome N, Oda Y, Iwamoto C, Ohuchida K, Hashizume M. Tissue classification of liver pathological tissue

21. Ogawa S, Shinoda K, Hasegawa M, Kato S, Ishikawa M, Komagata H, Kobayashi N. Demosaicking method for multispectral images based on spatial gradient and inter-channel correlation. In: International symposium on multispectral colour science; 2016. p. 157–66.

22. Yanagi Y, Shinoda K, Hasegawa M, Kato S, Ishikawa M, Komagata H, Kobayashi N. Optimal transparent wavelength and arrangement for multispectral filter array. In: IS & T international symposium on electronic imaging, IPAS-024; 2016. p. 1–5.

23. Shinoda K, Ogawa S, Yanagi Y, Hasegawa M, Kato S, Ishikawa M, Komagata H, Kobayashi N. Multispectral filter array and demosaicking for pathological images. APSIPA ASC. 2015:697–703.

24. Yanagi Y, Shinoda K, Hasegawa M, Kato S, Ishikawa M, Komagata H, Kobayashi N. Optimal multispectral filter array considering transparent wavelength and arrangement. IEICE Trasn Info Syst. 2016;J99-D(8):794–804. (Japanese)

25. Shinoda K, Kobayashi N, Katoh A, Komagata H, Ishikawa M, Murakami Y, Yamaguchi M, Abe T, Hashiguchi A, Sakamoto M. An efficient wavelet-based ROI coding for multiple regions. IEICE Trans on Fundamentals. 2015;E98-A(4):1006–20.

26. Shinoda K, Kawase M, Hasegawa M, Ishikawa M, Komagata H, Kobayashi N. Joint optimization of multispectral filter arrays and demosaicking for pathological images. IIEEJ Transactions on Image Electronics and Visual Computing. 2018;6(1):13–21.

27. Shinoda K, Kawase M, Hasegawa M, Ishikawa M, Komagata H, Kobayashi N. Optimal spectral sensitivity of multispectral filter array for pathological images. Image Electronics and Visual Computing Workshop. 2017;1P-10

Fundamental Technologies for Integration and Pathology in MCA

5

Yoshinobu Sato and Yoshito Otake

Abstract

This chapter describes fundamental technologies on integrating function and pathology with macro-normal anatomy to construct multidisciplinary computational anatomy (MCA) models. AI-based segmentation with Bayesian U-net and cross-modality image synthesis (image-to-image translation) with CycleGAN are firstly developed to obtain static macro-anatomy models from different modalities of clinical images, and then two approaches are described for MCA modeling: (1) construction of templates of high-fidelity multi-scale anatomy models and registration of the templates to clinical images to obtain patient-specific models for in-silico simulations, and (2) functioning and pathology anatomy modeling by integrating dynamic behaviors and disease progression. These two MCA modeling approaches are applied to musculoskeletal functional anatomy including muscle fiber arrangements and muscle-bone attachments, skeletal motions including rig cage motions and hip joint postures, and liver deformations due to fibrosis progression.

Keywords

Deep learning · Skeletal motion · Disease progression · Muscle anatomy · Liver fibrosis

5.1 Introduction

The multidisciplinary computational anatomy (MCA) research extends the computational models of the static macro-anatomy of healthy adult human bodies so as to integrate multiscale (macro to micro), temporal (dynamic and longitudinal), functional (physiological), and pathological (disease) aspects in the models. In this chapter, the functional and pathological aspects are especially focused on. Nevertheless, the multiscale and temporal aspects are closely related to function and pathology. For example, physiological units, which play a key role in organ function, are typically micro-scale structures, dynamic behaviors of the anatomy are often directly related to organ function and pathology, disease progression is a temporal (longitudinal) phenomenon, and so on. Consequently, this chapter addresses multiscale and temporal aspects as well.

Segmentation of organs at the macro anatomy level is a prerequisite in our MCA approaches [1, 2]. Firstly, artificial intelligence (AI)-based segmentation is applied to obtain static macro-anatomy models from various modalities of clinical images [3, 4]. Then, two MCA approaches are addressed for integrating function and pathology. (1) High-fidelity anatomy modeling for in-silico functional simulations [5–8]: Cadaveric data are used to acquire micro-scale and physical-attachment information of anatomy which is necessary for functional simulations but hardly or partially imaged in clinical images. We develop methods for constructing templates of high-fidelity anatomy models from the cadaveric data and their personalization using patient images in combination with AI-based segmentation and template registration. The usefulness of the methods is demonstrated for musculoskeletal anatomy. (2) Functioning and pathological anatomy modeling [9–13]: Focusing on the musculoskeletal anatomy, dynamic anatomy is investigated while the musculoskeletal system is functioning, for example, the hip joint anatomy during standing and the rib cage anatomy during breathing. In addition, anatomy-based pathological predictive models are constructed, with which biopsy-proven pathological data are associated, in application to liver fibrosis staging.

In the following, AI-based segmentation with Bayesian U-net and cross-modality image synthesis with CycleGAN are firstly addressed, and then the two MCA modeling

Y. Sato (✉) · Y. Otake
Division of information science, Nara Institute of Science and Technology, Nara, Japan
e-mail: yoshi@is.naist.jp

approaches, high-fidelity anatomy modeling and functioning/pathological anatomy modeling, are described.

5.2 AI-Based Segmentation and Cross-Modality Image Synthesis

The CT segmentation tasks, which were originally solved using traditional methods [1, 2], have been re-implemented using AI, that is, deep learning. As an example, Fig. 5.1a shows an overview and typical results of muscle segmentation using our novel Bayesian U-net [3]. The segmentation error, measured by ASD (average symmetric surface distance), in 19 individual muscle segmentation of the hip and thigh area was significantly reduced from 1.56 mm by the previous hierarchical multi-atlas method [2] to 0.99 mm by Bayesian U-net [3]. In addition, prediction of segmentation accuracy became possible using an uncertainty metric measured by Bayesian U-net. Correlation coefficients of predicted and actual segmentation accuracy (Dice measure) of 19 muscles were 0.716 in average. Further, the method was successfully applied to public domain CT data from TCIA (The Cancer Imaging Archive) database [14] as well as CT data of the same hospital as training data. Segmentation accuracy of muscles has shown to be sufficiently high in large-scale testing, and now automatically segmented regions can be practically used for functional and pathology modeling with minimum operator interventions.

In order to apply segmentation methods developed for CT data to a different modality of data, for example, MRI data as assumed here, manual traces on MRI data are typically needed as training data. However, manual tracing is highly time-consuming. To avoid it, one approach is to synthesize CT images from MR images, and then existing CT segmentation tools are applied to synthesized CT images. Figure 5.1b shows a diagram and typical results of our MR-CT image synthesis (image-to-image translation) and segmentation tool using CycleGAN [15] without using paired CT and MRI data [4]. In our method, just a bunch of unpaired CT and MRI data having a similar field of views are used. When more than two hundred cases of CT and MRI data are used as training data in combination with our newly introduced gradient consistency loss, significant accuracy improvement in image synthesis and final segmentation was observed compared with using around 20 cases of training data.

5.3 High-Fidelity Anatomy Modeling for Functional Simulations

One aim of the MCA research is to provide a framework for reconstructing patient-specific multiscale anatomy models, which are directly applicable to in-silico functional simulations, from clinical images. One approach is to construct the model templates, and then non-rigidly register the templates to clinical image data for reconstructing patient-specific models. The templates include high-fidelity anatomical information such as micro-scale structures and physical connections which are unable or difficult to be obtained only from patient clinical images. Because nonrigid registration

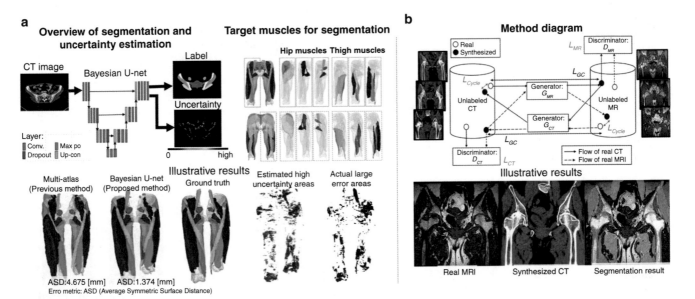

Fig. 5.1 AI-based segmentation and image synthesis (image translation) for macro-anatomy modeling from different modality data. (**a**) Overview, target muscles for segmentation, and illustrative results of our novel Bayesian U-net segmentation and uncertainty estimation [3]. Using Bayesian U-net, much more accurate results were obtained. In addition, estimated high uncertainty areas were well-correlated with actual large error areas. (**b**) Method diagram and illustrative results of our MR-CT image synthesis (image-to-image translation) using CycleGAN [15] without using paired CT and MRI data [4]

of the template to the whole image can be unstable and inaccurate, we perform the registration for each segmented macro-level anatomical region, by which patient-specific information on microstructures and physical connections, partly included in the patient images, is properly reconstructed in the models. We investigated this approach with applications to the musculoskeletal system.

Bones and muscles are the main anatomical structures in the musculoskeletal system, whose patient-specific macro anatomy is provided by AI-based segmentation described in 6.2. Detailed anatomical information is useful for functional simulation about microarchitectures (such as muscle fibers and trabecular bones) inside macrostructures, connective tissues (such as tendons, ligaments, and fasciae), physical connections among tissues, and so on, which are often not or only partially captured in routinely available or less invasive clinical images of patients. In the following, firstly, details of construction of the musculoskeletal high-fidelity model templates are described, and then reconstruction of the patient-specific models from routinely available clinical 3D data is demonstrated.

5.3.1 Constructing High-Fidelity Model Templates from Cadaver Cryosection Data

In order to construct musculoskeletal high-fidelity model templates, muscle fiber arrangement templates are modeled from the Visible Korean (VK) dataset, high-resolution 3D volume data (0.1 or 0.2 mm^3 voxel size) of the entire bodies of the cadavers, which are serially cryosectioned at 0.1 or 0.2 mm and stored as high-resolution optical images [16]. Figure 5.2a shows typical results of muscle fiber tractography from the VK 3D data. We firstly reduced artifacts caused by temporal variations of the lighting condition during a long-term data acquisition process of the VK data, and then performed structure tensor computation and fiber tracking [5]. Muscles are the power source of human movements. Muscles move the bones across the joint by their contraction. Their fiber directions are biomechanically important because muscle contraction happens along these directions. Muscle contraction force is transmitted to the bones through the muscle-bone attachment areas (origin and insertion for the fixed and moving bones, respectively). Therefore, muscle-bone attachment areas need to be determined in order to make muscle fiber arrangements utilized in biomechanical simulations.

Figure 5.2b shows the probabilistic atlases of the muscle attachment areas of the gluteal muscles used in our approach of the attachment area estimation. Unlike previous works based on cross-validation using two cadavers [17], the probabilistic atlases of the attachment areas were constructed by

3D digitization of 8 cadavers (later extended to 20 cadavers), and the patient-specific attachment areas were estimated using the probabilistic atlas and patient bone shapes [6]. Figure 5.2c shows the fiber templates of gluteus maximus, medius, and minimus muscles reconstructed from the VK data [7, 8]. Global fiber arrangement models based on B-splines are fitted to muscle fiber segments extracted from the VK data so that the fibers are constrained to start from the estimated origin areas and end to the insertion while covering the segmented muscle region. The size of the B-spline grid matrix to represent the deformation of the fiber arrangement was gradually increased so as to stabilize the deformation computation. Although the extracted muscle fiber segments are scattered and incomplete, the global model ensures that all fibers start from the origin area, reach the insertion area, and densely cover the whole muscle volume.

5.3.2 Patient-Specific Musculoskeletal High-Fidelity Modeling from Clinical Images

The constructed high-fidelity model templates are nonrigidly registered with patient CT/MR images. The bone and muscle regions are firstly segmented, followed by the attachment area estimation. In the formulation of the registration, shape similarity between segmented muscle regions of the patient images and the deformed templates are combined with vector field similarity between the fiber orientations estimated from the clinical images and those in the deformed fiber arrangement model templates.

Figure 5.2d shows validation results of patient-specific fiber arrangement reconstruction from clinical images in comparison with two previous methods (grid fitting and CFD) [18, 19] which only use the outer shapes of muscles to generate the fiber arrangements [8]. Our study demonstrated fully automated reconstruction of detailed patient-specific muscle fiber arrangements and muscle-bone attachment areas from routinely available clinical 3D data. The accuracy of reconstructed fiber arrangements was evaluated using the VK data of two cadavers [8]. One VK dataset (VK-1) includes high-resolution optical images (0.1 mm^3 voxel), CT (1 mm^3 voxel), and MR (1 mm^3 voxel) data, and the other (VK-2) only high-resolution optical images (0.2 mm^3). Firstly, the templates were constructed from VK-2. Then, they were non-rigidly registered with the segmented regions of CT and MR data of VK-1 to reconstruct patient-specific models. Finally, the reconstructed patient-specific models were validated using the templates constructed from VK-1. These results suggest a possibility that the current string approximation of muscle models in biomechanical simulations [20] can be replaced by volumetric high-fidelity musculoskeletal models [21].

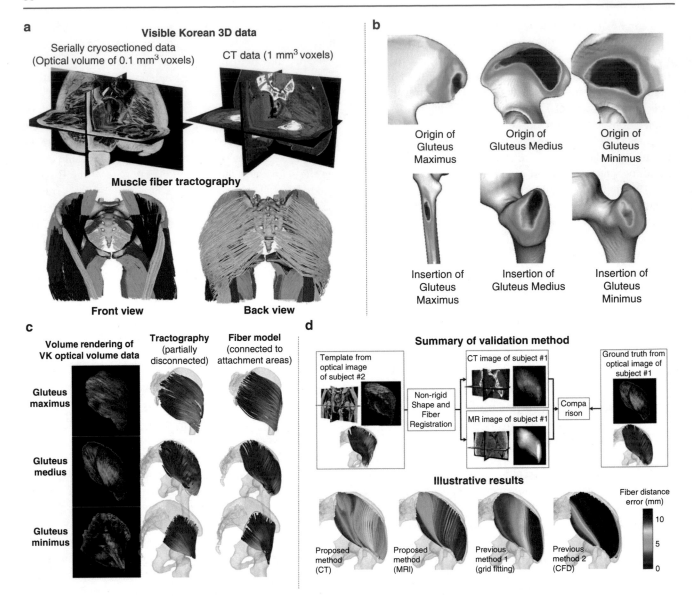

Fig. 5.2 Multidisciplinary musculoskeletal functional anatomy modeling. (**a**) Results of muscle fiber tractography from the VK 3D data [5]. (**b**) Probabilistic atlases of the attachment areas of gluteal muscles. They were constructed by 3D digitization of 8 cadavers (later extended to 20 cadavers), and the patient-specific attachment areas were estimated using the probabilistic atlas and patient bone shapes [6]. (**c**) Fiber templates of gluteus maximus, medius, and minimus muscles reconstructed from the VK data [7, 8]. The color in the fiber rendering cor-responds to the orientation at each line segment. X, Y, and Z components of the orientation vector were assigned to R, G, and B components. (**d**) Validation results of patient-specific fiber arrangement of the gluteus medius reconstructed from clinical images in comparison with two previous methods (grid fitting and CFD) [18, 19] which only use the outer shapes of muscles to generate the fiber arrangements [8]. Colors in the illustrative results denote the fiber distance error from the ground truth

5.4 Functioning and Pathological Anatomy Modeling

5.4.1 Functioning Anatomy Modeling

CT and MR scanners are typically used to acquire images of stationary anatomy at supine/prone positions. The musculoskeletal system is functioning in various positions. Especially, modeling of joint and muscle configurations at weight-bearing positions and during typical movements would be important for functional modeling. For the purpose of functional skeletal anatomy modeling, we show a couple of new approaches of combining dynamic 2D X-ray imaging with stationary 3D imaging. Figure 5.3 shows X-ray image-based chest 3D dynamic analysis of rib motion using anatomical and biomechanical (uniaxial rotation) constraints [9] combined with a robust and fast 2D-3D registration [22]. As another function modeling, supine and standing skeletal posture modeling was analyzed using a large-scale database of

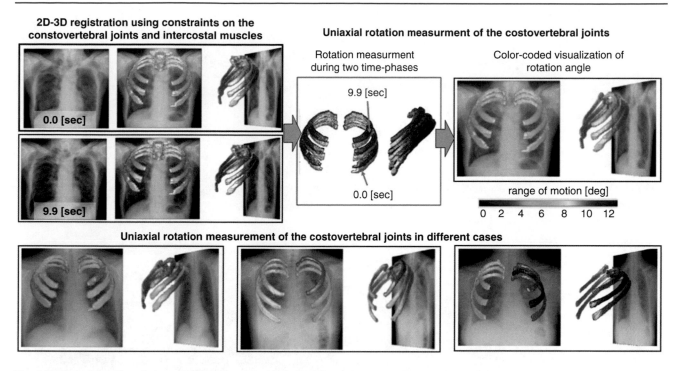

Fig. 5.3 X-ray image-based chest 3D dynamic analysis of rib motion using anatomical and biomechanical (uniaxial rotation) constraints [9] combined with a robust and fast 2D-3D registration [22]

CT data and X-ray images. In order to perform this task, we have established fully automated processes of CT segmentation and subsequent 2D-3D registration of uncalibrated X-ray images with segmented CT data. We confirm that there was no significant difference between known and unknown calibration parameters [10]. More recently, 3D skeletal postures can be estimated only from 2D X-ray images without CT [11] by using training data obtained from automated 2D-3D registration-based large-scale posture analysis [10]. This approach is now extended so as to deal with muscle analysis from 2D X-ray imaging, which will provide an AI-enhanced MCA framework of dynamic functioning anatomy modeling.

5.4.2   Pathological Anatomy Modeling

Regarding pathological anatomy modeling, the live fibrosis progression modeling is addressed. The dataset of multiparametric MR data of 56 patients of fibrosis stages F0 to F4, which were definitely diagnosed by biopsy pathology test, were collected. In addition, blood test data and stiffness measurements obtained from MR elastography data were associated with each patient. Our aim is to construct a prediction model of the fibrosis stage from the liver shape, compare to that from blood test or stiffness measurement, and eventually combine shape, blood test, and stiffness into a prediction scheme.

Initially, we used principal component analysis (PCA) to derive the shape feature vectors [12]. Then, we used partial least squares (PLS) instead, which is regarded as a supervised dimensionality reduction scheme to optimize classification or regression with known labels (in this case, known fibrosis stages) compared with PCA as an unsupervised one [13]. We confirmed that the SVM classification accuracy was largely improved by using PLS coefficients in comparison with PCA. Although shape-based classification was inferior to stiffness and blood test, its performance was close to them regarding classification between F0-F1 and F2-F4. Figure 5.4 shows shape changes along the normal vector of the SVM discrimination plane. In these changes, not only commonly known shape changes such as right robe shrinkage and left lobe enlargement but also enlargements in the posterior part of the right lobe and the caudate lobe were observed, which were found in a recent clinical study [23]. This is regarded as an example of MCA models due to disease progression.

5.5   Summary

We have described recent research achievements on fundamental technologies for integrating function and pathology of tissues, organs, and systems in the MCA models. We extended macro anatomy modeling so as to incorporate function and pathology information. We developed a framework of high-fidelity anatomy modeling for functional simulations

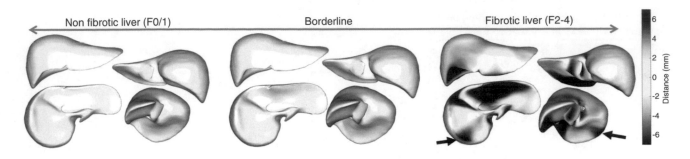

Fig. 5.4 Disease progression anatomy model using shape changes along the normal vector of the SVM discrimination plane [13]. Arrows indicate shape changes in the fibrotic liver suggested by a recent clinical paper [23]

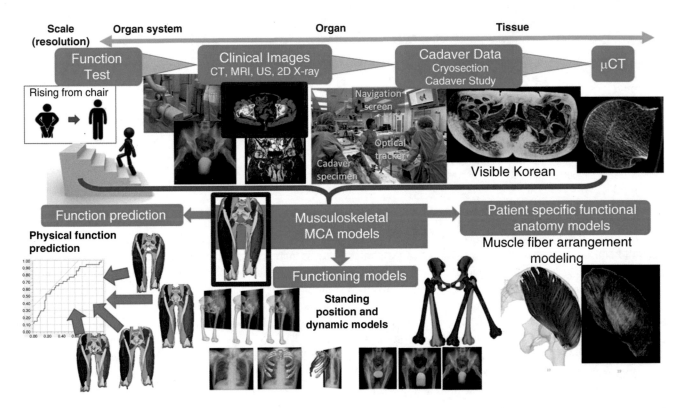

Fig. 5.5 Overview of musculoskeletal MCA modeling. Macro-level computational anatomy modeling from CT images is indicated by red frames, which are the starting point of the MCA modeling

using the template including information on micro-scale anatomy and physical connection constructed from cadaveric data, with application to musculoskeletal multi-scale modeling. In addition, disease progression was addressed with application with liver fibrosis modeling. AI-based segmentation and image synthesis for macro-anatomy modeling from different modalities of data, and 2D-3D integration technologies for functioning anatomy modeling were also described and enhanced the significance of the MCA models.

In the MCA project, a new discipline of computational modeling of human anatomy has been established, which integrates multiscale anatomy, function (physiology), pathol-

ogy (disease), and temporal changes (growth/progression). The MCA models are constructed from different modalities of data and used to reconstruct patient-specific multiscale and dynamic anatomy models, in which functional and pathological aspects are integrated, from clinical image data.

We have particularly investigated the MCA modeling for the musculoskeletal system, in which muscle fiber arrangements and bone-muscle attachment areas are modeled, and its patient-specific reconstruction schemes from clinical images are formulated. Further, musculoskeletal anatomy modeling during functioning such as standing and joint motion has been incorporated in the MCA modeling. Fig. 5.5 shows an overview of musculoskeletal multiscale MCA modeling, where our ongoing study about physical function prediction from muscle anatomy models is included as organ-system level modeling. Musculoskeletal analysis is addressed in a variety of research fields, including biomechanics, robotics, rehabilitation, orthopedics, ergonomics, prosthesis design, sports science, and so on. This new discipline developed in this research provides a framework for personalized functional musculoskeletal anatomy modeling, which has not been available before. A new integrated research filed of personalized human body sciences will emerge centered at this newly developed discipline. As future work, the developed MCA models should be shared among research communities and utilized for in-silico functional analysis and simulations of the human body.

References

1. Okada T, Linguraru MG, Hori M, Summers RM, Tomiyama N, Sato Y. Abdominal multi-organ segmentation from CT images using conditional shape–location and unsupervised intensity priors. Med Image Anal. 2015 Dec 1;26(1):1–8.
2. Yokota F, Otake Y, Takao M, Ogawa T, Okada T, Sugano N, Sato Y. Automated muscle segmentation from CT images of the hip and thigh using a hierarchical multi-atlas method. Int J Comput Assist Radiol Surg. 2018 Jul 1;13(7):977–86.
3. Hiasa Y, Otake Y, Takao M, Ogawa T, Sugano N, Sato Y. Automated muscle segmentation from clinical CT using Bayesian U-net for personalized musculoskeletal modeling. IEEE Trans Med Imaging. 2019 Sep 10;39(4):1030–40.
4. Hiasa Y, Otake Y, Takao M, Matsuoka T, Takashima K, Carass A, Prince JL, Sugano N, Sato Y. Cross-modality image synthesis from unpaired data using CycleGAN. In: International workshop on simulation and synthesis in medical imaging. Cham: Springer; 2018 Sep 16. p. 31–41.
5. Otake Y, Miyamoto K, Ollivier A, Yokota F, Fukuda N, O'Donnell LJ, Westin CF, Takao M, Sugano N, Chung BS, Park JS, Sato Y. Reconstruction of 3d muscle fiber structure using high resolution cryosectioned volume. In: International workshop and challenge on computational methods and clinical applications in musculoskeletal imaging. Cham: Springer; 2017 Sep 10. p. 85–94.
6. Fukuda N, Otake Y, Takao M, Yokota F, Ogawa T, Uemura K, Nakaya R, Tamura K, Grupp RB, Farvardin A, Armand M, Sugano N, Sato Y. Estimation of attachment regions of hip muscles in CT image using muscle attachment probabilistic atlas constructed from measurements in eight cadavers. Int J Comput Assist Radiol Surg. 2017 May;12(5):733–42.
7. Otake Y, Yokota F, Fukuda N, Takao M, Takagi S, Yamamura N, O'Donnell LJ, Westin CF, Sugano N, Sato Y. Patient-specific skeletal muscle fiber modeling from structure tensor field of clinical CT images. In: International conference on medical image computing and computer-assisted intervention. Cham: Springer; 2017 Sep 10. p. 656–63.
8. Otake Y, Takao M, Fukuda N, Takagi S, Yamamura N, Sugano N, Sato Y. Registration-based patient-specific musculoskeletal modeling using high fidelity cadaveric template model. In: International conference on medical image computing and computer-assisted intervention. Cham: Springer; 2018 Sep 16. p. 703–10.
9. Hiasa Y, Otake Y, Tanaka R, Sanada S, Sato Y. Recovery of 3D rib motion from dynamic chest radiography and CT data using local contrast normalization and articular motion model. Med Image Anal. 2019 Jan 1;51:144–56.
10. Uemura K, Takao M, Otake Y, Koyama K, Yokota F, Hamada H, Sakai T, Sato Y, Sugano N. Change in pelvic sagittal inclination from supine to standing position before hip arthroplasty. J Arthroplast. 2017;32(8):2568–73.
11. Jodeiri A, Zoroofi RA, Hiasa Y, Takao M, Sugano N, Sato Y, Otake Y. Fully automatic estimation of pelvic sagittal inclination from anterior-posterior radiography image using deep learning framework. Comput Methods Prog Biomed. 2020 Feb 1;184:105282.
12. Hori M, Okada T, Higashiura K, Sato Y, Chen YW, Kim T, Onishi H, Eguchi H, Nagano H, Umeshita K, Wakasa K. Quantitative imaging: quantification of liver shape on CT using the statistical shape model to evaluate hepatic fibrosis. Acad Radiol. 2015 Mar 1;22(3):303–9.
13. Soufi M, Otake Y, Hori M, Moriguchi K, Imai Y, Sawai Y, Ota T, Tomiyama N, Sato Y. Liver shape analysis using partial least squares regression-based statistical shape model: application for understanding and staging of liver fibrosis. Int J Comput Assist Radiol Surg. 2019 Dec 1;14(12):2083–93.
14. https://www.cancerimagingarchive.net/. As of July 30, 2020.
15. Zhu JY, Park T, Isola P, Efros AA. Unpaired image-to-image translation using cycle-consistent adversarial networks. In Proceedings of the IEEE international conference on computer vision 2017 (pp. 2223–2232).
16. Park JS, Chung MS, Hwang SB, Lee YS, Har DH, Park HS. Visible Korean human: improved serially sectioned images of the entire body. IEEE Trans Med Imaging. 2005 Feb 28;24(3):352–60.
17. Carbone V, Fluit R, Pellikaan P, van der Krogt M, Janssen D, Damsgaard M, Verdonschot N. TLEM 2.0? A comprehensive musculoskeletal geometry dataset for subject-specific modeling of lower extremity. J Biomech. 2015;48(5):734–41.
18. Kohout J, Clapworthy GJ, Zhao Y, Tao Y, Gonzalez-Garcia G, Dong F, Wei H, Kohoutová E. Patient-specific fibre-based models of muscle wrapping. Interface Focus. 2013 Apr;3(2):20120062.
19. Blemker SS, Delp SL. Three-dimensional representation of complex muscle architectures and geometries. Ann Biomed Eng. 2005;33(5):661–73.
20. Rajagopal A, Dembia CL, DeMers MS, Delp DD, Hicks JL, Delp SL. Full-body musculoskeletal model for muscle-driven simulation of human gait. IEEE Trans Biomed Eng. 2016 Oct;63(10):2068–79.
21. Modenese L, Kohout J. Automated generation of three-dimensional complex muscle geometries for use in personalised musculoskeletal models. Ann Biomed Eng. 2020 Mar;17:1–12.
22. Otake Y, Armand M, Armiger RS, Kutzer MD, Basafa E, Kazanzides P, Taylor RH. Intraoperative image-based multiview 2D/3D registration for image-guided orthopaedic surgery: incorporation of fiducial-based C-arm tracking and GPU-acceleration. IEEE Trans Med Imaging. 2011 Nov 18;31(4):948–62.
23. Ozaki K, Matsui O, Kobayashi S, Sanada J, Koda W, Minami T, Kawai K, Gabata T. Selective atrophy of the middle hepatic venous drainage area in hepatitis C-related cirrhotic liver: morphometric study by using multidetector CT. Radiology. 2010;257(3):705–14.

Part III

Basic Principles of MCA: Application Systems and Applied Techniques Based on MCA Model

Pre-/Intra-operative Diagnostic and Navigational Assistance Based on Multidisciplinary Computational Anatomy

6

Kensaku Mori

Abstract

The progress of such medical imaging devices as computed tomography (CT) or magnetic resonance (MR) scanners is continuously changing the environments of the clinical fields where medical diagnosis and treatment are performed. Medical imaging devices have become indispensable tools for safely and accurately performing medical procedures. Furthermore, although CT or MR obtains mm-level details of human anatomical structure, the μm-level structure is required in the pathological diagnosis for a final diagnosis. Handling multi-modality and multi-scale images reflect the nature of medical diagnosis and treatments. Multidimensionality exists both in the human body and in medical images obtained from the human body. For example, when we think about the resolution scale, we realize that there are many scales in medical images, such as the human-body scale, the organ scale, the organ internal-structure scale, and the cellular scale. This chapter presents a research overview of "Pre-, intra-, and post-operative diagnosis and navigational assistance based on multidisciplinary computational Anatomy," which was conducted under the MCA Project's support. In pre-, intra-, and post-operative diagnosis and navigational assistance, it is crucial to use macro- and microanatomical structure information appropriately. Seamless integration of this anatomical information is crucial for MCA-based clinical assistance. We show the overview of MCA-based clinical procedure assistance. Then we will show some examples of MCA-based clinical assistance systems that are focusing on macro- and microanatomical structures.

K. Mori (✉)
Graduate School of Informatics, Nagoya University, Information Technology Center, Nagoya, Aichi, Japan
e-mail: kensaku@is.nagoya-u.ac.jp

6.1 Introduction

The progress of such medical imaging devices as computed tomography (CT) or magnetic resonance (MR) scanners is continuously changing the environments of the clinical fields where medical diagnosis and treatment are performed. Medical imaging devices have become indispensable tools for safely and accurately performing medical procedures. Furthermore, although CT or MR obtains mm-level details of human anatomical structure, the μm-level structure is required in the pathological diagnosis for a final diagnosis. Handling multi-modality and multi-scale images reflect the nature of medical diagnosis and treatments. A micro-CT scanner is one device that enables the capture of volumetric images in μm-resolution. This means a computer needs to understand seamlessly human anatomy from macro- to micro-levels.

Multidimensionality exists both in the human body and in medical images obtained from the human body. For example, when we think about the resolution scale, we realize that there are many scales in medical images, such as the human-body scale, the organ scale, the organ internal-structure scale, and the cellular scale. One typical example in medical images is tumor growth along the time axis in the time scale. It is also possible to define the meta-anatomy scale that is a logical scale that a set of anatomical names can express. In endoscopic diagnosis and surgery, medical doctors utilize medical images of different modalities and resolutions, including CT, endoscopic ultrasound, microendoscopic images at a millimeter, hundreds of micrometers, and a few micrometers resolution, respectively.

Multi-scale image registration and navigation are interesting topics in the Multidisciplinary Computational Anatomy (MCA) project [1]. Such techniques will allow us to observe human anatomy from multilevels, including from mm to μm. The images of many different scales must be integrated. Meta-annotation at the macro- and micro-levels is also essential. For example, we can address human anatomy at the following scales: body, organ, internal organ structure, micro-anatomy structure, and cell. Tumor structure and growth analysis are good examples of multidisciplinary computational anatomy over time. Endoscopic diagnosis and surgery also require multi-scale navigation. In endoscopic procedure guidance, we need millimeter accuracy for arriving at a target location; micrometer accuracy is required for analyzing cancer structures and treating them. Multi-modal images are also required for clinical decisions. The MCA aims to develop several methodologies to analyze the medical image from the time, function, pathological axis and deploy these technologies in the clinical field.

MCA also can obtain many benefits from the multilayer neural network, also known as deep learning. Especially in the image recognition area, the convolutional neural network has made remarkable progress in conjunction with software frameworks and computing devices like GPUs. This progress has also enabled us to perform advanced image analysis, including image classification, object detection, image segmentation with high accuracy. Deep learning also had many effects on medical image analysis in MCA. As described before, the goal of multidisciplinary computation anatomy is to integrate various information from the viewpoints of the spatial axis, the time axis, the functional axis, and the pathological axis. These integrations required advanced medical image analysis. The deep learning technique enabled us to perform advanced image analysis.

Furthermore, medical diagnostic assistance devices' real clinical application using machine learning techniques has now started. AI-assisted endoscopic diagnosis assistance system is one example. Utilization of the power of deep learning accurate such integration.

This chapter presents a research overview of "Pre-, intra- and post-operative diagnosis and navigational assistance based on multidisciplinary computational Anatomy," which was conducted under the MCA Project's support. In pre-, intra-, and post-operative diagnosis and navigational assistance, it is essential to use macro- and microanatomical structure information appropriately. Seamless integration of these anatomical information is crucial for MCA-based clinical assistance. We show the overview of MCA-based clinical procedure assistance. Then we will show some examples of MCA-based clinical assistance systems that are focusing on macro- and microanatomical structures.

6.2 MCA and Clinical Procedure Assistance

In the clinical field, multidimensionality is always considered in diagnostic and surgical processes. This multidimensionality is truly multidiscipline. We propose that multi-dimensional navigation of the patient's anatomy space with the fusion of information would enable us to optimize diagnosis and provide aid to surgery. Here we define multi-dimensional navigation of the space of the patient's anatomy as technologies navigating inside a space formed of multidimensional and multimodal images interpreted in a multidisciplinary way. We call this *multidiscipline navigation* or multi-dimensional navigation (Fig. 6.1).

When we overview research of the medical image processing fields, most work focuses only on image information integration of images having almost the same resolution. For

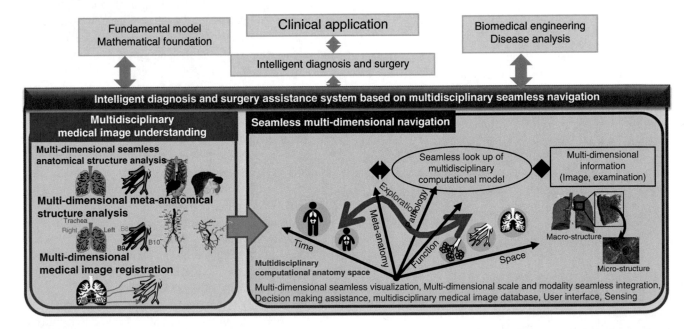

Fig. 6.1 Concept of pre-, intra-, and post-operative diagnosis and navigational assistance based on multidisciplinary computational anatomy

example, CT and MR image integration has been conducted for many years. Information obtained from one image is overlaid on another image in such work. CT images have good performance in depicting bone, while MR images have good performance in depicting soft tissues. Integration of two images compensates for information lacking in one image. Although this kind of integration is a multidisciplinary computational anatomy field, it is just a part of the project. There is a big gap between these researches and our project goal to develop a *total understanding of the human body based on medical images* or *advanced intelligent diagnostic and systems to aid surgery*. Integrating multidisciplinary medical images is the key technology to advance medical assistance based on medical images.

As stated above, MCA aims to enable us to develop an intelligent diagnostic and surgical assistance system, which enables us to analyze multidimensional medical images taken in the process of diagnosis and surgery based on the multidisciplinary computational anatomical model and enable us to navigate the space spanned by the multidimensional medical images seamlessly. Examples of MCA-based medical assistance systems include: (a) multidimensional seamless registration, (b) multidimensional seamless anatomical structure analysis, (c) multidimensional meta-anatomical structure understanding and annotation, (d) multidimensional seamless visualization, and (e) decision assistance for diagnosis and surgery.

6.3   Macroscopic Anatomy and its Therapeutic Application

6.3.1   Macroscopic Anatomical Structure Analysis

One of MCA's fundamental processes is to analyze anatomical structure from four viewpoints of the spatial axis, the time axis, the function axis, and the pathological axis. Anatomical structure segmentation from medical images is essential in MCA's anatomical structure analysis. Atlas-based segmentation was utilized to segment anatomical structures from CT or MR images. This method firstly constructs anatomical structure atlas, and then input images were registered with atlases [2]. Likelihood maps of organ regions were computed from the registration results with the input images and the atlas images. Organ regions were then determined by the graph-cut method combined with the MAP (Maximum a posteriori) estimation. Progress of FCN (fully convolutional network) has changed the scene of multi-organ segmentation. U-Net is one of the famous FCN methods used for multi-organ segmentation from medical images [3]. In this process, we prepare many original CT images and label images of organ regions on original CT

images. CT or MR images are volumetric mages consisting of many axial slice images. We manually trace organ regions on axial images to generate training datasets for the FCNs.

We have created an organ label image database of more than 600 abdominal CT scans cases. The database contains organ region label information of the stomach, the liver, the pancreas, the spleen, the artery, and the spleen [4]. This training dataset was utilized for training the cascaded U-Nets. The U-Net segmentation architecture is illustrated in Fig. 6.2. The cascaded U-Net's former step is designed to roughly segment organ regions. The later stage segments each organ precisely.

Furthermore, organ position information is utilized to improve segmentation accuracy [5]. This is based on the fact that each organ's locations in human anatomy are almost fixed. Figure 6.3 shows multi-organ segmentation results from abdominal CT images using the proposed U-Net-based framework. Cascaded segmentation workflow can achieve high accuracy in multi-organ segmentation.

6.3.2   Blood Vessel Recognition

We can see that the anatomical textbook illustrates blood vessel branching structure with anatomical name annotations. Organ regions extraction from CT images can identify organ shape by using CNN. However, anatomical organs' names are also important information to recognize patient anatomy. Surgeons describe patient anatomy by using anatomical names. We call anatomical name information meta-anatomical information. Meta-anatomical information is another axis that expands the MCA space. Meta-anatomical information is one of the key ideas of seamless understanding of human anatomy, which is essential in MCA.

In anatomical structure extraction using the CNN, we can extract organ regions from CT images by the CNN. However, a blood vessel network is extracted as one region, although blood vessels have many branches. Each blood vessel has its anatomical name. Anatomical name recognition of blood vessel branches by computers is essential for advanced human anatomy recognition and multidisciplinary recognition of a human body's total understanding. An automated anatomical labeling procedure is developed for assigning anatomical names for each branch extracted from CT images. We have utilized the machine learning method to recognize each branch of blood vessels or bronchi's anatomical names combined with rule-based logic [6]. The convolutional neural network approach is recently developed for automated anatomical labeling of blood vessels. Firstly, we extract blood vessels from CT images and obtain blood vessels' graph structures. We compute feature vectors for each blood vessel branch. Graph neural network (GNN) is applied to learn each branch's anatomical names [7]. We use trained

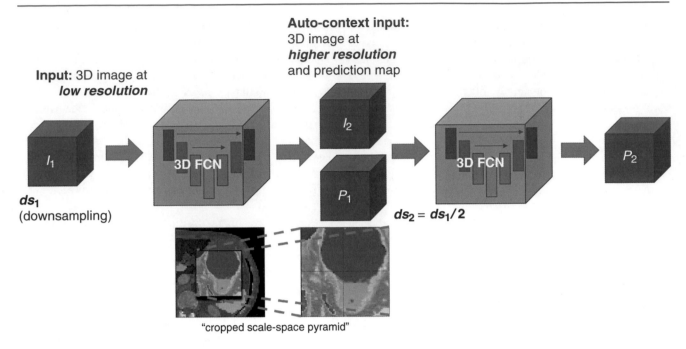

Fig. 6.2 Cascaded U-Net for multi-organ segmentation

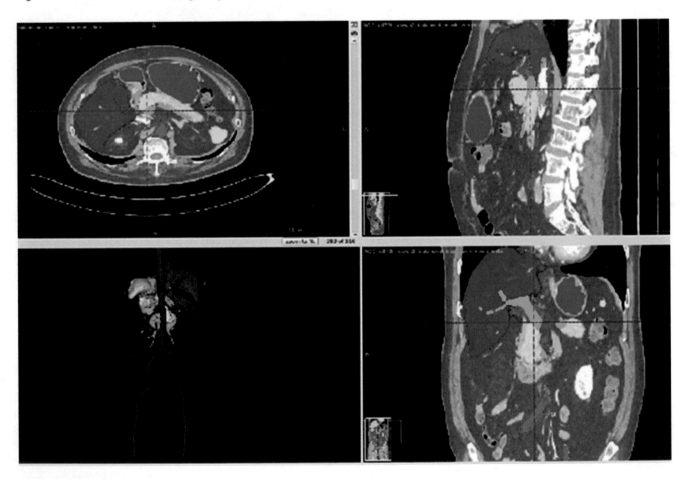

Fig. 6.3 Examples of multi-organ segmentation from CT images

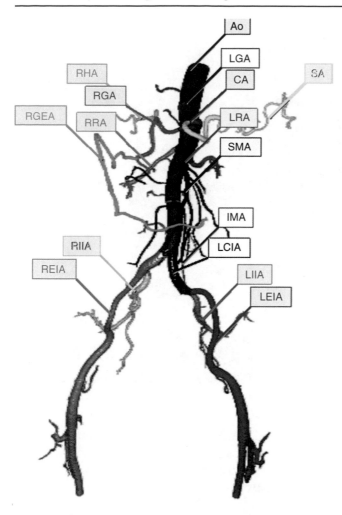

Fig. 6.4 Examples of automated anatomical labeling

GNN for inferring anatomical names of blood vessel branches. Figure 6.4 shows one example of GNN-based anatomical name recognition results.

6.3.3 Surgical Navigation

Macroscopic anatomical structure segmentation can be utilized for surgical navigation. Anatomical structure can be displayed according to the posture of the real laparoscope. Laparoscopic surgery is one of the less invasive surgical procedures. In laparoscopic surgery, a surgery inserts a laparoscopic into a patient's abdominal cavity. Surgical operations are conducted by watching laparoscopic camera videos. Although the laparoscopic procedure is less invasive, it requires higher operation skills. One of the reasons is the limited view of the laparoscopic camera. A laparoscopic surgery navigation system that assists a surgeon is desired to be expected to overcome such problems. In laparoscopic surgery navigation, the critical point is to show anatomical

structure information during laparoscopic surgery. Macroscopic anatomical structure information can be used for laparoscopic surgery assistance. For example, early gastric cancer laparoscopic surgery requires operating blood vessels related to the stomach with precisely understanding patients' anatomical structures around the stomach. It is possible to navigate laparoscopic surgical procedures by recognizing anatomical structures, including blood vessel structures or solid organ structures, from pre-operative CT images and presenting such information according to laparoscope movement. One example is displaying anatomical laparoscopic surgical navigation information based on a laparoscope's real-time tracking information by the optical or electromagnetic tracker. Such a navigation system requires coordinate system registration among the preoperative CT images, the patient, the laparoscopic camera, and the laparoscope body (Fig. 6.5). Coordinate system registration is typically performed based on markers attached to a patient body surface or anatomical structure that can be identified on a body surface and preoperative CT images [8]. Anatomical structures can now be easily segmented from abdominal CT images by the deep learning model using U-Net. Blood vessel names can also be identified on CT images by utilizing a graph-convolution network, a deep learning method.

The CNN is also useful for surgical scene recognition. For example, the CNN can recognize areas or blood vessels where a surgeon operates. Anatomical structures observed by a laparoscope can be identified by FCN (Fig. 6.6). Hayashi and his colleagues develop the method that displays 3D anatomical structures extracted from CT images based on surgical scene recognition results [9]. Laparoscopic surgical navigation systems introduced in the previous paragraph utilized positional trackers, including optical or electromagnetic trackers. However, the scene recognized system does not require such tracking systems that are typically expensive and are hard to set up in real clinical scenes. Although it is impossible to perform precise synchronization with laparoscope movement precisely, it is possible to automatically display important anatomical structure information at the critical point scenes during surgery. This system is one of the new applications of CNN. Figure 6.7 shows an example of such a system.

6.3.4 3D-Printed Anatomical Model

3D printing is one of the advanced techniques to reproduce 3D anatomical structures. 3D printing systems fabricate physical anatomical models based on anatomical structure information. A surgeon can touch and observe the 3D shape of organs using a 3D organ model fabricated by a 3D printer.

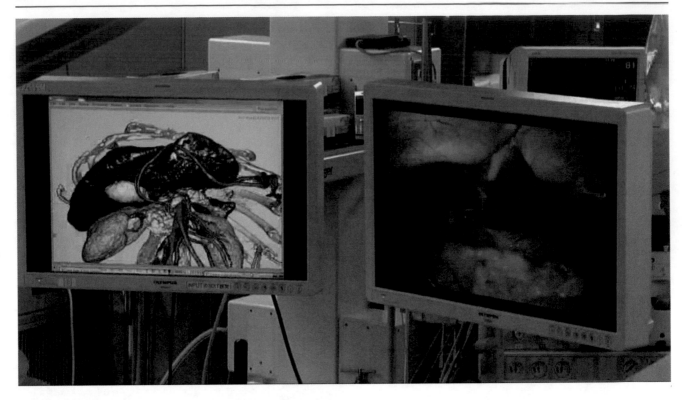

Fig. 6.5 Laparoscopic surgery navigation based on multi-organ segmentation

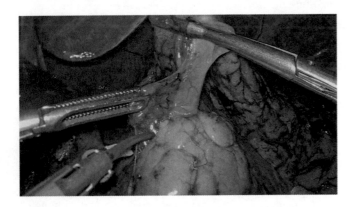

Fig. 6.6 Blood vessel detection on laparoscopic video frames

Intuitive observation becomes possible by bringing the 3D printed model to surgeons' operating rooms. Although it is possible to render organ shape in a three-dimensional way on a computer screen, it is still on a 2D computer display. The surgical navigation system presented in the previous section is a typical example of a 2D-based 3D display. If we use 3D printed organ mode, a surgeon can understand 3D structure information very intuitively. CNN-based MCA enables us to automate 3D printing data generation [10]. The CNN-based organ region segmentation method can generate organ-shape information (STL) used in a 3D printer. Figure 6.8 shows an example of a 3D printed liver model based on CNN-based organ region segmentation from the liver's preoperative CT images.

6.4 Colonoscopy Assistance Based on Macro- and Microanatomical Structure Analysis

Colonoscopy diagnosis assistance based on macro- and microanatomical structure recognition.

MCA project also conducted several developments of methods that assist colonoscope diagnosis assistance. Computer assistance of colonoscopy consists of two processes: (a) colonic polyp detection and (b) colonic polyp classification. In colonic polyp detection assistance, a computer detects colonic polyps in colonic video scenes and gives a physician some warning. Mori and his colleagues have created an automated polyp detection method based on CNN. The CNN is trained to detect frames where colonic polyps are observed by giving training images. C3D network is utilized to implement this colonic polyp detection function [11]. If colonic polyps are found in colonic video frames, the system makes a warning sound and changes colonoscopy videos' border color into yellow for giving some attention to a physician (Fig. 6.9).

In colonic polyp classification, a computer classifies each colonoscopic polyp into neoplastic or non-neoplastic polyps based on microscopic structure analysis, including cell patterns or tiny blood vessels. This method classifies super magnified images of colonic polyp surfaces. The super magnified colonoscope is a new endoscope that can take both conven-

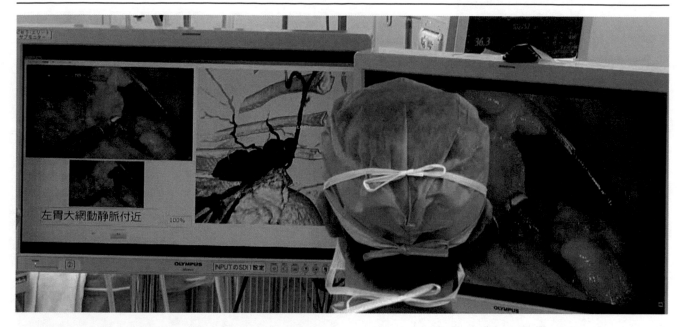

Fig. 6.7 Surgical assistance information display based on surgical scene analysis

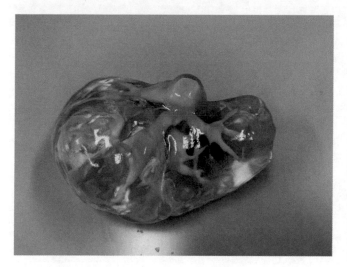

Fig. 6.8 3D-printed liver model based on organ segmentation results by 3D U-Net

function is executed when we push the button located at the manipulation handle of a super magnified endoscope. If each category's likelihood is lower than a given threshold, the system asks a physician to retake an image. This becomes possible because computer diagnosis is executed in real-time during endoscopy. We iterate this computer-assisted diagnosis process at a different position on a colonic polyp, and then finally, a physician decides polyps' pathological types. Figure 6.8 shows an example of colonic polyp qualitative diagnosis.

This colonoscopy assistance system assists a physician in the anatomical structure viewpoints of macro (polyp detection) and micro (cell structure) levels. It is an excellent example of the seamless integration of macro- and micro-level computational anatomy. These systems are approved as official medical devices by the PMDA (Pharmaceuticals and Medical Devices Agency), Japan.

6.5   Micro-Scale Anatomical Structure Segmentation

6.5.1   Micro-CT Images

In MCA, it is important to establish a framework that can seamlessly handle macro- and microanatomical structures. Microscopic anatomical structure segmentation is also one of the vital tasks in MCA. However, volumetric image scanners used in the clinical scenes are hard to obtain microscopic volumetric images of anatomical structures Typical resolution of CT scanners used in the clinical field are about 0.5 mm × 0.5 mm × 0.5 mm per voxel. It is necessary to use

tional colonoscopic images and super magnified colonoscopic images. We have developed a method that can classified images were taken by the super magnified endoscope into neoplastic images or non-neoplastic images [12]. Hilger-tensor features are computed from super magnified endoscopic images, and we classify such features by support vector machi (SVM) technique to obtain classification results [13]. The system displays classification results with their likelihood in real time during endoscopic procedures. It is possible to show the classification results of two classes and four classes (Fig. 6.10).

Furthermore, procedure recommendation functions such as biopsy are implemented in such a system. This assistance

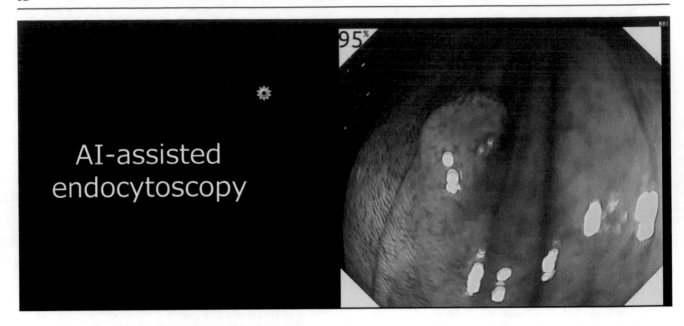

Fig. 6.9 Example of automated colonic polyp frame detection

Fig. 6.10 Example of automated colonic polyp pathological type classification if super magnified colonoscopic image

high-resolution CT devices to capture microscopic anatomy. One solution is to use micro-CT devices that can be easily operated on a desktop. Such devices can capture microscopic CT images in micrometer resolution of a specimen. Figure 6.11a shows a picture of a desktop-type micro-CT imaging device and an example of a CT image of an inflated lung specimen taken by the micro-CT scanner. Furthermore, the three-dimensional rendering result is shown in Fig. 6.11. From Fig. 6.11, we see the micro-CT can capture the micro-anatomical lung structures' image, including the alveoli as sponge structure.

We have conducted microscopic anatomical structure analysis segmentation and microanatomical structure esti-

mation from clinical CT images. Also, we can observe invasions of lung cancer in the alveoli-level by using micro-CT images. Micro-CT scanning enables us to analyze the micro-anatomical structure of the lung.

6.5.2 Micro-Anatomical Structure Segmentation from Micro-CT Images of Lung Specimen

As we described in the previous section, micro-CT imaging enables us to diagnose diseases, including lung cancer at the alveoli level. Although the CNN can be utilized to

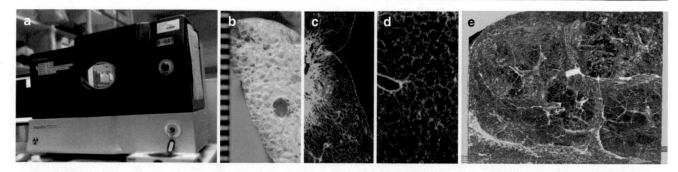

Fig. 6.11 Example of micro-CT images of inflated lung specimen; (**a**) desktop-type micro-CT scanner, (**b**) lung specimen, (**c**, **d**) examples of micro-CT images, (**e**) example of 3D rendering of micro-CT image

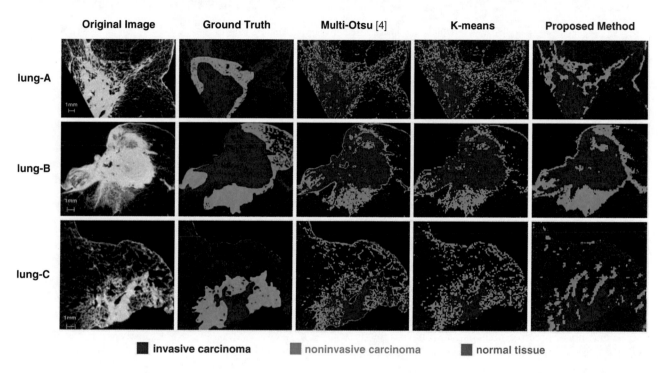

Fig. 6.12 Examples of micro-CT image segmentation

analyze such images to segment micro-CT images into invasive cancer regions, non-invasive cancer regions, and normal regions, a lot of training data (manual trace of each region) is required to make a CNN network to output desired segmentation results. Manual trace of each region on micro-CT images is a tough task. A spherical clustering-based method was developed for micro-CT image segmentation to overcome such problems, one of the unsupervised approaches. In the clustering method, it is essential to find suitable features to be used for clustering. Moriya et al. have developed a variational autoencoder-based approach for clustering [14]. Micro-CT images of the inflated lung specimen were segmented into invasive, noninvasive, and normal regions by this method. Figure 6.12 shows examples of spherical k-means clustering of micro-CT images.

6.5.3   Super-Resolution of Clinical CT Images Based on Micro-CT Image Database

Image generation method based on deep learning including GAN (generative adversarial network) has made remarkable progress in the computer vision field The GNA is useful not only in natural images (typical scenes captured by conventional camera) but also in medical images. The CycleGAN is especially known as the method that mutually converts images of two different domains [15]. If we can design an appropriate loss function, the CycleGAN can be used for domain conversion and super-resolution of input images. Zheng et al. has proposed a method for super-resolution of clinical CT images taken at the clinical field [16]. This method utilizes micro-CT images and clinical (macro)-CT images for training. Clinical CT images are converted to

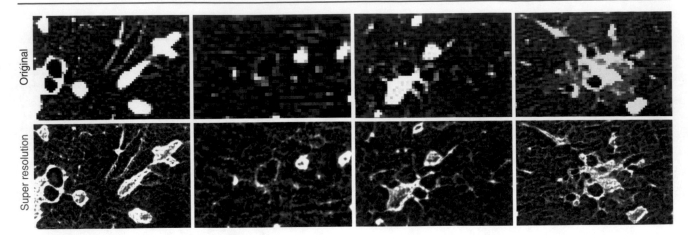

Fig. 6.13 Example of super-resolution of clinical CT image based on micro-CT image database

micro-CT-level high-resolution images by the trained CycleGAN. The method shown in utilizes the loss function that employs the SSIM index in Cycle loss to obtain stable super-resolution results. Figure 6.13 shows an example of super-resolution of clinical CT images. This method us to achieve seamless integration of macro- and microanatomical structure information for us.

6.6   Conclusion

This chapter presented a research overview of "Pre-, intra- and post-operative diagnosis and navigational assistance based on multidisciplinary computational anatomy," which was conducted under the MCA Project's support. In pre-, intra-, and post-operative diagnosis and navigational assistance, it is vital to use macro- and microanatomical structure information appropriately. Seamless integration of this anatomical information is a key point of MCA-based clinical assistance. We showed the overview of MCA-based clinical procedure assistance. Also, we presented some examples of MCA-based clinical assistance systems that are focusing on macro- and microanatomical structures.

References

1. http://www.tagen-compana.org
2. Wolz R, Chu C, Misawa K, Fujiwara M, Mori K, Rueckert D. Automated abdominal multi-organ segmentation with subject-specific atlas generation. IEEE Trans Med Imaging. 2013 Sep;32(9):1723–30. https://doi.org/10.1109/TMI.2013.2265805.
3. Ronneberger O, Fischer P, Brox T. U-net: convolutional networks for biomedical image segmentation. Lect Notes Comput Sci. 2015;9351:234–41.
4. Roth HR, Oda H, Zhou X, Shimizu N, Yang Y, Hayashi Y, Oda M, Fujiwara M, Misawa K, Mori K. An application of cascaded 3D fully convolutional networks for medical image segmentation. Comput Med Imaging Graph. 2018;66:90–9.
5. Shen C, Wang C, Roth HR, Oda M, Hayashi Y, Misawa K, Mori K. Spatial information-embedded fully convolutional networks for multi-organ segmentation with improved data augmentation and instance normalization. Proc SPIE 11313, Medical Imaging 2020.: Image Processing. 2020;1131316:1–7.
6. Kitasaka T, Kagajo M, Nimura Y, Hayashi Y, Oda M, Misawa K, Mori K. Automatic anatomical labeling of arteries and veins using conditional random fields. Int J Comput Assist Radiol Surg. 2017;12(6):1041–8.
7. Hibim Y, Hayashi Y, Oda M, Kitasaka T, Misawa K, Mori K. Study on automated labeling of abdominal arteries using spectral-based convolutional graph neural networks. Technical Report of IEICE. (in print)
8. Hayashi Y, Misawa K, Hawkes DJ, Mori K. Progressive internal landmark registration for surgical navigation in laparoscopic gastrectomy for gastric cancer. Int J Comput Assist Radiol Surg. 2016;11(5):837–45.
9. Hayashi Y, Misawa K, Mori K. Development of positional tracker-less surgical navigation system based on laparoscopic scene analysis using deep learning. J JSCAS. (in print). 2020;
10. Hayashi Y, Shen C, Igami T, Nagino M, Mori K. Extraction of blood vessel regions in liver from CT volumes using fully convolutional networks for computer-assisted liver surgery. Int J Comput Assist Radiol Surg. 2020;15(Supp 1):S152–3.
11. Misawa M, Kudo S-e, Mori Y, Cho T, Kataoka S, Yamauchi A, Ogawa Y, Maeda Y, Takeda K, Ichimasa K, Nakamura H, Yagawa Y, Toyoshima N, Ogata N, Kudo T, Hisayuki T, Hayashi T, Nakamura K, Baba T, Ishida F, Ito H, Holger R, Mori K. Artificial intelligence-assisted polyp detection for colonoscopy: initial experience. Gastroenterology. 2018;154(8):2027–9.
12. Misawa M, Kudo S, Mori Y, Takeda K, Maeda Y, Kataoka S, Nakamura H, Kudo T, Nakamura K, Hayashi T, Katagiri A, Baba T, Ishida F, Inoue H, Nimura Y, Oda M, Mori K. Accuracy of computer-aided diagnosis based on narrow-band imaging endocytoscopy for diagnosing colorectal lesions: comparison with experts. Int J Comput Assist Radiol Surg. 2017;12(5):757–66.
13. Itoh H, Nimura Y, Mori Y, Misawa M, Kudo S-E, Hotta K, Ohtsuka K, Saito S, Saito Y, Ikematsu H, Hayashi Y, Oda M, Mori K. Robust endocytoscopic image classification based on higher-order symmetric tensor analysis and multi-scale topological statistics. Int J Comput Assist Radiol Surg. 2020;15(12):2049–59.

14. Moriya T, Oda H, Mitarai M, Nakamura S, Roth HR, Oda M, Mori K. Unsupervised segmentation of micro-CT images of lung cancer specimen using deep generative models. MICCAI 2019, LNCS. 2019;11769:240–8.

15. Zhu J-Y, Park T, Isola P, Efros AA. Unpaired image-to-image translation using cycle-consistent adversarial networks. arxiv 170310593. 2017; In ICCV 2017

16. Zheng T, Oda H, Moriya T, Sugino T, Nakamura S, Oda M, Mori M, Takabatake H, Natori H, Mori K. Multi-modality super-resolution loss for GAN-based super-resolution of clinical CT images using micro CT image database. Proc SPIE 11313, Medical Imaging 2020: Image Processing. 2020;1131305:1–7.

Cancer Diagnosis and Prognosis Assistance Based on MCA

7

Noboru Niki, Yoshiki Kawata, Hidenobu Suzuki, Mikio Matsuhiro, and Kurumi Saito

Abstract

We are developing a detection/diagnosis system for lung cancer and chest diseases. These research use synchrotron radiation large-field microscopic CT images, high-definition CT images and pathology/clinical information, and long-term low-dose CT images and genetic information. We present high-performance computer-aided diagnosis system by analyzing the pathological condition from these multiscale image information. These results are described.

Keywords

Lung cancer · Synchrotron radiation CT · CAD · CT images · Genetic information

7.1 Introduction

Medical imaging is one of the major tools that have enriched medical science, disease diagnosis and treatment. The most widely used imaging modality in clinical practice for cancer detection, oncologic diagnosis, and treatment guidance is computed tomography (CT), which allows clinicians to assess the characteristics of human tissue noninvasively.

Recent advances in CT imaging technologies allow the high-throughput extraction of informative imaging features to quantify the differences that oncologic tissues exhibit. A key challenge is to transform a myriad of spatially and temporally quantified features into medical knowledge: the process of integrating diverse information (demographic, clinico-pathological, and quantitative imaging) to provide personalized clinical predictions that can accurately estimate cancer probability and predict patients' outcomes for risk-adaptive treatment.

Tumor heterogeneity is a key challenge for an era of precision medicine [1, 2]. The complex structures in tumors are exhibited on different spatial scales from whole-body to molecular imaging. Molecular characterization using genomic and proteomic technologies has been critical information on the development of precision medicine approaches. However, these techniques of tumor sampling, often invasive biopsy-based molecular assays, does not always reflect the entire tumor cell characteristics [3]. Advanced medical imaging technologies in clinical oncology have been expanded from a primary diagnostic tool to an important role in the context of precision medicine. The main reason is that the noninvasive imaging represents the entire tumor status, and provides complementary information to biopsy-based assays of tumor tissues. The quantitative CT imaging becomes increasingly attractive field [4–10]. The underlying hypothesis of this research area is that the advanced computational approaches discover imaging biomarkers associated with cancer probabilities, clinicopathological prognostic factors, and gene expression levels from large amounts of image-based features. If this hypothesis is proven through external and independent validation cohorts of patients, we can non-invasively infer biological characteristics of diseases, possibly representing cancer probability and prognostic information, from the quantitative CT imaging.

We describe a system for detecting and diagnosing lung cancer and chest diseases using high-resolution 3D CT images and low-dose 3D CT images information. Fig. 7.1

N. Niki (✉)
Emeritus Professor, Tokushima University, Tokushima, Japan
e-mail: niki@tokushima-u.ac.jp

Y. Kawata · H. Suzuki · M. Matsuhiro
Division of Science and Technology, Graduate School of Technology, Industrial and Social Sciences, Tokushima University, Tokushima, Japan
e-mail: niki@tokushima-u.ac.jp

K. Saito
Graduate School of Advanced Technology and Science, Tokushima University, Tokushima, Japan

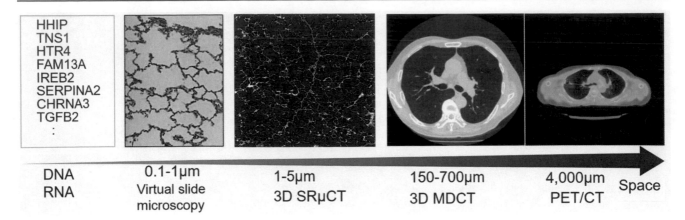

HHIP
TNS1
HTR4
FAM13A
IREB2
SERPINA2
CHRNA3
TGFB2
:

DNA RNA | 0.1-1μm Virtual slide microscopy | 1-5μm 3D SRμCT | 150-700μm 3D MDCT | 4,000μm PET/CT | Space

Fig. 7.1 Multiscale image information

shows multiscale image information. We develop high-performance computer-aided diagnosis system by analyzing the pathological condition from the multiscale image information. The synchrotron radiation large-field microscopic CT images, high-resolution CT images and pathology/clinical information, long-term low-dose CT images and genetic information are used for this progress. The main contents are as shown below.

1. Quantitative elucidation and understanding of three-dimensional microstructures of normal adult lungs and COPD lungs: This research has enabled us to add new knowledge to the conventional three-dimensional microstructure of normal adult lungs. We are elucidating the pathology of COPD lungs.
2. Development of computer-aided detection and diagnosis system for lung cancer, pulmonary embolism, and osteoporosis: We discuss the detection of lung cancer using low-dose CT images, the diagnosis and prognosis of lung cancer using high-resolution CT images, pulmonary embolism using non-contrast CT images, and the detection of osteoporosis using long-term low-dose CT images.
3. Finding genes associated with emphysematous lesion progression and detection of airway lesions: For emphysematous lesions, genes associated with progression are identified using the secular change in the low absorption region of long-term low-dose CT images of heavy smokers. Genetic information predicts the progression of emphysematous lesions in heavy smokers and presents the risk of emphysematous lesions. Airway lesions are detected by using high-resolution CT images to measure narrowing of the airway lumen and airway wall thickening.

7.2 Three-Dimensional Lung Microstructure

Revealing minute changes in organs will be achieved by developing imaging technologies with improved spatial and density resolutions. The recognition of abnormalities related to the lobular anatomy has become increasingly important in the lung diagnosis at clinical routines of CT examinations [11, 12]. This work analyzes the normal 3D microstructure of the lobular anatomy and advances the abnormal 3D microstructure. SRμCT opens up the 3D analysis of the fine microstructures of the lung. We developed SRμCT system with a spatial resolution of around 3 μm using a 36-megapixel CMOS digital camera to analyze the secondary pulmonary lobule microstructure [13, 14]. The volumetric SRμCT images were measured at the synchrotron radiation facility Super Photon ring-8 GeV (SPring-8) beamline BL20B2. Using the developed SRμCT system, we reconstructed a volumetric image of a human lung specimen with contrast agents. The normal microstructures of the secondary lobule of the lung with the basic structure of the peripheral lung were analyzed. These are 5–30 mm polyhedrons surrounded by interlobular septa. (1) Terminal bronchiole, respiratory bronchiole, alveolar duct, and alveolar sac/alveolar system, (2) arteriole, alveolar capillary, and venule system, and (3) the cooperative structure between the bronchial system and vascular system is elucidated. True 3D microanatomy was demonstrated. Figure 7.2 (a and b) shows terminal bronchioles, respiratory bronchioles, alveolar ducts, alveolar sacs, and alveoli of normal and emphysematous regions. Figure 7.2 (c and d) shows arterioles, alveolar capillary beds, and pulmonary arterial systems of normal and emphysematous regions. These analysis results added new knowledge to the three-dimensional fine structure of the normal adult lung,

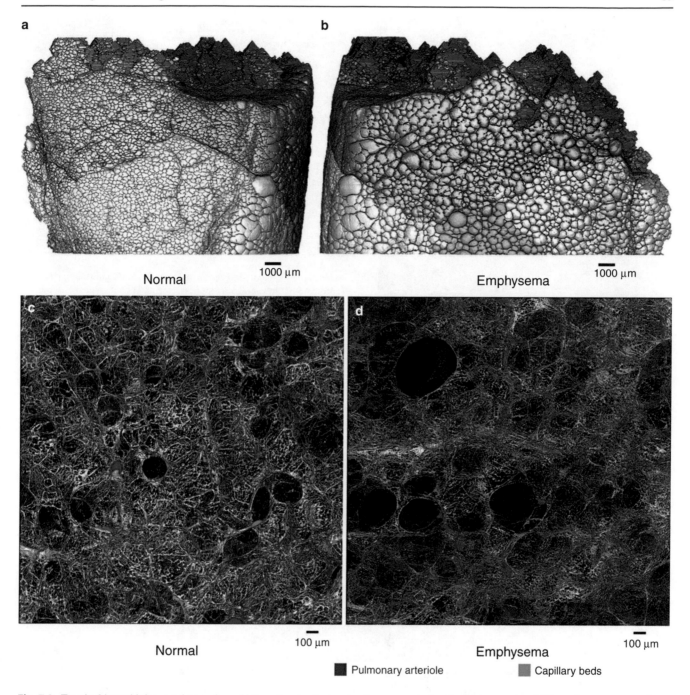

Fig. 7.2 Terminal bronchioles, respiratory bronchioles, alveolar ducts, alveolar sacs, and alveoli of (**a**) normal and (**b**) emphysematous regions. Arterioles, alveolar capillary beds, and pulmonary arterial systems of (**c**) normal and (**d**) emphysematous regions

which has been regarded as common sense shown in the schematic diagram of Netter [15]. We are conducting imaging experiments on COPD lungs. On the progress of COPD, (1) terminal bronchioles, respiratory bronchioles, alveolar ducts, alveolar sac, and alveolar system, (2) arterioles, alveolar capillaries, and venule system, and (3) the cooperative structure between bronchial system and vascular system, we aim to fully understand how it will be destroyed on a three-dimensional micro level.

7.3   Pulmonary Vascular System and Lymph Node System

Extraction of blood vessels is the basis of organ analysis. Contrast CT images are relatively easy to extract peripheral blood vessels, classify arteries and veins, and separate vascular contacts. These difficulties increase with non-contrast CT images. Figure 7.3 shows a blood vessel extraction result of contrast thoracoabdominal CT images [16]. Image enhance-

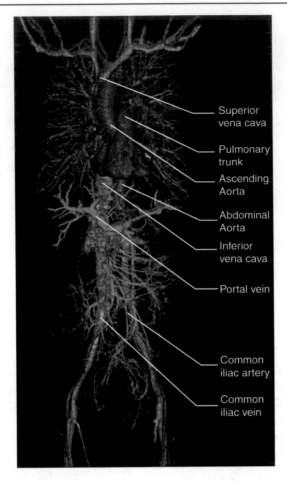

Fig. 7.3 Blood vessels segmentation result of contrast thoracoabdominal CT images [16]

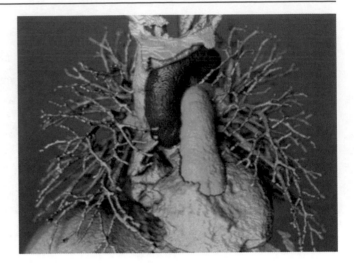

Fig. 7.4 Segmentation result of the aorta and main pulmonary artery from non-contrast CT images [24]. Red and green colors show the aorta and main pulmonary artery

ment is used to extract peripheral blood vessels. Extraction of the inferior vena cava is difficult due to insufficient staining of the contrast agent [17]. Diameters of the aorta and main pulmonary artery are effective indices for pulmonary hypertension diagnosis [18]. For segmentation of these blood vessels from non-contrast CT images, it is necessary to isolate the area where the aorta and pulmonary artery are in contact. The segmentation methods have implemented several methodological solutions: multi-atlas [19], three-dimensional level-set [20], cylinder-tracking [21], optimal surface graph cuts [22], and probabilistic atlas of curvature-based centerline [23, 24]. Figure 7.4 shows a segmentation result of the aorta and main pulmonary artery [24]. This approach constructs deformable probabilistic atlas of curvature-based centerlines of the aorta and main pulmonary artery. The centerlines in approximately cylindrical parts of the blood vessels are selected by a rigid registration of an optimized atlas. The atlas is represented by linear combination of standard probabilistic model [25] and principal components. The cost function for the atlas optimization consists of prior probability and vascular direction. Local optimal solution of the cost function is found out by Quasi-Newton method [26]. The boundaries of the blood vessels are restored

from the 3D Euclidean distances on the centerlines refined by B-spline interpolation [27]. Table 7.1 shows comparative evaluation of automatic and manual segmentation results for 64 normal cases and 19 pulmonary embolism cases [23]. The metrics for the quantitative evaluation are Jaccard coefficient [28] and Dice coefficient [29]. This algorithm achieved highly accurate segmentation for normal cases, but also pulmonary embolism cases with pathological deformations of the main pulmonary artery. We analyze lymph node metastasis of lung cancer using contrast CT images. Figure 7.5 shows the results of lymph node extraction. Lymph node metastasis can be confirmed by size and dense staining.

7.4 Detection and Diagnosis of Early Lung Cancer Using Low-Dose CT Images and High-Resolution CT Images

7.4.1 Malignancy Prediction of Suspicious Nodule on Low-Dose CT Images

Lung cancer is the leading cause of cancer-related mortality worldwide. Screening for lung cancer with low-dose computed tomography (CT) has led to increased recognition of small lung cancers and is expected to increase the rate of detection of early-stage lung cancer [30]. The effectiveness of low-dose CT lung cancer screening has been reported by the NLST and NELSON studies [31, 32]. In NELSON, the ratio of positive CT screening is as low as 2.1%. Major concerns in the implementation of the CT screening of large populations include determining the appropriate management of pulmonary nodules found on a scan. The guidelines for management of lung nodules detected from screening CT scans are based on estimations of the individual risk of malignancy. The ability to

Table 7.1 Performances of aorta and main pulmonary artery segmentation [23]

	Aorta		Main pulmonary artery	
	Normal	Pulmonary embolism	Normal	Pulmonary embolism
Jaccard coefficient	0.916±0.013	0.911±0.023	0.902±0.019	0.914±0.017
Dice coefficient	0.956±0.007	0.953±0.012	0.948±0.009	0.955±0.009

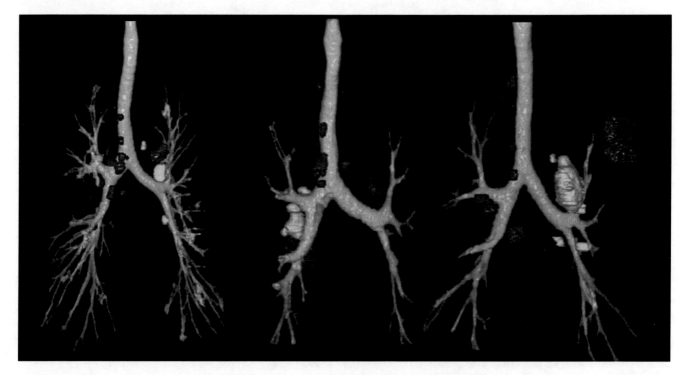

Fig. 7.5 Thoracic lymph node extraction results. Pathological N1 lung cancer cases (**a**), (**b**), and (**c**). Yellow color shows ipsilateral lymph nodes. Blue colors show mediastinal and contralateral lymph nodes, respectively. Brown color shows lung cancers

identify patients with a high rate of malignancy becomes crucial to guide treatment decisions and to develop risk-adapted treatment strategies. Considerable research efforts have been performed to enable the prediction of cancer likelihood in lung nodules based on CT image analyses for optimal therapeutic management to maximize patient survival and preserve lung function. There may be still room for the development of quantitative approaches to estimate the risk of malignancy. It will be essential to evaluate the discrimination ability of a widely-accepted lung cancer prediction model and to elucidate which benign nodules are hard to discrimination from malignant nodules. We have been developing a high-performance computer-aided detection system for lung cancer CT screening [33–35]. We investigated whether the computer-aided CT image features can improve the discrimination ability of lung cancer prediction models for nodules in whom malignancy is suspected. The CT image features included nodule morphology, a percentage of solid volume, internal and marginal intensity features: first-order features, and grey level co-occurrence

(GLCM) texture features. We built an integrated lung cancer prediction model with the computer-aided CT image features. When applying the prediction model to nodules with a maximum diameter of 2 cm or less in whom malignancy was suspected, we found that the benign nodules of granuloma are able to discriminate from malignant nodules diagnosed as stage 0, IA1, IA2, IA3, IB (Fig. 7.6).

7.4.2 Prognosis Prediction of Lung Cancer on Thin-Section CT Images

Cancer treatment includes multiple options depending on nodule aggressiveness, and has created the need for prognostic characterization. The tumor-node-metastasis (TNM) staging system for NSCLC is an internationally accepted system for determining the disease stage and is used for formulating a prognosis and guiding management. Patients with pathological stage IA disease have the most favorable prognosis

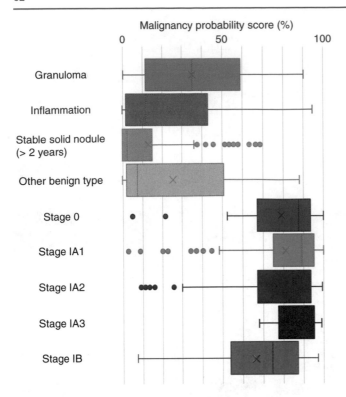

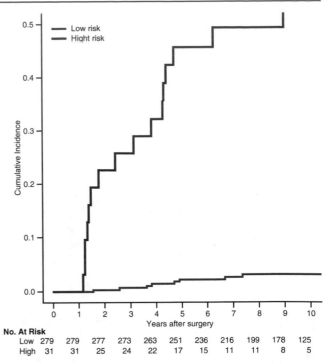

Fig. 7.7 Cumulative incidence curves for relapse using the relapse risk score from CT image features (low and high risk)

Fig. 7.6 Box plot showing the distribution of malignancy probability scores obtained by using the computer-aided CT image features in the discrimination of the benign nodules ($n = 349$; granuloma (29 nodules), inflammation (167 nodules), stable solid nodule (130 nodules), and other benign type (23 nodules)) from the malignant nodules ($n = 345$; stage 0 (35 nodules), IA1 (143 nodules), IA2 (114 nodules), IA3 (16 nodules), and IB (37 nodules)). The boxes represent the 25th and 75th centiles, with the medians indicated by the vertical lines. The whiskers represent the fifth to 85th centiles

and are treated with surgical resection. The ability to identify those patients with a high rate of recurrence is crucial to develop risk-adapted treatment strategies. Considerable research efforts have been performed to enable the stratification of lung cancer aggressiveness based on preoperative CT image analyses for optimal therapeutic management to maximize patient survival and preserve lung function. There may be still room for the development of quantitative approaches to stratify the risk of relapse in patients with early-stage lung cancer. We investigated potential usefulness of the computer-aided 3D image features, which extracted from inside nodule and marginal region, for stratifying the early-stage lung cancer. We measured 44 image features that describe nodule characteristics. The features can be divided into six groups; (a) nodule shape, (b) pleural attachment status, (c) intensity inside nodule, (d) texture inside nodule, (e) intensity in marginal region of nodule, and (f) texture in marginal region of nodule. We built a risk model using the generalized additive proportional hazard model based on the association between the total features and the duration of individual patient relapse-free survival (RFS). We used the multivariate generalized additive proportional hazard model to select the most

useful features from retained features in the pre-selection and avoidance of multicollinearity steps. When applying the risk prediction model to 310 adenocarcinoma lesions with stage IA, we found that quantitative image features in combination with surrounding characteristics of lung cancer are associated with an increased risk of relapse in patients with stage IA lung adenocarcinoma. These results were shown in Fig. 7.7. The risk model may help identify the most aggressive subgroup of patients with early-stage lung adenocarcinomas and could guide better clinical outcomes for these individuals.

7.5 Risk of Emphysematous Lesions by Long-Term Low-Dose 3D CT Images and Genetic Information and Airway Lesion Detection by High-Resolution 3D CT Images

Chronic obstructive pulmonary disease (COPD) is a major public health problem that is predicted to become the third leading cause of death worldwide by 2030 [36]. Smoking is a well-known risk factor in the development of COPD [37]. COPD has emphysematous lesions and airway lesions. Figure 7.8 shows the secular change in the low attenuation area and the pack year analysis results using long-term low-dose CT images of heavy smokers for emphysematous lesions [38]. There are slow group and fast group of progress in heavy smokers. Based on this result, an association analysis of single nucleotide polymorphisms (SNPs) with a CT image-based emphysema progression in heavy smokers has

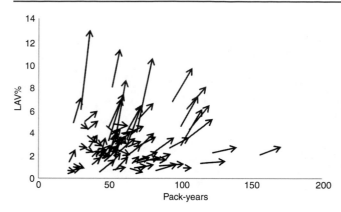

Fig. 7.8 Secular change in the low attenuation area and the pack year analysis results using long-term low-dose CT images of heavy smokers for emphysematous lesions [38]

been performed [39]. The procedure of this association analysis is as follows; (1) thoracic organs and low attenuation volume (LAV) segmentation from low-dose CT images, (2) data selection using CT image-based features, (3) evaluation of CT image-based emphysema progression, and (4) association analysis of SNPs with emphysema progression. COPD related genes are searched by 125 papers and 10 kinds of related genes of Japanese are selected; rs7733088 in HTR4 [40], rs7671167 in FAM13A [41], rs13118928 in HHIP [42], rs13180 in IREB2 [43], rs3923564 in SFTPD [44], rs7937 in EGLN2 [45], rs2736100 in TERT [46], rs401681 in CLPTM1L [47], rs1333040 in CDKN2B-AS1 [48], and rs10849605 in RAD52 [49]. We discover SNPs related to the progress of the low absorption volume [39]. The excavation results are shown in Table 7.2 [39].

Table 7.2 Excavation results of COPD related single nucleotide polymorphisms [39]

SNP	Genetic models	*p-value*	OR
rs7733088	Dominant (AA+AG)vs.GG	0.771	0.850
	Recessive AAvs.(AG+GG)	0.308	1.750
rs7671167	Dominant (CC+CT)vs.TT	0.716	0.828
	Recessive CC vs.(CT+TT)	0.389	0.495
rs13118928	Dominant (AG+GG)vs.AA	0.875	0.926
	Recessive GG vs.(AG+AA)	0.477	1.620
rs13180	Dominant (CT+TT)vs.CC	0.726	1.210
	Recessive TTvs.(CT+CC)	**0.028**	**3.490**
rs3923564	Dominant (AG+GG)vs.AA	**0.016**	**0.292**
	Recessive GGvs.(AG+AA)	0.786	0.842
rs7937	Dominant (CC+CT)vs.TT	0.864	0.917
	Recessive CCvs.(CT+TT)	0.889	1.110
rs2736100	Dominant (AC+CC)vs.AA	0.808	1.140
	Recessive CCvs.(AC+AA)	0.565	1.430
rs401681	Dominant (CT+TT)vs.CC	0.601	0.775
	Recessive TTvs.(CT+CC)	0.663	0.736
rs1333040	Dominant (CC+CT)vs.TT	0.510	1.380
	Recessive CC vs.(CT+TT)	0.719	0.777
rs10849605	Dominant (CC+CT)vs.TT	0.592	1.300
	Recessive CC vs.(CT+TT)	0.923	0.921

Table 7.3 Measurement result of wall thickness and inner diameter

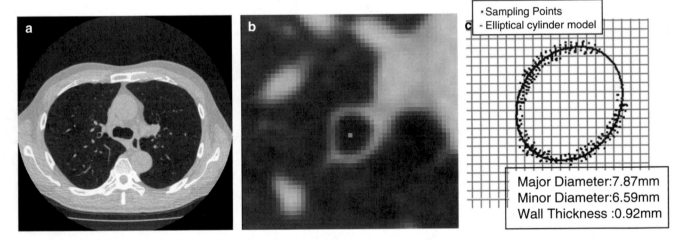

Design Value (mm)	Wall Thickness	1			
	Inner Diameter	6	4	3	2
Measurement Error(%)	Wall Thickness	0.4	0.2	0.3	0.2
	Inner Diameter	−0.3	−0.2	−0.1	0.3

Fig. 7.9 The result of applying low-dose CT to the right **b**1 bronchus in a COPD case (**a**) CT image including b1 bronchi, (**b**) Enlarged b1 bronchi, (**c**) Measurement result

In COPD diagnosis using 3D CT images, airway lesions are diagnosed by evaluating the bronchus with an inner diameter of 2 to 4 mm [50]. Bronchus with an inner diameter of 1 to 2 mm can be evaluated by using ultra-high-resolution 3D CT images [51]. We are developing a highly accurate bronchial measurement method using elliptic cylinder model fitting from high-resolution 3D CT images. Table 7.3 shows the evaluation results of ultra-high-resolution 3D CT images of the bronchial phantom. Figure 7.9 shows the evaluation results of 3D CT images of the bronchi.

7.6 Conclusion

We have researched and developed a detection and diagnosis system for lung cancer and chest diseases. With the basic idea of "diagnosing normal morphology and pathology on a micro to macro scale," we have consistently analyzed organs and pathologies using high-quality large-scale image databases. We have obtained remarkable research results by increasing the objectivity to "ambiguity" and achieving high-performance through advanced analysis. The above

Fig. 7.10 An overview of our Chest CAD system

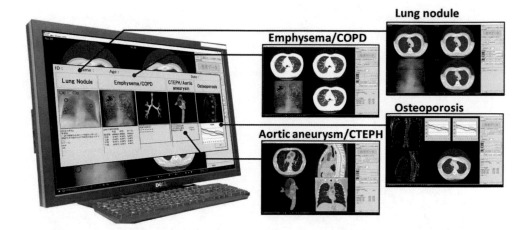

methods are integrated and systematized. Figure 7.10 shows an overview of the system. The lung cancer and chest diseases diagnosis workflow can be visualized by recording the disease read procedure, read location, and read time. These methods have also been successfully applied to colorectal cancer and kidney cancer.

Acknowledgments This research was supported by JSPS Kakenhi 26108007. We would like to express our deep gratitude to all our collaborators.

References

1. Maitland ML, Schilsky RL. Clinical trials in the era of personalized oncology. CA Cancer J Clin. 2011;61:365–81.
2. Aerts HJ. The potential of radiomic-based phenotyping in precision medicine. JAMA Oncol. 2016;2:1636–42.
3. Kern SE. Why your new cancer biomarker may never work: recurrent patterns and remarkable diversity in biomarker failures. Cancer Res. 2012;72:6097–101.
4. Rutman AM, Kuo MD. Radiogenomics: creating a link between molecular diagnostic and diagnostic imaging. Eur J Radiol. 2009;70:232–41.
5. Lambin P, Riso-Velazquez E, Leijenaar R, et al. Radiomics: extraction more information from medical images using advanced feature analysis. Eur J Cancer. 2012;48:441–6.
6. Lambin P, van Stiphout RG, Starmans MH, et al. Predicting outcomes in radiation oncology-multifactorial decision support systems. Nat Rev Clin Oncol. 2013;10:27–40.
7. Aerts HJ, Velazquez ER, Leijenaar RT, et al. Decoding tumour phenotype by noninvasive imaging using a quantitative radiomics approach. Nat Commun. 2014 Jun 3;5:4006. https://doi.org/10.1038/ncomms5006.
8. Aerts HJ. The potential of radiomic-based phenotyping in precision medicine: a review. JAMA Oncol. 2016;2:1636–42.
9. Lambin P, Leijenaar RTH, et al. Radiomics: the bridge between medical imaging and personalized medicine. Nat Rev Clin Oncol. 2017;14:749–62.
10. O'Connor JP, Aboagye EO, et al. Imaging biomarker road map for cancer studies. Nat Rev Clin Oncol. 2017;14:169–86.
11. Itoh H, Nakatsu M, Yoxtheimer LM, et al. Structural basis for pulmonary functional imaging. Eur J Radiol. 2001;37:143–54.
12. Itoh H, Nishino M, Hatabu H. Architecture of the lung morphology and function. J Thorac Imaging. 2004;19:221–7.
13. Umetani K, Okamoto T, Saito K, Kawata Y, Niki N. 36M-pixel synchrotron radiation micro-CT for whole secondary pulmonary lobule visualization from a large human lung specimen. Eur J Radiol Open. 2020;7:100262.
14. Umetani K, Itoh H, Kawata Y, Niki N. Large lung specimen imaging full-field micro-CT using a high-megapixel single lens reflex camera and synchrotron radiation. Proc. International Workshop on Advanced Image Technology (IWAIT). 2018;22:1–4.
15. Frank H, Netter M. Atlas of human anatomy. 6th ed. Winsland House I, Singapore: Elsevier Pte. Ltd; 2015.
16. Niki N, Kwata Y, Suzuki H, et al. Cancer diagnosis and prognosis assistance based on multidisciplinary computational anatomy–progress overview FY2014-FY2018–. Proceedings of the fifth international symposium on the project "multidisciplinary computational anatomy". 2019;107–114.
17. Maklad AS, Matsuhiro M, Suzuki H, et al. Blood vessel-based liver segmentation using the portal phase of an abdominal CT dataset. Med Phys. 2013;40(11):113501(17pp.). https://doi.org/10.1118/1.4823765.
18. Truong QA, Massaro JM, Rogers IS, et al. Reference values for Normal pulmonary artery dimensions by noncontrast cardiac computed tomography, the Framingham heart study. Circ Cardiovasc Imaging. 2012;5(1):147–54. https://doi.org/10.1161/CIRCIMAGING.111.968610.
19. Išgum I, Staring M, Rutten A, et al. Multi-atlas-based segmentation with local decision fusion–application to cardiac and aortic segmentation in CT scans. IEEE Trans Med Imaging. 2009;28(7):1000–10. https://doi.org/10.1109/TMI.2008.2011480.
20. Kurugol S, Come CE, Diaz AA, et al. Automated quantitative 3D analysis of aorta size, morphology, and mural calcification distributions. Med Phys. 2015;42(9):5467–78. https://doi.org/10.1118/1.4924500.
21. Xie Y, Liang M, Yankelevitz DF, et al. Automated measurement of pulmonary artery in low-dose non-contrast chest CT images. Proc SPIE Medical Imaging. 2015;9414:9414G-1–9. https://doi.org/10.1117/12.2081992.
22. Gamechi ZS, Arias-Lorza AM, Pedersen JH, et al. Aorta and pulmonary artery segmentation using optimal surface graph cuts in non-contrast CT. Proc of SPIE Medical Imaging. 2018;105742D:1–7. https://doi.org/10.1117/12.2293748.
23. Suzuki H, Kawata Y, Niki N, et al. Automated assessment of aortic and main pulmonary arterial diameters using model-based blood vessel segmentation for predicting chronic thromboembolic pulmonary hypertension in low-dose CT lung screening. Proc

SPIE Medical Imaging. 2018;10575:105750X-1–6. https://doi.org/10.1117/12.2293295.

24. Niki N, Kwata Y, Suzuki H, et al. Cancer diagnosis and prognosis assistance based on multidisciplinary computational anatomy–progress overview FY2016–Proceedings of the third international symposium on the project "multidisciplinary computational anatomy". 2017;87–94.

25. Kamber M, Shinghal R, Collins DL, et al. Mode-based 3-D segmentation of multiple sclerosis lesions in magnetic resonance brain images. IEEE Trans Med Imaging. 1995;14(3):442–53. https://doi.org/10.1109/42.414608.

26. Nocedal J, Wright S. Quasi-Newton methods. In: Numerical Optimization. Germany, Berlin: Springer-Verlag; 1999. p. 192–218.

27. Hughes TJR, Cottrell JA, Bazilevs Y. Isogeometric analysis: CAD, finite elements, NURBS, exact geometry and mesh refinement. Comput Methods Appl Mech Eng. 2005;194(39–41):4135–95. https://doi.org/10.1016/j.cma.2004.10.008.

28. Jaccard P. The distribution of flora in the alpine zone. New Phytol. 1912;11(2):37–50. https://doi.org/10.1111/j.1469-8137.1912.tb05611.x.

29. Dice LR. Measures of the amount of ecologic association between species. Ecology. 1945;26(3):297–302. https://doi.org/10.2307/1932409.

30. Kaneko M, Eguchi K, Ohmatsu H, et al. Peripheral lung cancer: screening and detection with low-dose spiral CT versus radiography. Radiology. 1996;201:798–802.

31. National Lung Screening Trial Research Team, Aberle DR, Adams AM, et al. Reduced lung-cancer mortality with low-dose computed tomographic screening. N Engl J Med. 2011;365(5):395–409. https://doi.org/10.1056/NEJMoa1102873.

32. de Koning HJ, van der Aalst CM, de Jong PA, et al. Reduced lung-cancer mortality with volume CT screening in a randomized trial. N Engl J Med. 2020;382(6):503–13. https://doi.org/10.1056/NEJMoa1911793.

33. Kanazawa K, Kawata Y, Niki N, et al. Computer-aided diagnosis for pulmonary nodules based on helical CT images. Comput Med Imaging Graph. 1998;22(2):157–67. https://doi.org/10.1016/s0895-6111(98)00017-2.

34. Kanazawa K, Kubo M, Niki N, et al. Computer aided screening system for lung cancer based on helical CT images. Proc. International Conference on Visualization in Biomedical Computing. 1996;223–228.

35. Kageyama T, Kawata Y, Niki N, et al. Differential diagnosis of pulmonary nodules using 3D CT images. Proc. SPIE Medical Imaging. 2020:113142J-1–6.

36. World Health Organization (WHO). World Health Statistics. 2008.

37. Global Initiative for Chronic Obstructive Lung Disease (GOLD). Global strategy for the diagnosis, management and prevention of chronic obstructive pulmonary disease: 2018 report. 2018.

38. Suzuki H, Matsuhiro M, Kawata Y, et al. Longitudinal follow-up study of smoking-induced emphysema progression in low-dose CT screening of lung cancer. Proc of SPIE Medical Imaging. 2014;9035:90352M-1–6. https://doi.org/10.1117/12.2044007.

39. Suzuki H, Matsuhiro M, Kawata Y, et al. Association analysis of SNPs with CT image-based phenotype of emphysema progression in heavy smokers. Proc SPIE Medical Imaging. 2020;11314:113142D-1–7. https://doi.org/10.1117/12.2549431.

40. Wilk JB, Shrine NRG, Loehr LR, et al. Genome-wide association studies identify CHRNA5/3 and HTR4 in the development of airflow obstruction. Am J Respir Crit Care Med. 2012;186(7):622–32. https://doi.org/10.1164/rccm.201202-0366OC.

41. Cho MH, Boutaoui N, Klanderman BJ, et al. Variants in FAM13A are associated with chronic obstructive pulmonary disease. Nat Genet. 2010;42(3):200–2. https://doi.org/10.1038/ng.535.

42. Kim WJ, Oh YM, Lee JH, et al. Genetic variants in HHIP are associated with FEV1 in subjects with chronic obstructive pulmonary disease. Respirology. 2013;18(8):1202–9. https://doi.org/10.1111/resp.12139.

43. Zhou H, Yang J, Li D, et al. Association of IREB2 and CHRNA3/5 polymorphisms with COPD and COPD-related phenotypes in a Chinese Han population. J Hum Genet. 2012;57:738–46. https://doi.org/10.1038/jhg.2012.104.

44. Chen W, Brehm JM, Manichaikul A, et al. A genome-wide association study of chronic obstructive pulmonary disease in Hispanics. Annals of American Thoracic Society. 2015;12(3):340–8. https://doi.org/10.1513/AnnalsATS.201408-380OC.

45. Castaldi PJ, Cho MH, Zhou X, et al. Genetic control of gene expression at novel and established chronic obstructive pulmonary disease loci. Hum Mol Genet. 2014;24(4):1200–10. https://doi.org/10.1093/hmg/ddu525.

46. Wei R, Cao L, Pu H, et al. TERT polymorphism rs2736100-C is associated with EGFR mutation–positive non–small cell lung cancer. Clin Cancer Res. 2015;21(22):5173–80. https://doi.org/10.1158/1078-0432.CCR-15-0009.

47. Wang Y, Broderick P, Webb E, et al. Common 5p15.33 and 6p21.33 variants influence lung cancer risk. Nat Genet. 2008;40(12):1407–9. https://doi.org/10.1038/ng.273.

48. Liu KY, Muehlschlegel JD, Perry TE, et al. Common genetic variants on chromosome 9p21 predict perioperative myocardial injury after coronary artery bypass graft surgery. J Thorac Cardiovasc Surg. 2010;139(2):483–488.e2. https://doi.org/10.1016/j.jtcvs.2009.06.032.

49. Timofeeva MN, Hung RJ, Rafnar T, et al. Influence of common genetic variation on lung cancer risk: meta-analysis of 14 900 cases and 29 485 controls. Hum Mol Genet. 2012;21(22):4980–95. https://doi.org/10.1093/hmg/dds334.

50. Oguma T, Hirai T, Niimi A, et al. Limitations of airway dimension measurement on images obtained using multi-detector row computed tomography. PLoS One. 2013;8(22):e76381. https://doi.org/10.1371/journal.pone.0076381.

51. Tanabe N, Shima H, Sato S, et al. Direct evaluation of peripheral airways using ultra-high-resolution CT in chronic obstructive pulmonary disease. Eur J Radiol. 2019;120:108687. https://doi.org/10.1016/j.ejrad.2019.108687.

Function Integrated Diagnostic Assistance Based on MCA Models

8

Hiroshi Fujita, Takeshi Hara, Xiangrong Zhou,
Atsushi Teramoto, Naoki Kamiya, Daisuke Fukuoka,
and Chisako Muramatsu

Abstract

This article describes the summary of our research achievement on the function integrated diagnostic assistance based on multidisciplinary computational anatomy (MCA) models. The main purpose of our research was to develop computer-aided diagnosis (CAD) systems based on such anatomical models for organ and tissue functions. During FY2014 to FY2018, we made research achievements on each of sub projects such as the fundamental techniques for constructing anatomical models, the analyses of functional images, and the work of other related CAD applications; the promising results indicated the success of our project, which may relate to improved clinical practice in the future.

Keywords

Multidisciplinary computational anatomy · Functional imaging · Computer-aided detection/diagnosis · Deep learning · PET/CT image

H. Fujita (✉) · T. Hara · X. Zhou
Department of Electrical, Electronic and Computer Engineering, Faculty of Engineering, Gifu University, Gifu, Japan

A. Teramoto
Graduate School of Health Sciences, Fujita Health University, Toyoake, Aichi, Japan

N. Kamiya
School of Information Science and Technology, Aichi Prefectural University, Nagakute, Aichi, Japan

D. Fukuoka
Faculty of Education, Gifu University, Gifu, Japan

C. Muramatsu
Faculty of Data Science, Shiga University, Hikone, Shiga, Japan

8.1 Introduction

In the previous project named as "computational anatomy", we had developed computational anatomy models of various organs in CT images [1]. On the other hand, in our recent project, named as "multidisciplinary computational anatomy (MCA)", we not only continued to improve model construction and application but also focused on establishment of computational models for *functional imaging*. These models can be effectively combined to process multidisciplinary information. Our roles in this MCA-based project were to investigate image analysis methods based on the fusion of anatomic and functional information and to establish methodologies of computer-aided detection/diagnosis (CAD) systems for organ and tissue functions.

The following sections describe the overall achievements of our project for each of the key topics. Section 8.2 explains the construction of anatomical models, which is the basis for the development of functional models. Section 8.3 describes the achievements on (1) the analyses of functional images in FDG-PET CT and PET/CT imaging, (2) these of muscle functions in CT imaging, and (3) on estimating the knee extension strength in ultrasound imaging. Section 8.4 provides several works of other related CAD applications, followed by Section 8.5 for a summary of the overall study.

8.2 Basic Techniques for Building Anatomical Models

8.2.1 Purpose

Localization, segmentation, and registration are major challenges in medical image analysis, and required as the fundamental pre-processing steps in our project, which aims to merge the multi-modality medical images and fuse the information of human anatomy and functional information as well as temporal change. Therefore, the purpose of this study was to accomplish these fundamental pre-processing steps based on abstract representations learned from the 3D CT and PET images to improve the robustness and efficiency.

8.2.2 Method

The proposed schemes used 2D and 3D deep convolutional neural networks (CNNs) to learn the hierarchical image feature space to accomplish (1) segmentation of the anatomical structures (including multiple organs) on a CT image (generally in 3D) by using a pixel-wised annotation, (2) localization of the bounding boxes that tightly surround the interesting regions in CT and PET images, and (3) registration that arranges the CT and PET images scanned separately into the same spatial space. All of these methods were based on deep learning (both supervised and unsupervised approaches), and the network structures of the CNNs used for these different tasks were similar and partially shared. The details of these methods can be referred in [2–6].

8.2.3 Results

A shared dataset consisting of 240 3D CT scans and a humanly annotated ground truth were used for training and testing the anatomical structures (17 types of major organ) segmentation and localization methods by using a four-fold cross validation. The coincidence (Jaccard index: JI) between the segmentation or localization result and human annotation was used as the evaluation criteria. In our experiments of organ localization and segmentation, we confirmed that 77.5% of bounding boxes were localized accurately and mean JI was 78.8% (Fig. 8.1). These performances were comparable or better in accuracy and had an advantage in computational efficiency and robustness comparing to the conventional methods without deep learning approach.

Another dataset consisting of 170 whole-body PET/CT scans was used for training and testing the image registration method by using un-supervised deep learning technique. The effectiveness of the registration process was evaluated by the normalized cross-correlation (NCC) and mutual information (MI) between CT and PET images using a ten-fold cross validation. The experimental results demonstrated that the mean values of NCC and MI improved from 0.40 to 0.58 and from 1.98 to 2.34, respectively, via the image registration process. The usefulness of our scheme for PET/CT image registration was confirmed by the promising results (Fig. 8.2).

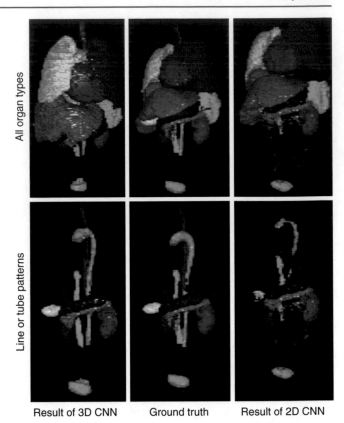

Fig. 8.1 An example of organ segmentation in a 3D CT case [2, 3]. Left: segmentation result of multiple organs by using a 3D deep CNN [3], Middle: a ground truth of the target organs. Right: segmented result by our previous method using a 2D deep CNN with 3D voting [2]

8.2.4 Conclusion

We proposed novel approaches to realize semantic segmentation, organ localization, and image registration based on the abstract representations learned from the 3D CT and PET images. Comparing to our previous works that were based on the image signal directly, the robustness and efficiency of these fundamental pre-processing were improved significantly. We confirmed that learning an image representation by high-level hierarchical features was the most critical step for model construction required by MCA.

8.3 Analysis of Functional Imaging

8.3.1 Automated Evaluation of Tumour Activities on FDG-PET CT Images

8.3.1.1 Purpose

Standardized uptake value (SUV) is a semi-quantitative indicator of FDG-PET. The maximum value of SUV is often employed for tumour evaluations, but it is difficult to evalu-

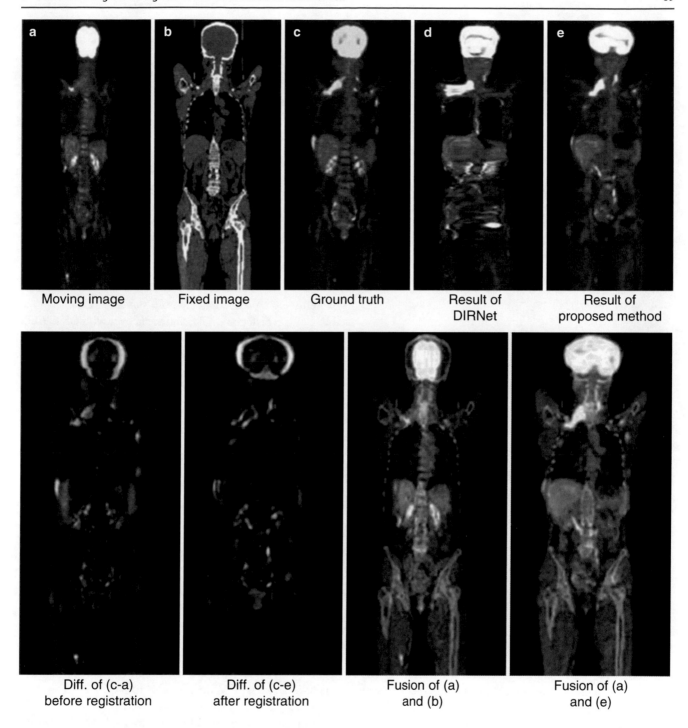

a	b	c	d	e
Moving image	Fixed image	Ground truth	Result of DIRNet	Result of proposed method

Diff. of (c-a) before registration	Diff. of (c-e) after registration	Fusion of (a) and (b)	Fusion of (a) and (e)

Fig. 8.2 An example of PET/CT image registration result by using 3D deep CNNs [5, 6]

ate the size and spread of the tumour because the maximum value is obtained for only one voxel within the tumour region. Metabolic volume (MV) and total lesion glycolysis (TLG) have been used to evaluate the response to tumour treatments in recent years; MV is an indicator of the volume that reflects the size and spread of tissues with high metabolism and TLG is an indicator that expresses the degree and volume of glucose metabolism. In order to determine the

MV and TLG, segmented regions of objective tumours are required that indicate high glucose metabolism. Currently, an automated detection method for pulmonary nodules in PET/CT images using CNN is being used. Z-score is a statistical index used to present deviation from a mean value and to show a quantitative abnormality in various measured features in medical fields by comparing with feature values in normal groups. Z-score images have been widely used in statistical image analysis methods to indicate the region abnormality pixel-by-pixel after a normal database was constructed.

We developed a new approach for application in the torso regions for tumour analysis using FDG-PET/CT images. In the statistical image analysis in the torso region, the organ segmentation technique is important, because the function of the organs is normalized based on the segmented regions. We adopted a novel approach using a deep learning technique for 3D CT images. The purpose of this study was to examine the differences in Z-scores obtained by the two organ segmentation results of the bounding box and the segmentation approaches for the normalization step by using typical tumours in lung regions, because tumours with various accumulations and sizes were evaluated in the FDG-PET/CT images. Furthermore, we developed a system to measure MV, TLG, and effective dose based on the automated segmented results.

8.3.1.2 Methods

The overall scheme of our procedure consisted of the following five steps: (1) segmentation of the left and right lungs (CT), (2) anatomical standardization (CT and FDG), (3) creation of normal model (FDG), (4) creation of Z-score images, and (5) automated segmentation of tumours.

The segmentation of lung regions was performed based on our deep learning approach. Two-hundred CT cases were used for the initial training of the networks which include manual segmentation results of 10 organs: lung (left and right), heart, liver, stomach, spleen, kidney (left and right), pancreas, and bladder. All regions were used for the training, but the results of left and right lungs were employed in this study.

After the determination of the organ regions, landmarks (LMs) were set to perform the anatomical standardization. Two methods for setting LMs, the bounding box method and the surface method, were used to compare the anatomical standardization results. The bounding box method is based on our localization approach for multiple organs. In this method, we set LMs on the apexes and the side of the smallest rectangular parallelepiped surrounding the segmented organs. The surface method is based on the segmentation results of organs. In this method, we set LMs on the surface of the organ. To translate the locations of LMs on CT onto PET images, the image resolutions were unified on PET images. CT and PET images were registered based on the location records from the PET/CT device during the image acquisition [7, 8].

The normal model was created by using nine normal cases excluding training cases for the deep learning of organ segmentations. All LMs were aligned and deformed into one lung shape based on the thin-plate spline (TPS) method as a non-rigid image deformation technique. LMs on patient images were also aligned into one lung shape based on the TPS method.

All normal cases were anatomically standardized by the TPS method. The results of anatomical standardization indicated that all pixels of PET images were aligned at the standard locations that present the functions in the organ. The mean (M) and standard deviation (SD) can express the confidence interval of the normal function based on the accumulation, that is, SUV in this study. The M and SD from the normal cases indicated the ranges of normal activities of glucose metabolism in FDG-PET images.

We evaluated 9 normal and 21 abnormal cases. Initially, the regions of tumours were segmented by one of the authors using a graph-cut method. The initial regions were verified by a radiologist, independently: 25 obvious and 10 suspected tumours on PET images were marked as the correct regions, which were employed as the gold standard (GS). Two features of CG distance (CGD) and area cover with GS (ACGS) were obtained from the GS and automatically detected regions. The regions were detected correctly when the CGD and ACGS were three voxels or less and 0.2 and more, respectively.

8.3.1.3 Results

The true positive (TP) fraction (%), the number of false positives (FPs), and average volume of FP (voxels) were 77, 1.10, and 77.61 for surface method, 69, 1.05, and 94.0 for bounding box method, and 69, 1.00, and 184.52 for single SUV method without Z-score, respectively, in the case of employing the combinations of obvious and suspicious 35 regions. We found that TP fraction was 7% decreased when applied from obvious to combined regions. Fig. 8.3 shows two examples from obvious 25 cases with GS regions in red regions. The MV, TLG, and effective doses were also obtained in each detected region based on the determined area.

8.3.1.4 Conclusion

LMs on the organ surface will be useful for statistical analysis of torso images. Various values from the analysis will be helpful indices for quantitative analysis of FDG-PET images. Performance studies including human observers are required to prove the usefulness of the automated methods.

8.3.2 Automated Detection of Lung Nodule in PET/CT Images

8.3.2.1 Purpose

Automated detection of lung nodules using PET/CT images that we previously developed shows good sensitivity;

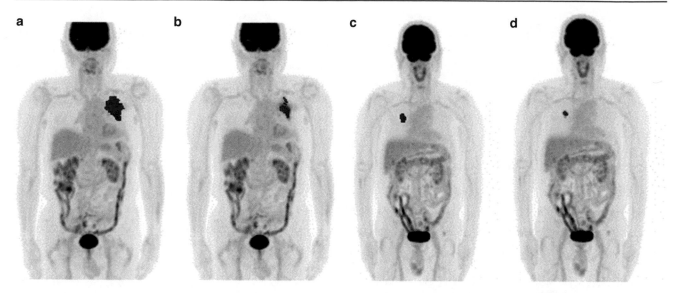

Fig. 8.3 Two examples of obvious cases in which automatic segmentation was performed by the surface method and the gold standard verified by a radiologist [7]. (**a**) Case 1 GS, (**b**) Case 1 automatic detection, (**c**) Case 2 GS, and (d) Case 2 automatic detection.

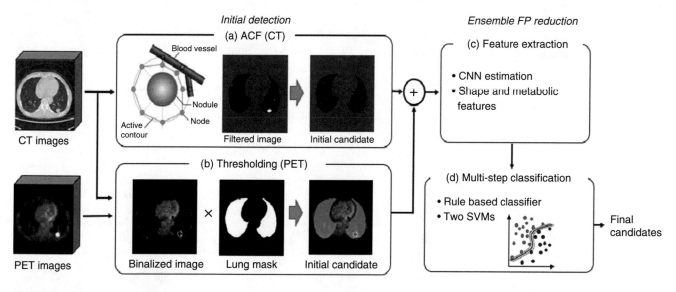

Fig. 8.4 Proposed overall scheme for detecting lung nodules in PET/CT images [9]

however, there was a challenge to reduce the FPs. In this study, we proposed an improved FP reduction method for the detection of lung nodules in PET/CT images using CNN [9].

8.3.2.2 Methods

The outline of our overall scheme for the detection of lung nodules is shown in Fig. 8.4. First, initial nodule candidates were identified separately on the PET and CT images using the algorithm specific to each image type. Subsequently, candidate regions obtained from them were combined. FPs contained in the initial candidates were eliminated by an ensemble method using convolutional neural network and hand-crafted features.

8.3.2.3 Results and Conclusion

We evaluated the detection performance using 104 PET/CT images collected by a cancer-screening programme. As a result of evaluation, the sensitivity of detection was 90.1%, with 4.9 FPs/case. Our ensemble FP reduction method eliminated 93% of the FPs; our proposed method using CNN technique eliminates approximately half the FPs existing in the previous study. These results indicate that our method may be useful in the computer-aided detection of pulmonary nodules using PET/CT images. As a continuation of this research, we also conducted a study on the malignancy analysis of pulmonary nodules using hand-crafted features and a machine learning algorithm, and showed that cases requiring biopsy can be accurately classified only by PET/CT images [10].

8.3.3 Recognizing Skeletal Muscle Regions for Muscle Function Analysis

8.3.3.1 Purpose

We have been working on the model-based skeletal muscle recognition from the previous project. In this project, we aimed to advance muscle recognition techniques using the models for function analysis in whole-body CT images. We started an international collaborative work with Bern University, Switzerland, and we worked on recognition of spinal erector muscle using a deep learning technique.

8.3.3.2 Method

Skeletal muscle recognition was achieved using whole-body CT images and torso CT images. Figure 8.5 shows the progress of research on muscle recognition for functional analysis from FY 2014 to FY 2018. Each colour represents a research area of the MCA project. The blue areas (Fig. 8.5a, b-1 and c-1) are on the *spatial axis*, and the green areas (Fig. 8.5b-2 and d) are on the *functional axis*. The orange area (Fig. 8.5c-2) is on the *pathology axis*.

We generated muscle shape models for sternocleidomastoid [11], supraspinatus, psoas major, and iliac muscles [12] (Fig. 8.5a).

Realizing complex muscle recognition requires automatic recognition of bone anatomical features that are deeply related to the muscle. Therefore, skeletal muscle attachment sites corresponding to the anatomical names were recognized from the scapula based on accurate bone recognition

and position/distribution characteristics (Fig. 8.5b-1). In addition, muscle bundle modelling was performed to properly position of the model and to create a precise model for functional analysis [13] (Fig. 8.5b-2).

For muscle function and muscle disease analysis of whole-body CT images, the whole-body muscle recognition using shape model (Fig. 8.5c-1) and a 3D texture analysis of the muscles were proposed (Fig. 8.5c-2). We modelled the body cavity region from the whole-body CT image, acquired the surface muscle regions, and recognized the deep muscles by applying the site-specific models. Based on the automatic recognition results, we segmented the skeletal muscles and achieved not only the muscle volume but also muscle image characteristics in whole-body CT images [14]. Here, we performed a 3D texture analysis using Haralick's features and found features that indicate a possible imaging classification of neurogenic and myopathic diseases, including amyotrophic lateral sclerosis (ALS).

As a challenging study, we achieved automatic recognition of the spinal erector muscle [15], which has a large and complex structure. By the conventional model-based method, the recognition rate in the 3D volume had a problem, although recognition rate was high in the 12th thoracic spine cross section. Therefore, in this study, recognition was performed using the deep learning method. In addition, in order to improve recognition results of the skeletal muscles by deep learning for functional analysis, modelling of attachment sites on bones [16] and muscle bundles by site [17] was realized (Fig. 8.5b-2 and d).

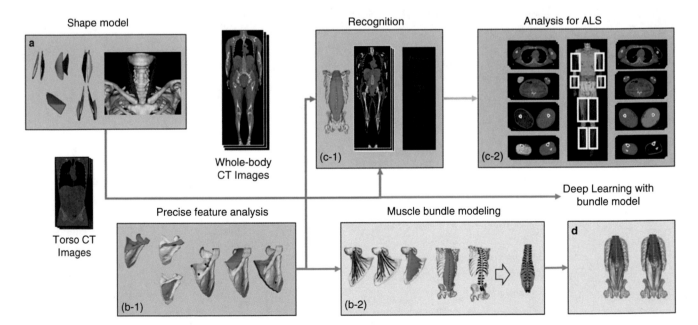

Fig. 8.5 Schematic diagram of the muscle recognition for functional analysis. (**a**) Model-based recognition of muscle, (**b**) precise analysis of scapula for muscle recognition, (**c**) segmental recognition for the whole-body muscle analysis, and (**d**) spinal erector muscle recognition using deep learning with muscle bundle model

8.3.3.3 Results

The effectiveness of the proposed method was evaluated using non-contrast torso CT and whole-body CT images. In the automatic recognition of the muscle with the shape model, concordance rates of 60.3% and 65.4% were obtained in sternocleidomastoid region using 20 cases of torso CT images and 10 whole-body CT images, respectively.

In the automatic recognition of the precise bone features, we used the trunk CT images of 26 cases, and the average concordance rate of 80.7% was obtained for the infraspinous fossa. As the result, feature recognition on the shoulder blades became possible in seven areas together with feature recognition based on six anatomical names, which we have already recognized.

In the automatic recognition and analysis of the whole-body CT images was successful in 36 of 39 cases using the shape models at inner muscle region including psoas major and iliac muscle. We succeeded in analysis of ALS cases in four limbs using the recognition result.

For the erector spinal muscle, mean recognition accuracies of erector spinal muscle and attached area were 89.9% and 65.5%, respectively, in terms of Dice coefficient using 11 cases. The muscle bundle running of the spinal column erector muscle was drawn, and the region of the muscle group constituting the spinal column erector muscle was acquired. As a result, an average Dice coefficient of 65.2% was obtained. This makes it possible to obtain muscle running information as structural information and to realize analysis of the structure in the muscle region recognized by deep learning.

8.3.3.4 Conclusion

The proposed method was realized recognition and analysis of skeletal muscles in torso and whole-body CT images. We developed the method to implement skeletal muscle modelling from micro to macros by modelling the distribution shape of muscle fibres and muscle fibre bundles. At the same time, by comparing and integrating the deep learning method with a model-based approach, we achieved skeletal muscle recognition for skeletal muscle function analysis.

8.3.4 Knee Extension Strength Using Ultrasound Images

8.3.4.1 Purpose

The word "Locomotive syndrome" has been proposed to describe the state of requiring care by musculoskeletal disorders and its high-risk condition. The goal of this study is to evaluate the relationships among ageing, muscle strength, and sonographic findings. This study aimed to develop a system for measuring the knee extension strength using the ultrasound images of rectus femoris muscles and quadriceps femoris muscles obtained with non-invasive ultrasound diagnostic equipment [18, 19].

8.3.4.2 Methods

First, we extracted the muscle area from the ultrasound images of the rectus femoris muscles. Edges from an original image were detected using Canny edge detector and the edges close to the boundary of the muscle area were manually selected. The boundary lines of the upper and lower ends of the muscle area were determined by approximating the selected edges with curves. The area between these boundaries was defined as the muscle area. Image features such as angular second moment and entropy were obtained from grey level co-occurrence matrix (GLCM), and the average value, standard deviation, skewness, and kurtosis were calculated from the echogenicity histograms. In addition, the vertical length at the centre of the muscle area was considered as the thickness of the muscle. We combined these features and physical features, such as the patient's height, and build a regression model of the knee extension strength from the training data. Finally, we substituted the features calculated from the test data into the regression model, and estimated their knee extension strength [19].

To assess muscle quality, texture analysis using GLCM was performed, in which the mean, skewness, kurtosis, inverse difference moment, sum of entropy, and angular second moment were included. The knee extension force in the sitting position and thickness of the quadriceps femoris muscle were also measured [18].

8.3.4.3 Results

We employed 168 B-mode ultrasound images of rectus femoris muscles scanned on both legs of 84 subjects (19–86 years old) for evaluation. As a result, the correlation coefficient value between the measured and estimated values and the root mean squared error were 0.82 and 7.69 kg, respectively [19]. To assess muscle quality, one hundred forty-five healthy volunteers were classified into six groups on the basis of sex and age. The quadriceps femoris thickness, skewness, kurtosis, inverse difference moment, angular second moment, and muscle strength were significantly smaller in elderly participants versus those in the younger and middle-aged groups ($p < 0.05$). In contrast, the mean and sum of entropy were significantly smaller in the younger group than in the middle-aged and elderly groups [18].

8.3.4.4 Conclusion

We developed a system for estimating the knee extension strength using ultrasound images. The system has a potential in quantitatively assessing the muscular morphologic changes due to ageing and could be a valuable tool for early detection of musculoskeletal disorders.

8.4   Other CAD Applications

8.4.1   Automated Malignancy Analysis in CT Images Using GAN

8.4.1.1 Purpose
In the detailed diagnosis of lung nodule using CT images, it is often hard task for radiologist to classify between benign and malignant nodules. In this study, we investigated the automated classification of pulmonary nodules in CT images using a CNN [20]. We used a generative adversarial network (GAN) to generate additional images when only small amounts of data are available, which is a common problem in medical research, and evaluated whether the classification accuracy is improved by generating a large amount of new pulmonary nodule images using the GAN.

8.4.1.2 Methods
The volume of interest centred on the pulmonary nodule was extracted from CT images, and two-dimensional images were created with the axial section. The CNN was trained using nodule images generated by the Wasserstein GAN, and then fine-tuned using the actual nodule images to allow the CNN to distinguish between benign and malignant nodules. CNN model was based on AlexNet; it consisted of five convolution layers, three pooling layers, and three fully connected layers.

8.4.1.3 Results
CT images of 60 cases with pathological diagnosis confirmed by biopsy were analysed. They consisted of 27 benign and 33 malignant nodules. As a result of evaluation, a correct identification rate of 67% for benign nodules and 94% for malignant nodules was obtained. Thus, almost all cases of malignancy and two-thirds of the benign cases were classified correctly. This performance was higher than that of existing method without using GAN generated images.

8.4.1.4 Conclusion
Evaluation results indicated that the proposed method may reduce the number of biopsies in patients with benign nodules, which are difficult to differentiate on CT images, by more than 60%. Furthermore, the use of GAN technology clearly improved the accuracy of the classification. In addition, we also developed a method to generate three orthogonal cross sections with GAN and analyse malignancy [21]. Furthermore, we tried to synthesize the volume data using 3D GAN and analyse with 3D CNN [22]. These results show that GAN helps improve classification accuracy, even in medical datasets with relatively few images.

8.4.2   Automated Classification of Cytological Images Using CNN

8.4.2.1 Purpose
In the detailed diagnosis of lung lesion, pathological examination is performed based on the morphological shape of tissue or cell. However, pathologists have to examine the huge number of images per patient; it is difficult to maintain the diagnostic accuracy avoiding reading errors. In order to assist the image diagnosis of thoracic regions, we have developed the automated classification using cytological images using CNN [23–25].

8.4.2.2 Methods
The microscopic images of specimens stained by using the Papanicolaou method were first collected. Then, they were given to the input layer of CNN. As for the CNN architecture for the classification of lung cancer types [23], we introduced original CNN for consisted of three convolutional layers, three pooling layers, and two fully convolutional layers. The probabilities of three types of cancers (adenocarcinoma, small cell cancer, and squamous cell cancer) were estimated from the output layer. Training on the CNN was conducted by using our original database. As for the automated classification between benign and malignant cells, we introduced pretrained model of VGG-16 network [24].

8.4.2.3 Results
For classification of the three cancer types, the classification accuracy of adenocarcinoma, squamous cell carcinoma, and small cell carcinoma was 89.0%, 60.0%, and 70.3%, respectively; the overall accuracy was 71.1%. Regarding the classification of benign and malignant cells, overall accuracy was 79.2%.

8.4.2.4 Conclusion
The classification accuracies were comparable to those of a cytotechnologist or pathologist. It is noteworthy that CNN is able to understand cell morphology and placement of cancer cells solely from images without prior knowledge and experience of biology and pathology. These results indicate that CNN is useful for classification of cytological images. In a recent study, we introduced the GAN technique used in the previous section to generate high-resolution cytological images and showed that the classification accuracy was improved [25].

8.4.3   Miscellaneous

8.4.3.1 CAD with Radiogenomics
When genetic testing is performed in daily clinical practice in near future, such information can be used for early detec-

tion of Alzheimer's disease (AD) and subtype classification of cancer. For the detection of AD, a normal ageing model of MR brain images was developed, and z-score maps were generated to identify the difference from the normal model [26]. The statistical parametric mapping was used for three-dimensional anatomical standardization. The relationship between the genotypes and cerebral atrophy revealed that anatomical locations of the cerebral atrophy differ, and the transition in the disease state from mild cognitive impairment to AD differs depending on the genotypes [27]. The result indicated genomic information can be used to identify patients susceptible to AD who should be subjected to periodic screening.

For differential diagnosis of breast cancer, radiomic features were determined to classify patients with triple negative breast cancer (TNBC). The area under the ROC curve of 0.70 indicated a possibility of the radiomic features in classification of TNBC, which can promote precision medicine of breast cancer [28].

8.4.3.2 Detection of Cardiovascular Disease Using Funduscopy

Fundus photograph is useful for observing condition of circulatory system. Measurement of retinal arteriolar-to-venular (AVR) diameter ratio can suggest a risk of hypertensive retinopathy. Microaneurysm (MA) is one of the early findings of diabetic retinopathy. However, detection of subtle change in vessel diameter and tiny aneurysms is not easy. We developed an automated AVR measurement method [29] and a fusion model of conventional [30, 31] and CNN-based [32] MA detection methods in this study. The results indicate the potential utility of the proposed method for early detection of hypertensive retinopathy and diabetic retinopathy.

8.4.3.3 Detection and Classification of Teeth for Automatic Dental Record Filing

Dental record plays an important role in dental diagnosis and personal identification. Automatic dental record (chart) filling can help dentists in improving diagnostic efficiency. It can also be used to establish systemic records for forensic identification. We have developed an automatic method to identify individual tooth and classify their tooth types in dental panoramic radiographs and dental CTs for dental chart filing [33, 34]. A CNN model using surrounding information and a relation network which takes the anatomic relationship of teeth into account provided improved detection and classification accuracies. The results indicate the potential usefulness of the proposed method for automatic dental chart filing.

8.5 Summary

In this article, we have introduced: (1) anatomical standardization of CT images using automatic organ extraction and construction of a glucose metabolism model of PET/CT images, (2) establishment of a tumour detection, differentiation, and analysis technology using morphological and functional information in the pulmonary region, (3) automatic skeletal muscle recognition technology and construction of a composite model of skeletal muscle in whole-body CT images, and (4) development of an imaging diagnosis support system in related fields. These technologies we have developed for various functional diagnosis support systems that incorporate artificial intelligence (AI) such as deep learning into multidisciplinary computational anatomy (MCA) models are expected to be useful in the future development of clinical systems and we hope it will contribute to saving many patients.

Acknowledgments The authors would like to thank all those who have contributed to this project. Y. Hatanaka from the University of Shiga Prefecture for research on fundus image CAD, Y. Uchiyama from Kumamoto University for research brain MRI CAD and radiogenomics CAD, K. Azuma from University of Occupational and Environmental Health, T. Katafuchi from Gifu University of Medical Science, T. Matsubara from Nagoya Bunri University, M. Matsuo from Gifu University, T. Miyati from Kanazawa University, M. Kanematsu from Gifu University, presently Gifu Prefectural General Medical Center, H. Jiang from Northeast University, China, S. Li from the University of Western Ontario, Canada, S. Wang from the University of South Carolina, USA, X. Zhang from Guangxi University, China, G. Zheng of University of Bern, Switzerland, presently Shanghai Jiao Tong University, China. This research was supported in part by a Grant-in-Aid for Scientific Research on Innovative Areas (No. 26108005) and International Activities Supporting Group (No. 15 K21716), Japanese Government.

References

1. Fujita H, Hara T, Zhou X, Muramatsu C, Kamiya N, Zhang M, et al. Model construction of computational anatomy: progress overview FY2009–2013. Proc. of Fifth International Symposium on the Project "Computational Anatomy". 2014:25–35.
2. Zhou X, Takayama R, Wang S, Hara T, Fujita H. Deep learning of the sectional appearances of 3D CT images for anatomical structure segmentation based on an FCN voting method. Med Phys. 2017;44:5221–33. https://doi.org/10.1002/mp.12480.
3. Zhou X, Yamada K, Kojima T, Takayama R, Wang S, Zhou XX, et al. Performance evaluation of 2D and 3D deep learning approaches for automatic segmentation of multiple organs on CT images. Proc. SPIE, Medical Imaging 2018:Computer-aided diagnosis. 2018:10575:105752C. https://doi.org/10.1117/12.2295178.
4. Zhou X, Kojima T, Wang S, Zhou XX, Hara T, Nozaki T, et al., Automatic anatomy partitioning of the torso region on CT images by using a deep convolutional network with a majority voting. Proc. SPIE, Medical Imaging 2019:Computer-Aided Diagnosis. 2019;10950:109500Z. https://doi.org/10.1117/12.2512651.

5. Kang H, Jiang H, Zhou X, Yu H, Hara T, Fujita H, et al. An optimized registration method based on distribution similarity and DVF smoothness for 3D PET and CT images. IEEE Access. 2019;8:1135–45. https://doi.org/10.1109/ACCESS.2019.2961268.

6. Yu H, Jiang H, Zhou X, Hara T, Yao Y-D, Fujita H, et al. Unsupervised 3D PET-CT image registration method using a metabolic constraint function and a multi-domain similarity measure. IEEE Access. 2020;8:63077–89. https://doi.org/10.1109/ACCESS.2020.2984804.

7. Hara T, Kobayashi T, Ito S, Zhou X, Katafuchi T, Fujita H. Quantitative analysis of torso FDG-PET scans by using anatomical standardization of normal cases from thorough physical examinations. PLoS One. 2015;10(5):e0125713. https://doi.org/10.1371/journal.pone.0125713.

8. Takeda K, Hara T, Zhou X, Katafuchi T, Kato M, Ito S, et al. Normal model construction for statistical image analysis of torso FDG-PET images based on anatomical standardization by CT images from FDG-PET/CT devices. Int J Comput Assist Radiol. 2017;12(5):777–87. https://doi.org/10.1007/s11548-017-1526-4.

9. Teramoto A, Fujita H, Yamamuro O, Tamaki T. Automated detection of pulmonary nodules in PET/CT images: ensemble false-positive reduction using a convolutional neural network technique. Med Phys. 2016;43(6):2821–7. https://doi.org/10.1118/1.4948498.

10. Teramoto A, Tsujimoto M, Inoue T, Tsukamoto T, Imaizumi K, Toyama H, et al. Automated classification of pulmonary nodules through a retrospective analysis of conventional CT and two-phase PET images in patients undergoing biopsy. Asia Ocean J Nucl Med Biol. 2019;7(1):29–37. https://doi.org/10.22038/AOJNMB.2018.12014.

11. Kamiya N, Ieda K, Zhou X, Chen H, Yamada M, Kato H, et al. Automated segmentation of sternocleidomastoid muscle using atlas-based method in X-ray CT images: preliminary study. Med Imag Inform Sci. 2017;34(2):87–91. (in Japanese). https://doi.org/10.11318/mii.34.87.

12. Kamiya N, Zhou X, Azuma K, Muramatsu C, Hara T, Fujita H. Automated recognition of the iliac muscle and modeling of muscle fiber direction in torso CT images. Proc. SPIE, Medical Imaging 2016:Computer-Aided Diagnosis. 2016;9785:97853K. https://doi.org/10.1117/12.2214613.

13. Kamiya N. Muscle segmentation for orthopedic interventions. In: Zheng G, Tian W, Zhuang X, editors. Intelligent orthopaedics: artificial intelligence and smart image-guided technology for orthopaedics (advances in experimental medicine and biology, 1093). Singapore: Springer; 2018. p. 81–91.

14. Kamiya N, Ieda K, Zhou X, Yamada M, Kato H, Muramatsu C, et al. Automated analysis of whole skeletal muscle for early differential diagnosis of ALS in whole-body CT images: preliminary study. Proc. SPIE, Medical Imaging 2017:Computer-Aided Diagnosis. 2017;10134:1013442. https://doi.org/10.1117/12.2251584.

15. Kamiya N, Li J, Kume M, Fujita H, Shen D, Zheng G. Fully automatic segmentation of paraspinal muscles from 3D torso CT images via multi-scale iterative random forest classifications. Int J Comput Assist Radiol Surg. 2018;13:1697–706. https://doi.org/10.1007/s11548-018-1852-1.

16. Kamiya N, Kume M, Zheng G, Zhou X, Kato H, Chen H, et al. Automated recognition of erector spinae muscles and their skeletal attachment region via deep learning in torso CT images. In: Vrtovec T, Yao J, Zheng G, Pozo J, editors. Proc. the 6th MICCAI workshop on computational methods and clinical applications in musculoskeletal imaging. MSKI2018, vol. 11404; 2019. p. 1–10. https://doi.org/10.1007/978-3-030-11166-3_1.

17. Kamiya N. Deep learning technique for musculoskeletal analysis. In: Lee G, Fujita H, editors. Deep learning in medical image analysis - challenges and applications - (advances in experimental medicine and biology, 1213). Springer Nature Switzerland AG; 2020. p. 165–76.

18. Watanabe T, Murakami H, Fukuoka D, Terabayashi N, Shin S, Yabumoto T, et al. Quantitative sonographic assessment of the quadriceps femoris muscle in healthy Japanese adults. J Ultrasound Med. 2017;36(7):1383–95. https://doi.org/10.7863/ultra.16.07054.

19. Murakami H, Watanabe T, Fukuoka D, Terabayashi N, Hara T, Muramatsu C, et al. Development of estimation system of knee extension strength using image features in ultrasound images of rectus femoris. Proc. SPIE 9790, Medical Imaging 2016: Ultrasonic Imaging and Tomography. 2016;9790:979012. https://doi.org/10.1117/12.2214843.

20. Onishi Y, Teramoto A, Tsujimoto M, Tsukamoto T, Saito K, Toyama H, et al. Automated pulmonary nodule classification in computed tomography images using a deep convolutional neural network trained by generative adversarial networks. Biomed Res Int. 2019;6051939:1–9. https://doi.org/10.1155/2019/6051939.

21. Onishi Y, Teramoto A, Tsujimoto M, Tsukamoto T, Saito K, Toyama H, et al. Multiplanar analysis for pulmonary nodule classification in CT images using deep convolutional neural network and generative adversarial networks. Int J Comput Assist Radiol Surg. 2020;15(1):173–8. https://doi.org/10.1007/s11548-019-02092-z.

22. Onishi Y, Teramoto A, Tsujimoto M, Tsukamoto T, Saito K, Toyama H, et al. Investigation of pulmonary nodule classification using multi-scale residual network enhanced with 3DGAN-synthesized volumes. Radiol Phys Technol. 2020;13(2):160–9. https://doi.org/10.1007/s12194-020-00564-5.

23. Teramoto A, Tsukamoto T, Kiriyama Y, Fujita H. Automated classification of lung cancer types from cytological images using deep convolutional neural networks. Biomed Res Int. 2017;4067832:1–6. https://doi.org/10.1155/2017/4067832.

24. Teramoto A, Yamada A, Kiriyama Y, Tsukamoto T, Yan K, Zhang L, et al. Automated classification of benign and malignant cells from lung cytological images using deep convolutional neural network. Inform Med Unlocked. 2019;16:100205. https://doi.org/10.1016/j.imu.2019.100205.

25. Teramoto A, Tsukamoto T, Yamada A, Kiriyama Y, Imaizumi K, Saito K, et al. Deep learning approach to classification of lung cytological images: two-step training using actual and synthesized images by progressive growing of generative adversarial networks. PLoS One. 2020;15(3):e0229951. https://doi.org/10.1371/journal.pone.0229951.

26. Kai C, Uchiyama Y, Shiraishi J, Fujita H. Quantitation of cerebral atrophy due to normal aging: principal component analysis with MR images in patients' age groups. Nihon Hoshasen Gijutsu Gakkai Zasshi. 2018;74(12):1389–95. (in Japanese). https://doi.org/10.6009/jjrt.2018_jsrt_74.12.1389.

27. Kai C, Uchiyama Y, Shiraishi J, Fujita H, Doi K. Computer-aided diagnosis with radiogenomics: analysis of the relationship between genotype and morphological changes of the brain magnetic resonance images. Radiol Phys Technol. 2018;11(3):265–73. https://doi.org/10.1007/s12194-018-0462-5.

28. Kai C, Ishimaru M, Uchiyama Y, Shiraishi J, Shinohara N, Fujita H. Selection of radiomic features for the classification of triple negative breast cancer based on radiogenomics. Nihon Hoshasen Gijutsu Gakkai Zasshi. 2019;75:24–31. (in Japanese). https://doi.org/10.6009/jjrt.2019_jsrt_75.1.24.

29. Hatanaka Y, Tachiki H, Ogohara K, Muramatsu C, Okumura S, Fujita H. Artery and vein diameter ratio measurement based on improvement of arteries and veins segmentation on retinal images. Conf Proc IEEE Eng Med Biol Soc. 2016;2016:1336–9. https://doi.org/10.1109/EMBC.2016.7590954.

30. Inoue T, Hatanaka Y, Okumura S, Ogohara K, Muramatsu C, Fujita H. Automatic microaneurysm detection in retinal fundus images by density gradient vector concentration. J Inst Image Electron Eng Jpn. 2015;44(1):58–66. (in Japanese). https://doi.org/10.11371/iieej.44.58.

31. Hatanaka Y, Inoue T, Ogohara K, Okumura S, Muramatsu C, Fujita H. Automated microaneurysm detection in retinal fundus images based on the combination of three detectors. J Med Imag Health Inform. 2018;8(5):1103–12. https://doi.org/10.1166/jmihi.2018.2419.

32. Miyashita M, Hatanaka Y, Ogohara K, Muramatsu C, Sunayama W, Fujita H. Automatic detection of microaneurysms in retinal image by using convolutional neural network. Med Imag Tech. 2018;36(4):189–95. (in Japanese). https://doi.org/10.11409/mit.36.189.

33. Muramatsu C, Hayashi T, Hara T, Katsumata A, Fujita H. Computer-aided diagnosis with dental images. In: Mazzoncini de Azevedo- Marques P, Mencattini A, Salmeri M, Rangayyan RM, editors. Medical image analysis and informatics: computer-aided diagnosis and therapy. Boca Raton: CRC Press;2017. p. 103–127.

34. Miki Y, Muramatsu C, Hayashi T, Zhou X, Hara T, Katsumata A, et al. Classification of teeth in cone-bean CT using deep convolutional neural network. Comput Biol Med. 2017;80:24–9. https://doi.org/10.1016/j.compbiomed.2016.11.003.

Part IV

Basic Principles of MCA: Clinical and Scientific Application of MCA

Clinical Applications of MCA to Surgery

9

Kenoki Ohuchida, Chika Iwamoto, and Makoto Hashizume

Abstract

Now, we developed multidisciplinary computational anatomy (MCA) model, in which we constructed the database based on the spatial axis, the time axis, the functional axis, and the pathological axis that are important components in actual clinical practice. We have developed a highly intelligent diagnosis and treatment system using a mathematical model that can cope with the individuality of various diseases and each patient. Here, we outlined the application of the MCA model in cancer treatment and surgical treatment.

Keywords

Laparoscopic surgery · Robot-assisted surgery · Navigation surgery · Simulation surgery

9.1 Introduction

Currently, major revolutions happen in clinical setting. In particular, treatment options for cancer are dramatically increasing. In medical treatment, a wide variety of anti-cancer agents have been developed. Their therapeutic effects have been reliably demonstrated in the medical field, leading to the progress of cancer medicine based on evidence-based medicine. In the surgical fields, the treatment with minimally invasiveness has surely become widespread due to endoscopic surgery. However, not only in cancer but also in all diseases, there are individual differences in the clinical condition, so that an average therapeutic effect derived from simple statistical analysis cannot be expected in all patients.

In surgical treatment, the size, position, and mobility of the tumor, body shape, organ size, fat mass, and degree of adhesion of the patient vary widely among patients. Therefore, today, it is hard to say that all minimally invasive treatments are carried out without compromising the security and oncologic safety for all patients. As a result, sub-optimal treatment may be selected or given to some patients. Today, to avoid this, we need skilled doctors with the sufficient experience. However, only some of the skilled doctors have the ability and only a limited number of patients can enjoy their ability because human resources and ability are limited.

To break down such medical situation, it is indispensable to accelerate research with medical-engineering collaboration for development of surgical simulation, surgical navigation, and surgical robots based on objective medical images. da Vinci Surgical System (Intuitive Surgical) was created from medical-industrial collaboration research, and more than 5000 units have been used in all over the world to date. In Japan, regulatory approval was received in 2009, and it has already been installed at many facilities. Robot-assisted surgery, which can maintain a high level of minimally invasive, curative, and safe treatment, has become an insured medical treatment and has become indispensable.

We also previously developed small master–slave surgery assisted robot, lesion diagnosis system with virtual endoscopy, navigation system for endoscopic surgery, Active running robotic endoscopy and contributed to the development of computer surgery through medical-engineering collaboration. However, in the current situation, these technologies have not been put to practical use and brought about a change in surgical medical treatment.

Now, we developed multidisciplinary computational anatomy (MCA) model, in which we constructed the database based on the spatial axis, the time axis, the functional axis, and the pathological axis that are important components in actual clinical practice. We have developed a highly

K. Ohuchida (✉) · C. Iwamoto · M. Hashizume
Department of Surgery and Oncology, Kyushu University, Fukuoka, Japan

Center for Advanced Medical Innovation, Kyushu University, Fukuoka, Japan
e-mail: kenoki@surg1.med.kyushu-u.ac.jp

intelligent diagnosis and treatment system using a mathematical model that can cope with the individuality of various diseases and each patient. Here, we outlined the application of the MCA model in cancer treatment and surgical treatment.

9.2 Application of MCA Model in the Research of Pancreatic Cancer

Pancreatic cancer is a solid cancer and consists of a microenvironment characterized by abundant stromal components. To develop a truly effective treatment for pancreatic cancer, it is necessary to more accurately reproduce and understand the physical condition of pancreatic cancer including microenvironment. In MCA, various types of images and models enable us to analyze such tumor microenvironment from morphological, spatial, temporal, and functional aspects, although such solid analysis was difficult to perform with conventional molecular biology analysis alone. Now, we clarified the pathophysiology of pancreatic cancer microenvironment. Pancreatic cancer arises from a precancerous lesion called PanIN in the early stage of carcinogenesis. PanIN is classified into PanIN-1, PanIN-2, and PanIN-3 according to the degree of atypia. Among them, the lesion called PanIN-3 initially remained in the lumen as a noninvasive cancer, but when its malignancy further increased, it ruptured the basement membrane, invaded the surrounding stroma, and became invasive cancer. Then, it progresses inside the pancreas and invades adipose tissue and other organs outside the pancreas. Pathological observation of the local invasion process revealed that pancreatic cancer invasion was always accompanied by a strong stromal reaction. In addition, it is well known that the center of pancreatic tumor is accompanied by a high degree of fibrosis called desmoplasia, but similarly, in most cases, a stromal reaction is associated with the tip of the infiltrate at the margin of the pancreatic tumor. Conventionally, it was quite difficult to analyze a series of carcinogenic processes over time, but in 2003, a mouse with genetic modification in which activated Kras was expressed specifically in a pancreas was reported [1] and the situation was changed. This mouse exhibits a pancreatic carcinogenesis pattern which is histologically very similar to that of human pancreatic cancer. Furthermore, a pancreatic carcinogenesis model (KPC mouse) with Kras activation and p53 inactivating mutations was also reported, and this model produces invasive cancer with marked interstitial hyperplasia and fibrosis similar to human pancreatic cancer tissue [2]. And in this model, the carcinogenic process also occurred PanIN lesions like human pancreatic cancer.

In the MCA projects, we crossed this KPC mouse with a luciferase mouse and created a KPCL mouse. In this mouse, the pancreatic tumors emit light with administration of luciferin, and it became possible to evaluate the size of the tumor and the presence or absence of metastatic foci in a live image (Fig. 9.1). Also, using this model, we can analyze the time-dependent spatial distribution of luciferase-positive precancerous cells in pathological images. When we slice all specimens fixed as a paraffin block, the distribution of PanIN, which is a precancerous lesion, can be visualized as a 3D image, and we can perform spatial analysis and observe the changes in their distribution. Furthermore, since it is possible to perform functional evaluation when we stain a marker for the proliferative ability such as Ki67, leading to the integration of spatial, temporal, pathological, and functional evaluations. Such integrated analyses of PanIN can make us deepen the comprehensive understanding of the lesions.

On the other hand, it is extremely difficult to analyze the carcinogenic process of human pancreatic cancer. However, there is a subtype of pancreatic cancer called intraductal papillary mucinous neoplasm (IPMN). Since this IPMN is a lesion with cysts, it is easy to detect it early by ultrasound or CT, and it is often found in the state of precancerous lesions. However, even if it can be detected early, it is difficult to evaluate the malignant potential, so that we often perform a surgical operation with extremely high invasiveness even though it is a precancerous lesion. Therefore, it is important to analyze the tissues of various carcinogenic processes, such as precancerous lesions, non-invasive cancer lesions, and microinvasive cancer lesions. To date, we examined these resected IPMN tissues using micro CT, which is one of major technologies of MCA. Figure 9.2 showed a case of total pancreatectomy because IPMN with cystic lesions is distributed throughout the pancreas. In this case, the usual evaluation method with observation of 5 mm sliced section planes is not sufficient to evaluate the important lesions such as micro-infiltrates. To improve this situation, we use the micro CT to image the whole excised sample, reconstruct this cystic lesion as a 3D image, and identify the minute nodular lesions. In the future, micro CT will surely extract the lesions that are particularly important in the carcinogenic process, and construct a 3D image at the micro level without slicing all block samples, staining it, and capturing the pathological tissue image with a high-resolution scanner. The rapid clinical application of such system is expected.

Also, this IPMN is often followed up due to the high invasiveness of pancreatectomy, and its time-dependent change is evaluated based on CT and MRI images. When we integrate these images bases on the method derived from MCA, it is possible to examine time-dependent morphological changes of macroscopic pancreatic tumors (Fig. 9.2). And now, we reconstruct a 3D image of resected tissues at the pathological level. Soon, based on these results, we will combine the preoperative tumor image with the postoperative pathological image of the resected tumor.

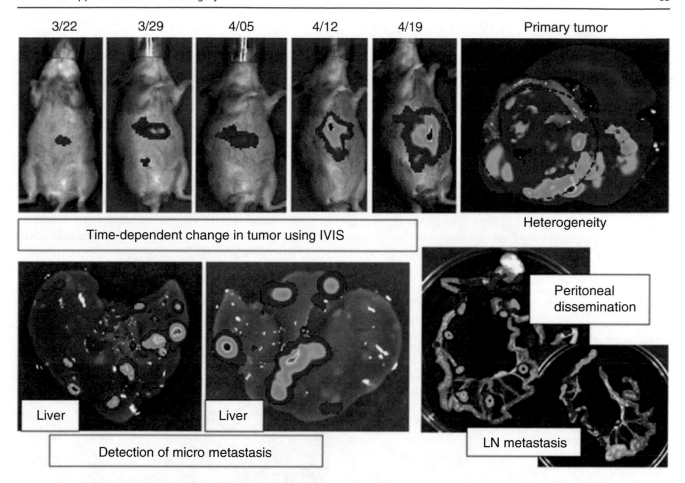

Fig. 9.1 Time-dependent change and heterogeneity in pancreatic tumor and metastatic lesions in KPCL mouse model

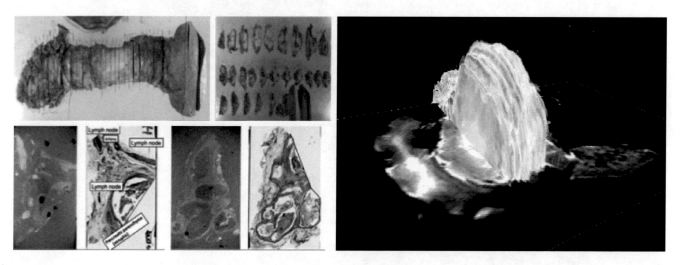

Fig. 9.2 Micro CT images of total pancreas with IPMN and 3D-fusion of microMRI and pathological images

Pancreatic tumors consist of various cell populations such as vascular endothelial cells, inflammatory cells, and fibroblasts as well as cancer cells. There are also some clonal populations of cancer cells, the so-called heterogeneity exists, and it has been reported that a small number of cell populations called cancer stem cells, which are less than sev-

eral percent, are involved in tumor formation, maintenance, and metastasis. Furthermore, it has been suggested that the individuality of pancreatic tumors may be closely related to the individuality of surrounding stromal cells in addition to the individuality of cancer cells themselves. Conventional molecular biological analysis has revealed some of the individuality of the cancer cells themselves and of the stromal cells in the surrounding microenvironment that have strong interactions with the cancer cells. However, the overall picture is not yet fully understood, and individualized treatment has not yet been put to practical use. Therefore, to understand the invasion and metastasis process of pancreatic cancer, we used MCA models and analyzed them from a different perspective from the conventional molecular biology method. In this project, we used MCA method with time-dependent images of micro CT and micro MRI, and reconstructed 3D images to analyze the time-dependent spatial heterogeneity in the tumor. Furthermore, the tumor was fixed with formalin together with the surrounding tissue, paraffin blocks were created, and all blocks were sliced to reconstruct a 3D image at the pathological resolution. Then, we can observe a non-uniform distribution at the cell level (Fig. 9.3). It was also possible to extract only the stroma such as collagen by special staining and reconstruct its 3D image. Using this MCA-based method, we can evaluate the heterogenous distribution of the stroma in the tumor microenvironments.

We also performed the analysis based on MCA method using human pancreatic cancer tissues. Using the analysis along the functional axis in the MCA model, we investigated the three-dimensional distribution of cancer cells with proliferation ability, and also investigated its correlation with the distribution of the tubular structure of pancreatic cancer cells corresponding to the degree of cell differentiation (Fig. 9.3). Further, we reconstructed the three-dimensional pathological tissue image of normal pancreas, and evaluated the three-dimensional expansion of the luminal structure, which was previously thought to be a small branch located near the pancreatic duct. Then, we identified the pancreatic duct gland (Fig. 9.3), which was recently reported to be the specific site where precancerous cells exit.

We created an organoid from a resected tissue of pancreatic cancer using a 3D model, and made it possible to analyze cancer cells in a more biological environment. So far, we have used this 3D model of pancreatic cancer cells and Pancreatic stellate cells (PSCs) and observed pre-invasion of PSCs before the invasion of pancreatic cancer cells (Fig. 9.4). At that time, such invading PSCs contracted the collagen gel layer and changed the fiber direction, revealing that PSCs have the role as the leading cells in cancer invasion [3]. Also, the disruption of basement membrane is the first step in the progression of cancer. Our findings using 3D organoid model suggested that there is a leading PSC that directly acts on the basement membrane of cancer and leads to its destruction [4]. In a pancreatic cancer tissue, a differentiated tubular adenocarcinoma has a basement membrane on the side oppo-

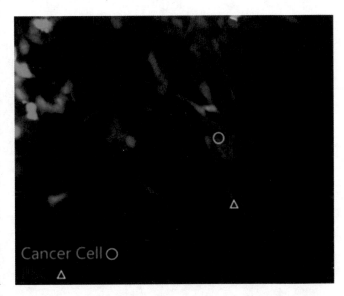

Fig. 9.4 The red PSC pre-invaded and lead the invasion of pancreatic cancer cells

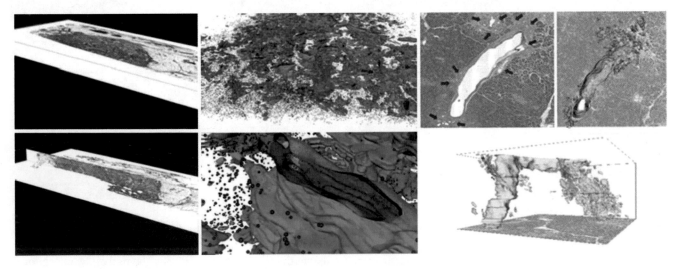

Fig. 9.3 3D reconstruction of pathological images of Ki67-staining cancer cells and pancreatic duct gland

site to the lumen, like a normal pancreatic duct. These data suggest that the destruction and regeneration of the basement membrane may be involved not only when non-invasive cancer becomes invasive cancer, but also when invasive cancer cells itself invade while maintaining its tubular structure.

The process of invasion or metastasis of pancreatic cancer is deeply related to the surrounding microenvironment, and its mechanism is expected to vary depending on its localization. In the future, it is necessary to clarify the cancer microenvironment from the biochemical, biomechanical, and biophysical aspects as well as molecular biological aspect. MCA-based methods are expected to be one of the effective methods to support these researches.

9.3 Clinical Application of MCA Model in Surgical Treatment

We promoted clinical application of MCA methods in the clinical field of oncologic surgery. In particular, we focused on pancreas, because the development of both preoperative simulation and navigation was behind due to the difficulty in segmentation of the pancreas itself. The morphology of the pancreas and the main blood vessels and organs around it, and their relative position differ from patient to patient. Understanding individual anatomy is needed not only to perform the surgery for pancreatic disease but also to perform lymph node dissection around the pancreas, which is important for gastric cancer surgery. Particularly in endoscopic surgery, not only the position and shape of the pancreatic tumor, but also the shape of the pancreas itself, the running structures of blood vessels around it, and their relative position are deeply involved in the difficulty of the surgical operation. Now, we can understand the three-dimensional surgical anatomy based on the MCA model that we recently constructed, and operate intuitively based on the image information necessary for endoscopic surgery. At the same time, it is possible to measure the distances and angles of the organs and tissues and accumulate them as objective data.

Understanding such individual anatomy intuitively and accumulation of objective data are useful in assessing the degree of difficulty and selecting an appropriate approach or the surgical technique according to the individual anatomy of the patient in endoscopic surgery that deals with the periphery of the pancreas, contributing to establishing the standard procedures based on the patient individuality.

Moreover, the pancreatic duct in the pancreas can be easily extracted using the MCA method. In particular, in the case where the pancreatic duct is occluded due to cancer with the dilated peripheral pancreatic duct or in the IPMN case where the pancreatic duct is dilated, the expanded pancreatic duct is extracted based on MCA method and we reconstructed a virtual pancreatic duct endoscopic images to observe the branch duct and the lesion in the pancreatic duct from the lumen of the pancreatic duct. The virtual pancreatic duct endoscopic images make us comprehend the spread of lesions in the pancreas like a real pancreatic duct endoscopic image (Fig. 9.5).

In addition, the decision of the resection line of pancreas is one of the important processes in pancreatic surgery. In the present pancreas simulation based on MCA model, the pancreatic cut surface at the planned pancreatic cut line can be shown together with the pancreatic duct stump by constructing a 3D image of the pancreatic parenchyma and pancreatic duct (Fig. 9.5).

3D printers have already spread widely, and we can get a real 3D model based on the constructed 3D image (Fig. 9.6). The real full-scale 3D model can be created from the same information as the 3D image on the monitor, but we can intuitively observe the part which we want to see directly without using the mouse, and imagine the situations encountered during surgery more realistically. It is also possible to combine materials of two colors, and the real 3D model colored by painting makes it easier to understand the anatomy and is more useful from an educational perspective.

With the technology created from MCA methods, it is now possible to intuitively understand the relative position between the tumor and the main pancreatic duct from various

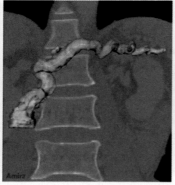

Fig. 9.5 The images of virtual pancreatic duct endoscopy and cross section of pancreatic duct

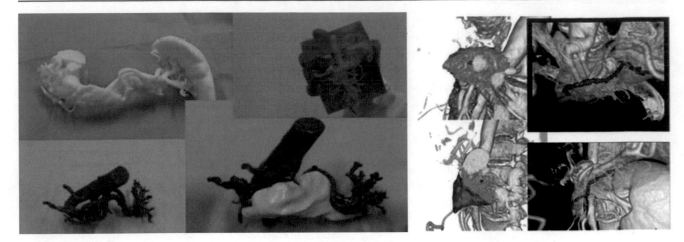

Fig. 9.6 3D printer model around the pancreas and simulation of the distances between tumor and main pancreatic duct

angles based on the tumor extraction and the main pancreatic duct extraction and we can actually measure the shortest distance between them (Fig. 9.6). In recent years, laparoscopic surgery for benign tumors of the pancreas and low-grade tumors has been increasing, and in particular pancreas-preserving surgery removing only the tumor called enucleation is increasing. The pancreatic endocrine tumor is one of the lesions to which such pancreatic preservation surgery is applied, and enucleation is often selected depending on the size and localization of the tumor. It is important to obtain accurate information such as the absolute distance between the tumor and the main pancreatic duct to determine this surgical procedure. Therefore, the clinical application of the MCA method in this field is highly expected.

Using MCA models, we investigated the deformation simulation of the pancreatic parenchyma. Although there are individual differences, the pancreas is an organ that is naturally soft and easily deformed. Deformation of the pancreas is an important factor in surgical operations related with the pancreas, especially in laparoscopic surgery. To simulate deformation of the pancreas, we used MCA models and calculated various factors such as its hardness, shape, and fixability with surrounding tissues. We also created a flexible model using a 3D printer and reconstructed the pancreas of various hardness by changing the hardness of the material. Now, we use this flexible model and simulate an actual procedure and feedback the results. Therefore, this flexible model is not only useful for preoperative planning, but also for further development of a deformation model, itself. Currently, as one of the results of MCA, it is possible not only to automatically extract organs and blood vessels, but also to automatically recognize the extracted organs and blood vessels and label their names. This system is also useful for medical education. Although such labeling of organ names may not be useful to surgeons who are familiar with anatomy, this automatic recognition function allows the AI to automatically extract individual anatomical features for each patient, which can reduce the burden on the surgeon. Furthermore, when we link this function with information based on the experience of expert surgeons, it will lead to the development of a surgical supporting system that suggests an appropriate surgical method according to individual anatomy.

9.4 MCA-Based Surgical Processing Model (SPM)

Although the judgment for surgical procedures is important during surgery, it is difficult to carry out it based on the objective findings because it is usually performed based on the subjective evaluation of the surgeon [5, 6]. Therefore, the attempts have been made to clarify them by time-series analysis of events during surgery, which is called Surgical Processing Model (SPM). To establish SPM, AI can play an active role, especially in surgical technique training. The objective assessment of the most advanced surgical procedure with medical-engineering collaboration is most compatible with engineering approaches and contributes to the development of SPM. It is also useful to provide training tasks according to surgeons, such as optimization of training curriculum. AI-based Measure of the Hand Motions developed by Uemura et al. [7] is one of AI-based researches for evaluation of surgical procedures. They compared the behaviors of beginners and experts in operating forceps during surgery, and defined parameters that can be evaluated as an expert. Using these parameters, we verified whether an expert and a beginner can be separated from different subjects [8, 9]. Using the AI optimized by this verification, they could perform surgical technique evaluation automatically. Furthermore, using MCA models, we analyzed objectively hard-to-commit technology the so-called craftsmanship that relies on the experience and intuition of a skilled surgeons through analysis of surgical workflow and applied and developed an objective and quantitative evaluation method for SPM. Then, MCA-related joint research was conducted between Japan and France to establish a global standard for surgical technique evaluation methods and training methods for achieving safe and accurate surgery. In Japan, we mainly performed data collection and SPM verification, and in France, they performed surgical skill analysis based on the data such as motion data and SPM analysis data, and verified its usefulness. Based on these results, we verbalized the surgical technique, arranged it in order and found out the surgical technique pattern of an expert operator.

9.5 Conclusion

The surgical navigation system, simulation system, robot system, AI system, and SPM, which were developed at MCA, are extremely important for future medical development. The database of surgeon's judgment in not only actual clinical surgery but also surgical training will lead to further development of MCA-based medicine for risk management of surgery, precision treatment, and individualized medicine.

References

1. Hingorani SR, Petricoin EF, Maitra A, Rajapakse V, King C, Jacobetz MA, et al. Preinvasive and invasive ductal pancreatic cancer and its early detection in the mouse. Cancer Cell. 2003;4:437–50.
2. Hingorani SR, Wang L, Multani AS, Combs C, Deramaudt TB, Hruban RH, et al. Trp53R172H and KrasG12D cooperate to promote chromosomal instability and widely metastatic pancreatic ductal adenocarcinoma in mice. Cancer Cell. 2005;7:469–83.
3. Koikawa K, Ohuchida K, Ando Y, Kibe S, Nakayama H, Takesue S, Endo S, Abe T, Okumura T, Iwamoto C, Moriyama T, Nakata K, Miyasaka Y, Ohtsuka T, Nagai E, Mizumoto K, Hashizume M, Nakamura M. Basement membrane destruction by pancreatic stellate cells leads to local invasion in pancreatic ductal adenocarcinoma. Cancer Lett. 2018 Jul 1;425:65–77.
4. Koikawa K, Ohuchida K, Takesue S, Ando Y, Kibe S, Nakayama H, Endo S, Abe T, Okumura T, Horioka K, Sada M, Iwamoto C, Moriyama T, Nakata K, Miyasaka Y, Ohuchida R, Manabe T, Ohtsuka T, Nagai E, Mizumoto K, Hashizume M, Nakamura M. Pancreatic stellate cells reorganize matrix components and lead pancreatic cancer invasion via the function of Endo180. Cancer Lett. 2018 Jan 1;412:143–54.
5. Uemura M, Sakata K, Tomikawa M, Nagao Y, Ohuchida K, Ieiri S, Akahoshi T, Hashizume M. Novel surgical skill evaluation with reference to two-handed coordination. Fukuoka Igaku Zasshi. 2015 Jul;106(7):213–22.
6. Obata S, Ieiri S, Uemura M, Jimbo T, Souzaki R, Matsuoka N, Katayama T, Hashizume M, Taguchi T. An endoscopic surgical skill validation system for pediatric surgeons using a model of congenital diaphragmatic hernia repair. J Laparoendosc Adv Surg Tech A. 2015 Sep;25(9):775–81.
7. Uemura M, Tomikawa M, Miao T, Souzaki R, Ieiri S, Akahoshi T, Lefor AK, Hashizume M. Feasibility of an AI-based measure of the hand motions of expert and novice surgeons. Comput Math Methods Med. 2018;4:9873273.
8. Busch C, Nakadate R, Uemura M, Obata S, Jimbo T, Hashizume M. Objective assessment of robotic suturing skills with a new computerized system: a step forward in the training of robotic surgeons. Asian J Endosc Surg. 2019 Oct;12(4):388–95.
9. Jimbo T, Ieiri S, Obata S, Uemura M, Souzaki R, Matsuoka N, Katayama T, Masumoto K, Hashizume M, Taguchi T. A new innovative laparoscopic fundoplication training simulator with a surgical skill validation system. Surg Endosc. 2017 Apr;31(4):1688–96.

Clinical Applications of MCA to Diagnosis

10

Shoji Kido, Shingo Mabu, Tohru Kamiya, Yasushi Hirano, Rie Tachibana, and Kunihiro Inai

Abstract

In multidisciplinary computational anatomy (MCA), its scheme will be expanded in spatial, time series, functional, and pathological axes. Therefore, we have expected computer-aided diagnosis (CAD) applications based on this scheme are able to support diagnosis for wide range of clinical images including not only radiological images, but also pathological images and autopsy images. From these axes of views, we have developed robust CAD methods for pathological lungs such as diffuse lung diseases (DLD), lung nodules, and also colon polyps. In addition, we have obtained three dimensional (3D)-scanned images of whole lungs as new pathological images to assist diagnosis of clinical images.

Keywords

Computer-aided Diagnosis (CAD) · Unsupervised Learning · Semi-supervised Learning · Transfer Learning Temporal Subtraction · Bayesian Optimization · Electronic Cleansing · Generative Adversarial Network (GAN) · Autopsy Imaging (Ai) · Digital Autopsy

S. Kido (✉)
Department of Artificial Intelligence Diagnostic Radiology, Osaka University Graduate School of Medicine, Osaka, Japan
e-mail: kido@radiol.med.osaka-u.ac.jp

S. Mabu
Graduate School of Sciences and Technology for Innovation, Yamaguchi University, Yamaguchi, Japan

T. Kamiya
Department of Mechanical and Control Engineering, Faculty of Engineering, Kyushu Institute of Technology, Fukuoka, Japan

Y. Hirano
Medical Informatics and Decision Sciences, Yamaguchi University Hospital, Yamaguchi, Japan

R. Tachibana
Department of Information Science & Technology, National Institute of Technology, Oshima College, Komatsu, Japan

K. Inai
Division of Molecular Pathology, School of Medical Sciences, Fukui University, Fukui, Japan

10.1 Introduction

In the research of medical imaging, CAD applications targeting for breast and colon cancers have achieved a measure of success. However, despite strong expectations from clinicians for CAD applications, they are not as widespread in clinical practice as originally envisioned. One of the reasons for this is that the current CAD limits the target to specific organs and diseases for the purpose of screening. Clinicians pointed out that CAD applications had to be useful for various diseases in daily clinical practice. Therefore, we have been working on the development of CAD applications those can handle multiple organs and multiple diseases. However, clinicians pointed out that the developed CAD applications were complicated to use in daily clinical practice, because they needed to execute different algorithms for each organ and disease. One of the reasons was that these CAD applications were not based on generic anatomical models.

In spatial, time series, functional, and pathological axes, MCA models are expanded from computational anatomical (CA) models. In the spatial axis, we have expanded the CAD scheme from pathological level (cell size) to CT level (organs). In the time series axis, we have expanded the CAD scheme from living images to postmortem images. We called these images as life-time images which include living images and autopsy images. In the functional and pathological axes, we have also developed our CAD applications. MCA models have enormous quantity of information compared with conventional CA models. Therefore, we have expected clinical applications based on MCA are able to support diagnosis for wide range of clinical images including not only conventional radiological images, but also pathological (microscopic or macroscopic) images, and autopsy images.

From these points of view, we have developed clinical CAD applications based on MCA. One is the classification method of DLD opacity patterns on high-resolution CT (HRCT) images using a convolutional neural network (CNN). These CAD applications can classify normal and DLD opacities such as consolidation, ground-glass opacity (GGO), honeycombing emphysema, diffuse nodular, which reveal typical DLD opacities. However, for realizing high accuracy, a large number of images with correct annotations are necessary for training CNN. It is quite tough and impractical task for radiologists to annotate a large number of images. Therefore, we have proposed some unsupervised learning algorithms without using annotated training data for distinguishing DLD opacities. Another solution for small number of images to effectively train classifiers is transfer learning, where deep neural networks trained in a certain domain are fine-tuned in a different domain. In addition, we have developed some CAD applications for lung nodules on chest CT images. One is an image registration technique and a detection algorithm of abnormalities on temporal subtraction images for lung nodules. The second is an optimization of network architecture for deep learning for classification algorithm of lung nodules based on 3D-CNN. And, third is a deep learning-based method for classification of lung nodules based on medical findings. Moreover, for CAD application to detect colon polyps that are submerged in residual materials by virtual subtraction of the residual materials from CT-colonography (CTC), we have developed a deep-learning scheme for performing electronic cleansing (EC). As a new challenge, we have collected surface images of whole lungs by use of 3D-scanner. The surface images are new modality images. For obtaining such images, we have made inflated fixed lungs from human resected lungs. And, we have registered 3D-scanned images of autopsy lungs with micro-CT images as digital autopsy lungs. From the new point of view, these images will contribute to autopsy imaging.

10.2 Methods

10.2.1 Unsupervised Learning, Semi-Supervised Learning, and Transfer Learning for Efficient Training of Classifiers for DLD Opacities

Deep learning has been widely applied to medical image analysis. One of the reasons behind the success of deep learning is the availability of large application-specific annotated datasets. However, it is difficult to prepare large-scale datasets of various organs and diseases because annotation can only be performed by qualified field experts and it is tough work for them to give annotations to thousands of images.

Therefore, we proposed some unsupervised learning algorithms without using annotated training data for distinguishing DLD opacities. In [1], unsupervised class labeling was realized by the combination of evolutionary data mining and genetic algorithm (GA). This method aims at regions-of-interest-based (ROI-based) clustering, where (1) we used feature values of {mean, variance, skewness, kurtosis} of CT values and three eigenvalues of Hessian matrix [2], then (2) frequent attribute (feature) patterns are extracted by evolutionary data mining, and (3) ROIs having the similar attribute patterns are assigned to the same cluster. The experimental results showed that the clustering accuracy was 47.7% when six types of DLD opacities (consolidation, GGO, honeycombing, emphysema, nodular and normal) were clustered.

Another unsupervised learning method was proposed in [3], where ROI-based clustering was also executed. The flow of the method is shown in Fig. 10.1. First, as a preprocessing, CT images are divided into 32 × 32 [pixels] ROI

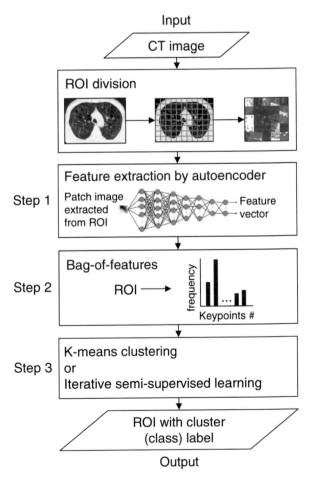

Fig. 10.1 Flow of unsupervised and semi-supervised learning. The unsupervised and semi-supervised learning methods consist of three steps. At step 3, the unsupervised learning implements k-means clustering and the semi-supervised learning implements iterative semi-supervised learning

images. Then, the feature extraction is executed by combining step (1) autoencoder and step (2) bag-of-features, and each ROI is represented by a histogram of key-point features. At step (3), the generated histograms are clustered by k-means clustering. This method showed the clustering accuracy of 72.8%. Then, this unsupervised learning method was extended to semi-supervised learning by changing the part of k-means clustering to iterative semi-supervised learning. The iterative semi-supervised learning contains support vector machine (SVM) classification, self-training, and active learning to gradually improve the classifier. At the first training iteration, only 1% of the data were annotated by human and used to train SVM. Then, the remaining 99% testing data were classified by the trained SVM. After that, the testing data with more than 99% classification confidence were regarded as new training data (self-training), and quite a small number of data with the lowest confidence were annotated by human as new training data. The experimental results showed that the proposed method obtained 98.5% classification accuracy at 507th iteration when 50% of the data were annotated by human.

Another solution to effectively train classifiers is transfer learning, where deep neural networks trained in a certain domain are fine-tuned in a different domain. In [4], aiming the efficient learning of CNN, the effects of transfer learning were analyzed. We used cifar-10 and cifar-100 datasets [5] to pre-train CNN. Cifar-10 and cifar-100 have patch images with 32×32 [pixels] that is the same size as ROI images. After training CNN using cifar-10 or cifar-100 datasets, some fully-connected (FC) layers were added for fine-tuning. Here, additional one FC layer, two FC layers, and three FC layers were compared to find the best structure. The experimental results showed that CNN pre-trained with cifar-100 and fine-tuned with one additional FC layer obtained the best results. Therefore, it was clarified that the pre-training dataset with more (various) classes and the compact structure for fine-tuning contribute to better accuracy in this problem.

10.2.2 Image Registration Techniques and Detection of Abnormalities on Temporal Subtraction Images

To detect lung nodules and GGO on thoracic multi-detector row computed tomography (MDCT) images, CAD systems are used in clinical fields. Without the CAD systems, it is a difficult task to detect abnormalities for radiologists, since subtle lesions such as small lung nodules tend to be low in contrast on the CT images. The CAD system from medical images provides as a second opinion to radiologists on visual screening [6]. In this study, we describe a new method for the detection of automatic temporal changes in thoracic multi-detector row CT (MDCT) images which is obtained different time series and detection of GGO based on machine learning technique from The Lung Image Database Consortium (LIDC) data sets.

The temporal subtraction image which is obtained by subtraction of a previous image from a current one from a same subject can be used for enhancing interval changes on medical images. Radiologists detect these changes, e.g. size and location of regions on both scan images by manually comparing on visual screening. To make a temporal subtraction images, image registration techniques with high accuracy are necessary, since subtraction artifacts with low accuracy registration images are remained caused misregistration. In the past 5 years, we have developed non-rigid image registration techniques and detection of abnormalities on thoracic MDCT images.

To obtain a temporal subtraction image with high accuracy, we have proposed a non-rigid registration method [7]. There are 2 steps, global matching and local matching. Global matching step includes rigid transformation, affine transformation, and high order polynomial transformation. On the other hand, local matching is a step for the nonlinear transform such as a 3D-elastic transformation and voxel matching technique [8]. The voxel matching is a non-rigid registration technique for reducing the subtraction artifacts. Finally, temporal subtraction image is obtained by subtraction of warping previous image from current one. Figure 10.2 shows an experimental result. Fig. 10.2a, b, c, and d are current, previous, temporal subtraction image with our previous method [8] and proposed method [7] from a same patient, respectively. Small lung nodules on both images (arrow) exist on previous and current images and abnormal shadow is enhanced on subtraction image. Most subtraction artifacts are removed on both temporal subtraction images (Fig. 10.2c and d) significantly. Using our new temporal subtraction technique, radiologist can easily detect abnormal shadows on a current image.

To detect abnormal areas such as GGO and lung nodules, we have proposed automatic detection methods for small-sized of lung nodules using on temporal subtraction images based on artificial neural networks [9] and current image set based on CNN [10]. In [9], we extract statistical features based on intensity, gradient, and shape of segmented regions from a temporal subtraction image. After that, we classify true positives and false positives using Artificial Neural Network (ANN) classifier, class featuring information compression (CLAFIC), Mahalanobis distance, and Fisher's linear discriminant (FLD) classifier. From the employed 31 small sized lung nodules, the FROC curve of ANN classifier is the closest to the left top and the detection performance of ANN classifier is 5.1 false positives per CT at 80.7% sensi-

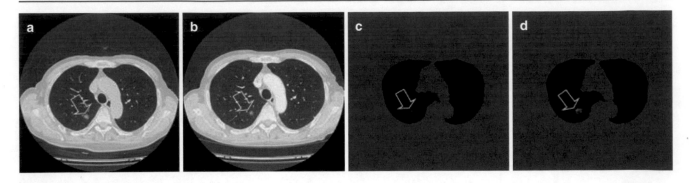

Fig. 10.2 Temporal subtraction image based on non-rigid image registration technique. (**a**) Current image, (**b**) Previous image, (**c**) Subtraction image [8], (**d**) Subtraction image [7]

tivity. On the other hand, detection of GGOs on current image sets [11] based on CNN [10], we obtained a TP of 86.05% with 4.8/scan FP. From the both of experimental results, our new technique may be useful for radiologists in the detection of abnormalities on visual screening.

10.2.3 Optimization of Network Architecture for Deep Learning

As is well known, the deep learning is a powerful tool for many applications such as segmentation, classification, and image enhancement in the field of medical image analysis. However, determination of the network structure to obtain the best performance is a very time-consuming work, because there are too many hyperparameters that represent the network structure. In many cases, pre-defined networks that are published via Internet sites are used, but these networks are usually defined for some specific purposes other than medical purpose. These networks do not necessarily have the best performance for the medical purpose. Furthermore, most of these networks are built for 2D images, but not for 3D images. Therefore, networks for 3D images have to be manually constructed from scratch.

Although the grid search and the random search can be used for optimizing hyperparameters, they are inefficient because each search point is evaluated separately. On the other hand, the Bayesian optimization [12] is efficient and the search point can be located arbitrarily. In the Bayesian optimization, better hyperparameters are decided sequentially by iteration of the exploration phase and the exploitation phase. The current search point (=hyperparameters) is evaluated in the former phase, and the next search point is decided by reference to the results of the previous explorations in the latter phase. As the result of the Bayesian optimization for deep learning, the optimized network structure and the optimized weights are obtained. Figure 10.3 shows the training phase of the Deep Learning including the Bayesian optimization.

As an example of the Bayesian optimization, we performed a classification of lung nodules into benign and malignant by using a 3D-CNN. The hyperparameters to be optimized are number of convolution layers, filter size in convolution layers, numbers of channels in convolution layers, type of pooling layers, and number of nodes in fully-connected layers. The AUC of the ROC curves by the automatically optimized CNN, manually constructed CNN, SVM (RBF kernel), SVM (Polynomial kernel), and Random forest are 0.816, 0.722, 0.749, 0.736, and 0.680, respectively. It is shown that the automatically optimized CNN has the highest performance [13]. Currently, the automatic hyperparameter optimization methods are realized by the python libraries [14] and frameworks [15].

10.2.4 Classification of Lung Tumors into Benign and Malignant Based on Medical Findings Using Deep Learning

Generally, CAD systems based on deep learning provide only the final decision such as category of the tumors or likelihood of malignancy of tumors. The function of these systems is enough for supporting radiologists' decision, but not for explaining the reasons of malignancy to patients and ensuring the validity of the CAD systems. We have been developing the deep learning-based system that provides both the likelihoods of existence of medical findings and likelihood of malignancy of tumors based on them. The medical findings that we used are utilized in clinical practice. First, we calculated likelihoods of four medical findings: clearness of the marginal region of tumor, circularity of tumor, existence of air-bronchogram, and existence of notch. Each likelihood of these medical findings was calculated by using a 3D-CNN trained and constructed for each medical finding. The accuracy of each medical finding is 87%, 68%, 73%, and 76%, respectively, where existence of each medical finding was decided by thresholding of each likelihood.

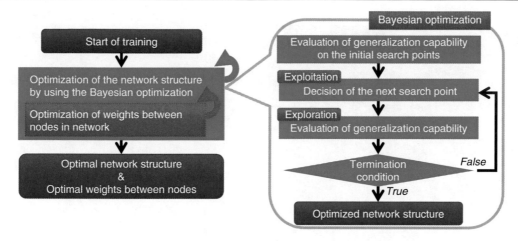

Fig. 10.3 Processing flow of the training phase for the Deep Learning including the Bayesian optimization

Next, the tumors were classified into malignant and benign tumors by using a Multi-Layer Perceptron (MLP). The MLP we used consisted of one input layer, one hidden layer, and one out layer. The likelihoods of the existence of medical findings were fed as input values of the MLP. The final accuracy of malignancy was 79%. As the comparison with the normal CNN that output only the likelihood of malignancy of tumors, the accuracy was almost the same. The proposed method was superior to the normal CNN in terms that the proposed method was able to provide the likelihoods of the existence of medical findings.

10.2.5 EC for CTC

EC enables computer-aided detection systems to detect polyps that are submerged in residual materials by virtual subtraction of the residual materials from CTC images. Previously, we developed a deep-learning scheme for performing EC in dual-energy CTC to overcome the limitations of EC in single-energy CTC [16]. Although the deep-learning-based method can reduce EC artifacts in single-energy CTC, it requires a huge amount of manually annotated volumetric images that are difficult to generate. Moreover, it was computationally expensive because it was designed for analyzing 54 cut-plane images at each voxel.

Generative adversarial network (GAN) is a novel machine-learning algorithm that can directly translate an input image into an output image without the segmentation of a target object. Therefore, we developed a GAN-based EC method for improving the quality of EC images in CTC [17].

Based on the pix2pix GAN method [18] that was developed for 8-bit 2D photos, we developed a 3D-vox2vox GAN for performing the EC on 16-bit volumetric CTC images using 3D-convolution kernels (Fig. 10.4a and b). Figure 10.4a shows an overview of the 3D-vox2vox GAN scheme. The 3D-vox2vox GAN consists of a generator and a discriminator network. The generator network, which is based on encoder-decoder architecture like 3D U-Net, produces fecal-material-cleansed images from an input image with tagged fecal materials. The generator consists of seven 3D-convolutional/deconvolutional layers. The discriminator network, which is based on the 3D-PatchGAN [18], attempts to differentiate cleansed images that are generated by the generator from the images without tagged fecal materials. It consists of three-stride convolutional layers, two convolution layers, and a sigmoid function. After 3D-vox2vox GAN is trained, the resulting generator is used to perform the cleansing of a new unseen CTC volume. As the generator is a fully convolutional network, the new case can be cleansed rapidly regardless of its size (Fig. 10.4b).

In our pilot study, 200 volumes of interest (VOIs) with 128^3 voxels were extracted from the CTC scans of an anthropomorphic phantom (Phantom Laboratory, Salem, NY) with 20 and 40 mg/ml contrast agent. The phantom was scanned using a CT scanner (SOMATOM Definition Flash, Siemens Healthcare) with a single-energy mode, 120 kVp, 68 mA, 0.6-mm slice thickness, and 0.6-mm reconstruction interval. A paired training dataset was sampled from the VOIs of CT scans acquired with and without a contrast agent. After the 3D-vox2vox GAN was trained, the resulting generator was used to cleanse a new unseen CTC volume that was scanned using a CT scanner (LightSpeed 16, GE Medical Systems) with a single-energy mode, 120 kVp, 60 mA, 2.5-mm slice thickness, and 1.25-mm reconstruction interval. For a clinical CTC volume with a size of 512 × 512 × 604 voxels, the test of the 3D-GAN EC scheme took approximately 1 minute using a single GPU (NVIDIA® GeForce GTX 1080 Ti). The test of the deep-learning-based EC scheme took approximately 54 hours using four GPUs (NVIDIA® GeForce GTX 1080). Figure 10.4c and d shows an example of the virtual endoscopic views before and after the 3D-vox2vox

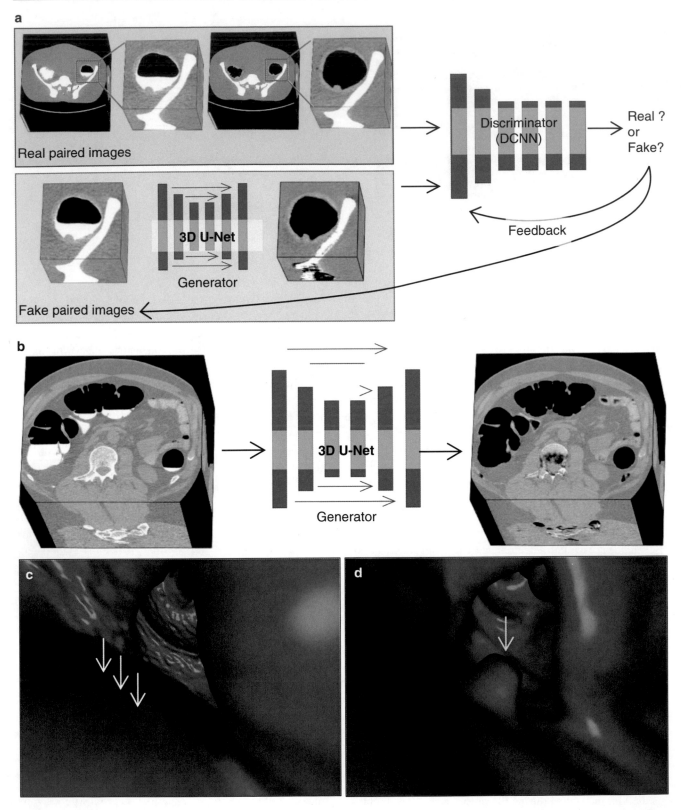

Fig. 10.4 Overview of the 3D-vox2vox GAN EC scheme in the CTC. (**a**) Training based on 3D-vox2vox GAN. (**b**) Testing based on 3D-vox2vox GAN. (c, d) An example of virtual endoluminal view before/after EC. (**c**) The original uncleansed CT image. The white arrows on a virtual endoluminal image indicate tagged residual fecal materials. (**d**) Virtual endoluminal image after cleansing by the 3D-vox2vox GAN EC Scheme. A yellow arrow indicates that a polyp is clearly visible

GAN EC method. The 3D-vox2vox GAN was able to make a polyp visible as shown in Fig. 10.4d.

Deep-learning-based EC can reduce EC artifacts but requires manually annotated training images and intensive calculation time. Our novel EC scheme based on 3D-vox2vox GAN does not require manually annotated training images and is faster in the removal of residual fecal materials than is the deep-learning-based EC. Therefore, the 3D-vox2vox GAN-based EC scheme has the potential in providing an effective solution for EC of CTC images for clinical use. We have further extended the 3D-vox2vox GAN-base EC scheme to incorporate a progressive training method [19] and developed a self-supervised 3D-GAN EC scheme. This EC scheme can generate EC images of higher quality than those obtained by use of only a supervised training dataset.

10.2.6 Establishment of 3D-Whole Lungs for Medical Image Investigation

Recent progression of 3D-image technique allowed us to visualize the stereo views of surface structures of many architectures; however, 3D-observation of human organs is rarely performed yet. In addition, there is no education tool

for health care practitioners to observe the internal structure of organs in association with their surface structure. We present a novel technique how to use the autopsy-resected whole lung in combination with autopsy imaging (Ai) [20] for future medical image analyses.

Human lungs resected by hospital autopsy were transbronchially prefixed by 15% formalin for 30 min followed by the fixation for 7 to 14 days using modified Heitzman's fixative (50% v/v polyethylene glycol 400, 20% ethanol, 4% formalin in distilled water) with 10 cm H2O [21] constant pressure in a custom-designed continuous infusion device (Zek Tech, Osaka, Japan). Then, the fixed lungs were continuously inflated for additional 4 to 7 days by room air with equal pressure via a tube enclosed in the bronchus (Fig. 10.5a). The fixation and influx periods were adjusted by the organ condition such as size, weight, and in the presence or absence of pulmonary diseases.

Figure 10.5b shows the representative 3D-lung images established by an Artec Eva 3D-scanner and Artec Studio Ver.11 software (Data Design, Nagoya, Japan) according to the manufacture's instruction, which can easily access the surface information such as shape, lobes, folds, degree of anthracosis of the lung. Following the completion of 3D-fixation, the lungs were performed conventional CT and

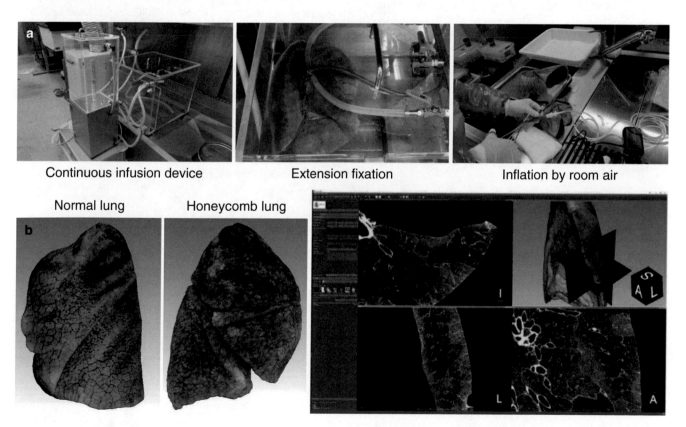

Fig. 10.5 Procedure making 3D-digital lung content from resected lung. (**a**) Establishment of a 3D-whole lung preparation. (**b**) Representative 3D-surface images captured by a 3D-scanner. (**c**) Internal micro-CT images from the arbitrary surface of the 3D-normal lung image

micro-CT scanning (Ai-organ) using a hospital CT and an industrial micro-CT apparatus, respectively, and the scanned images were superimposed. The data sets were embedded in the 3D-lung image by the registration techniques. Figure 10.5c represents a representative digital content working on the 3D-slicer ver.4.6.2. This content can arbitrarily rotate the 3D-lung image and is capable to overview the outside and inside of the lung. Additionally, this content made possible to compare the images of conventional CT with those of micro-CT images on the same cross-section, indicating that this content would become an e-textbook for understanding the internal structure of both normal and pathological lungs.

We established the methodology to make 3D-whole lung specimens from the resected lungs at autopsy in order to preserve the anatomical and radiological data sets. Our new methods allow us to establish further image analyses as well as new medical educations [22].

10.3 Conclusion

We have developed CAD applications for pathological lungs such as DLD and nodules, and also for colon polyps. In addition, we have developed new modality images which registered 3D-scanned images of whole lungs with micro-CT images as digital autopsy lungs. The CAD applications we have developed and the digital autopsy lungs are able to assist diagnosis for wide range of clinical images.

Acknowledgements This research was supported by research grants of Grant-in-Aid for Scientific Research on Innovative areas, MEXT, Japanese Government.

References

1. Mabu S, Obayashi M, Kuremoto T, Hashimoto N, Hirano Y, Kido S. Unsupervised class labeling of diffuse lung diseases using frequent attribute patterns. Int J Comput Assist Radiol Surg. 2017;12(3):519–28.
2. Rui X, Hirano Y, Tachibana R, Kido S. A bag-of-features approach to classify six types of pulmonary textures on high-resolution computed tomography. IEICE Trans Inf Syst. 2013;96(4):845–55.
3. Mabu S, Kido S, Hirano Y, Kuremoto T. Unsupervised and semi-supervised learning for efficient opacity annotation of diffuse lung diseases. In: SPIE 11050, International Forum on Medical Imaging in Asia; 2019. p. 110501D-1–6.
4. Mabu S, Atsumo A, Kido S, Kuremoto T, Hirano Y. Investigating the effects of transfer learning on ROI-based classification of chest CT images: a case study on diffuse lung diseases. J Signal Process Syst. 2020;92(3):307–13.
5. CIFAR-10 and CIFAR-100 datasets. Accessed: 15 July 2017 https://www.cs.toronto.edu/~kriz/cifar.html
6. Doi K. Current status and future potential of computer-aided diagnosis in medical imaging. British J Rad. 2005;78:S3–S19.
7. Lu K, Li T, Kim M, Aoki K. Extraction of GGO candidate regions on thoracic CT images using Supervoxel-based graph cuts for healthcare systems. Mobile Networks and Applications. 2018;23(6):1669–79.
8. Itai K, Ishikawa K, Doi. Development of a voxel matching technique for substantial reduction of subtraction artifacts in temporal subtraction images obtained from thoracic MDCT. J Digit Imaging. 2010;23(1):31–8.
9. Yoshino M, Tan L, Kim M, Aoki T, Hirano K. Automatic classification of lung nodules on MDCT images with the temporal subtraction technique. Int J Comput Assist Radiol Surg. 2017;12(10):1789–98.
10. Hirayama M, Lu T, Kim T, Hirano K. Extracting of GGO regions from chest CT images using deep learning. In: Proceedings of the 17th International Conference on Control, Automation and Systems; 2017. p. 351–5.
11. Lung Image Database Consortium (LIDC) – Cancer Imaging Program http://imaging.cancer.gov/programsandresources/informationsystems/lidc
12. Kushner HJ. A new method of locating the maximum point of an arbitrary multipeak curve in the presence of noise. J Fluids Eng. 1964;86(1):97–106.
13. Hirano Y, Ito T, Kido S et al. Automated construction of the optimal structure for 3D CNN by Using the Bayesian Optimization. In: RSNA 2018. Radiological Society of North America, Chicago; 2018.
14. https://github.com/fmfn/BayesianOptimization/releases
15. https://preferred.jp/en/projects/optuna/
16. Tachibana R, Näppi JJ, Ota J, Kohlhase N, Hironaka T, Kim SH, Regge D, Yoshida H. Deep learning electronic cleansing for single- and dual-energy CT colonography. Radiographics. 2018;38:2034–50.
17. Tachibana R, Näppi J J, Yoshida H. The next step in electronic cleansing for CT colonography: unsupervised machine learning. Radiological Society of North America 2018 Scientific Assembly and Annual Meeting, Chicago; 2018.
18. Isola P, Zhu J-Y, Zhou T, Efros AA. Image-to-image translation with conditional adversarial networks. 2017 IEEE Conference on Computer Vision and Pattern Recognition (CVPR); 2017. p. 1125–1134.
19. Tachibana R, Näppi J, Hironaka T, Yoshida H. Electronic cleansing in CT colonography using a generative adversarial network. Proc SPIE Med Imaging. 2019;10954:1095419.
20. Inai K, Noriki S, Kinoshita K, et al. Postmortem CT is more accurate than clinical diagnosis for identifying the immediate cause of death in hospitalized patients: a prospective autopsy-based study. Virchows Arch. 2016;469:101–9.
21. Ichikado K, Johkoh T, Ikezoe J, et al. Acute interstitial pneumonia: high-resolution CT findings correlated with pathology. Am J Roentgenol. 1997;168:333–8.
22. Tozawa K, Saito A, Kido S, et al. Super resolution of a lung CT volume using a generative adversarial network. In: Computer assisted radiology and surgery, 32th international congress and exhibition (CARS2018). Berlin: Int J Comput Assist Radiol Surg; 2018. p. 170–1.

Application of MCA across Biomedical Engineering

Etsuko Kobayashi, Qingchuan Ma, Daeyoung Kim, Kazuaki Hara, Junchen Wang, and Ken Masamune

Abstract

Many surgical robots have been investigated recently. To create safe and effective control systems for surgical robots, intraoperative biological information is required. However, this information is limited, making it difficult to predict accurate biological responses. If patient-specific models are constructed by combining multidisciplinary computational anatomy with intraoperative biological information, the robots could be controlled by accurately predicting biological responses to the surgery. This approach would imply the realization of safer and more sophisticated treatments. This formed the background for the proposed highly intelligent surgical robot that can approach a surgical area using multidisciplinary computational anatomy and limited intraoperative biological information. In this chapter, we will present about our research activities based on this idea. We will present about (i) study on stapler device control for pancreatic tissue damage suppression, and (ii) navigation and robotic system for oral and maxillofacial surgery.

Keywords

Surgical robot · Minimally invasive surgery · Intraoperative measurement · Oral and maxillofacial surgery

E. Kobayashi (✉) · Q. Ma · K. Hara
School of Engineering, The University of Tokyo, Tokyo, Japan
e-mail: etsuko@bmpe.t.u-tokyo.ac.jp

D. Kim
Faculty of Health and Medical Science, Teikyo Heisei University, Tokyo, Japan

J. Wang
School of Mechanical Engineering and Automation, Beihang University, Beijing, China

K. Masamune
Institute of Advanced Biomedical Engineering and Science, Tokyo Women's Medical University, Tokyo, Japan

11.1 Introduction

Endoscopic surgery is currently a popular form of minimally invasive surgery involving the use of forceps, an endoscope, and a trocar. However, although endoscopy has numerous merits, it has the disadvantage of limited working flexibility. To overcome this obstacle, many surgical robots have been studied.

Most of these robots position surgical instruments based on preoperative or intraoperative images or surgeon input and can support surgery, such as precise positioning and scaling of the manipulation. However, if these robots are controlled using the properties of the affected organs, safer and more effective surgical robots can be created.

To acquire the properties of the organs, intraoperative biological information is required. However, this information is limited, making it difficult to predict accurate biological responses. Therefore, we aimed to construct a patient-specific model by combining multidisciplinary computational anatomy with intraoperative biological information. Using this approach, the surgical robot could be controlled by accurately predicting the biological response to the surgery.

Based on this idea, we will present about (i) study on stapler device control for pancreatic tissue damage suppression and (ii) navigation and robotic systems for oral and maxillofacial surgery.

11.2 Study on Stapler Device Control for Pancreatic Tissue Damage Suppression

In distal pancreatectomy, stapler devices are often used. However, pancreatic juice leakage because of the mechanical damage of pancreas occurs in some cases. We believe that if the relationship between the physical properties of the pancreas, the compression conditions by the stapler during surgery, and pancreatic juice leakage is known, pancreatic juice leakage can be prevented and safe surgery can be realized. To achieve highly safe stapling, the following objectives were performed: (1) analysis of pancreatic injury based on mechanics and pathology and (2) application to the robotic stapler.

11.2.1 Analysis of Pancreatic Injury Based on Mechanics and Pathology

To analyze the relation between compression by stapler and mechanical damage, the following two studies are essential; damage evaluation system and gathering the mechanical properties for simulation studies. Therefore, we conducted the following three studies. (1) Compression force and pancreatic juice leakage measurement system for isolated pig's pancreas, (2) System development for in vivo and in situ organ elasticity measurements, and (3) Compression and pathology observation study for human isolated pancreas.

1. Compression force and pancreatic juice leakage measurement system for isolated pig's pancreas.

In this research, we developed a force and fluorescence measurement system, and compression experiments were performed at several compression speeds [1].

For force measurement, a commercialized compression testing machine (EZ Test, Shimadzu Scientific) was used. To measure the pancreatic juice leakage, a special fluorescence probe (a chymotrypsin probe, which reacts only with pancreatic juice) was used [2]. For fluorescence measurement, an electron-multiplying charge-coupled device camera, an exci-

tation light, and lowpass (490 nm) and bandpass (520 nm) filters were used (Fig. 11.1a).

We isolated the pancreas of a pig; the size of the organs was approximately $250 \times 100 \times 25$ mm^3. We compressed the pancreas using our system at three compressing speeds of 500, 100, and 10 mm/min and evaluated the pancreatic juice leakage. The size of the indenter was 10×65 mm^2. Compression was performed until the pancreas reached 2 mm in thickness (same as with the stapler).

In the fluorescence experiment, 150 µL of chymotrypsin probe was sprayed on the surface of the pancreas before and after compression. The fluorescence intensity of each pixel at each second was measured for 5 min. We then compared the maximum increase rate of the fluorescence intensity at each second before and after compression. We considered the maximum increase rate because (i) the isolated pancreas was already damaged, and some fluorescence was observed before compression and (ii) this rate is related to the concentration of the pancreatic juice.

The results showed that in some areas of the pancreas, the maximum increase rate of fluorescence intensity was higher than before in the pancreas with compression when the reaction force decreases as the membrane breaks (Fig. 11.1b). After the compression test, pathological observation was also performed. The results indicated that the force and fluorescence measurement system we developed can evaluate pancreatic juice leakage. Using this system, we can investigate the proper compression parameters for robotic pancreatectomy.

2. System development for in vivo and in situ organ elasticity measurements.

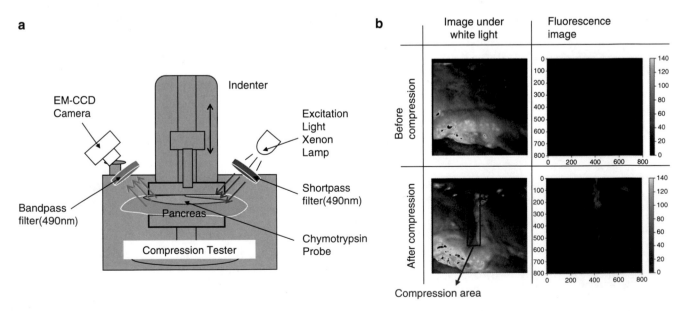

Fig. 11.1 Compression force and pancreatic juice leakage measurement system (**a**) System configuration of the system and (**b**) Images acquired by the system

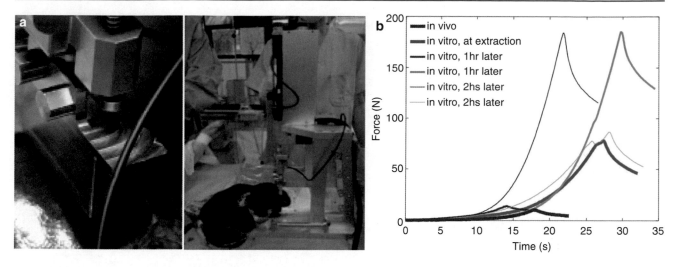

Fig. 11.2 Instrumentation system for organ elasticity measurement (**a**) In vivo (left) and in situ (right) measurements using pig liver and (**b**) Load cell signals were recorded in vivo and in situ

An organ's mechanical properties strongly influence the force control system. The relationship between in vivo and in situ properties is not yet clear because these properties are often measured after isolation or during intraoperative procedures using different methods and parameters. Hence, a measurement instrument that allows consistent measurement is needed.

We are developing an instrumentation system for organ elasticity measurement which can measure elasticity during operation and after isolation [3]. Its manipulator has a servomotor (ENC-258101G 1000/3CH, CITIZEN) and an indenter plate with a load cell (UNLRS-FG200, unipulse). The load cell signal ranges from 0 to 200 N. A photocoupler detects the indenter at the adjustable limit of its descending movement. The controller box includes power supplies and an Arduino Uno R3 microcontroller board (Arduino Srl) with interface modules. Signals are recorded in 10-bit resolution and 0.1-s sampling intervals. From the encoder pulse of the servomotor, the position of the indenter plate is also recorded.

The system is programmed to run an automatic sequence of descend, stop, and ascend phases of the indenter's movement: the descending movement speed is approximately 15 mm/min, and it stops at the 15-mm point above the base plate so that the organ is pressed to a thickness of 10 mm.

Figure 11.2a shows system testing on the liver of a pig, and Fig. 11.2b shows recorded data. We repeated the measurements until 2 h after extraction of the liver. The load cell signals were median-filtered, where the filter length was 0.3 s. The figure plots are aligned with the time when the filtered signal rose above zero (at which time the indenter was supposed to touch the organ surface). The peak of each plot indicates that the indenter stopped compressing at 15-mm thickness of the liver.

The elasticity right after death was less than that in vivo. We suggest that the elasticity decreases owing to the removal of blood during isolation. As time goes by, elasticity decreases owing to rigor mortis. The difficulty of calibration causes differences in elasticity measurements. Because this is a single case, more investigation is required.

3. Compression and pathology observation study for human isolated pancreas.

In a first step, we performed a compression experiment with the swine pancreas in situ, and in vivo and measured the elasticity of the pancreas and the changes in the pancreas that occurred upon rupture. With the swine pancreas, we secure a number of experiments that cannot be obtained with the human pancreas. The swine pancreas is different in shape and size from the human pancreas; however, it is necessary to find the correlation between them. In this study, we started to measure the elasticity of the human pancreas from cadavers.

We compressed the pancreas with the measurement system we developed and a commercialized stapler. With the system, to obtain the viscoelasticity, we compressed the pancreas by 5 mm with 1 mm/min speed to avoid any damage to the pancreas. The reaction force from the gauge was measured. We preserved the pancreas under the clamped condition in formalin and prepared slides for pathological analysis. We compressed the proximal part (pancreas body) and distal part (pancreas tail).

Figure 11.3 shows the first results of the experiments. As we compressed the pancreas, the stress increased exponentially, and relaxation was observed when pausing (shown in Fig. 11.3a). We plotted the results ($n = 10$) as shown in

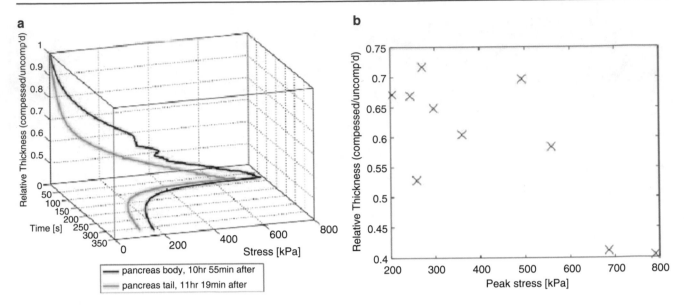

Fig. 11.3 Elasticity of the human pancreas. (**a**) Change of stress in the compressed pancreas and (**b**) Correlation between peak stress and relative thickness (*n* = 10)

Fig. 11.3b, and we observed a correlation between peak stress and relative thickness except in two cases. The pancreas was thin and had some separation from the test table according to our observation based on the picture of the prepared slide. There is little study that has investigated the mechanical properties of the human pancreas. We will continue this work and gather the human pancreas mechanical properties. Continuation of this study is expected to become an important finding for predicting pancreatic damage.

11.2.2 Application to the Robotic Stapler and Future Work

We have developed a pancreatic stapling device (PSD) as shown in Fig. 11.4. A conventional reload for the Covidien Endo GIA Ultra Universal Stapler (Endo GIA) can be attached to the PSD. A coreless DC motor with gear head (1/36) and optical encoder (NC-185801 Citizen Chiba Precision Co., LTD.) is connected to a leadscrew and slider (KR1502A THK). The original rod of the Endo GIA is used to transfer the force from the slider to the stapler mechanism. A load cell is included to measure the force the slider exerts on the rod and optionally control that force [4].

The purpose of our research is to achieve a safe pancreatectomy. To do this, it is important to clarify the relationship between the reaction force during compression by the stapler, mechanical parameters of the pancreas, and damage to the pancreas. However, it is very difficult to know the mechanical parameters of the living human pancreas accurately. Therefore, it is necessary to estimate the real value from the parameters that can be obtained by the experiments,

such as comparison of in vivo and in situ experimental results and comparison of human and pig experiments. For the estimation, we can apply modeling methods studied in the research project of MCA. Finally, we would like to realize safe treatment with the robot stapler we developed.

11.3 Navigation and Robotic System for Oral and Maxillofacial Surgery

Many patients undertake oral and maxillofacial surgery (OMS) for diseases treatment and esthetical purposes. The surgeon is under a heavy physical and mental burden when conducting precise OMS procedures such as drilling the screw hole onto the mandible because of the limited observation capability for surgical region and long-time surgical procedures. Therefore, a surgical navigation method and a surgical assistance robot have been proposed to guide the operation and relieve the surgeon's workload. We have developed marker-based/markerless AR navigation systems [5] and a compact OMS robot for precise positioning [6]. These two systems have been seamlessly integrated to further develop an autonomous surgical system [7], which could exchange the roles between surgeons and surgical systems by making the robot the primary operator and the surgeon the surveillant.

In this section, we describe a summary of the OMS robot and a major improvement by updating both the hardware and the software of the previous navigation system to develop a fundamentally new system. The technical details and performance evaluations of the new system and prospects for future work are introduced.

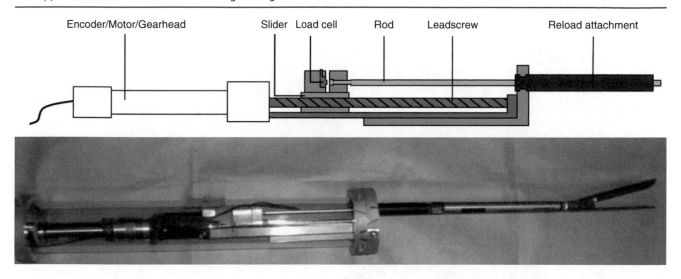

Fig. 11.4 Pancreatic stapling device with reload attached

11.3.1 OMS Robot for Precise Positioning and Drilling with patient's Specific Safety Mechanism

Preoperative surgical simulations are utilized in orthognathic surgery with a patient 3D dataset obtained from X-ray CT images. Although the simulations of drill placements are used as quantitative information, it is still difficult to reflect the precise and numerical preoperative plan in actual surgery because of human factors at the operation of the pointer or devices in their hands. In addition, esthetically oral and maxillofacial surgery involves operation on the patient's oral cavity, which also makes the surgery more difficult.

To solve the above problems, we have been developing a compact and lightweight six-DOF surgical robot for precise positioning in OMS. Although there have been various surgical robots developed specifically for oral and maxillofacial surgery, the size and weight of these conventional robots prevent their widespread use in clinical practice. Therefore, a small and lightweight six-DOF robot was proposed in this study to achieve a better positioning performance, as shown in Fig. 11.5. This robot adopted a three-DOF orientation mechanism, which used the remote center of motion (RCM) design to change the end-effector's rotation angle. It also has a three-DOF position mechanism, which used a novel parallel mechanism to change the end effector's axial coordinate. Six motors (1628 024B, Faulhaber, Schönaich 71,101, Germany) were used to actuate all movements. All mechanical parts of the robot were made of aluminum to reduce the weight. A robot adaptor was used to work as the power unit and facilitate the data exchange between the workstation and the robot. To mechanically protect the subject from any unintended injury, a driving range limit mechanism (DRLM) was proposed to restrict the movement range of the drill tip. The

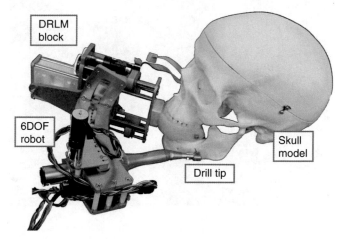

Fig. 11.5 Overview of the maxillofacial robot

DRLM is a hard block fabricated by a 3D printer whose data are preacquired from the patient's anatomical information.

11.3.2 Markerless Navigation System for OMS

Previously proposed navigation systems need to place a real maker frame onto the patient's mouth region to identify the pose of patient's head, which may either affect the navigation accuracy because of marker's drift or make the marker become obstacle to the operation procedure. Our proposed navigation system uses the pose of teeth to reflect the pose of skull without using the real marker. This navigation system uses the nature of teeth and proposed a novel working concept without using the real marker, which is able to identify the pose of teeth to reflect the pose of skull.

The main process of this navigation system includes offline phase and online phase. During the offline phase, the specific teeth group was segmented from the CT images as a STL file. Then the aspect graph of the teeth model was built by setting a large number of virtual cameras around the 3D object. An oversampling method was used to create a 5-layer image pyramid of the obtained 2D images cluster for increasing the online matching speed [5]. During the online phase, an exhaustive search was conducted from the top to the bottom layer of the image pyramid to find the best match between the real camera image and the model image. Finally, the position, rotation, scaling of the matched image was converted into the six-DOF pose as the final result. The matched results were displayed as a VR image with the real-time pose and latency shown on the monitor (Fig. 11.6).

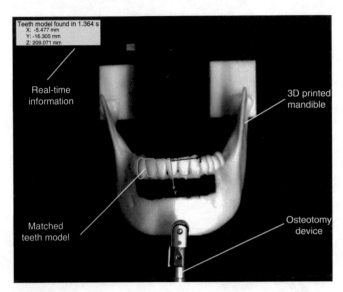

Fig. 11.6 VR interface of matching model in new generation marker-less navigation system

11.3.3 Integration of Navigation System for Compact Robot

The aforementioned robot and navigation method were integrated to work as one functional system, where the navigation method can dynamically detect the pose of teeth and combined with preoperative plan to guide the movement of robot. The overview of the newly developed autonomous OMS robot is shown in Fig. 11.7. The robot was suspended onto the operating table with a mechanical arm. The pose of the robot could be freely adjusted to fit the surgical region and ensure the teeth within the field of view (FOV) of camera. Once reaching the target region, the mechanical arm could be manually locked and the pose of the robot remains unchanged during the surgery. The robot used an osteotomy device (CORE System, Stryker Corporation, Kalamazoo, MI 49002 USA) as the end effector and its header can be changed based on specific surgical tasks.

A monograph camera (UI-3370 CP, IDS, Obersulm74182, Germany) with a resolution of 2048 × 2048 pixels was fixed onto the robot base by an aluminum frame for obtaining the real-time image. The navigation program was integrated with robot control software. The main process of navigation could be divided into the offline phase and online phase. The STL file of the osteotomy device was obtained by using a high-precision 3D scanning system (Artec Space Spider, Artech3D, Luxembourg). Then the aspect graph of the teeth model was built by setting a large number of virtual cameras around the 3D object. An oversampling method was used to create a 5-layer image pyramid of the obtained 2D images cluster for increasing the online matching speed [5]. During the online phase, an exhaustive search was conducted from the top to the bottom layer of the image pyramid to find the

Fig. 11.7 Overview of the current prototype of the autonomous surgical robot

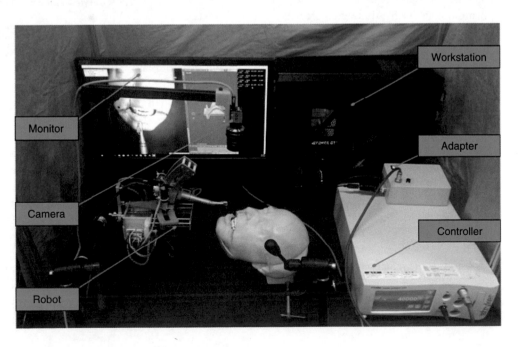

best match between the real camera image and the model image. Finally, the position, rotation, scaling of the matched image was converted into the six-DOF pose as the final result. The STL file of the osteotomy device was also obtained by using a high-precision 3D scanning system (Artec Space Spider, Artech3D, Luxembourg). Then a process similar to that used for the teeth model was implemented to build the aspect group for online matching. The overview of the integrated system is shown in Fig. 11.7 [7].

A drilling experiment was designed to verify the usability and positioning accuracy of the new system on five 3D-printed mandible models. The open-source 3D group planning software Blender (Version 2.79, Blender.org) was used to make preoperative plan. A circular motion path was designed to imitate the complexity of the actual drilling operation for the robot trajectory, where a first hole was located in the center, four holes in the first circle, and eight holes in the second circle. The five tested mandible models were further scanned by the aforementioned 3D scanning system to obtain the 3D image of the holes after the experiment. Then the barycenter of the hole in each scanned mandible model STL file of the scanned model was analyzed in Blender by using a 3D imaging processing method. Finally, the coordinate comparison was made between the planned holes and the actual holes to statistically evaluate the drilling accuracy.

The actual drilling results and corresponding scanning results are shown in Fig. 11.8. The statistical results indicate that this OMS robot maintained a high precision with average positioning accuracy in x- and y-axes of 1.20 ± 0.24 mm

and 0.04 ± 0.07 mm, respectively. Nevertheless, as listed in Table 11.1, Trial 1 and Trial 2 had relatively larger deviations than the others, indicating that the pose difference would influence the final matching result if it significantly varied from the reference pose. Given that the surgical region in the second circle had relatively larger values than the first circle in terms of surface curvature and offset from the center, we also calculated the radius of the holes from the center by combining the x- and y-axes values to evaluate the influence of the surface condition and the moving distance on the drilling accuracy. The results showed that the average radii of the first and second circles were 5.09 ± 0.41 mm and 10.27 ± 0.27 mm, respectively.

11.3.4 Prospects for the Next OMS Robot and Navigation System

This research part proposed an autonomous OMS robot that could detect the pose of the skull without a marker and automatically finish the operation under the surveillance of the surgeon. Phantom experiments showed that this robot could conduct the drilling operation with high accuracy based on a preoperative plan. In the next step of this research part, we will investigate and quantify the surgeon's surgical planning mechanism by statistically analyzing patient's preoperative and postoperative CT images, which could be used to develop a machine learning-based planning program for integration with the current robot to achieve a more intelligent surgical system.

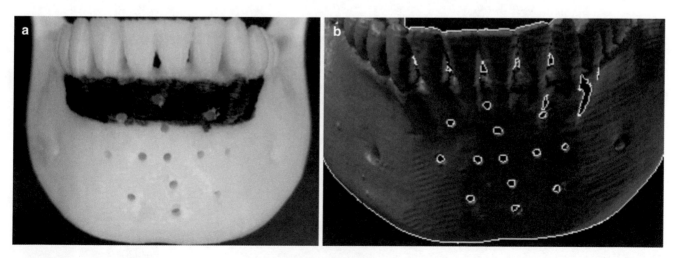

Fig. 11.8 Drilling results of the tested model (**a**) and its 3D scanning image (**b**)

Table 11.1 Errors of x- and y-axes in five mandible drilling trials (Mean ± SEM)

Axis	Trail 1	Trail 2	Trail 3	Trail 4	Trail 5
X (mm)	2.97 ± 0.19	3.57 ± 0.23	0.06 ± 0.19	−0.33 ± 0.22	−0.31 ± 0.26
Y (mm)	−0.15 ± 0.14	−0.19 ± 0.15	0.07 ± 0.11	−0.18 ± 0.12	0.67 ± 0.15

11.4 Conclusion

We proposed an intelligent, autonomous surgical robot to approach the affected surgical area based on multidisciplinary computational anatomy and limited intraoperative biological information. As an example of research, (i) study on stapler device control for pancreatic tissue damage suppression, and (ii) navigation and robotic system for oral and maxillofacial surgery are introduced.

References

1. Kobayashi E, Tsuchiya S, Akagi Y, Tomii N, Nakagawa K, Inai K, et, al, A novel reaction force-fluorescence measurement system for evaluating pancreatic juice leakage from an excised swine pancreas during distal pancreatectomy, J Hepatobiliary Pancreat Sci, in press, 2020.
2. Yamashita S, Sakabe M, Ishizawa T, Hasegawa K, Urano Y, Kokudo N. Visualization of the leakage of pancreatic juice using a chymotrypsin activated fluorescent probe. Br J Surg. 2013;100:1220–8.
3. Kim D, Kobayashi E, Sato R, Kiguchi K, Sakuma I. Comparison study of elasticity of swine liver in vivo and in vitro. Proceedings of EMBC. 2017;66
4. Stroo JL, Kim D, Kobayashi E, Ando T, Joung S, Toyoda M, et al. Development of powered pancreas stapler for stapling condition evaluation. Proceedings of the 9th Asian Conference on Computer Aided Surgery. 2013:78–9.
5. Wang J, Suenaga H, Yang L, Kobayashi E, Sakuma I. Video see-through augmented reality for Oral and maxillofacial surgery. International Journal of Medical Robotics and Computer Assisted Surgery. 2017;13 https://doi.org/10.1002/rcs.1754.
6. Hara K, Ma Q, Hideyuki S, Kobayash E, Sakuma I, Masamune K. Orthognathic surgical robot with a workspace limitation mechanism. IEEE/ASME Transactions on Mechatronics. 2019;24:2652–60. https://doi.org/10.1109/TMECH.2019.2945605.
7. Ma Q, Kobayash E, Suenaga H, Hara K, Wang J, Nakagawa K, et al. Autonomous surgical robot with camera-based Markerless navigation for Oral and maxillofacial surgery. IEEE/ASME Transactions on Mechatronics. 2020;25:1084–94. https://doi.org/10.1109/TMECH.2020.2971618.

Part V

New Frontier of Technology in Clinical Applications Based on MCA Models: Lifelong Human Growth

Three-Dimensional Analyses of Human Organogenesis

12

Tetsuya Takakuwa

Abstract

The three-dimensional (3D) observations are required for analyzing complex morphogenetic processes that occur during human embryonic development. Serial histological sections were utilized not only for histological two-dimensional observation, but also for designing three-dimensional (3D) plaster or wax models, which enable the 3D changes to be visible, since the late nineteenth century. Additionally, imaging modalities, such as magnetic resonance imaging and phase-contrast X-ray computed tomography, have been applied to embryology. High-resolution 3D datasets with an adequately large number of samples, covering a broad range of developmental periods with various methods of acquisition, are key features for the research. These datasets have the advantage of morphology, morphometry, and quantitative analysis using 3D coordinates. In particular, an adequate sample size is required for quantitative analysis using statistical methods and multidisciplinary computational anatomy (MCA) based analysis, which are expected to be useful analyzing methods for many unresolved tasks, such as quantitative movement (differential growth), branching morphogenesis, and information concerning physical and structural property. As a future perspective, analysis targets using digital imaging data with MCA based method may shift from embryonic period to early-fetal period (9-12 weeks after fertilization), which can apply to prenatal diagnosis using ultrasound. This data will timely contribute to improvements in prenatal diagnostics by detailing and comparing suitable markers for estimating developmental growth.

T. Takakuwa (✉)
Human Health Science, Graduate School of Medicine, Kyoto University, Kyoto, Japan
e-mail: tez@hs.med.kyoto-u.ac.jp

Keywords

Three-dimensional analyses · human organogenesis · imaging modalities · MRI

12.1 Three-Dimensional Analysis of Human Development Using Histological Sections

The three-dimensional (3D) analysis of human development is required for analyzing complex morphogenetic processes that occur during human embryonic development. Serial histological sections were utilized not only for histological two-dimensional observation, but also for designing three-dimensional (3D) plaster or wax models, which enable the 3D changes to be visible, since the late nineteenth century [1, 2]. Such history of human embryology including morphometrics will be described in Chap. 15.

12.2 3D Analysis of Human Development Using Imaging Modalities in High Resolution

Remarkable progress has been made in non-destructive imaging technologies, such as magnetic resonance imaging (MRI) and phase-contrast X-ray computed tomography (CT), which have all been applied to embryology [3, 4]. The imaging modalities are selected based on their destructive versus non-destructive features, the size of the samples, and the desired resolution. The technological and historical aspects to acquire embryo imaging will be mentioned in detail in Chap. 15.

Using MRI and ptychographic X-ray CT, our group proceeded the 3D analysis of human embryonic development, including organogenesis as research [5] (Fig. 12.1).

M. Hashizume (ed.), *Multidisciplinary Computational Anatomy*, https://doi.org/10.1007/978-981-16-4325-5_12

Fig. 12.1 Three-dimensional analysis of human embryonic development, including organogenesis using digital datasets: outline

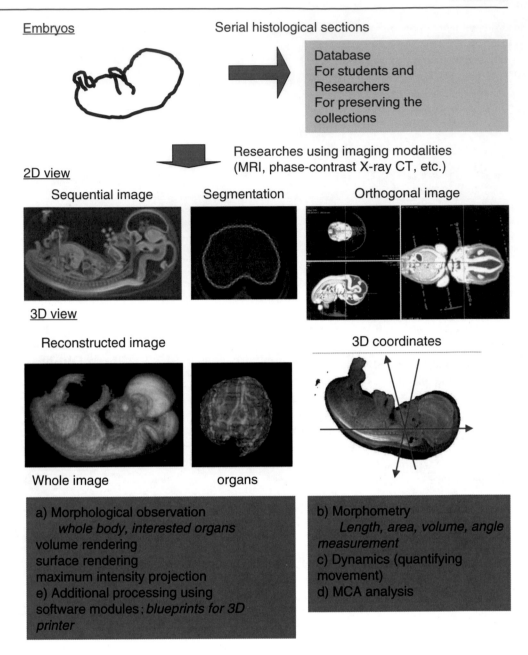

High-resolution 3D datasets with an adequately large number of samples, covering a broad range of developmental periods with various methods of acquisition, are key features for this research. These data sets have the advantage of morphology, morphometry, and quantitative analysis using 3D coordinates. In particular, an adequate sample size is required for quantitative analysis using statistical methods and multidisciplinary computational anatomy (MCA) based analysis.

12.2.1 Morphological Observations

The digital data had the following gross merits, which increased the efficiency of observations and accelerated the speed of morphological observations: (1) The data enabled us to analyze classical morphology and morphometry efficiently. (2) The complete 3D external and internal views and their reconstructions are easily obtained. (3) The obtained images can be resliced and rotated freely on the screen, by which the 3D shapes of the objects and their spatial relationships with the adjacent organs and tissues are easily recognizable. (4) The images from different embryos are simultaneously comparable on the screen. (5) Volume-rendering data can be utilized for the MCA based analysis, which will be described in another chapter.

12.2.2 Morphometry

Name Classical embryology using histological techniques provided little morphometric data [6, 7]. For the measurement of spatial distances and angles between anatomical landmarks of interest, 3D reconstruction from serial histological sections was required. The procedure for such measurements was quite laborious with a number of possible issues that could arise, such as non-rigid deformation, tissue discontinuity, and accumulation of scale change [8]. The digital data from MRI and CT have merit for morphometry. Precise morphometric data, length, angle, area, and volume of target regions and organs can be measured on the screen using digitalized data from MRI and CT images. Such morphometric data are useful for demonstrating the development features at each stage and for screening abnormally developed samples.

12.2.3 Quantitative Analysis Using 3D Coordinates

A 3D coordinate can be given for each landmark by examining the position of the voxel on 3D digital data. Application of 3D coordinates of anatomical landmarks, especially MCA based analysis, is expected to be useful analyzing methods for many tasks, which remain to be dissolved as follows:

12.2.4 Quantitative Movements (Differential Growth)

Both the external and internal structures of embryos rapidly change in size and shape during the period of organogenesis. Many dynamic events are traditionally described as migration in which the position of structures changes from one region of the embryo to another. Gasser (2006) [9] recently demonstrated most of the positional changes of the developing structures, such as the sclerotome formation from the somite, the spinal ganglion formation from the neural crest, and the endocrine glands formation from the pharyngeal endoderm, which can be explained by differential growth.

For understanding the positional change of landmarks of interest and their relationships during development, the digitized data was advantageous for comparing structures of interest between different stages with identical magnification, superimposed on the same screen. The 3D positional change of interesting landmarks and their relationships during development were demonstrated [5, 9–11], which indicated that many dynamic events can be explained by differential growth using 3D coordinates (Fig. 12.2).

12.2.5 Shared Mechanism for Human Organogenesis (Branching Morphogenesis)

Organs, such as bronchi of the lungs, urinary collecting trees of the kidneys, the milk ducts of the mammary gland, develop from branched tubes during embryonic and fetal development. Because of many similarities between these branched tube structures, the shared mechanism was assumed. Recently, Hannezo et al. (2017) [12] proposed a unifying theory to solve this issue. Namely, the certain tips stop growing in a random manner. For example, the branched mammary gland structures stop growing when the tips of the structure impinge on neighboring branches. In the kidney, this cessation has been proposed to occur when nephrons form near the end of the collecting ducts. The unifying theory mainly comes from experimental in vitro model or in vivo animal models. Analysis using human samples with 3D coordinates is awaited [13] (Figure 12.3a,b). Several 3D analyses for shared mechanism for human organogenesis based on MCA models will be described in Chap. 14.

12.2.6 Information Concerning Physical and Structural Property

Imaging modalities data contain information not only regarding 3D morphology, but also regarding physical property. Diffusion tensor images MRI has also been applied to fetal brain [14] and cardiac muscles in mice [15, 16]. The method is applicable to various organs and tissues that are anisotropic in nature (Figure 12.3c). Mesenchymal tissues consisting of fibers and membranous structures, such as the muscles, tendons, arteries, and bones, may be candidates for application of this method. These tissues have not been analyzed vigorously because of technical reason and their large target size.

Phase-contrast X-ray computed tomography with Zeff imaging methods can be used to recognize and differentiate heavy metals, such as iron, aluminum, nickel, and copper [17]. The 3D dynamics of such elements during human embryonic development are not currently known. Hematogenesis of the embryos may be also detectable using iron as a trace marker. This information with 3D distribution may provide new insight of human development.

Fig. 12.2 Three-dimensional analyses of quantitative movements (differential growth). (**a**) Lateral view of embryos between Carnegie stage (CS) 19 and CS 23 showing the tympanic cavity and ear canal (upper) and craniofacial morphogenesis (lower). The ear canal (ec), external ear (Ex), eye (Ey), first cervical vertebra (C1), internal ear (int), pituitary gland (Pg), and tympanic cavity (tc). The black line indicates the reference axis connecting the middle point of the bilateral Ey and Pg (X-axis). The red line indicates the reference line connecting Pg and C1. Blue arrow indicates the frontal side of the face. Note that the Exm and Int are observed at similar position on the red segment. There is a gross change of angle between the black and red lines used. The change of angle may result from the formation of the mandibular apparatus and the structures at the base of the skull. (**b**) Three-dimensional graph showing the relationship between the right external ear (Exm) and internal ear (Int) during development. Axes and anatomical landmarks are shown on frontal view of volume-rendering images. Abbreviations: Carnegie stage (CS), external ear (Exm), first cervical vertebra (C1). internal ear (int), pituitary gland (Pg)

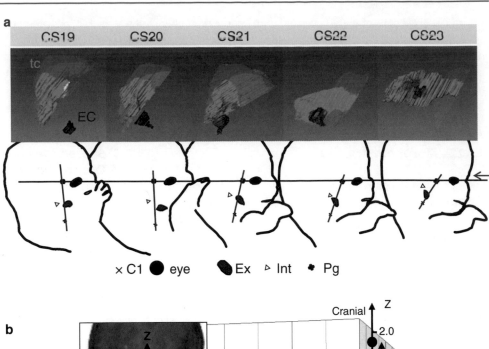

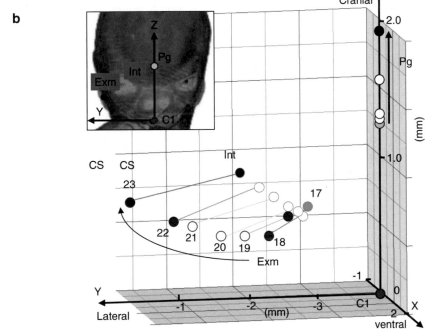

12.3 Perspective

12.3.1 Shift from Embryonic Period to Early-Fetal Period (9-12 Weeks after Fertilization)

The number of morphological studies on the early-fetal period (9-12 weeks after fertilization) is less as compared to that on the first 8 weeks after fertilization (at the end of Carnegie stage [CS] 23) [2] due to several reasons. First, many researchers have been attracted to the dynamic morpho-genesis in rather earlier developmental stages. Establishment of CS may contribute in encouraging studies for the early-fetal period. Second, it is difficult to apply histological analysis for the entire body of the fetus with a size larger than that at CS 23. Therefore, studies conducted on fetal period are mainly confined to localized histological analysis. The 3D datasets of larger samples corresponding to early-fetal period can be acquired with MRI in high resolution, which are worth analyzing as they can reveal the 3D development of the entire body and organs. Such morphometric data are also valuable for connecting and comparing the sonography data.

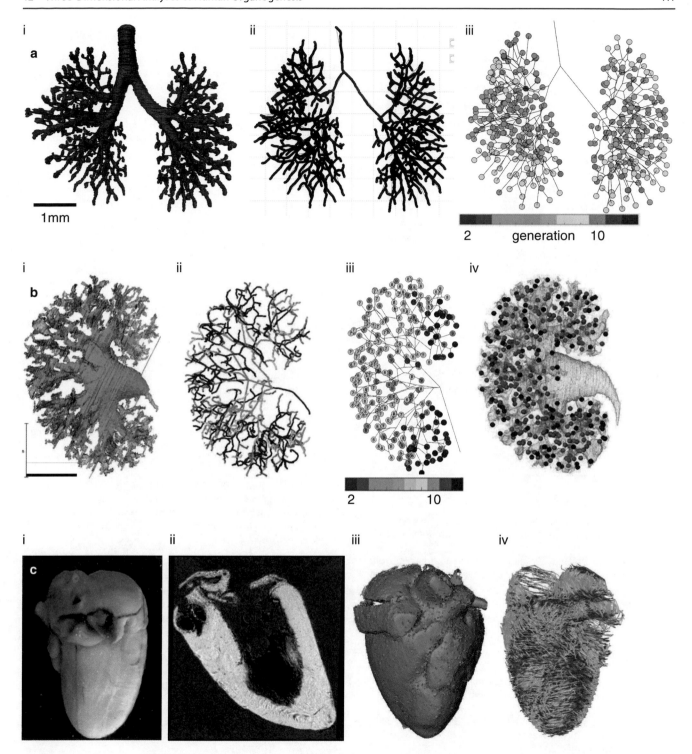

Fig. 12.3 Three-dimensional analysis using digitalized data acquired with imaging modalities. (**a**) Branching morphogenesis of the human bronchi during embryonic period. (i) The bronchial tree was extracted from ptychographic X-ray computed tomography image. (ii) Centerline of the bronchial tree was processed. (iii) Generation number of each branch segments was indicated by colors. (**b**) Branching morphogenesis of the human urinary collecting system (UCS) at Carnegie stage 23. (i) The UCS was extracted. (ii) Centerline of the bronchial tree was processed. The tree was illustrated by rainbow colors from the proximal to peripheral branches. (iii) Generation number of each branch segments was indicated by number and colors. (iv) Position of nascent nephrons connected (green) and not connected (red) to UCS are shown. (**c**) The fetal heart ex vivo (crown-rump length = 94 mm). (i) The frontal gross view. (ii) Magnetic resonance imaging (MRI)-T1 image. (iii) The 3D volume-rendering image. (iv) Diffusion tensor images MRI

12.3.2 Application to Prenatal Diagnosis

The 3D information obtained in classical embryology since the late nineteenth century has been used as the basis of prenatal diagnosis using ultrasound (US). The use of US for prenatal diagnostics has rapidly increased in the past 30 years [18]. Moreover, 3D sonography performed with high-frequency transvaginal transducers has expanded as 3D sonoembryology, which provides a basis for assessing normal human development and can also be useful in detecting developmental anomalies [18, 19].

Prenatal diagnosis using US enables a shift in diagnostics from the second trimester to the first trimester of gestation. At present, an embryo at 9 weeks after fertilization or younger can be assessed via morphological and morphometrical analyses, which corresponds to a CS of 15-16. While abnormal embryos younger than 12 weeks after fertilization are observed by chance for clinical indications, systematic screening using sonographic parameters results in the detection of abnormalities during the late first trimester (12-13 weeks after fertilization). Analysis using the digital data with MCA based method, during early-fetal period could timely contribute to improvements in prenatal diagnostics by detailing and comparing suitable markers for estimating growth and development [20].

References

1. Morgan LM. A social biography of Carnegie embryo no. 836. Anat Rec B New Anat. 2004;276:3–7. https://doi.org/10.1002/ar.b.20002.
2. O'Rahilly R, Muller F. Developmental stages in human embryos: including a revision of Streeter's horizons and a survey of the Carnegie collection. Carnegie Institution of Washington: Washington; 1987.
3. Smith BR. Visualizing human embryos. Sci Am. 1999;280:76–81. https://doi.org/10.1038/scientificamerican0399-76.
4. Momose A, Takeda T, Itai Y, Hirano K. Phase-contrast X-ray computed tomography for observing biological soft tissues. Nat Med. 1996;2:473–5. https://doi.org/10.1038/nm0496-473.
5. Takakuwa T. 3D analysis of human embryos and fetuses using digitized datasets from the Kyoto collection. Anat Rec. 2018;301:960–9. https://doi.org/10.1002/ar.23784.
6. O'Rahilly R, Müller F. Developmental stages in human embryos: revised and new measurements. Cells Tissues Organs. 2010;292:73–84. https://doi.org/10.1159/000289817.
7. Levitan ML, Desmond ME. Expansion of the human embryonic brain during rapid growth: area analysis. Anat Rec. 2009;292:472–80. https://doi.org/10.1002/ar.20882.
8. Kajihara T, Funatomi T, Makishima H, Aoto T, Kubo H, Yamada S, et al. Non-rigid registration of serial section images by blending transforms for 3D reconstruction. Pattern Recogn. 2019;96:106956. https://doi.org/10.1016/j.patcog.2019.07.001.
9. Gasser RT. Evidence that some events of mammalian embryogenesis can result from differential growth, making migration unnecessary. Anat Rec B New Anat. 2006;289:53–63. https://doi.org/10.1002/ar.b.20092.
10. Kagurasho M, Yamada S, Uwabe C, Kose K, Takakuwa T. Movement of the external ear in human embryo. Head Face Med. 2012;8:2. https://doi.org/10.1186/1746-160X-8-2.
11. Katsube M, Yamada S, Yamaguchi Y, Takakuwa T, Yamamoto A, Imai H, et al. Critical growth processes for the midfacial morphogenesis in the early prenatal period. Cleft Palate-Craniofacial J. 2019;56:1026–37. https://doi.org/10.1177/1055665619827189.
12. Hannezo E, Scheele CLGJ, Moad M, Drogo N, Heer R, Sampogna RV, et al. A unifying theory of branching morphogenesis. Cell. 2017;171:242–55. https://doi.org/10.1016/j.cell.2017.08.026.
13. Ishiyama H, Ishikawa A, Kitazawa H, Fujii S, Matsubayashi J, Yamada S, et al. Branching morphogenesis of the urinary collecting system in the human embryonic metanephros. PLoS One. 2018;13:e0203623. https://doi.org/10.1371/journal.pone.0203623.
14. Huang H, Xue R, Zhang J, Ren T, Richards LJ, Yarowsky P, et al. Anatomical characterization of human fetal brain development with diffusion tensor magnetic resonance imaging. J Neurosci. 2009;29:4263–73. https://doi.org/10.1523/JNEUROSCI.2769-08.2009.
15. Mekkaoui C, Porayette P, Jackowski MP, Kostis WJ, Dai G, Sanders S, et al. Diffusion MRI tractography of the developing human fetal heart. PLoS One. 2013;8:e72795. https://doi.org/10.1371/journal.pone.0072795.
16. Angeli S, Befera N, Peyrat JM, Calabrese E, Johnson GA, Constantinides CA. High-resolution cardiovascular magnetic resonance diffusion tensor map from ex-vivo C57BL/6 murine hearts. J Cardiovasc Magn Reson. 2014;16:77. https://doi.org/10.1186/s12968-014-0077-x.
17. Yoneyama A, Hyodo K, Takeda T. Feasibility test of Zeff imaging using x-ray interferometry. Appl Phys Lett. 2013;103:204108. https://doi.org/10.1063/1.4831773.
18. Blaas HG. Detection of structural abnormalities in the first trimester using ultrasound. Best Pract Res Clin Obstet Gynaecol. 2014;28:341–53. https://doi.org/10.1016/j.bpobgyn.2013.11.004.
19. Pooh RK, Shiota K, Kurjak A. Imaging of the human embryo with magnetic resonance imaging microscopy and high-resolution transvaginal 3-dimensional sonography: human embryology in the 21st century. Am J Obste Gynecol. 2011;204(77):e1–e16. https://doi.org/10.1016/j.ajog.2010.07.028.
20. Kobayashi A, Ishizu K, Yamada S, Uwabe C, Kose K, Takakuwa T. Morphometric human embryonic brain features according to developmental stage. Prenat Diagn. 2016;36:338–45. https://doi.org/10.1002/pd.4786.

Skeletal System Analysis during the Human Embryonic Period Based on MCA

13

Tetsuya Takakuwa

Abstract

During human development, both external and internal morphological features change dramatically. External features, including those on the body and limbs, provide a good basis for determining the staging of each developing embryo. The application of three-dimensional (3-D) sonography with high-frequency transvaginal transducers has expanded and now fosters 3-D sonoembryology, which provides a basis for assessing normal human development and can also be useful in detecting developmental anomalies. The quantitative data of standard morphology for each Carnegie stage (CS) and early fetal period is required for the evaluation of the body and limbs in clinically obtained data, to allow for better prenatal morphological diagnosis. Analysis of the skeletal system during the human embryonic and early fetal period based on MCA will be described, including the rib cage, shoulder girdle, pelvis, and femur. The data obtained may contribute to such evaluations.

Keywords

Human embryonic period · Phase-contrast CT · Skeletal system · Rib cage · Femur · Shoulder girdle

13.1 Introduction

During human development, both external and internal morphological features change dramatically. External features, including those on the body and limbs, provide a good basis for determining the staging of each developing embryo. For

example, flexion and extension of the body combined with the posture of the upper and lower limbs have been integrated into the Carnegie stage (CS), which is universally accepted for determining the staging of human embryos [1]. Although such qualitative external changes are well described in the literature, 3-D quantitative changes in the body and limbs have not been well analyzed. The application of 3-D sonography with high-frequency transvaginal transducers has expanded and now fosters 3-D sonoembryology, which provides a basis for assessing normal human development and can also be useful in detecting developmental anomalies [2, 3]. Such technology could contribute to more accurate prenatal diagnoses as well as enable a shift in the diagnostic time window (from the second to first trimester). Under these circumstances, the quantitative data of standard morphology for each CS and early fetal period are required for evaluation of the body and limbs in clinically obtained data, to allow for better prenatal morphological diagnoses. In the present chapter, analysis of the skeletal system during the human embryonic and early fetal period based on MCA will be described, including the rib cage, shoulder girdle, pelvis, and femur.

13.2 Methods

All human embryo and early fetal specimens used are stored at the Congenital Anomaly Research Center of Kyoto University [4, 5]. The phase-contrast X-ray computed tomography (PXCT) and magnetic resonance imaging (MRI) were used for 3-D data acquisition [6, 7]. PXCT and MRI data from selected embryos were analyzed precisely as serial 2-D and reconstructed 3-D images using Amira software (version 5.5; Visage Imaging, Berlin, Germany). The 3-D coordinates were initially assigned by examining the voxel position on 3-D images, which were subjected to MCA-based analysis such as principal component (PC) analysis and Procrustes analysis.

T. Takakuwa (✉)
Human Health Science, Graduate School of Medicine, Kyoto University, Kyoto, Japan
e-mail: tez@hs.med.kyoto-u.ac.jp

13.3   Results

13.3.1   Rib Cage

The rib cage mainly consists of 12 pairs of ribs and vertebrae as well as the hypaxial muscles on the chest and upper abdominal parts of the body. In the early embryonic period until CS16, all visceral organs, such as the heart, lung, liver, and digestive tract, are covered with soft tissue. However, these organs are not protected by the rib cage because no cartilage or muscle formation is observed until CS16 [1, 8]. Such organs are covered by the rib cage until the end of the embryonic period. The rib cage becomes detectable with cartilage formation at CS17, expanding outward from the dorsal side of the chest-abdominal region [9]. The ribs elongate progressively to surround the chest, differentiating into the upper and lower rib cage regions by CS20. The ends of the corresponding ribs in the upper region elongate toward each other, leading to their joining and sternum formation between CS21 and CS23, whereas the lower region of the rib cage remains widely open. The rib cage can be divided anatomically and functionally into the upper and lower thoracic regions. The upper region is related to the pulmonary part of the respiratory system and upper limbs, while the lower thorax is anatomically related to the diaphragmatic part of the respiratory system and also closer to the abdominal cavity and locomotor apparatus [10, 11]. We aimed to analyze the morphogenesis of all ribs from the first to the twelfth rib pairs plus vertebrae to compare their differences and features according to the position along the cranial–caudal axis during the human embryonic period [12].

Seven rib cage landmarks, from the first to the twelfth vertebra, and the ribs (for a total of 84 landmarks) were located for each sample. A total of 384 sample data from 32 samples were subjected to process and principal component (PC) analyses, using MATLAB (R2017b, MathWorks, USA) software-assisted algorithms based on orthogonal coordinates of the voxels at each reference point. PC1 and PC2 accounted for 76.3% and 16.4% (sum, 92.7%) of the total variance, respectively, indicating that the change in shape was accounted for by two components (Fig. 13.1a). Changes in PC1 resulted in a circular form surrounding the trunk. A decrease in PC1 showed closing of the rib tips, while an increase in PC1 showed opening of the rib tips. Changes in PC2 showed the movement of the lateral projection and dorsal convexity of the ribs. An increase in PC2 showed posterior movement of the lateral projection at the middle part of the rib, which is related to the dorsal convexity of the ribs, while a decrease in PC2 showed anterior movement of the lateral projection at the middle part of the ribs.

The distribution of scatter plots of the PC1 and PC2 values for each rib showed a fishhook-like shape (Fig. 13.1b).

The distribution was fitted to a quartic equation as follows: $y = 0.072X^4 + 0.1535X^3 + 0.1785X^2 - 0.2976X - 0.7001$; $R^2 = 0.82$. PC1 and PC2 plots for each rib moved positions along the fitting curve according to the development of CS18 to CS23. The scatter plots moved in a wide range from the center right (quadrant IV) to the left, reached the left end (quadrant II) of the fitting curve, and moved slightly back in the upper ribs (1–7). However, the scatter plots moved in a narrower range from the center right to the center (quadrant IV) in the lower ribs. Movement was limited to the right (quadrant I) in the 11th and 12th ribs.

PC1 and PC2 values for each rib were plotted close to the fitting curve, for which the shape could be determined using a single parameter. We denote the fitting curve in the principal component subspace in Fig. 13.1b by $\mathbf{F}(Shape)$, where *Shape* is the arc length parameter along the fitting curve and provides a linear scale for shape representation. Thus, the right end of the fitting curve (2.0,1.8) was designated as origin O of $\mathbf{F}(Shape)$, while the left end of the curve (−2.3, 1.0) was 6.6. PC1 and PC2 plots by rib x can determine the nearest point X on the fitting curve. The distance XO along the fitting curve was defined as the value of rib x for $\mathbf{F}(Shape)$.

$\mathbf{F}(Shape) = 0.5$ showed that the paired ribs were on the dorsal side of the trunk with the opening of the rib tips (Fig. 13.1c). $\mathbf{F}(Shape) = 2.5$ showed that the paired ribs surrounded the dorsal side of the trunk with dorsal convexity of the ribs. The rib tips were separated. $\mathbf{F}(Shape) = 4.5$ showed that the paired ribs had a circular form that surrounded more than half of the trunk, with the lateral projection at the middle part of the ribs. $\mathbf{F}(Shape) = 6.5$ showed that the paired ribs had a circular form surrounding almost all of the trunk with the dorsal convexity of the ribs. The bilateral rib tips were almost closed. The change in $\mathbf{F}(Shape)$ at each rib pair is indicated. The development of each rib pair could be indicated as an increase in the $\mathbf{F}(Shape)$ scale in almost all conditions (Fig. 13.1d). However, the development of a subset of rib pairs (1st–8th) resulted in a plateau or even decrease in the $\mathbf{F}(Shape)$ scale at the end of the embryonic period (between CS22 and CS23).

Our data clearly demonstrated that human embryonic ribs all progress through common morphological forms irrespective of their position on the axis. The data suggested that in a parsimonious model, the common series of rib pairs can be controlled by a small number of factors.

13.3.2   Femur

The femur is a long bone that develops via endochondral ossification. In particular, the human femur first appears as mesenchymal condensation between CS16 and CS17. Chondrification occurs between CS17 and CS18 and subse-

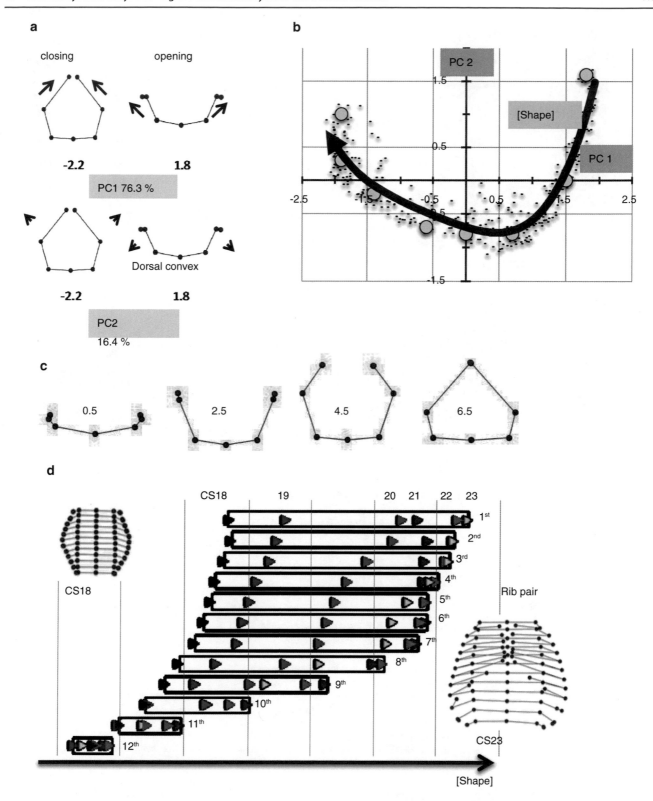

Fig. 13.1 Analysis of the rib cage based on MCA. (**a**) Changes in the ribs based on principal component analysis (PC) 1 and PC2. PC scores are indicated below each illustration. (**b**) Scatter plot of PC1 and PC2 values for all rib pairs ($n = 348$). (**c**) Rib morphologies for each F(Shape) value. (**d**) Changes in F(Shape) according to rib number. CS18, purple; blue, CS19; green, CS20; red, CS21; orange, CS22; yellow, CS23

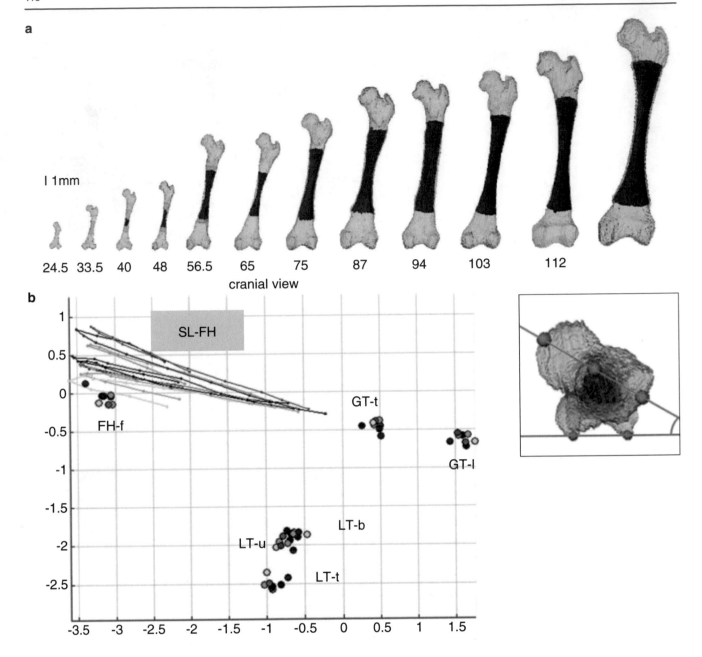

Fig. 13.2 Analysis of the femur based on MCA. (**a**) Reconstruction of the Femur during development. (**b, c**) Reconstructed Procrustes shape coordinates for the proximal (**b**) and distal (**c**) epiphysis of the fetal femur. FH-f: center of the femoral head fovea; GT-1: most lateral point of the greater trochanter; GT-t: top of the greater trochanter; IF: Intercondylar fossa; LC-b: lateral condyle (bottom); LC-p: lateral condyle (posterior); LE: lateral epicondyle; LT-b: bottom end of the lesser trochanter; LT-t: top of the lesser trochanter; LT-u: upper end of the lesser trochanter; MC-b: medial condyle (bottom); MC-p: medial condyle (posterior); ME: medial epicondyle; SL-FH: semi-landmarks from the upper end to the lower end of the femoral head along the plane passing through the midpoint of the femoral head, femoral neck, and greater trochanter; SL-L: semi-landmarks along the roundness of the lateral condyle from the upper end to the opposite side; SL-M; semi-landmarks along the roundness of the medial condyle from the upper end to the opposite side

quently proceeds to endochondral ossification between CS22 and CS23 [1, 8, 13]. The cartilage structure influences bone structure formation, as the cartilage structure acts as the blueprint replaced by the bone structure. How the morphological features of the cartilage structure may be replaced by those of the bone structure has not been fully demonstrated,

especially precise quantitative information regarding the 3-D form of the femur.

The morphogenesis and internal differentiation process of the femur were analyzed in 3-D from the fetus (CRL11–185 mm, $n = 62$) [14]. Procrustes analysis was performed to distinguish the change in shape from the change in size

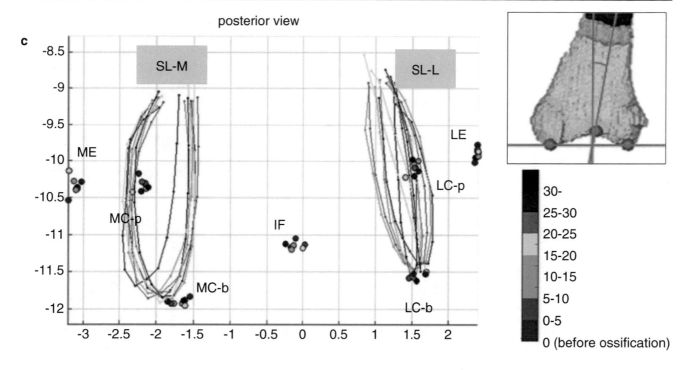

Fig. 13.2 (continued)

according to growth using defined landmarks ($n = 13$) and semi-landmarks ($n = 3$). Centroid sizes at both the proximal and distal epiphyses showed a strong positive correlation with the ossified shaft length (OSL) ($R^2 = 0.99$ and 0.99). The Procrustes shape coordinates for the proximal epiphysis indicated that each landmark on the greater and lesser trochanters and femoral head fovea was located in the same position irrespective of the OSL (Fig. 13.2a). In comparison, semi-landmarks at femoral head (SL-FH), which lined the femoral head, moved in accordance with the increase in OSL. Procrustes analysis indicated that changes in the femur shape after ossification were limited, and were mainly detected at the time of initial ossification and shortly thereafter. In contrast, femoral neck anteversion and torsion of the femoral head continuously changed during the fetal period. The Procrustes shape coordinates for the distal epiphysis indicated that each landmark was located in the same position irrespective of the OSL (Fig. 13.2b). semi-landmarks along the lateral and medial condyles (SL-L, SL-M) were located in different positions according to the OSL, although no obvious regularity was noted.

Torsion of the lower leg, including the femur (anteversion), was continuously observed during the fetal period and after birth [15–17]. Torsion of the femur may be affected by repetitive and persistent mechanical forces and the intrauterine position [3]. With respect to mechanical forces, muscle tension and local forces exert a rotary stress on the epiphysis. Remodeling at the metaphysis and epiphysis dur-

ing the growth of a long bone such as the femur is well known as the mechanism that maintains the shape [8]. Anatomical landmarks remained in the same relative position during subsequent endochondral ossification in the present study, indicating that the remodeling system during femur shaft growth in the longitudinal direction is elaborate.

13.3.3 Shoulder Girdle

The shoulder girdle (pectoral girdle) is the set of bones in the appendicular skeleton, which anchors the upper limb on each side to the axial skeleton [18, 19]. In humans, it consists of the clavicle and scapula. Well and complex movements of the shoulder girdle cause difficulty in describing the default position of the shoulder girdle. Information about the 3-D morphogenesis and position of the entire shoulder girdle except for the scapula height during the embryonic and fetal periods is limited [20, 21].

The 3-D reconstruction and morphometry in our study revealed that all landmarks on the shoulder girdle remained at a similar height except the inferior angle, which means that the scapula enlarges in the caudal direction and reaches the adult position during the fetal period (Fig. 13.3a-b). The position of the shoulder girdle during the embryonic and fetal periods was unique (Fig. 13.3a and c). In contrast to the constant position of the clavicle, the scapula body was

Fig. 13.3 Analysis of the shoulder girdle based on MCA. (**a**) Ventral, lateral, and cranial views of the 3-D reconstruction of the scapula (purple) and clavicle (green) in the embryonic (CS19) and fetal (CRL 225 mm) period. Th1: first thoracic vertebra; ac: acromion; cp: Coracoid process. (**b**) Position of the scapula along *z*-axis. Solid purple circles indicate the superior angle of the scapula (spa), open purple circles indicate the inferior angle of the scapula (ifa), and blue stars indicate the center of the glenoid cavity (glc). (**c**) Relationship between the scapula and clavicle from the cranial view. Solid yellow circles indicate the angles between the glc-Th1 and ACJ-Th1 segments from the cranial view (∠T1Scc-∠T1Clc)

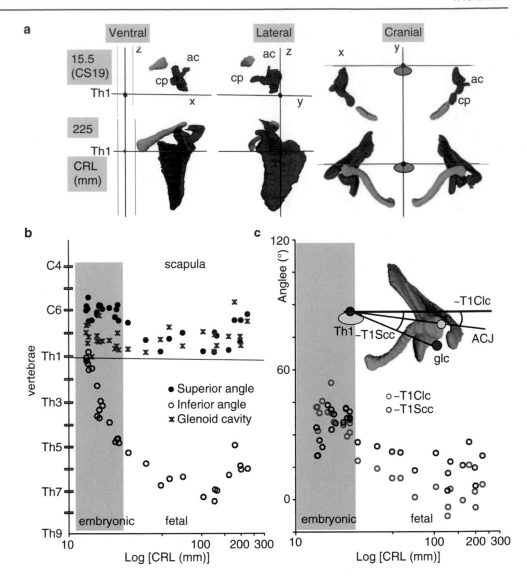

rotated internally and upward at the initiation of the morphogenesis. Thus, the scapula body of both seems almost parallel. The internal rotation of the scapula was changed externally, while the upward rotation remained unchanged. Compared with the adults, the scapula was still rotated internally and upward during the fetal period. The shoulder girdle is located in the ventral part of the body (vertebrae) during the initial morphogenesis, which changes the position to the lateral side of the vertebrae during the late embryonic period and fetal growth. The position during the fetal period may be consistent with that in adults. Such unique positioning of the shoulder girdle may contribute to the stage-specific posture of the upper limb, which is an important external feature for determining the staging, especially between CS18 and CS23 [1]. For example, the axial skeleton (vertebrae) becomes straight and the upper limb extends vertically to the axial skeleton at CS19. The shoulder (humerus head) becomes externally evident, and the joint flexed with the elbow is pronated at CS23. The posture may be explained, in part, by the

unique positional change of the scapula during the embryonic period.

13.4 Conclusion

Skeletal system analysis based on MCA may provide a useful standard for morphogenesis and morphometry of the skeletal system, which can serve as the basis to better understand embryonic and early fetal development and aid in differentiating normal and abnormal development.

References

1. O'Rahilly R, Muller F. Developmental stages in human embryos: including a revision of Streeter's horizons and a survey of the Carnegie collection. Carnegie Institution of Washington: Washington; 1987.

2. Blaas HG. Detection of structural abnormalities in the first trimester using ultrasound. Best Pract Res Clin Obstet Gynaecol. 2014;28:341–53. https://doi.org/10.1016/j.bpobgyn.2013.11.004.

3. Pooh RK, Shiota K, Kurjak A. Imaging of the human embryo with magnetic resonance imaging microscopy and high-resolution transvaginal 3-dimensional sonography: human embryology in the 21st century. Am J Obste Gynecol. 2011;204(77):e1–e16. https://doi.org/10.1016/j.ajog.2010.07.028.

4. Nishimura H, Takano K, Tanimura T, Yasuda M. Normal and abnormal development of human embryos: first report of the analysis of 1,213 intact embryos. Teratology. 1968;1:281–90.

5. Yamaguchi Y, Yamada S. The Kyoto collection of human embryos and fetuses: history and recent advancements in modern methods. Cells Tissues Organs. 2018;205:314–9. https://doi.org/10.1159/000490672.

6. Yoneyama A, Yamada S, Takeda T. Fine biomedical imaging using X-ray phase-sensitive technique. In: Gargiulo G, editor. Advanced biomedical engineering. Rijeka: InTech; 2011. p. 107–28.

7. Toyoda S, Shiraki N, Yamada S, Uwabe C, Imai H, Matsuda T, et al. Morphogenesis of the inner ear at different stages of normal human development. Anat Rec. 2015;298:2081–90. https://doi.org/10.1002/ar.23268.

8. O'Rahilly R, Müller F. The skeletal system and limbs. In: O'Rahilly R, Müller F, editors. Human Embryology & Teratology. 3rd ed. New York: Wiley-Liss; 2001. p. 357–94.

9. Okuno K, Ishizu K, Matsubayashi J, Fujii S, Sakamoto R, Ishikawa A, et al. Rib cage morphogenesis in the human embryo: a detailed three-dimensional analysis. Anat Rec. 2019;302:2211–23. https://doi.org/10.1002/ar.24226.

10. Bastir M, García-Martínez D, Recheis W, Barash A, Coquerelle M, Rios L, et al. Differential growth and development of the upper and lower human thorax. PLoS One. 2013;8:e75128. https://doi.org/10.1371/journal.pone.0075128.

11. García-Martínez D, Recheis W, Bastir M. Ontogeny of 3D rib curvature and its importance for the understanding of human thorax development. Am J Phys Anthropol. 2016;159:423–31. https://doi.org/10.1002/ajpa.22893.

12. Matsubayashi J, Okuno K, Fuji S, Ishizu K, Yamada S, Yoneyama A, et al. Human embryonic ribs all progress through common morphological forms irrespective of their position on the axis. Dev Dyn. 2019;248:1257–63. https://doi.org/10.1002/dvdy.107.

13. O'Rahilly R, Gardner E. The timing and sequence of events in the development of the limbs in the human embryo. Anat Embryol. 1975;148:1–23. https://doi.org/10.1007/BF00315559.

14. Suzuki Y, Matsubayashi J, Ji X, Yamada S, Yoneyama A, Imai H, et al. Morphogenesis of the femur at different stages of normal human development. PLoS One. 2019;14:e0221569. https://doi.org/10.1371/journal.pone.0221569.

15. Felts WJ. The prenatal development of the human femur. Am J Anat. 1954;94:1–44. https://doi.org/10.1002/aja.1000940102.

16. Guidera KJ, Ganey TM, Keneally CR, Ogden JA. The embryology of lower-extremity torsion. Clin Orthop Relat Res. 1994;302:17–21.

17. Gulan G, Matovinović D, Nemeć B, Rubinić D, Ravlić-Gulan J. Femoral neck anteversion: values, development, measurement, common problems. Coll Antropol. 2000;24:521–7.

18. Browne D. Congenital deformities of mechanical origin. Arch Dis Child. 1955;30:37–41. https://doi.org/10.1136/adc.30.149.37.

19. Standring S. Gray's anatomy. 41st ed. Amsterdam: Elsevier; 2005.

20. Schuenke M, Schulte E, Schumacher U, Ross LM, Lamperti ES. THIEME atlas of anatomy. General anatomy and musculoskeletal system. 1st ed. Stuttgart: Thieme; 2006.

21. Lewis WH. The development of the arm in man. Am J Anat. 1902;1:145–86.

MCA-Based Embryology and Embryo Imaging

Shiori Nakano, Ryota Kodama, Yutaka Yamaguchi,
Tetsuya Takakuwa, and Shigehito Yamada

Abstract

The study of human embryology has a long history owing to its development in the human embryo collections that were first established in the nineteenth century. The first established large collection of human embryos was the Carnegie Collection, followed by several other major collections. After the Carnegie stages of development were defined based on morphological features of developing embryos, researchers have conducted morphological measurements and analyses to discover new insights using the stored specimens efficiently. At present, conducting analysis using nondestructive methods has been prioritized, and novel imaging techniques are adopted to preserve the specimens and have promoted the use of 3D imaging modalities. Visualizing tissues and organs in three dimensions has helped understand and characterize complex morphogenic changes in the body. The use of 3D imaging modalities started in the twentieth century using the histological sections for reconstruction, and now, 3D image datasets are also used. This chapter describes how the collections have been made, to provide new insights into human embryonic development, along with details of novel 3D imaging techniques for morphological analyses and their methods of application.

Keywords

Human embryo · Embryo collection · 3D imaging techniques · Morphological analyses

14.1 History of Human Embryology, Embryo Collection, and Morphometrics

Research on human embryology began in the nineteenth century using human embryo specimens derived from maternal deaths, abortion, or surgery [1, 2]. Although animal experimental biology has developed in the past decades, human embryo specimens are still mainly used for research owing to the ethical restrictions on research using human embryos. In the past, some human embryo collections such as the Carnegie Collection, Blechschmidt Collection, and Kyoto Collection (Table 14.1) have contributed to the establishment and progress of human embryology. Resources on human development based on the results of research using human embryo collections are available and useful for students and researchers [9–12]. The details of the three collections mentioned above will be stated in the following text.

14.1.1 Human Embryo Collections

The Carnegie Collection is the oldest human embryo collection, established in 1887 by Franklin P. Mall. The collection grew at a rate of about 400 specimens per year, and the number of samples attained over 8000 by the 1940s. Mall and his colleagues prepared serial sections and created hundreds of 3D models of the embryos using the wax plate technique. Eventually, more than 700 wax-based models were created, and their drawings are still used as schemes in textbooks on human embryology. During this era, several members strongly supported their research; Osborne O. Heard worked

S. Nakano · Y. Yamaguchi
Congenital Anomaly Research Center, Kyoto University Graduate School of Medicine, Kyoto, Japan

R. Kodama · S. Yamada (✉)
Congenital Anomaly Research Center, Kyoto University Graduate School of Medicine, Kyoto, Japan

Human Health Sciences, Kyoto University Graduate School of Medicine, Kyoto, Japan
e-mail: shyamada@cac.med.kyoto-u.ac.jp

T. Takakuwa
Human Health Sciences, Kyoto University Graduate School of Medicine, Kyoto, Japan

Table 14.1 Comparison among major human embryo collections (Based on [3])

Collection	Place	Number	Characteristics	Establishment	Collaboration with the Kyoto Collection
Carnegie	Washington DC, USA	About 10,000	Human fixed specimens and histology	1887	[4–6]
Belchschmidt	Göttingen, Germany	About 120	Human histology	1948	[7, 8]
Kyoto	Kyoto, Japan	About 44,000	Human fixed specimens and histology	1961	–

as an embryo modeler and James D. Didusch as a scientific illustrator. The research from the Carnegie Collections was compiled in the journal "*Contributions to Embryology of the Carnegie Institution of Washington*" published from 1915 to 1966. After Mall unexpectedly passed away in 1917, George L. Streeter, George W. Corner, and Ronan O'Rahilly took over the management of the collection. The greatest achievement of this collection was the establishment of "Carnegie stages," the definition of human-specific developmental stages [2], which is widely used today. The collection is now preserved at the Human Developmental Anatomy Center under the directorship of Elizabeth Lockett in Washington D.C. (Fig. 14.1), and many researchers continue to seek collaboration. Further details on the Carnegie Collection can be found in earlier publications [13, 14] and on the website (http://nmhm.washingtondc.museum/collections/hdac/carnegie_history.htm).

The Blechschmidt Collection is stored at the Center of Anatomy, University of Göttingen, Germany. This collection was created in 1948 and named after Erich Blechshmidt, who directed the Anatomical Institute from 1942 to 1973. The collection contains a histological collection of serial sections and model collection reconstructed from the serial sections. Approximately 120 human embryos are sectioned serially into almost 200,000 serial sections and stored in the histology collection. In recent years, two digitization projects have been performed by the Kyoto Collection [7] and the Digital Embryology Consortium [15]. The model collection comprised 64 large models, which were generated from1946 to 1979. Each model is a large-scale polymer plastic reconstruction from regularly spaced histological sections at an intermediate magnification [16]. These models are in a permanent exhibition housed at the center (Fig. 14.2).

The Kyoto Collection began in 1961 by Hideo Nishimura, and currently has over 44,000 human embryo specimens. In 1975, the Congenital Anomaly Research Center was established at Kyoto University. Since then, the collection has been stored at the center (Fig. 14.3) and provides a resource for international researchers of human embryology. One of the characteristics of the collection is that many abnormal embryos are included in the collection, such as embryos with holoprosencephaly [17]. External appearances of the embryos have been examined in detail, but most of the internal anomalies of the whole samples (meaning "not sectioned serially") remain unchecked. Recent advances in imaging technology have enabled detailed imaging of the internal

organs of the embryos, and multiple imaging techniques have been applied to this collection. The Kyoto Collection and the applied imaging techniques will be introduced in Sect. 14.2 of this chapter.

14.1.2 Classical Morphometrics Using Human Embryos and Fetuses

Morphogenesis occurs dramatically during embryonic and fetal periods, and microscopic observation of the histological sections was the only way to analyze the morphogenetic changes until computer-assisted techniques became available. The morphological shape of the embryo and fetus changes in three dimensions(3D); therefore, visualization by 3D reconstruction using wax plates was significant [18]. The resource for the reconstruction was serial sections. Now, the digitization of the histological glass slides is used as one solution for preserving the collections, which could also decrease the maintenance cost for administrators of human embryo collections. In 2014, the Digital Embryology Consortium, an international partnership, was established to digitize, preserve, and disseminate the major embryology histological collections for researchers [15].

The next step of morphological analyses is morphometrics; one of the early studies on embryonic and fetal measurement was published by Mall, using the Carnegie Collection [19]. In the early era, the researchers measured the volume, length, and weight for quantitation. In the late twentieth century, quantitative analyses of human embryos were still performed based on such measurements [20, 21]. Magnetic resonance (MR) imaging and X-ray computed tomography (CT) have been developed, and 3D reconstruction and analyses have become possible. Therefore, to apply 3D analyses to human embryology, 3D imaging techniques are required to obtain high-resolution images of human embryos and fetuses, which can be applied to MCA.

14.2 Imaging of Human Embryo and Fetus

As mentioned above, the classical reconstruction method was wax modeling, and the technique has evolved into digitized materials and computer graphics (CGs) [22]. Stained serial sections were digitized into the color images of TIFF, JPEG, or PNG format, and then they are reconstructed into

Fig. 14.1 The Carnegie Collection. (Top left) Many wax-plate reconstructions are stored on the shelf. There are several shelves in the center. (Top right) Vascular system in an embryo is reconstructed. (Bottom) Reconstructions of the embryo at Carnegie stage 12. The left one is surface reconstruction, and the right one is a model with central nervous system (CNS) and primitive gut. The model seen in the back of the desk is another embryo reconstruction, unrelated to the two in the front row

3D models using 3D software (Fig. 14.4). Previously, 3D software were scarce and expensive, but now many software programs are available, some of which are free. For example, the embryos in the Kyoto Collection were first reconstructed using Cosmozone-2SA [23], next NIH image [24], DeltaViewer [22], Volocity [5], Amira [25], and 3D Slicer (https://www.slicer.org/). Reconstruction from serial sections requires registration and is time consuming. In contrast, 3D imaging allows rapid 3D rendering and digital resectioning in arbitrary planes and is also convenient for mor-

phometric analyses. However, the serial section created so far has a lot of information and is a valuable sample; therefore, it should be used effectively. There are some problems with the serial sections coming from the process of sectioning, fixation, and staining; sections are stretched, bent, tore during sectioning, folded during fixation, and stained unevenly. These problems have disturbed smooth reconstruction, and a new automatic alignment method is now established using nonrigid registration [26] (Fig. 14.4). The method is now being improved, and it is expected that smooth

Fig. 14.2 The Blechschmidt Collection. (Top) The models are in a permanent exhibition housed in the basement room of the center. (Bottom) Reconstructions from the same embryos (CRL 6.3 mm): (Bottom right) surface reconstruction, (Bottom left) CNS, primitive gut, and primitive urinary system

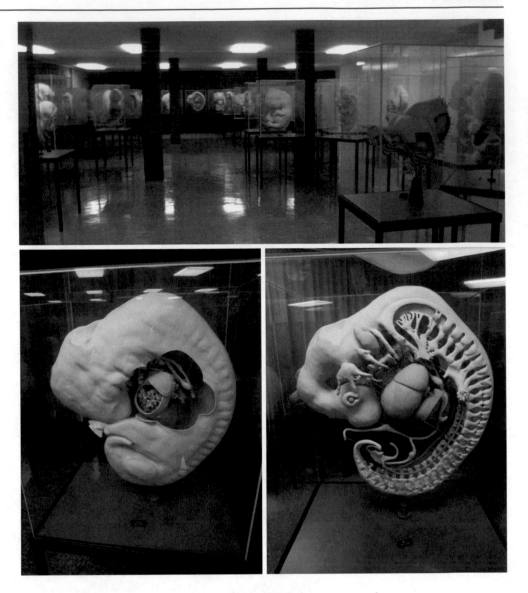

3D reconstruction from serial sections will become possible in the near future.

For unsectioned specimens, 3D imaging methods are available. Multiple 3D-imaging modalities have been applied to human embryology.

14.2.1 Episcopic Fluorescence Image Capture

Episcopic fluorescence image capture (EFIC) is a 3D-imaging method that relies on the embedding of the embryo in paraffin [27] (Weninger and Mohun 2002), followed by sectioning of the paraffin block using a sliding microtome. Immediately after cutting each section, the surface of the block is imaged using tissue autofluorescence. The block is accurately placed in the same photo position on the microtome after cutting every time; therefore, the obtained 2D image stacks are completely registered [28]. EFIC is a destructive way to image

because the sample is sectioned. The resolution is approximately 2 μm/pixel in the section plane, depending on the magnification of the microscope [5]. This method has been applied to human embryos of the Kyoto Collection [5].

14.2.2 Magnetic Resonance Imaging, MR Microscopy

Magnetic Resonance (MR) imaging is a powerful tool to image not only human patients but also chemical-fixed samples (Fig. 14.5). The resolution of the MR device for clinical use is now 200 μm/pixel, which is sufficient for imaging of the human fetus. MR devices specialized for shooting small objects are called MR microscopes and have been previously applied to developmental embryology in a number of animal models [29–31]. It is a noninvasive and nondestructive imaging technique; therefore, it is extremely effective for imaging

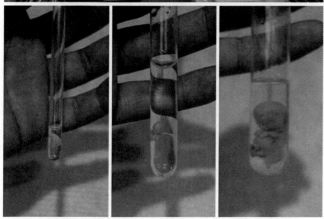

Fig. 14.3 The Kyoto Collection. (Top) Storage lookers in the specimen room. Formalin-fixed embryos in bottles are stored in bottles. (Center) Microscopic room. Many sets of serial sections (approximately 1000 embryos) are stored. (Bottom) Human embryo specimens in glass tubes for MRI scan. (Bottom left) 5 mm tube, (Bottom center) 10 mm tube, (Bottom right) 20 mm tube

precious human embryos. Imaging of human embryos by MR microscopy was reported using superconducting magnets ranging from 1.0 T to 9.4 T [32–34], and over 1200 human embryos from the Kyoto Collection were imaged

using a super-parallel MR microscope operated at 2.34 T [35–38].

14.2.3 X-Ray Computed Tomography

X-rays are electromagnetic waves characterized by amplitude and phase. When X-rays pass through a sample, the phase shifts, and the amplitude decreases. Conventional X-ray imaging (radiography) is based on absorption-contrast (amplitude imaging), and phase-contrast X-ray imaging is based on phase imaging [39]. Fetal bones, which become hard by calcification, can be imaged by conventional X-ray imaging, whereas embryonic bones cannot be imaged owing to their softness, and X-rays are hardly absorbed by the embryo. In contrast, to detect the phase shift, X-rays must pass through the sample. Therefore, phase-contrast imaging is applied to soft samples such as embryos and is not applied to hard samples such as fetuses. The sensitivity of the phase shift is approximately 1000 times larger than that of absorption [39]. In the detection of phase shift, conversion of phase shift into X-ray intensity is required by interferometry or diffractometry, and the X-ray intensity can be measured using an X-ray camera. Devices based on this principle were developed [40, 41]. Human fetuses have been imaged by conventional X-ray CT [42], and human embryos have already been imaged by phase-contrast X-ray CT [43] (Fig. 14.6).

14.3 Morphometrics of Human Embryos Using 3D Imaging

It is said that processing image data obtained by 3D imaging is easy, but morphometry for MCA using such 3D data cannot be performed without segmenting regions of interest (ROI). Computer-assisted segmentation technology is in progress, and manual segmentation is the most reliable at this stage. Some images of human embryos and fetuses in the Kyoto Collection have been manually segmented, and they are used and analyzed for MCA analyses. Early papers focused on two dimensions [44], but soon extended the focus to three dimensions [45]. Analyses of dynamic morphological changes from the embryonic stage to the fetal period are in progress by the method of MCA, and some of the results have already been published as follows: changes in the position of the eyes on the face [46], complex morphology of the collecting duct of the kidney [47], morphological changes on the surface of the brain [48], and skeletal system (face [49], femur [50], rib [51]). For 3D analyses, it is necessary to set reference points that indicate homologous positions for different specimens. During the development of humans and other animals, there are problems in which the reference points move significantly and the shape changes greatly as a

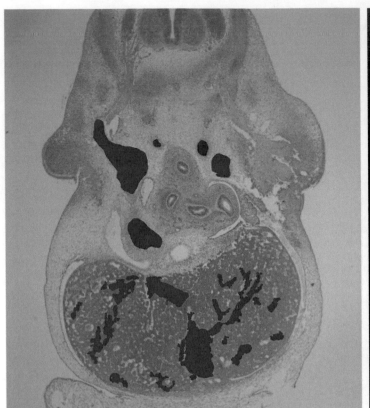

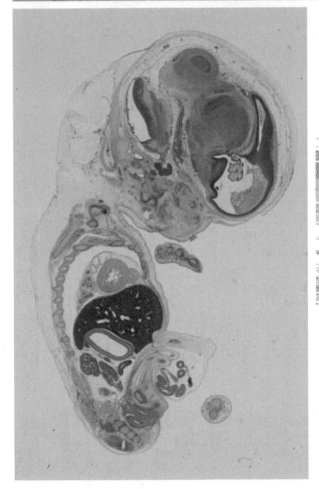

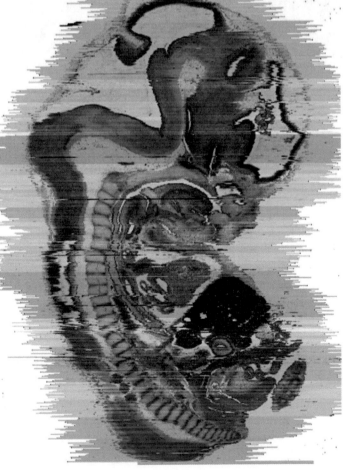

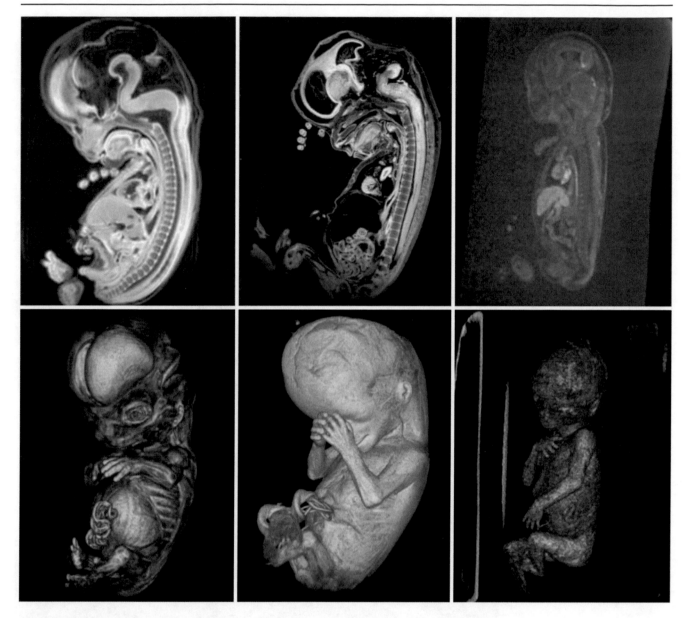

Fig. 14.5 X-ray CT. (Top) Conventional X-ray CT for a fetus and (Center and Bottom) phase-contrast X-ray CT for an embryo. (Center Left) Surface reconstruction and (Center Right) midsagittal section; (Bottom Left) coronal section in the head region, and (Bottom Right) transverse section in the chest region

result, or new points appear owing to the change in the shape. Regarding the former, one solution is shown [52]. To resolve the latter is important, now and in the future, to understand the developmental changes as a continuous phenomenon from the MCA perspective.

Fig. 14.4 3D models of human embryos from serial sections. (Top) Manual segmentation with manual registration. (Top left) Blood vessels are marked with manual segmentation: arteries (red) and veins (blue). (Top right) Reconstruction by manual registration. (Bottom) Automatic registration of serial images from histological sections. (Bottom left) Original image of the human embryo, midsagittal section. (Bottom right) midsagittal resection of 3D volume obtained from 2D stack in transverse section

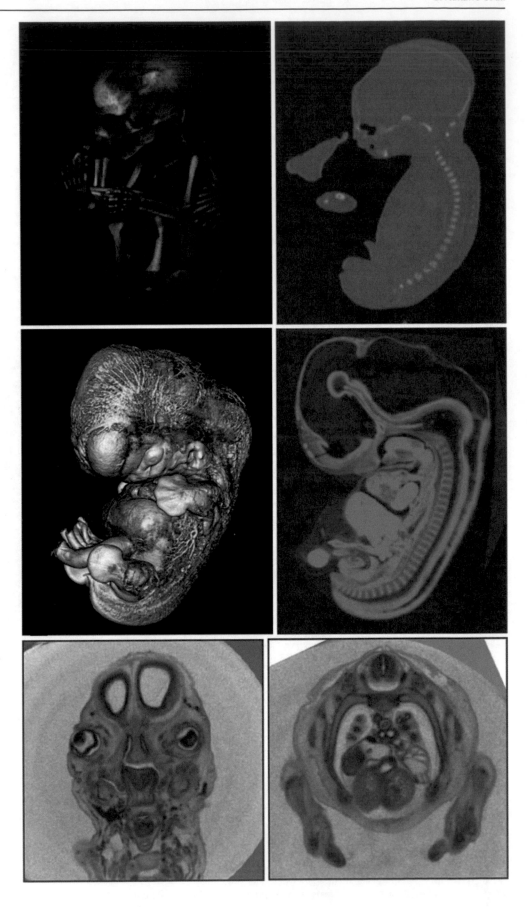

Fig. 14.6 X-ray CT. (Top) Conventional X-ray CT for a fetus and (Center and Bottom) phase-contrast X-ray CT for an embryo. (Center Left) Surface reconstruction and (Center Right) midsagittal section; (Bottom Left) coronal section in the head region, and (Bottom Right) transverse section in the chest region

References

1. Morgan LM. A social biography of Carnegie embryo no. 836. Anat Rec B New Anat. 2004;276(1):3–7. https://doi.org/10.1002/ar.b.20002.

2. O'Rahilly R, Müller F. Developmental stages in human embryos: including a revision of streeter's horizons and a survey of the Carnegie collection. Carnegie Institution of Washington: Washington, D.C; 1987.

3. Yamada S, Hill MA, Takakuwa T. Human Embryology. In: Wu B, editor. New discoveries in embryology. Rijeka, Croatia: IntechOpen; 2015. https://doi.org/10.5772/61453.

4. Dhanantwari P, Lee E, Krishnan A, Samtani R, Yamada S, Anderson S, et al. Human cardiac development in the first trimester: a high-resolution magnetic resonance imaging and episcopic fluorescence image capture atlas. Circulation. 2009;120(4):343–51. https://doi.org/10.1161/CIRCULATIONAHA.108.796698.

5. Yamada S, Samtani RR, Lee ES, Lockett E, Uwabe C, Shiota K, et al. Developmental atlas of the early first trimester human embryo. Dev Dyn. 2010;239(6):1585–95. https://doi.org/10.1002/dvdy.22316.

6. Krishnan A, Samtani R, Dhanantwari P, Lee E, Yamada S, Shiota K, et al. A detailed comparison of mouse and human cardiac development. Pediatr Res. 2014;76(6):500–7. https://doi.org/10.1038/pr.2014.128.

7. Miyazaki R, Makishima H, Manner J, Sydow HG, Uwabe C, Takakuwa T, et al. Blechschmidt collection: revisiting specimens from a historical collection of serially sectioned human embryos and fetuses using modern imaging techniques. Congenit Anom (Kyoto). 2018;58(5):152–7. https://doi.org/10.1111/cga.12261.

8. Ueno S, Yamada S, Uwabe C, Manner J, Shiraki N, Takakuwa T. The digestive tract and derived primordia differentiate by following a precise timeline in human embryos between Carnegie stages 11 and 13. Anat Rec (Hoboken). 2016;299(4):439–49. https://doi.org/10.1002/ar.23314.

9. Gasser RF, Cork RJ, Stillwell BJ, McWilliams DT. Rebirth of human embryology. Dev Dyn. 2014;243(5):621–8. https://doi.org/10.1002/dvdy.24110.

10. Kerwin J, Yang Y, Merchan P, Sarma S, Thompson J, Wang X, et al. The HUDSEN atlas: a three-dimensional (3D) spatial framework for studying gene expression in the developing human brain. J Anat. 2010;217(4):289–99. https://doi.org/10.1111/j.1469-7580.2010.01290.x.

11. de Bakker BS, de Jong KH, Hagoort J, de Bree K, Besselink CT, de Kanter FE, et al. An interactive three-dimensional digital atlas and quantitative database of human development. Science. 2016;354(6315):aag0053. https://doi.org/10.1126/science.aag0053.

12. Belle M, Godefroy D, Couly G, Malone SA, Collier F, Giacobini P, et al. Tridimensional visualization and analysis of early human development. Cell. 2017;169(1):161–73 e12. https://doi.org/10.1016/j.cell.2017.03.008.

13. Brown DD. The Department of Embryology of the Carnegie Institution of Washington. BioEssays. 1987;6(2):92–6. https://doi.org/10.1002/bies.950060213.

14. O'Rahilly R. One hundred years of human embryology. In: KALTER H, editor. Issues and reviews in Terratology. New York: Plenum Press; 1988.

15. Hill MA. Two web resources linking major human embryology collections worldwide. Cells Tissues Organs. 2018;205(5–6):293–302. https://doi.org/10.1159/000495619.

16. Blechschmidt E. Reconstruction method by using synthetic substances; a process for investigation and demonstration of developmental movements. Z Anat Entwicklungsgesch. 1954;118(2):170–4.

17. Matsunaga E, Shiota K. Holoprosencephaly in human embryos: epidemiologic studies of 150 cases. Teratology. 1977;16(3):261–72. https://doi.org/10.1002/tera.1420160304.

18. Born J. Ueber die Nasenhöhlen und den Tränennasengang der Amphibien. Morphologisches Jahrbuch. 1876;2:577–646.

19. Mall FP. On measuring human embryos. Anat Rec. 1907;1(6):129–40.

20. Desmond ME, O'Rahilly R. The growth of the human brain during the embryonic period proper. 1. Linear axes. Anat Embryol (Berl). 1981;162(2):137–51. https://doi.org/10.1007/BF00306486.

21. Diewert VM. A morphometric analysis of craniofacial growth and changes in spatial relations during secondary palatal development in human embryos and fetuses. Am J Anat. 1983;167(4):495–522. https://doi.org/10.1002/aja.1001670407.

22. Yamada S, Itoh H, Uwabe C, Fujihara S, Nishibori C, Wada M, et al. Computerized three-dimensional analysis of the heart and great vessels in normal and holoprosencephalic human embryos. Anat Rec (Hoboken). 2007;290(3):259–67. https://doi.org/10.1002/ar.20427.

23. Shiota K, Nakatsu T, Irie H. Computerized three-dimensional reconstruction of the brain of normal and holoprosencephalic human embryos. Birth Defects Orig Artic Ser. 1993;29(1):261–71.

24. Miura T, Komori M, Takahashi T, Shiota K. Computerized three-dimensional reconstruction of human embryos and their organs using the "NIH image" software. Kaibogaku Zasshi. 1995;70(4):353–61.

25. Kishimoto H, Yamada S, Kanahashi T, Yoneyama A, Imai H, Matsuda T, et al. Three-dimensional imaging of palatal muscles in the human embryo and fetus: development of levator veli palatini and clinical importance of the lesser palatine nerve. Dev Dyn. 2016;245(2):123–31. https://doi.org/10.1002/dvdy.24364.

26. Kajihara T, Funatomi T, Makishima H, Aoto T, Kubo H, Yamada S, et al. Non-rigid registration of serial section images by blending transforms for 3D reconstruction. Pattern Recogn. 2019;96:106956. https://doi.org/10.1016/j.patcog.2019.07.001.

27. Weninger WJ, Mohun T. Phenotyping transgenic embryos: a rapid 3-D screening method based on episcopic fluorescence image capturing. Nat Genet. 2002;30(1):59–65. https://doi.org/10.1038/ng785.

28. Rosenthal J, Mangal V, Walker D, Bennett M, Mohun TJ, Lo CW. Rapid high resolution three dimensional reconstruction of embryos with episcopic fluorescence image capture. Birth Defects Res C Embryo Today. 2004;72(3):213–23. https://doi.org/10.1002/bdrc.20023.

29. Bone SN, Johnson GA, Thompson MB. Three-dimensional magnetic resonance microscopy of the developing chick embryo. Investig Radiol. 1986;21(10):782–7. https://doi.org/10.1097/00004424-198610000-00003.

30. Smith BR, Effmann EL, Johnson GA. MR microscopy of chick embryo vasculature. J Magn Reson Imaging. 1992;2(2):237–40. https://doi.org/10.1002/jmri.1880020220.

31. Smith BR, Johnson GA, Groman EV, Linney E. Magnetic resonance microscopy of mouse embryos. Proc Natl Acad Sci U S A. 1994;91(9):3530–3. https://doi.org/10.1073/pnas.91.9.3530.

32. Smith BR, Linney E, Huff DS, Johnson GA. Magnetic resonance microscopy of embryos. Comput Med Imaging Graph. 1996;20(6):483–90. https://doi.org/10.1016/s0895-6111(96)00046-8.

33. Smith BR. Visualizing human embryos. Sci Am. 1999;280(3):76–81. https://doi.org/10.1038/scientificamerican0399-76.

34. Haishi T, Uematsu T, Matsuda Y, Kose K. Development of a 1.0 T MR microscope using a Nd-Fe-B permanent magnet. Magn Reson Imaging. 2001;19(6):875–80. https://doi.org/10.1016/s0730-725x(01)00400-3.

35. Matsuda Y, Ono S, Otake Y, Handa S, Kose K, Haishi T, et al. Imaging of a large collection of human embryo using a super-

parallel MR microscope. Magn Reson Med Sci. 2007;6(3):139–46. https://doi.org/10.2463/mrms.6.139.

36. Matsuda Y, Utsuzawa S, Kurimoto T, Haishi T, Yamazaki Y, Kose K, et al. Super-parallel MR microscope. Magn Reson Med. 2003;50(1):183–9. https://doi.org/10.1002/mrm.10515.

37. Yamada S, Uwabe C, Nakatsu-Komatsu T, Minekura Y, Iwakura M, Motoki T, et al. Graphic and movie illustrations of human prenatal development and their application to embryological education based on the human embryo specimens in the Kyoto collection. Dev Dyn. 2006;235(2):468–77. https://doi.org/10.1002/dvdy.20647.

38. Shiota K, Yamada S, Nakatsu-Komatsu T, Uwabe C, Kose K, Matsuda Y, et al. Visualization of human prenatal development by magnetic resonance imaging (MRI). Am J Med Genet A. 2007;143A(24):3121–6. https://doi.org/10.1002/ajmg.a.31994.

39. Momose A, Fukuda J. Phase-contrast radiographs of nonstained rat cerebellar specimen. Med Phys. 1995;22(4):375–9. https://doi.org/10.1118/1.597472.

40. Becker P, Bonse U. The skew-symmetric two-crystal X-ray interferometer. J Appl Crystallogr. 1974;7(6):593–8. https://doi.org/10.1107/S0021889874010491.

41. Yoneyama A, Takeda T, Tsuchiya Y, Wu J, Thet Thet L, Koizumi A, et al. A phase-contrast X-ray imaging system—with a 60×30mm field of view—based on a skew-symmetric two-crystal X-ray interferometer. Nucl Instrum Methods Phys Res, Sect A. 2004;523(1):217–22. https://doi.org/10.1016/j.nima.2003.12.008.

42. Morimoto N, Ogihara N, Katayama K, Shiota K. Three-dimensional ontogenetic shape changes in the human cranium during the fetal period. J Anat. 2008;212(5):627–35. https://doi.org/10.1111/j.1469-7580.2008.00884.x.

43. Shigehito Y, Takashi N, Ayumi H, Akio Y, Tohoru T, Tetsuya T. Developmental anatomy of the human embryo – 3D-imaging and analytical techniques; 2012. https://doi.org/10.5772/32104.

44. Katsube M, Yamada S, Miyazaki R, Yamaguchi Y, Makishima H, Takakuwa T, et al. Quantitation of nasal development in the early prenatal period using geometric morphometrics and MRI: a new insight into the critical period of binder phenotype. Prenat Diagn. 2017;37(9):907–15. https://doi.org/10.1002/pd.5106.

45. Katsube M, Yamada S, Yamaguchi Y, Takakuwa T, Yamamoto A, Imai H, et al. Critical growth processes for the Midfacial morphogenesis in the early prenatal period. Cleft Palate Craniofac J. 2019;56(8):1026–37. https://doi.org/10.1177/1055665619827189.

46. Kishimoto M, Saito A, Takakuwa T, Yamada S, Matsuzoe H, Hontani H, et al. A spatiotemporal statistical model for eyeballs of human embryos. IEICE Trans Inf Syst. 2017;E100.D(7):1505–15. https://doi.org/10.1587/transinf.2016EDP7493.

47. Ishiyama H, Ishikawa A, Kitazawa H, Fujii S, Matsubayashi J, Yamada S, et al. Branching morphogenesis of the urinary collecting system in the human embryonic metanephros. PLoS One. 2018;13(9):e0203623. https://doi.org/10.1371/journal.pone.0203623.

48. Kasahara K, Saito A, Takakuwa T, Yamada S, Matsuzoe H, Hontani H, et al. A spatiotemporal statistical shape model of the brain surface during human embryonic development. Advanced Biomedical Engineering. 2018;7:146–55. https://doi.org/10.14326/abe.7.146.

49. Katsube M, Rolfe SM, Bortolussi SR, Yamaguchi Y, Richman JM, Yamada S, et al. Analysis of facial skeletal asymmetry during foetal development using muCT imaging. Orthod Craniofac Res. 2019;22(Suppl 1):199–206. https://doi.org/10.1111/ocr.12304.

50. Suzuki Y, Matsubayashi J, Ji X, Yamada S, Yoneyama A, Imai H, et al. Morphogenesis of the femur at different stages of normal human development. PLoS One. 2019;14(8):e0221569. https://doi.org/10.1371/journal.pone.0221569.

51. Okuno K, Ishizu K, Matsubayashi J, Fujii S, Sakamoto R, Ishikawa A, et al. Rib cage morphogenesis in the human embryo: a detailed three-dimensional analysis. Anat Rec (Hoboken). 2019;302(12):2211–23. https://doi.org/10.1002/ar.24226.

52. Saito A, Tsujikawa M, Takakuwa T, Yamada S, Shimizu A. Level set distribution model of nested structures using logarithmic transformation. Med Image Anal. 2019;56:1–10. https://doi.org/10.1016/j.media.2019.05.003.

Modeling of Congenital Heart Malformations with a Focus on Topology

15

Ryo Haraguchi, Wataru Ueki, Yoshiaki Morita, and Taka-aki Matsuyama

Abstract

Congenital heart disease (CHD) involves structural abnormalities in blood vessels connected to the heart or in the heart itself, which are present since birth. Morphological abnormalities in CHD lead to abnormalities in blood flow and circulatory function. Education and training on cardiac morphology for medical students and cardiologists have traditionally been based on a hands-on experience with congenitally malformed cardiac specimens; however, archives are no longer widely available due to multiple reasons. This chapter presents several digitalization of congenitally malformed heart specimens to address this issue. It also introduces several studies on shape modeling for CHD, with a focus on topology.

Keywords

Congenital heart malformation · Modeling · Topology Ontology

15.1 Modeling for Congenital Heart Malformation Focused on Topology

15.1.1 Introduction

The heart functions by cycling through contraction and expansion, so the morphology and function of the heart are closely related [1]. Congenital heart disease (CHD) involves structural abnormalities in blood vessels connected to the heart or in the heart itself, which are present since birth [2].

A CHD results when the heart, or blood vessels near the heart, fails to develop normally before birth [3]. The heart has a complex three-dimensional (3D) structure, and diseases such as CHD increase that complexity.

Atrial septal defects (ASDs) constitute 8% to 10% of CHDs in children [4]. An ASD is a "hole" in the wall that separates the left and right atria. It allows oxygen-rich blood to leak into the oxygen-poor blood chambers of the heart [3]. Thus, morphological abnormalities in CHD lead to abnormalities in blood flow and circulatory function. In other words, some CHDs alter the topology of blood circulation.

Education and training on morphology for medical students and professionals specializing in pediatric cardiology and surgery has traditionally been based on a hands-on experience with congenitally malformed cardiac specimens [5]. However, the frequency of autopsies is decreasing globally. There are multiple reasons for this, including improved treatments, the development of imaging methods, and the lack of pathologists. Archived specimens are damaged by repeated use, and there is concern that such specimens, including those from patients with CHD, will be lost. Therefore, immediate action is required, and we believe that the digitization of specimens is an appropriate and effective way to preserve them [6]. A 3D computational shape model of the heart would be useful for medical education, and the combination of such models with biophysical simulations would contribute to our understanding, diagnosis, and treatment of complex diseases, including cardiac arrhythmias [6, 7].

This chapter presents several digitalization of congenitally malformed heart specimens to address the above-mentioned issues. It also introduces several studies on shape modeling for CHD, with a focus on topology.

R. Haraguchi (✉)
University of Hyogo, Kobe, Japan
e-mail: haraguch@ai.u-hyogo.ac.jp

W. Ueki · Y. Morita
National Cerebral and Cardiovascular Center, Suita, Japan

T.-a. Matsuyama
Showa University, Tokyo, Japan

15.1.2 Acquisition of Three-Dimensional Images of Congenital Heart Malformations

Heart specimens with congenital heart malformation can be digitalized in 3D using two methods. The first involves stacking a large number of microscopic images, which is known as 3D reconstruction: the target part is excised and sectioned by a pathologist, stained, and microscopic images are acquired. The second method involves the acquisition of volumetric data using a medical imaging modality such as computed tomography (CT) or magnetic resonance imaging (MRI).

The former method allows the acquisition of detailed images with high spatial resolution and visualization of detailed pathological conditions by staining. However, it is generally difficult to reconstruct volumetric data by slice-by-slice registration. Also, the specimen is irreversibly lost because of the cutting and slicing required. In contrast, 3D data acquired using CT or MRI can be obtained without damaging the specimen. However, these techniques cannot provide information as detailed as that provided by staining, and their spatial resolution is also inferior to that of microscopic imaging.

15.1.2.1 Acquisition of Microscopic Images

Virtual microscopy is used in many hospital pathology departments. It involves the automatic acquisition of microscopic images of various resolutions of specimens mounted on slides and stained. However, it is typically difficult to reconstruct volumetric data from a series of such microscopic images. We have reconstructed volumetric data for the pulmonary vein (Takayasu disease) [8] and atrioventricular node (normal) by manual registration [9, 10]. Kugler et al. demonstrated automatic volume reconstruction of the pancreas of a KPC mouse [11].

15.1.2.2 Acquisition of Macroscopic Images

CT and MRI techniques can provide obtain volume data directly and non-invasively, but they have limited ability to reflect the physical characteristics of each part of the specimen. The acquisition of volume data from isolated hearts using imaging devices is also known as ex vivo imaging.

Howard et al. [12] used MRI to image two human hearts deemed not viable for transplantation. Neither patient had any history of a heart condition that would suggest an atrial defect, yet an ASD, a type of CHD, was found in each heart. The authors presented images, videos, and 3D reconstructions to provide a clear view of the anatomy of ASDs. Hill et al. [13] imaged 12 human hearts deemed not viable for transplantation, using endoscopic video cameras inserted into the cardiac chambers. The Visible Heart Lab at the University of Minnesota created a heart database, which it has made publicly available on the Internet [14]. Kiraly et al. [5] scanned about 400 human cardiac specimens using high-resolution micro-CT/MRI to establish a virtual museum of congenital heart defects.

Clinical 3D MRI sequences provide good image contrast with high spatial resolution. However, the optimal MRI sequence for imaging of autopsied human heart specimens fixed in formalin, which replaces water in cardiac tissue, is uncertain. We compared the visibility of a formalin-fixed heart specimen using various 3D MRI sequences to determine the optimal sequence for obtaining macroscopic images and concluded that magnetization-prepared rapid acquisition with gradient echo (MPRAGE) is the most adequate sequence for obtaining macroscopic images of human-autopsied heart specimens with CHDs [15] (Fig. 15.1).

15.1.2.3 Archiving of Images

Archiving of images improves their accessibility and increases the value of stored data. As noted above, the Visible Heart Lab at the University of Minnesota maintains a heart database that is publicly available on the Internet [14], and Kiraly et al. have reported the establishment of a virtual museum of congenital heart defects, which contained about 400 specimens of hearts with CHD [5]. Additionally, the National Institutes of Health's 3D Heart Library provides a repository for digital reproductions of human heart anatomies, including CHD [16].

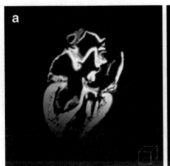 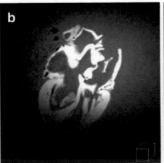 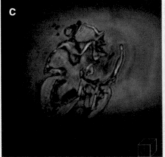 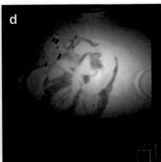

Fig. 15.1 Comparison of the visibility of a formalin-fixed human heart specimen using various 3D MRI sequences. (**a**) MPRAGE, (**b**) FLASH, (**c**) T2-SPACE, (**d**) True-FISP

15.1.3 Analysis of Heart Morphology and Topology

15.1.3.1 Segmentation and Visualization from Volume Data

Segmentation is important for effective visualization. A variety of segmentation methods for microscopic and macroscopic images have been proposed. The following discussion introduces the system we have developed for the visualization of microscopic images [9, 10] and macroscopic images [6].

Formalin-fixed paraffin-embedded tissues taken from around the atrioventricular node were serially sectioned, and sections obtained at 200 intervals were mounted on glass slides and stained with Masson's trichrome. Next, microscopic images were obtained using an Aperio's slide scanner. The 30 slices were segmented into seven regions, and the slices were aligned manually. An iPad application was developed to import the images and display them in 3D. We used Unity™ as the development framework. There is a marked difference between the resolution of the slice image (0.50 μm) and the slice spacing (400 μm). Therefore, the interpolation method and the mesh generation method have a considerable influence on the quality and reproducibility of the 3D display. A comparison of various methods revealed that the Poisson surface reconstruction method [17] it optimal in terms of display quality and operability. We implemented a transparency change function and a gyro-sensor-based viewpoint change function in the iPad app, which enabled more intuitive operation.

We also developed a system for visualizing macroscopic images [6]. We obtained MRI data from an isolated heart specimen using MPRAGE sequences (FOV 320, voxel size of 1.0 × 1.0 × 1.0 mm). For this, we used a three clinical machine (MAGNETOM Verio, Siemens AG Healthcare Sector, Erlangen, Germany). Next, we performed an interactive volumetric segmentation using VoTracer2 software [18]. Another iPad application was developed to visualize the segmented MRI volume data. We used the Unreal Engine™ [19] as the software development framework. Figure 15.2a presents an overview of the iPad application. Users can freely zoom and change the viewpoint by means of touch and pinch-in/out operations; the application also has translucent display and cross-section display functions. Users can visualize MRI datasets obtained from macroscopic specimens.

15.1.3.2 Interactive Extraction of Graph-like Structure from Volume Data

CHD causes abnormalities in hemodynamics by altering the connections between the heart chambers and blood vessels. Consequently, insufficient blood reaches the lungs, resulting in the circulation of oxygen-poor blood [20]. A "graph" is a mathematical representation used to model the pairwise relationships between objects, making it suitable for representing the characteristics of CHD. Conventional computational organ models involving voxels, polygons, or statistical shapes, are poor representations of heart functions such as hemodynamics. Therefore, we are developing interactive software for extracting graph-like structures representative of heart morphology and function [21]. Figure 15.2b presents an overview of the system. The input to the system is an MRI dataset obtained from macroscopic specimens. The system generates supervoxels using the simple linear iterative clustering (SLIC) segmentation algorithm [22]. The output is

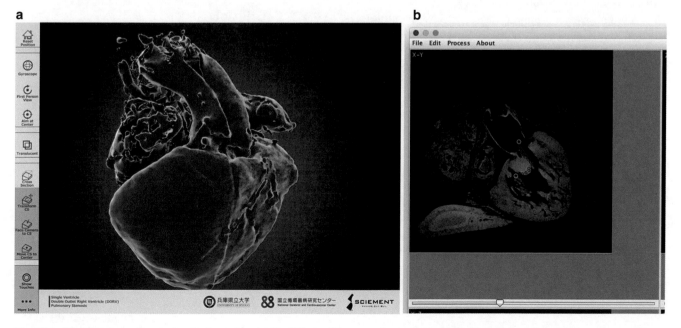

Fig. 15.2 Visualization and analysis of volume data, (**a**) iPad Viewer (**b**) system for extracting graph-like structures

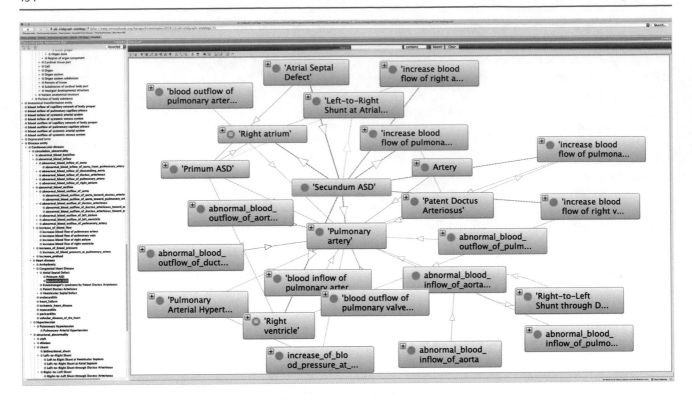

Fig. 15.3 Sample of an ontology of congenital heart malformation

the optimal path between two supervoxels specified by the user. We intend to enable the user to extract the graph-like structure embedded in the volume data.

15.1.4 Ontology of Congenital Heart Malformations

For a computer to process the graph structure data described above according to its semantics, an ontology is needed. By referencing such an ontology, we can link topology (morphological information) to the knowledge of the disease (conceptual information) in a machine-readable database. An ontology of congenital heart malformation may allow identification of a disease from topology or generation of morphological information from the name of the disease.

The Foundational Model of Anatomy (FMA) [23] is an ontology that expresses the anatomical structure of a healthy human body. The FMA ontology contains approximately 75,000 classes and over 120,000 terms, over 2.1 million relationship instances. Although the FMA is comprehensive, it contains only healthy anatomical structures and not diseased structures or functional information such as blood flow. Several ontologies of "diseases" or "abnormalities" are available [24, 25], but they do no include the relationship

between morphological and functional abnormalities in CHD. Therefore, we have developed a new ontology of congenital heart malformation by extending the FMA. We used Protégé software [26] as an ontology construction support tool. Figure 15.3 presents a sample description of structural and functional abnormalities and their relationships in cases of ASD.

15.1.5 Conclusion

The availability of heart specimens with congenital heart malformations has been reduced dramatically as a result of stricter data protection regulations, fewer autopsies, natural attrition, and improved treatments [5]. The digitalization of macroscopic and histological specimens could overcome this issue.

"Graph" is important in computational cardiac modeling because graphs can represent the morphology, function, and structure of the heart.

Acknowledgments I am grateful to Dr. Hatsue Ishibashi-Ueda (National Cerebral and Cardiovascular Center, Japan), and Mr. Hirofumi Seo (Sciment, Inc.) for their collaboration. This work was supported (in part) by a JSPS Grant-in-Aid for Scientific Research on Innovative Areas (Multidisciplinary Computational Anatomy) JSPS KAKENHI Grant Number 15H01133.

References

1. Haraguchi R, Katsuda T. Morphologic and functional modeling of the heart. In: Kobatake H, Masutani Y, editors. Computational anatomy based on whole body imaging, chap. 3.8.1. Tokyo: Springer; 2017. p. 230–5. https://doi.org/10.1007/978-4-431-55976-4.

2. Haraguchi R, Nakao M, Kurosaki KI, Iwata M, Nakazawa K, Kagisaki K, Shiraishi I. Heart modeling of congenital heart disease based on neonatal echocardiographic images. Advanced Biomedical Engineering. 2014;3:86–93. https://doi.org/10.14326/abe.3.86. http://jlc.jst.go.jp/DN/JST.JSTAGE/abe/3.86?lang=en&from=CrossRef&type=abstract

3. About Congenital Heart Defects. 2018. https://www.heart.org/en/health-topics/congenital-heart-defects/about-congenital-heart-defects

4. Sachdeva R. Congenital cardiovascular malformations. In: Allen HD, Driscoll DJ, Shaddy RE, Feltes TF, editors. Moss & Adams' heart disease in infants, children, and adolescents: including the fetus and young adult. 8th ed. Philadelphia: LWW; 2012. p. 672–808.

5. Kiraly L, Kiraly B, Szigeti K, Tamas CZ, Daranyi S. Virtual museum of congenital heart defects: digitization and establishment of a database for cardiac specimens. Quant Imaging Med Surg. 2019;9(1):115–26. https://doi.org/10.21037/qims.2018.12.05. http://qims.amegroups.com/article/view/23202/22468

6. Haraguchi R, Morita Y, Matsuyama TA, Ishibashi-Ueda H, Seo H. Construction of computational cardiac model with congenital heart diseases based on isolated human hearts – Progress overview FY2016. In: Proceedings of the 3rd International Symposium on multidisciplinary computational anatomy; 2017. p. 217–9. Nara.

7. Lopez-Perez A, Sebastian R, Ferrero JM. Three-dimensional cardiac computational modelling: methods, features and applications. Biomed Eng Online. 2015;14(1):35. https://doi.org/10.1186/s12938-015-0033-5. http://www.biomedical-engineering-online.com/content/14/1/35

8. Takagi Y, Ogo K, Miyaji K, Sakuma M, Ikeda Y, Nakanishi N, Ishibashi-Ueda H. Pathological characteristics of Takayasu arteritis with pulmonary hypertension. In: C63. Endothelium and pulmonary hypertension pathogenesis, A5020–A5020. American Thoracic Society, Denver. 2011. http://www.atsjournals.org/doi/abs/10.1164/ajrccm-conference.2011.183.1_MeetingAbstracts.A5020.

9. Haraguchi R, Seo H, Matsuyama TA, Iwata M, Hasegawa S, Ootou K, Ishibashi-Ueda H. Three-dimensional reconstruction and visualization of cardiac conduction system based on serial histological specimens. In: IEICE Technical Report MI2013–95. Okinawa; 2014. p. 215–8.

10. Haraguchi R, Seo H, Matsuyama TA, Morita Y, Iwata M, Hasegawa S, Ishibashi-Ueda H. Improvement of representation for three-dimensional reconstruction and visualization of serial histological specimens. In: IEICE Technical Report MI2014–80. Okinawa; 2015. p. 125–8.

11. Kugler M, Goto Y, Tamura Y, Kawamura N, Kobayashi H, Yokota T, Iwamoto C, Ohuchida K, Hashizume M, Shimizu A, Hontani H. Robust 3D image reconstruction of pancreatic cancer tumors from histopathological images with different stains and its quantitative performance evaluation. Int J Comput Assist Radiol Surg. 2019;14(12):2047–55. https://doi.org/10.1007/s11548-019-02019-8. http://link.springer.com/10.1007/s11548-019-02019-8

12. Howard SA, Quill JL, Eggen MD, Swingen CM, Iaizzo PA. Novel imaging of atrial septal defects in isolated human hearts. J Cardiovasc Transl Res. 2013;6(2):218–20. https://doi.org/10.1007/s12265-013-9451-6. http://link.springer.com/10.1007/s12265-013-9451-6

13. Hill AJ, Laske TG, Coles JA, Sigg DC, Skadsberg ND, Vincent SA, Soule CL, Gallagher WJ, Iaizzo PA. In vitro studies of human hearts. Ann Thorac Surg. 2005;79(1):168–77. https://doi.org/10.1016/j.athoracsur.2004.06.080. https://linkinghub.elsevier.com/retrieve/pii/S0003497504014146

14. Atlas of Human CARDIAC Anatomy website. 2014. http://www.vhlab.umn.edu/atlas/histories/histories.shtml

15. Ueki W, Morita Y, Shiotani M, Haraguchi R, Matsuyama TA, Tanida Y, Harumoto K, Kono AK, Fukuda T. Evaluation of 3D magnetic resonance imaging of autopsied human heart specimens for computational modeling of congenital heart diseases. In: 103rd Scientific Assembly and Annual Meeting, Radiological Society of North America (RSNA2017). Chicago; 2017. https://rsna2017.rsna.org/.

16. NIH 3D Heart Library website. 2014. https://3dprint.nih.gov/collections/heart-library

17. Kazhdan MM, Bolitho M, Hoppe H. Poisson Surface Reconstruction. In: Sheffer A, Polthier K, editors. Proceedings of the fourth Eurographics symposium on geometry processing, SGP '06, vol. 256. Switzerland: Eurographics Association, Aire-la-Ville, Switzerland; 2006. p. 61–70. http://dl.acm.org/citation.cfm?id=1281957.1281965.

18. Takashi I. VoTracer website. 2015. http://www.riken.jp/brict/Ijiri/VoTracer/

19. Epic Games Japan: Unreal Engine. 2017. https://www.unrealengine.com/ja/

20. NIH National Heart, Lung, and Blood Institute website. https://www.nhlbi.nih.gov/health-topics/congenital-heart-defects

21. Haraguchi R. Preliminary study for interactive extraction of graph-like structures from volume data. Technical Report of IEICE. 2016;116(298):29–30.

22. Achanta R, Shaji A, Smith K, Lucchi A, Fua P, Süsstrunk S. SLIC Superpixels compared to state-of-the-art Superpixel methods. IEEE Trans Pattern Anal Mach Intell. 2012;34(11):2274–82. https://doi.org/10.1109/TPAMI.2012.120. http://ieeexplore.ieee.org/document/6205760/

23. Foundational Model of Anatomy ontology. 2016. http://sig.biostr.washington.edu/projects/fm/AboutFM.html

24. SNOMEDInternational: SNOMED-CT. 2020. https://www.snomed.org/snomed-ct/get-snomed

25. (NLM), T.N.L.o.M.: Medical Subject Headings (MeSH). 2019. https://www.nlm.nih.gov/mesh/meshhome.html

26. Musen MA. The protege project. AI Matters. 2015;1(4):4–12. https://doi.org/10.1145/2757001.2757003. https://dl.acm.org/doi/10.1145/2757001.2757003

New Frontier of Technology in Clinical Applications Based on MCA Models: Tumor Growth

A Technique for Measuring the 3D Deformation of a Multiphase Structure to Elucidate the Mechanism of Tumor Invasion

Yasuyuki Morita

Abstract

Cancer is a leading cause of death worldwide and over 90% of cancer-related deaths are due to metastasis. Therefore, understanding the mechanism of metastasis is an important goal for treating cancer patients. Metastasis is initiated by the invasion of cancer cells from a primary lesion via the interstitial extracellular matrix (ECM). Metastasis involves biomechanical interactions between the ECM and a single cancer cell or cancer cell aggregation (cancer spheroid) as it makes its way through ECM collagen fibers. It is important to quantify the ECM deformation fields produced in this process to clarify the biomechanical interactions. We visualized the dynamic deformation of the ECM using a digital volume correlation (DVC) method. As a result, my research group quantified the three-dimensional ECM deformation caused by a single cancer cell or cancer spheroid. This work would be contributory to construct a fundamental knowledge of metastasis suppression when investigated using the multidisciplinary computational anatomy (MCA) techniques.

Keywords

Cancer · Metastasis · Extracellular matrix (ECM) Epithelial–mesenchymal transition (EMT) · Digital volume correlation

16.1 Introduction

Cancer has been the leading cause of death in Japan since 1981, followed by cardiac disorders and vascular brain disease [1]. The number of fatalities due to the last two disorders have remained stable or decreased with the development of treatments. By contrast, cancer has been increasing consistently since the end of World War II and this trend is predicted to continue, as in other developed countries [2]. Recent studies have revealed that 90% of cancer deaths are due to metastasis [3, 4].

Briefly, cancer progresses as follows (Fig. 16.1):

(i) The primary lesion gains necessary nutrients by attracting blood vessels.
(ii) Cancer cells proliferate and grow in the primary lesion.
(iii) Some cancer cells leave the lesion and invade the interstitial extracellular matrix (ECM).
(iv) These cells metastasize to other tissues through the intravasation and extravasation of blood or lymph vessels.

Cancer is a disorder of cellular function. When cancer cells proliferate only in a primary lesion [processes (i) and (ii) above], this is a benign growth, which we can cure. If cancer cells in the primary lesion acquire invasive capacity through epithelial–mesenchymal transition (EMT) and invade the ECM [process (iii)], they metastasize to other tissues through blood or lymph vessels [process (iv)]. This is malignant growth and is difficult to control. We can say that invasion of the ECM by cancer cells is the initiation of metastasis. This invasion is a biomechanical interaction, since cancer cells migrate through the ECM network of collagen fibers by pushing and pulling the fibers. Knowledge of the biomechanical interaction between cancer cells and the ECM is important for elucidating the biomechanical mechanism of metastasis. However, most studies of these relationships have studied two-dimensional (2D; cancer cells on hard plastic dishes) or 2.5-dimensional (2.5D; cancer cells on a soft gel) cultures [5–8]. The outcomes of such studies do not reflect the actual biomechanical interaction, since the primary metastasis microenvironment (the ECM) is a 3D-fiber-rich collagen network. Cellular behavior is very different in 2D and 3D environments [9]. Therefore, the bio-

Y. Morita (✉)
Faculty of Advanced Science and Technology, Kumamoto University, Kumamoto, Japan
e-mail: ymorita@kumamoto-u.ac.jp

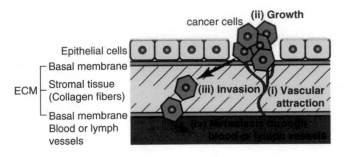

Fig. 16.1 The progression of cancer

mechanical interaction in a 3D environment must be elucidated, but few studies have examined this due to the difficulties involved [10, 11]. My research group hopes to contribute to clarifying the biomechanical mechanism of metastasis by visualizing the 3D deformation of the ECM caused by a single cancer cell or cancer cell aggregation (cancer spheroid). The standard treatments for metastatic cancer are chemotherapy and radiation therapy, although both have adverse effects. We believe that an understanding of the biomechanical interaction between cancer cells and the ECM will enable the development of new methods to control metastasis.

16.2 Biomechanical Interaction between a Single Cancer Cell and the ECM

The human cervical cancer cell line HeLa (RCB0007; RIKEN BRC, Tsukuba, Japan) was used as a typical epithelial cancer cell. EMT was induced in the cells using transforming growth factor-β1 (TGF-β1). EMT typically changes epithelial cancer cells into mesenchymal cancer cells, which have higher metastatic potential [12, 13]. Many triggers can induce EMT, such as SNAIL excitation [5, 14] and biochemical interaction with cancer-associated fibroblasts (CAFs) [15]. TGF-β is often used to induce EMT [6–8, 13, 16], because it has the broadest effects among the cytokines produced in the cancer environment, and influences many aspects of tumorigenesis [17]. We added 20 ng/mL TGF-β1 (PeproTech, Rocky Hill, NJ, USA) to the culture medium. Figure 16.2 shows the levels of markers for the cells treated with TGF-β1. TGF-β1 treatment upregulated the expression of fibronectin 1 (*Fn1*), a mesenchymal cell marker, by more than 11-fold. As a cellular adhesion molecule, *Fn1* is important in EMT interactions [18]; its high expression is likely to contribute greatly to the reconstruction of the ECM. In addition, the addition of TGF-β1 reduced the epithelial characteristics of the HeLa cells, shown by the slight decrease in expression of the epithelial cell marker keratin 19 (*Krt19*) [19]. qRT-PCR results indicated that the addition of TGF-β1-induced EMT in the HeLa cells. Consequently, the migration

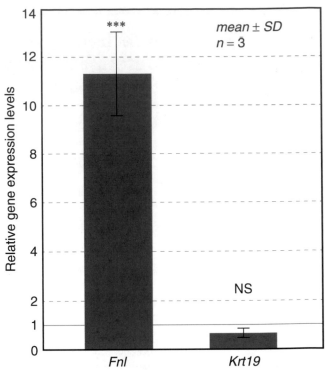

Fig. 16.2 mRNA expression in HeLa cells in the presence of TGF-β1. Fibronectin 1 (*Fn1*) is a typical mesenchymal cell marker, whereas keratin 19 (*Krt19*) is an epithelial cell marker. Data are normalized to the corresponding mRNA expression in cells not treated with TGF-β1 (defined as 1). All data are expressed as the mean ± standard deviation. ***$p < 0.001$. NS, not significant

speeds of HeLa cells in the gel roughly doubled from 80 to 180 nm/min with the addition of TGF-β1 (data not shown). The full-field 3D deformation of the collagen gel around a cancer cell was determined using the digital volume correlation (DVC) method [20, 21]. When the cancer cells were embedded in a 3D collagen gel, 7.0% carboxylate-modified fluorescent polystyrene microspheres (FluoSpheres F8821, 1.0-μm diameter, red dye; Thermo Fisher Scientific, Waltham, MA, USA) were dispersed randomly in the gel. The movement of the microspheres reflects gel deformation, since the carboxyl group of the microspheres binds covalently with collagen fibers [22]. Immunofluorescence staining with actin–green fluorescent protein (GFP; Life Technologies, Carlsbad, CA, USA) was used to examine the location and morphology of cancer cells in the gel using confocal laser scanning microscopy (A1Rsi; Nikon Instech, Tokyo, Japan). Figure 16.3 shows representative examples of the ECM deformation caused by HeLa cells. The maximum displacement and deformation of the ECM were greater in the HeLa cells with TGF-β1. Once the deformation field of the ECM around a cell was determined, the strain tensor of the ECM was computed using a displacement gradient technique [23]. Material constitutive equations describing the relationship between the strain and stress tensors were used

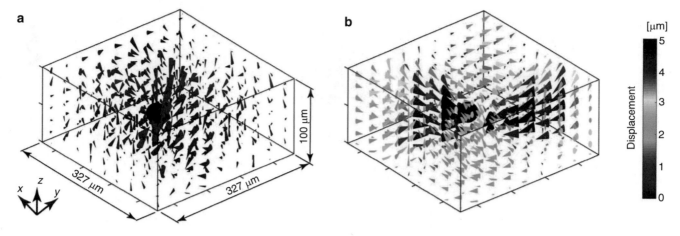

Fig. 16.3 ECM deformation field induced by a HeLa cell: (**a**) before and (**b**) after adding TGF-β1. The black spheres show the cell's location

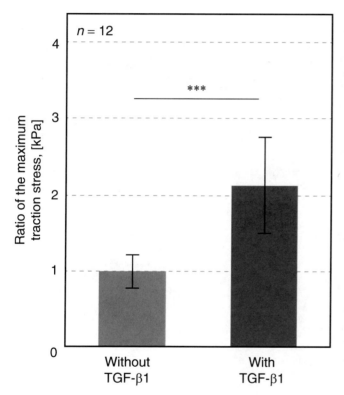

Fig. 16.4 Ratio of the maximum traction force exerted by a HeLa cell without and with TGF-β1. Data are normalized to the corresponding maximum traction force exerted by a Hela cell without TGF-β1 (defined as 1). ***$p < 0.001$

to determine the traction stress of the cell [24]. The ECM collagen gel can be considered a linear elastic material, since the strain was ≤10% [25, 26]. Figure 16.4 plots the ratio of the traction forces of the cancer cell. EMT was induced by the addition of TGF-β1 and the resulting traction forces roughly doubled. Therefore, the enhanced migration speed and traction forces induced by EMT contribute to the increased invasiveness and metastasis of cancer cells.

16.3   Biomechanical Interaction between Cancer Spheroids and the ECM

My group has also been investigating deformation of the ECM surrounding a cancer spheroid, which is a model of actual cancer tissue. In this study, the cancer spheroids consisted of aggregations of pancreatic adenocarcinoma cells (PANC-1). The spheroids were embedded in a collagen gel, and the deformation of the collagen ECM caused by the spheroid was determined using the DVC method [20, 21]. Figure 16.5 shows a representative ECM deformation field. The deformation increased with time. Once the deformation field around the spheroid was determined, the bulk strain of the ECM was calculated using a displacement gradient technique. Figure 16.6 shows the bulk strain of the ECM. The cancer spheroid formed a fluctuating deformation field in the surrounding ECM, implying that there was pushing and pulling deformation of the ECM (Fig. 16.6a). Finally, contraction dominated the deformation of the ECM (Fig. 16.6b) and this was particularly high around the tip of the invasive protrusion of the spheroid (dotted circles in Fig. 16.6b).

16.4   Conclusion

My research group quantified the 3D deformation fields in the ECM exerted by a single cancer cell or cancer spheroid. Now, we hope to elucidate the mechanism of cancer metastasis by studying the detailed biomechanical interaction between cancer cells and the ECM. Consequently, the knowledge of the biomechanical interaction will make it possible to predict metastatic pathways of the cancers, and then suppress and control cancer progression by observing the in-situ deformation field in the tissue around cancer with time using the multidisciplinary computational anatomy (MCA) techniques.

Fig. 16.5 ECM deformation field induced by a PANC-1 cancer spheroid. (**a**) initially (0 h) and (**b**) 2 h later

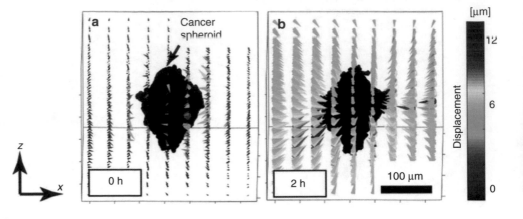

Fig. 16.6 Bulk strain field of ECM induced by a PANC-1 cancer spheroid: (**a**) initially (0 h) and (**b**) 2 h later

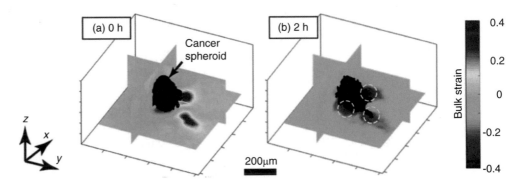

References

1. The Ministry of Health, Labour and Welfare in Japan, Annual reports of Vital Statics (in Japanese), 2018.
2. The National Institute of Population and Social Security Research in Japan, Reports of World's Vital Statics (in Japanese), 2019.
3. Steeg PS. Targeting metastasis. Nat Rev Cancer. 2016;16:201–18. https://doi.org/10.1038/nrc.2016.25.
4. Mehlen P, Puisieux A. Metastasis: a question of life or death. Nat Rev Cancer. 2006;6:449–58. https://doi.org/10.1038/nrc1886.
5. Ghosh D, Dawson MR. Microenvironment influences cancer cell mechanics from tumor growth to metastasis. In: Dong C, Zahir N, Konstantopoulos K, editors. Biomechanics in oncology, vol. 1092; 2018. p. 69–90. https://doi.org/10.1007/978-3-319-95294-9_5.
6. Nalluri SM, O'Connor JW, Virgi GA, Stewart SE, Ye D, Gomez EW. TGFβ1-induced expression of caldesmon mediates epithelial-mesenchymal transition. Cytoskeleton. 2018;75:201–12. https://doi.org/10.1002/cm.21437.
7. Mekhdjian AH, Kai FB, Rubashkin MG, Prahl LS, Przybyla LM, McGregor AL, et al. Integrin-mediated traction force enhances paxillin molecular associations and adhesion dynamics that increase the invasiveness of tumor cells into a three-dimensional extracellular matrix. Mol Biol Cell. 2017;28:1467–88. https://doi.org/10.1091/mbc.E16-09-0654.
8. Lam CRI, Tan C, Teo Z, Tay CY, Phua T, Wu YL, et al. Loss of TAK1 increases cell traction force in a ROS-dependent manner to drive epithelial-mesenchymal transition of cancer cells. Cell Death Dis. 2013;4:e848. https://doi.org/10.1038/cddis.2013.339.
9. Pedersen JA, Swartz MA. Mechanobiology in the third dimension. Ann Biomed Eng. 2005;33:1469–90. https://doi.org/10.1007/s10439-005-8159-4.
10. Koch TM, Munster S, Bonakdar N, Butler JP, Fabry B. 3D traction forces in cancer cell invasion. PLoS One. 2012;7:e33476. https://doi.org/10.1371/journal.pone.0033476.
11. Indra I, Undyala V, Kandow C, Thirumurthi U, Dembo M, Beningo KA. An in vitro correlation of mechanical forces and metastatic capacity. Phys Biol. 2011;8:015015. https://doi.org/10.1088/1478-3975/8/1/015015.
12. Voulgari A, Pintzas A. Epithelial-mesenchymal transition in cancer metastasis: mechanisms, markers and strategies to overcome drug resistance in the clinic. Biochim Biophys Acta-Rev Cancer. 1796;2009:75–90. https://doi.org/10.1016/j.bbcan.2009.03.002.
13. Lamouille S, Xu J, Derynck R. Molecular mechanisms of epithelial-mesenchymal transition. Nat Rev Mol Cell Biol. 2014;15:178–96. https://doi.org/10.1038/nrm3758.
14. McGrail DJ, Mezencev R, Kieu QMN, McDonald JF, Dawson MR. SNAIL-induced epithelial-to-mesenchymal transition produces concerted biophysical changes from altered cytoskeletal gene expression. FASEB J. 2015;29:1280–9. https://doi.org/10.1096/fj.14-257345.
15. Giannoni E, Bianchini F, Masieri L, Serni S, Torre E, Calorini L, et al. Reciprocal activation of prostate cancer cells and cancer-associated fibroblasts stimulates epithelial-mesenchymal transition and cancer stemness. Cancer Res. 2010;70:6945–56. https://doi.org/10.1158/0008-5472.CAN-10-0785.
16. Yoshie H, Koushki N, Molter C, Siegel PM, Krishnan R, Ehrlicher AJ. High throughput traction force microscopy using PDMS reveals dose-dependent effects of transforming growth factor-β on the epithelial-to-mesenchymal transition. J Vis Exp. 2019;148:e59364. https://doi.org/10.3791/59364.
17. Ikushima H, Miyazono K. TGFβ signalling: a complex web in cancer progression. Nat Rev Cancer. 2010;10:415–24. https://doi.org/10.1038/nrc2853.

18. Griggs LA, Hassan NT, Malik RS, Griffin BP, Martinez BA, Elmore LW, et al. Fibronectin fibrils regulate TGF-β1-induced epithelial-mesenchymal transition. Matrix Biol. 2017;60-61:157–75. https://doi.org/10.1016/j.matbio.2017.01.001.

19. Flozak AS, Lam AP, Russell S, Jain M, Peled ON, Sheppard KA, et al. Beta-catenin/T-cell factor signaling is activated during lung injury and promotes the survival and migration of alveolar epithelial cells. J Biol Chem. 2010;285:3157–67. https://doi.org/10.1074/jbc.M109.070326.

20. Morita Y, Kawase N, Ju Y, Yamauchi T. Mesenchymal stem cell-induced 3D displacement field of cell-adhesion matrices with differing elasticities. J Mech Behav Biomed Mater. 2016;60:394–400. https://doi.org/10.1016/j.jmbbm.2016.02.025.

21. Franck C, Hong S, Maskarinec SA, Tirrell DA, Ravichandran G. Three-dimensional full-field measurements of large deformations in soft materials using confocal microscopy and digital volume correlation. Exp Mech. 2007;47:427–38. https://doi.org/10.1007/s11340-007-9037-9.

22. Bloom RJ, George JP, Celedon A, Sun SX, Wirtz D. Mapping local matrix remodeling induced by a migrating tumor cell using three-dimensional multiple-particle tracking. Biophys J. 2008;95:4077–38. https://doi.org/10.1529/biophysj.108.132738.

23. Franck C, Maskarinec SA, Tirrell DA, Ravichandran G. Three-dimensional traction force microscopy: a new tool for quantifying cell-matrix interactions. PLoS One. 2011;6:e17833. https://doi.org/10.1371/journal.pone.0017833.

24. Fung YC. Biomechanics: mechanical properties of living tissues. New York: Springer Science & Business Media; 2013.

25. Arevalo RC, Kumar P, Urbach JS, Blair DL. Stress heterogeneities in sheared type-I collagen networks revealed by boundary stress microscopy. PLoS One. 2015;10:e0118021. https://doi.org/10.1371/journal.pone.0118021.

26. Motte S, Kaufman LJ. Strain stiffening in collagen I networks. Biopolymers. 2013;99:35–46. https://doi.org/10.1002/bip.22133.

Construction of Classifier of Tumor Cell Types of Pancreas Cancer Based on Pathological Images Using Deep Learning

17

Naoaki Ono, Chika Iwamoto, and Kenoki Ohuchida

Abstract

Recently, computer-aided diagnosis methods based on machine learning, mainly using Deep Leaning, have been studied and developed very rapidly. Especially image recognition based on Convolutional Neural Networks showed high accuracy in diagnosis problems when they are given the huge amount of training data. Those methods are not only helpful for classification but also useful for feature extraction from given images. Here we introduce a new classification method to find the features of tumor tissues from histopathology images by unsupervised clustering based on Information Maximization Self-Augmented Training. Moreover, to evaluate fibrosis and classify tumor cells, we used histopathological images with different staining methods as concatenated inputs. Using this approach, we can quantify integrated features based on multimodal imaging using deep learning. In this study, we analyzed pathological images of pancreas cancers and optimized to classify the patches of the images into the categorize with different features, which are consistent with annotation of the medical doctors. It can also provide a map to visualize the probability where cell types are categorized into specific classes according to the given pathological images.

Keywords

Deep Learning · Convolutional Neural Networks · Pathological images · Unsupervised Clustering

N. Ono (✉)
Data Science Center, Graduate School of Science and Technology, Nara Institute of Science and Technology, Nara, Japan
e-mail: nono@is.naist.jp

C. Iwamoto · K. Ohuchida
Department of Medicine and Surgery, Graduate School of Medical Sciences, Kyushu University, Fukuoka, Japan

17.1 Introduction

Pathologists inspect a huge number of visual samples every day and make a diagnosis, detect lesions, classify tumors and so on. It requires a huge amount of experience to train those skills, and it can be a bottleneck that the shortage of skilled pathologists. Computer-aided diagnosis has not been to replace human doctors, but it can provide useful tools to help diagnose and improve efficiency and accuracy of those human pathologists by showing, for example, possible areas of images to be focused on [1, 2].

There have been many studies to propose computational methods for image analysis and feature extraction of medical images [3]. Note that most studies of those image processing are based on supervised learning using heuristically designed models that need appropriate labels according to the task, the prediction from the model changes depending on how to label it. Since these heuristic annotations cost a lot of time for clinical doctors, it is desirable to construct a model that automatically classifies given images into certain categories, for example, to evaluate the types and stages of the tumor in order to decide the appropriate treatment plan. To address this problem, we introduce an approach to image classification using an unsupervised clustering model based on Information Maximizing Self Argument Training (IMSAT) [4]. We applied this model to evaluate pathological images of pancreas cancer and showed that this model could automatically learn the patterns of tumor cells and classification, which are consistent with the categorization made by medical doctors.

Pancreatic cancer is known as one of the worst prognosis cancers due to the difficulty of early detection, the fast progression stages and the frequency of distant metastasis, etc. Therefore, estimation of medical features of tumor cells, such as the proliferation rate, metastatic state, etc., will provide much advance to lead to better treatment. Generally, Hematoxylin and eosin (H&E) stain is a standard for histopathological imaging for diagnosis, and to visualize other detail status of tissues such as fibrosis, various staining meth-

Fig. 17.1 A schematic illustration of the architecture of the Deep Convolutional Neural networks

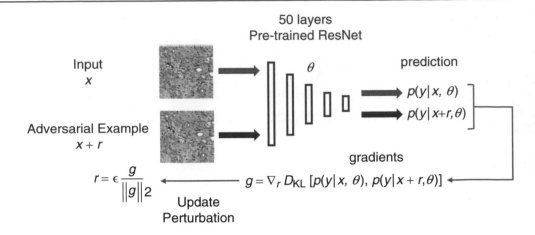

ods such as Masson's Trichrome (MT) staining are used for diagnosis. MT stain visualizes the interstitium of collagen fibers with blue. It is possible to detect the infiltrative area more clearly than HE staining because it can be used for discrimination of the degree of progress. In this study, we analyzed multi-stained pathological images of the pancreas obtained from the KPC model mouse [5].

Tumor tissues are generally heterogeneous, composed of a mixture of various types of cells, it implies that the tumor futures should be recognized from the local variation of morphological features of the cells, such as the shapes and distribution of cell types in each region in order to help a doctor's judgement. Applications based on deep learning are very promising for image recognition; however, it is still a black box and difficult to understand the process of the trained neural networks. In previous studies [6], we proposed a method to visualize the "latent space" learned by feature extraction. Based on those models, pixels of given images are transformed into a vector of variables that represents the informative features of the original images. These latent variables extracted from the model will be important to elucidate the behaviour of extracted features in the histopathological images.

In our study, we analyze the pancreatic histopathological image using Convolutional Neural Network (CNN) that trained two different dying methods, i.e., H&E and MT staining images registered with the help of Hontani and others [7], to evaluate a degree of fibrosis. Using this model, we extract morphological patterns that contribute to predicting the malignancy of pancreatic cancer.

Visual Recognition Challenge (ILSVRC) held in 2012, a model of deep learning named AlexNet won with overwhelming results. The diagnostic application of deep learning has been already shown successful results, for example, detections of tumors, identification of CoViD-19 from CT images, classification of skin melanomas [8] but the illnesses that can be dealt with are also limited due to the difficulty of obtaining a teacher label. Careful annotation by a specialist is necessary.

We constructed a model of cluster analysis using Information Maximizing Self-Augmented Training (IMSAT) to obtain clusters using extracted features by CNN. Although IMSAT is unsupervised learning, it can perform clustering with high accuracy. It is based on Self-Augmented Training and maximization of mutual information. When data is distorted by any affine transformation and perturbations, local representations in the latent space depart greatly from the predictions of original data points. Using SAT, it can be close the predictions of perturbed data points to original data points. KL divergence D_{KL} is used as the distance between original and distorted distribution. We choose Virtual Adversarial Training (VAT) [9] and typical data augmentation such as random contrast enhancement, rotation, flipping, and scaling as distortion in SAT. Adversarial perturbations are generated in VAT. These disturb the prediction distribution greatly and can be computed from multiplying random normalized vectors according to the gradient of the KL divergence of the prediction using the perturbed input to the original input in the latent space. The scheme of the VAT is illustrated in Fig. 17.1.

17.2 Methods of Unsupervised Image Classification

Applications of image analysis using deep learning have been rapidly developing. The convolutional neural network is a popular model of deep learning mostly applied for image classification. In the competition of ImageNet Large Scale

17.3 Results

To construct a model of IMSAT, we first need to determine the number of possible classifications. In this study, we tested different number of clusters $k = 4, 8, 16$, and 32, then evaluated comparing the images in each obtained cluster.

And we chose $k = 8$ since it was most consistent with annotations of medical doctors. When the number of clusters was small ($k = 4$), many different types of cells were included within the single category, and when k was larger than 8, some classes were almost blank, i.e., the model could not find samples to distinguish so detail.

Figure 17.2 shows the example of the patches classified into different categories using this model. Each row represents the clusters which are automatically optimized by the IMSAT method. It is clear that the features and density of the cells are different between the categories, and they can be annotated as dense tumor cells, gap between cells, acinar cells, blood vessels, etc.

Since the output of the discriminator is normalized by the softmax function, we can evaluate the probability to categorize each category at each site of the original image. Figure 17.3 shows an example of a heatmap of the probability where the cell types are categorized to dense tumor cells, represented by the 8 th (indexed by 7) row in Fig. 17.2. To provide this quantitative evaluation will greatly help interpretations by medical doctors.

Analyzing pancreatic cancer pathological images using IMSAT. Using CNN, it is possible to extract features based on the morphological characteristics of the tissue in addition to the brightness value. For pathological images with heterogeneous cell types, it is more interpretable to encode them into discrete representations. Therefore, as a result of clustering using IMSAT, we obtained clusters involved in differentiation. IMSAT is considered to be the most suitable model for pathological image analysis since it can provide a plausible heatmap of the probability distribution of cell types. There have been some models of unsupervised clustering model [10] that achieves higher accuracy in classification. It is considered that more cohesive clustering is possible by analysis. On the other hand, an evaluation index of the optimum number of clusters applicable to a clustering model based on a neural network has not yet been developed, and pathology.

17.4 Discussion

We constructed the unsupervised classification model using deep learning based on IMSAT in order to evaluate and quantify pathological images with different staining methods. This approach can be applied for quantitative evaluation

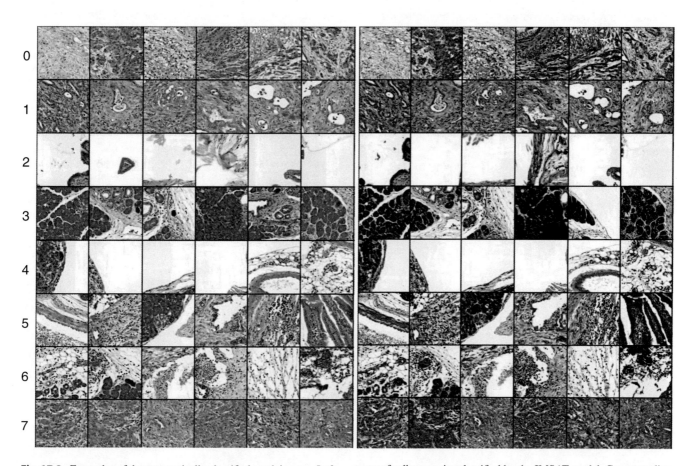

Fig. 17.2 Examples of the automatically classified patch images. Left: HE stained images, and Right: MT stained images are patches at the corresponding areas of the contact slices. Each row represents different types of cell categories classified by the IMSAT model. Corresponding patches in the left and the right were taken from the same position in the original images

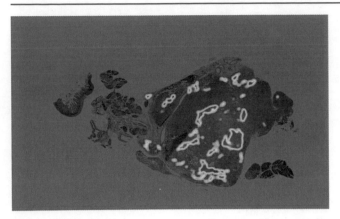

Fig. 17.3 An example of the heatmap to visualize the probability of dense tumor cell areas

level close to that of a pathologist, use high-magnification images. It is considered necessary to perform learning based on cell morphology. Already, by using supervised CNN for pathological image data sets by the magnification of 40 to 400 times, it is possible to identify cancer classification with high accuracy. Diseases that are difficult to annotate by clustering based on potential features in pathological images, such as those used in this study, using datasets cut out at various magnifications. It is possible to perform pathological image analysis considering structural malformations and cell malformations, and it is thought that a predictive model of cancer malignancy and progression can be constructed that will contribute to diagnostic support in the future.

of various types of multimodal analysis. Although integrated analysis of multidisciplinary computational anatomy often requires complex models in a high dimensional data space, feature extraction using IMSAT itself can be regarded as optimization of embedding from given training data into a reduced latent space that represents manifold of sample data. Moreover, this method can be incorporated with other models of deep learning such as autoencoders, style-GAN (Generative Adversarial Networks), etc. One of the difficulties of the implementation of IMSAT is its high computational cost since it requires a large amount of sample data. And when enough training sample is not available, the embedded latent space may not properly represents the latent space. Multiplying training samples using data augmentation or training the model using a similar data sample which is plentifully available and applying transfer learning will help to address this difficulty.

Immunostaining that can specifically bind to and visualize specific gene-related proteins is also widely used. This is because many biomarkers related to malignancy and progression can be detected by immunostaining. It has already been clarified in clinical studies that the expression level of specific proteins and transcription factors is predominantly correlated with clinical stage classification, lymph node metastasis, and tissue differentiation and is involved in predicting prognosis.

Another characteristic of the data set used this time is that it focuses on structural irregularities such as the cyclic structure formed by the cancer cell population and does not consider the degree of detailed irregularities. In the actual diagnosis of pancreatic cancer, in addition to structural malformations, evaluation is performed by evaluating nuclear cell malformations. In the future, if predictions are made at a

References

1. Van Ginneken B, Romeny BT, Viergever MA. Computer-aided diagnosis in chest radiography: a survey. IEEE Trans Med Imaging. 2001 Dec;20(12):1228–41.
2. Doi K, MacMahon H, Katsuragawa S, Nishikawa RM, Jiang Y. Computer-aided diagnosis in radiology: potential and pitfalls. Eur J Radiol. 1999;31(2):97–109.
3. Farjam R. Soltanian, Zadeh H, Jafari, Khouzani K, Zoroofi RA. An image analysis approach for automatic malignancy determination of prostate pathological images. Cytometry Part B: Clinical Cytology: The Journal of the International Society for Analytical Cytology. 2007 Jul;72(4):227–40.
4. Hu W, Miyato T, Tokui S, Matsumoto E, Sugiyama M. Learning discrete representations via information maximizing self-augmented training. arXiv preprint arXiv:1702.08720. 2017 Feb 28.CNN.
5. Lee JW, Komar CA, Bengsch F, Graham K, Beatty GL. Genetically engineered mouse models of pancreatic Cancer: the KPC model (LSL-Kras$^{G12D/+}$; LSL-Trp53$^{R172H/+}$; Pdx-1-Cre), its variants, and their application in Immunooncology drug discovery. Current protocols in pharmacology. 2016 Jun;73(1):14–39.
6. Asano K, Ono N, Iwamoto C, Ohuchida K, Shindo K, and Kanaya S, "Feature extraction and Cluster analysis of Pancreatic Pathological Image Based on Unsupervised Convolutional Neural Network," 2018 IEEE International Conference on Bioinformatics and Biomedicine (BIBM), Madrid, Spain, 2018, pp. 2738–2740, https://doi.org/10.1109/BIBM.2018.8621323.
7. Hontani H, et al. "Registration between histopathological images with different stains and an MRI image of pancreatic cancer tumor." International Forum on Medical Imaging in Asia 2019. Vol. 11050. International Society for Optics and Photonics, 2019.
8. Mikołajczyk A, Grochowski M. Data augmentation for improving deep learning in image classification problem. In 2018 international interdisciplinary PhD workshop (IIPhDW) 2018 May 9 (pp. 117–122). IEEE.
9. Miyato T, Maeda SI, Koyama M, Ishii S. Virtual adversarial training: a regularization method for supervised and semi-supervised learning. IEEE Trans Pattern Anal Mach Intell. 2018 Jul 23;41(8):1979–93.
10. Bridle JS, Heading AJR, MacKay DJC. Unsupervised classifiers, mutual information and 'Phantom targets. Adv Neural Inf Proces Syst. 1992;4:1096–101.

Part VII

New Frontier of Technology in Clinical Applications Based on MCA Models: Cranial Nervous System

Multi-Modal and Multi-Scale Image Registration for Property Analysis of Brain Tumor

18

Takashi Ohnishi

Abstract

Analyses of the relationships between physical properties and the microstructure of human tissue have been widely conducted. In particular, the relationships between acoustic parameters and the microstructure of the human brain fall within the scope of our research. In order to analyze the relationships between physical properties and microstructure of the human tissue, accurate image registration is required. To observe the microstructure of the tissue, a pathological (PT) image, which is an optical image capturing a thinly sliced specimen, has generally been used. However, spatial resolution and image features of PT images are markedly different from those of other imaging modalities. This study proposes a modality conversion method from PT to ultrasonic (US) images, including a downscaling process using a convolutional neural network (CNN). Namely, the constructed conversion model estimates the US signals from the patch image of the PT image. The proposed method was applied to PT images, and it was confirmed that the converted PT images were similar to the US images by visual assessment. Image registration was then performed with the converted PT and US images measuring the consecutive pathological specimens. Successful registration results were obtained for every pair of images. Analysis methods using PT and US images were also developed. First, several tissue densities were calculated from PT images, and useful parameters associated with tumor grade were investigated. A CNN-based nuclear density estimation method from acoustic characteristics was then developed. The estimated nuclear densities from the attenuation were highly correlated with those calculated from the PT image.

Keywords

Modality conversion · Pathological image · Ultrasonic image · Convolutional neural network

18.1 Introduction

To diagnose and determine the tumor region is an important process, but it is difficult in brain tumor resection. Physicians have to depend on their skills or experience in diagnosing brain tumors. For instance, color or stiffness plays an important role in tumor diagnosis. If we understand the physical properties of brain tumors more deeply, it can help establish a new quantitative index for diagnosing brain tumors, and it can also provide useful information for treatment. Our plan is to develop a diagnostic system for brain tumors using small ultrasonic probes. First, we measure stiffness or other correlated parameters from an ultrasound (US) signal. Then, a precise diagnosis can be made based on these parameters. To accomplish this system, we must analyze the relationship between US signals and tissue information.

In recent years, the physical properties of human tissue, such as mechanical, optical, and acoustic properties, have been widely measured. In addition, these properties have been compared with the microstructure of tissue, such as the distribution of the cell nuclei and the running directions of nerve fibers [1–3]. The microstructure of tissue can be acquired from pathological (PT) images, which are optical images of thinly sliced specimens. Methodologies of multimodal analysis using such PT images and other modal images have been widely developed. We have also been analyzing the relationships between acoustic characteristics and the microstructure of the human brain using PT images and microscopic ultrasonic (US) images. To compare the physical properties and the microstructure at the same location using multi-modal images, accurate image registration is required. Previous studies used landmark-based or semiauto-

T. Ohnishi (✉)
Center for Frontier Medical Engineering, Chiba University, Chiba, Japan

matic methods [4–7]. However, correction of local differences was too difficult because tissue characteristics in the PT image are not taken into consideration in these methods, which makes detection of the corresponding landmarks difficult. In this case, intensity-based registration may be more promising.

When intensity-based registration is performed, the spatial resolution of the PT image can be an obstacle because it is much higher than that of other image modalities such as computed tomography, magnetic resonance imaging, and US imaging. For example, the highest spatial resolution of the PT image is approximately 230×230 nm^2, whereas that of the US image measured by a US microscopic system is approximately 8×8 μm^2 at most. Therefore, when a pixel is selected on the US image during the registration process, the corresponding pixel value is calculated from 35×35 pixel regions in the PT image. In this situation, the spatial resolution of the PT image is generally adjusted to be almost the same as that of another image using an averaging and downsampling technique before image registration. However, such a simple downscaling processing eliminates microscopic patterns that each organ inherently possesses and leads to a decline in registration accuracy.

To enhance each structural component in the PT image and achieve highly accurate image registration, we introduced a modality conversion method combined with the downscaling process. This study tuned a conversion method assuming image registration between PT and US images. Additionally, as an initial study, a tissue density calculation method was constructed to conduct a relationship analysis of PT and US images. The obtained tissue densities were compared with the acoustic characteristics at corresponding regions.

18.2 Materials

Brain tumor samples were resected from six patients as normal clinical procedures. After the surgery, the resected tumor samples were further dissected into some pieces. These obtained pieces were named macro-specimens S1 to S6. This study was approved by the Ethical Review Board of our University, and informed consent was obtained from all six patients who participated in the study. Each resected macro-specimen then underwent formalin fixation, tissue processing, and paraffin embedding. Thinly sliced specimens with 8-μm thickness were then obtained from paraffin-embedded specimens using a microtome. These thinly sliced specimens were deparaffinized with xylene and cleaned with ethanol. For US measurement, the images of specimens in this status were captured. These specimens were further stained with hematoxylin-eosin (HE), and the PT images of the stained specimens were then captured.

For macro-specimen S1, sectioning by microtome was performed repeatedly, and 19 consecutive pathological specimens were obtained from the paraffin-embedded specimen. Both US measurement and PT image acquisition were performed on only the first pathological specimen. A pair of PT and US images acquired in this process was used to construct the conversion model. For the other pathological specimens, US and PT images were acquired from odd and even numbered pathological specimens, respectively. As for macro-specimens S2–6, one pathological specimen was obtained from each macro-specimen, and a pair of US and PT images was acquired in each macro-specimen just as the pair of PT and US images of S1.

For US measurement, two ultrasonic microscopic systems were used. One was a modified version of a commercial product (AMS-50SI, Honda Electronics Co., Ltd., Toyohashi, Japan) and was used for S1. The other was an in-house developed system that was used for S2–6. In both systems, a ZnO wave transducer (Fraunhofer IMBT, St. Ingbert, Germany) with a center frequency of 250 MHz was commonly used. This transducer was attached to the X-Y stage and scanned with 8-μm pitch in each direction. Echo amplitude, speed of sound, and attenuation were calculated from the acquired RF echo signal at each scan point and used as the pixel value of each image. Image size and pixel size were 300×300 to 800×800 pixels and 8.0×8.0 μm^2/pixel. The detailed calculation method for the acoustic characteristics has previously been described [8, 9].

For PT image acquisition, HE stained pathological specimens were digitalized with a virtual slide scanner (NanoZoomer S60, Hamamatsu Photonics K.K., Hamamatsu, Japan). The image and pixel sizes were approximately $12,000 \times 12,000$ pixels and 228×228 nm^2, respectively.

18.3 Modality Conversion from Pathological Image to Ultrasonic Image

18.3.1 Construction of the Conversion Model

The proposed method consists of two steps: model construction and actual registration steps. In the model construction step, landmark-based registration with PT and US images was conducted. The US image was moved to the coordinate system of the PT image in this registration process. If the imaged area of the original PT image was too large compared with that of the US image, a region of interest was set in the PT image. A rescaled PT image was generated using the simple average method and then binarized with the discriminant analysis method. The landmarks were detected by AKAZE feature detector [10] from the binarized PT image and US image. Outliers for the landmarks were removed by random sampling consensus [11]. These registration results had to be visually confirmed by the operator. The conversion model was constructed with the original PT image and the registered US image. Figure 18.1a shows the flow for conversion model construction using a convolutional neural network (CNN) [12]. Some patch images were extracted from the original PT image. The conversion model estimates a US signal from each small region on the patch image. Estimated US signals p_k were compared with actual US signals l_k. CNN was optimized until the loss function was minimized. The mean absolute error defined as follows was used as the loss function:

$$loss(\mathbf{P}, \mathbf{L}) = \frac{1}{N} \sum_{k=1}^{N} |p_k - l_k| \qquad (18.1)$$

Here, k and N represent the index of the patch image and the total number of patch images input into the CNN, respectively. These processes were repeated until the epoch number reached a predefined limit. In terms of the framework of the CNN, there were two convolution layers and two pooling layers followed by dropout and fully connected layers. CNN construction had to be conducted once before the actual registration.

In the actual registration step, the PT images for image registration were converted by the constructed model. Affine registration including shift, rotation, and scaling operations was then conducted. The normalized cross correlation (NCC) and Powell-Brent methods were used as a similarity measure between converted PT and US images and an optimization method, respectively.

18.3.2 Studies of Modality Conversion and Image Registration

In this study, two kinds of experiments were conducted. In the first experiment, the applicability of the conversion model was evaluated with PT and US images obtained from the same macro-specimen S1. A conversion model was constructed with a pair of PT and US images and applied to the other nine PT images. Image registration was then performed. The first US image was used as a reference image, and the other images were registered into the first US image. To evaluate the versatility of the conversion model, another experiment was conducted with images S2–6. A conversion model was constructed with the images of S2 and applied to the PT images of S3–6. The patch size for the conversion model was set to 32 × 32 pixels. Namely, the pixel size after conversion was 7.30 × 7.30 μm^2. The number of epochs, batch size, learning rate, and dropout rate for CNN were set to 20,000, 100, 1.0×10^{-3}, and 0.5, respectively.

Figure 18.1b shows a result of the conversion model construction. Black spots are clearly enhanced after conversion. On visual assessment, the converted PT image was similar to the US image when compared with the PT image. The constructed conversion model was applied to other PT images. It was confirmed that the features of all converted images were similar to those of the US images. In addition, image registration was performed with the US and converted PT images. All US images were registered into the neighboring PT images. Figure 18.1c shows the registration results with pathological specimens #3–5 of S1. All images including both US and PT images were successfully registered using the original US images and the converted PT images on visual assessment.

A conversion model was constructed with the image dataset of S2 and applied to the image datasets of S3–6. The resultant images are shown in Fig. 18.1d. Some structures in the converted PT image of S2 were slightly enhanced. For the image dataset of S3, the tendency for the conversion result was similar to that for S2. The effect of modality conversion was confirmed. However, it was less than that in the previous experiment. Although the effectiveness of the proposed method could not be visually confirmed, the histogram or spatial distribution of pixel values was similar to that of the US image. From these results, it can be expected that the proposed method can produce better registration than the simple downscale method.

18.4 Property Analysis

18.4.1 Investigation of Effective Properties

To identify an effective parameter for estimating tumor grade, tissue microstructure and acoustic characteristics were compared using the registered US and PT images. The densities of nuclei and red blood cells are important, and are associated with tumor grade as pathological diagnostic information. Therefore, a calculation method for tissue densities was developed. The tissue densities were calculated for high- and low-grade regions of PT images, and the property that could effectively estimate tumor grade was investigated.

The areas of nuclei and red blood cells were individually extracted from PT images using a deep learning method [13]. These extraction models were previously constructed for each target tissue using the original PT and annotation images. The regions of nuclei and blood cells were removed from PT images, and the color space for the residual regions was translated to the HSV color space. The residual region was divided into glass and cytoplasm regions by applying the maximum entropy threshold method [9] for the saturation channel. Finally, the tissue density was calculated from the

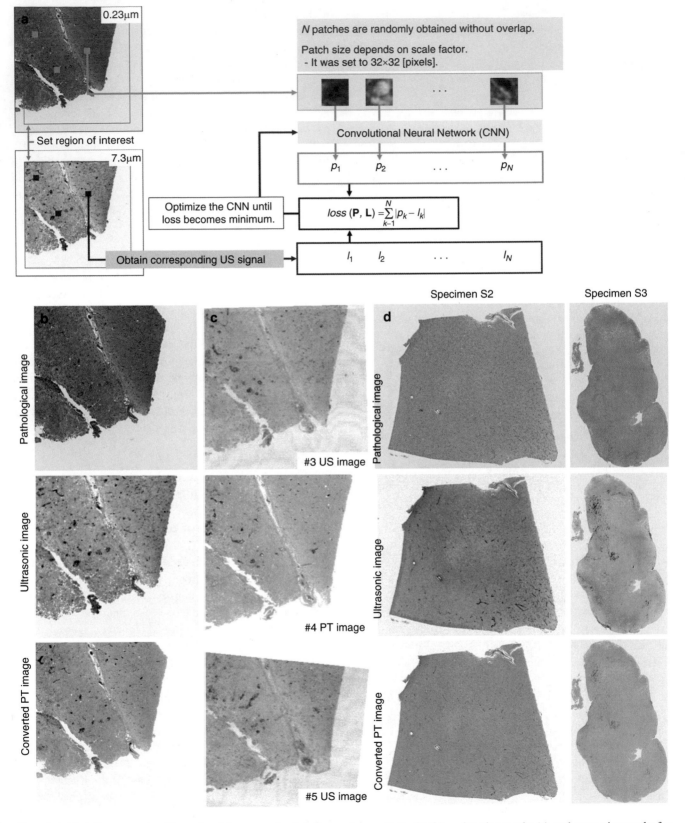

Fig. 18.1 Modality conversion. Flow of model construction (**a**), Conversion results for S1 (**b**), registration results (**c**), and conversion results for S2 and S3 (**d**)

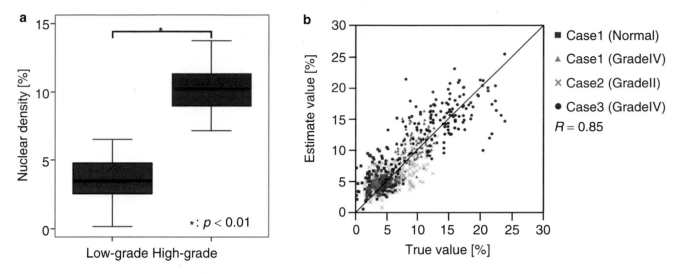

Fig. 18.2 Property analysis. Comparison of the density of nuclei between low- and high-grade regions (**a**) and the results of nuclear density estimation (**b**)

binary images of nuclei, red blood cells, and cytoplasm. The tissue density ρ was defined as follows:

$$\rho = \frac{A_{target}}{A_{all}} \times 100 \qquad (18.2)$$

Here, A_{target} and A_{all} represent the number of pixels in the target tissue region and the whole area of the patch image, respectively.

Each tissue density was compared between high- and low-grade regions using a t-test. The comparison results for the nuclear density are shown in Fig. 18.2a. The nuclear density between low- and high-grade regions showed a significant difference ($p < 0.01$). The nuclear density appears to be a useful parameter for estimating the tumor grade.

18.4.2 Pathological Properties from Acoustic Characteristics

From the results of subsection 18.4.1, it might be possible to assess the tumor grade indirectly if the nuclear density is obtained. As an initial study, a CNN was used to estimate the nuclear density from the acoustic characteristics. Small regions were extracted from US images and input into the CNN. The CNN estimated the nuclear density l_k. On the other hand, a corresponding small region was extracted from the PT image, and its nuclear density p_k was calculated, in a similar manner to that described in subsection 18.4.1. The CNN for the nuclear density estimation was constructed with 4 residual blocks and a fully connected layer. All convolution layers in the residual blocks consisted of a 7×7 kernel with 16 channels. Four pairs of PT and US images, S3–S6, were divided into 4884 pairs of patch images. The attenuation

image was used as a US image in this experiment, and 4096 pairs and 788 pairs of patch images were used for network training and testing, respectively. Training conditions, such as the loss function and optimization method, were the same as for the modality conversion.

Figure 18.2b shows a scatter plot, in which the horizontal and vertical axes describe the nuclear densities calculated from the PT image and estimated from the US image using the constructed CNN, respectively. This graph indicates that the CNN could roughly estimate the pathological characteristics from the acoustic characteristics. Although the estimation accuracy should be improved, the feasibility of this approach was shown, because the correlation coefficient was 0.85. Additionally, we would like to improve the CNN to identify the tumor grade from the acoustic characteristics.

18.5 Conclusion

To conduct image registration with pathological (PT) and ultrasonic (US) images, a CNN-based modality conversion method for PT images was proposed. On visual assessment, converted PT images were similar to the US images when compared with the original PT images. Therefore, highly accurate registration results can be obtained without additional intelligent and/or complicated registration methods. Because it was ensured that the registration results were promising for the analysis application, analysis methods were also developed, and some analyses to estimate the tumor grade from acoustic characteristics were conducted. First, the tissue densities were compared between low- and high-grade regions. The analysis results showed that nuclear density was an effective parameter for determining tumor grade. Additionally, a CNN-based nuclear density estimation method from US images was developed. The estimated nuclear densities from the attenuation were highly correlated with those calculated from PT images.

In the present study, two analysis methods were constructed to indirectly determine the tumor grade from the attenuation image through the nuclear density. A speed of sound image will be introduced into the nuclear density estimation as well in future studies. In addition, we will try to directly determine the tumor grade from the US characteristics using a CNN-based method. To evaluate the contribution to tumor grade estimation, the relationship with other tissue densities will be investigated again. To conduct such advanced analyses, we need to increase the size of the dataset.

In the Multidisciplinary Computational Anatomy project, a multi-modal analysis is one of the key methodologies to understand the human body over various axes defined in this project. Our achievement indicated the possibility that analysis by using ultrasonic and pathological images. Although further improvements are still required depending on the analysis purpose, the concepts of the modality conversion and the property analysis might be applicable and transferable for other researches.

References

1. Ban S, Min E, Baek S, Kwon HM, Popescu G, Jung W. Optical properties of acute kidney injury measured by quantitative phase imaging. Biomed Opt Express. 2018;9(3):921–32.
2. Nandy S, Mostafa A, Kumavor PD, Sanders M, Brewer M, Zhu Q. Characterizing optical properties and spatial heterogeneity of human ovarian tissue using spatial frequency domain imaging. Biomed Optics. 2016;21(10):101402-1–8.
3. Rohrbach D, Jakob A, Lloyd HO, Tretbar SH, Silberman RH, Mamou J. A novel quantitative 500-MHz acoustic microscopy system for ophthalmologic tissues. IEEE Trans Biomed Eng. 2017;64(3):715–24.
4. Choe AS, Gao Y, Li X, Compthon KB, Stepniewska I, Anderson AW. Accuracy of image registration between MRI and light microscopy in the ex vivo brain. Magn Reson Imaging. 2011;29(5):683–92.
5. Goubran M, Ribaupirre S, Hammond RR, Currie C, Burneo JG, Parrent AG, Peters TM, Khan AR. Registration of in-vivo to ex-vivo MRI of surgically resected specimens: a pipeline for histology to in-vivo registration. J Neurosci Method. 2015;241:53–65.
6. Elyas E, Papaevangelou E, Alles EJ, Erler JT, Cox TR, Robinson SP, Bamber JC. Correlation of ultrasound shear wave elastography with pathological analysis in a xenografic tumour model. Sci Rep. 2017;7(1):165.
7. Schalk SG, Postema A, Saidov TA, Demi L, Smeenge M, Rosette JJ, Wijkstra H, Mischi M. 3D surface-based registration of ultrasound and histology in prostate cancer imaging. Comput Med Imaging Graph. 2016;47:29–39.
8. Kobayashi K, Yoshida S, Saijo Y, Hozumi N. Acoustic impedance microscopy for biological tissue characterization. Ultrasonics. 2014;54(7):1922–8.
9. Ogawa T, Yoshida K, Kashio S, Ohnishi T, Haneishi H, Yamaguchi T. Multi-scale speed of sound analysis by comparing of histological image and ultrasound microscopic images at multiple frequencies. Proceedings of International Forum on Medical Imaging in Asia. 2019;11050:163. https://doi.org/10.1117/12.2521637.
10. Alcantarilla PF, Nuevo J, Bartoli A. Fast explicit diffusion for accelerated features in nonlinear scale spaces. Proceedings of British Machine Vision Conference. 2013:13.1–13.11.
11. Fischler MA, Bolles RC. Random sample consensus: a paradigm for model fitting with applications to image analysis and automated cartography. Commun ACM. 1981;24(6):381–95.
12. Lawrence S, Giles CL, Tsoi AC, Back AD. Face recognition: a convolutional neural-network approach. IEEE Trans Neural Netw. 1997;8(1):98–113.
13. Cui Y, Zhang G, Liu Z, Xiong Z, Hu J. A deep learning algorithm for one-step contour aware nuclei segmentation of histopathological images. Med Biol Eng Comput. 2019;57:2027–43.

Brain MRI Image Analysis Technologies and its Application to Medical Image Analysis of Alzheimer's Diseases

19

Koichi Ito

Abstract

Statistical analysis using large-scale brain magnetic resonance (MR) image databases has observed that brain tissues have presented age-related morphological changes. This result indicated that the age of a subject could be estimated from his/her brain MR image by evaluating morphological changes in healthy aging. We explore brain local features, which are useful for analyzing brain MR images. The brain local features are defined by volumes of brain tissues parcellated into local regions defined by the automated anatomical labeling atlas. Age is estimated by the machine learning approach with brain local features extracted from T1-weighted MR images. In addition, we consider using the convolutional neural network (CNN) to extract brain features, where any medical knowledge is not required to define local features. We evaluate the performance of the proposed approaches using large-scale MR image databases. We also consider applying the above approaches to identify Alzheimer's disease from brain MR images.

Keywords

Magnetic resonance · Brain · T1-weigted image · Age estimation · Feature extraction · Alzheimer's disease · Convolutional neural network

19.1 Introduction

Morphological changes in the human brain have followed a specific pattern of growth and atrophy in the process of brain development and healthy aging. Statistical analysis using magnetic resonance (MR) images such as T1-weighted images has demonstrated that age-related changes are found in gray matter (GM) volume, white matter (WM) volume and cerebrospinal fluid (CSF) [1–4]. GM volume monotonically decreases with age from 20s to 70s, WM volume shows small changes, and CSF monotonically increases with age from 20s to 70s in contrast with GM. Such volume changes make it possible to estimate the age of subjects from T1-weighted images. This fact might help in early identification and diagnostic support of age-related brain disorders since neurodegenerative diseases such as Alzheimer's disease (AD) have caused the accelerated aging process, i.e., accelerated brain atrophy. As mentioned above, this chapter introduces the temporal and pathological axes of the brain in multidisciplinary computational anatomy. This chapter also explores MCA-based medicine and artificial intelligence (AI) through a comparison of conventional medical image analysis-based and AI-based approaches.

19.2 Age Estimation Methods from T1-Weighted Images

Table 19.1 summarizes age estimation methods from T1-weighted images. In the early years of the study, global or local features extracted from T1-weighted images are used to estimate the age of subjects [5–9]. Lao et al. [5] defined brain morphological signature (BMS), which is a feature vector combined with GM, WM, and CSF, where each brain tissue is divided into 101 pieces based on the manually set region of interests (ROI). Neeb et al. [6] defined features based on voxel water content (VWC), reduced its dimension by principal component analysis (PCA) and estimated age using linear regression analysis of VWCs. Franke et al. [7] employed GM volume whose dimension is reduced by PCA and estimated age using relevance vector machine (RVM) [13]. Wang et al. [8] used surface features of the brain such as cortical thickness, mean curvature, Gaussian curvature, and surface area and estimated age using RVR. Kondo et al. [9] employ brain local features (BLF),

K. Ito (✉)
Graduate School of Information Sciences, Tohoku University, Sendai, Japan
e-mail: ito@aoki.ecei.tohoku.ac.jp

M. Hashizume (ed.), *Multidisciplinary Computational Anatomy*, https://doi.org/10.1007/978-981-16-4325-5_19

Table 19.1 Summary of age estimation methods from T1-weighted images

Method	Features	# of subjects (Age range [y/o])	MRI	MAE [y/o]
Lao et al. [5]	BMS	153 (56 ∼ 85)	1.5 T	–
Neeb et al. [6]	BWM	44 (23 ∼ 74)	1.5 T	6.3
Franke et al. [7]	PCA for GM	547 (19 ∼ 96)	1.5 T, 3 T	4.61
J. Wang et al. [8]	Surface of GM and WM	663 (7 ∼ 82)	1.5 T, 3 T	4.57
Kondo et al. [9]	BLF	1096 (20 ∼ 80)	0.5 T	4.24
Huang et al. [10]	2D CNN	1096 (20 ∼ 80)	0.5 T	4.0
J.H. Cole et al. [11]	3D CNN	2001 (18 ∼ 90)	1.5 T, 3 T	4.16
M. Ueda et al. [12]	3D CNN	1096 (20 ∼ 80)	0.5 T	3.2

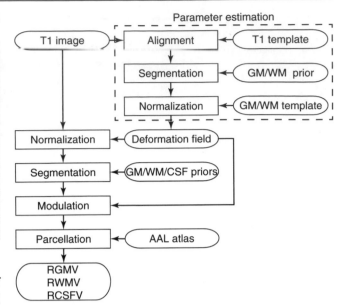

Fig. 19.1 Flow of preprocessing used in BLF extraction [9]

which are defined by regional volume from local regions of GM, WM, and CSF parcellated by the automated anatomical labeling (AAL) atlas [14, 15], and estimated age using RVR as well as other methods. In recent, a convolutional neural network (CNN) [16] was used to improve the accuracy of age estimation. Huang et al. [10] employed the 2-dimensional (2D) CNN to extract features and estimate the age from T1-weighted images automatically. Cole et al. [11] and Ueda et al. [12] used 3-dimensional (3D) CNN [17] to extract features from the whole T1-weighted images. We describe two approaches of brain MR image analysis: (i) brain local features and (ii) CNN in the following.

19.3 Brain Local Features

This section describes BLF extracted from T1-weighted images proposed in [9]. The preprocessing [18] is applied to T1-weighted images in order to extract local features of each brain tissue using statistical parametric mapping (SPM)[1] and voxel-based morphometry (VBM).[2] We empirically confirmed that SPM2 and VBM2 are effective for MR images acquired by a 0.5 T MR scanner, while SPM12 and CAT12 are effective for MR images acquired by 1.5 T and 3 T MR scanner. In the following, the process of using SPM2 and VBM2 is described. Note that the processes are almost the same in other versions of SPM and VBM. The preprocessing using SPM2 and VBM2, which is illustrated in Fig. 19.1, is described in the following. First, all the T1-weighted images are transformed into the Talairach stereotactic space by aligning each of the images to the template. The ICBM 152 template, which approximates the Talairach space [19], is used in this process. The deformation field is estimated using GM to prevent any contribution of non-brain voxels and per-

form optimal spatial normalization of brain tissues. The T1-weighted image is normalized using the estimated deformation field. Next, the normalized images are segmented into each brain tissue such as GM, WM, and CSF using the SPM2 default segmentation procedure, where a mixture model cluster analysis with a priori knowledge of the spatial distribution of tissues is used. The brain tissues are then modulated by the Jacobian determinants derived from spatial normalization to correct volume changes in spatial normalization. Finally, each brain tissue is parcellated into 1024 local regions defined by the 1024 AAL atlas [15]. The GM volume in each parcellated local region is calculated as the regional GM volume (RGMV). The regional WM volume (RWMV) and the regional CSF volume (RCSFV) are obtained in the same way. BLF is obtained as a feature vector combined with RGMV, RWMV, and RCSFV. RVR [13] is used to estimate an age using BLF.

19.4 Age Estimation Using 2D-CNN

Any technical knowledge of brain science is not required when developing brain image analysis methods using CNN-based approaches. If there are much training data, CNN-based approaches can be used to estimate the age from T1-weighted images. This section describes the early approach using 2D CNN proposed in [10]. The 2D CNN-based method consists of two steps: (i) preprocessing and (ii) age estimation using CNN. The step (i) is normalization that all the images are transformed into the standard space by aligning images into the ICBM template, which is the same as BLF as mentioned above. The step (ii) is age estimation

---

[1] https://www.fil.ion.ucl.ac.uk/spm/

[2] http://dbm.neuro.uni-jena.de/wordpress/vbm/

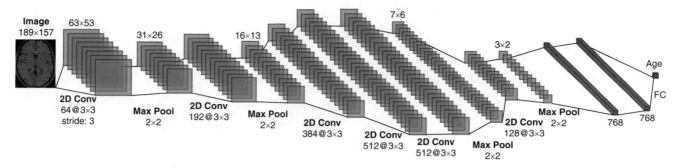

Fig. 19.2 Network architecture used in 2D CNN for age estimation [10]

using CNN, whose architecture is based on VGG-16. Figure 19.2 shows the network architecture of 2D CNN for age estimation. This architecture is inspired by VGG-16 [20], which is usually used in object recognition. The features of VGG-16 is the small size of convolution layers, i.e., 3 × 3 kernels, to reduce the number of parameters and enhance the accuracy of CNN. This architecture is designed according to the same idea. In the fully-connected layers, drop-out layers are introduced to prevent overfitting, where all the units in the fully-connected layers are randomly cut off during training. The network weights are trained to minimize the mean squared error (MSE) using the stochastic gradient descent (SGD) optimization with momentum. The loss function L is defined by

$$L = \frac{1}{N_n}\left(t_n - y_n\right)^2 + \frac{1}{2}\lambda W^2, \qquad (19.1)$$

where N is the number of the data, tn is the actual age of the n-th subject, y_n is the predicted age of the n-th subject, λ is the weight decay, and W is the weights to be trained. The loss function includes the regularization term to prevent overfitting in training. The regularization term is calculated by the weight decay and L_2 norm of all the weights. The learning rate η is determined for every epoch as follows:

$$\eta = \frac{\eta_0}{1 + \left(epoch \times decay_\eta\right)}, \qquad (19.2)$$

where η_0 is the initial value of learning rate, $decay_\eta$ is the learning rate decay, and epoch is the number of epochs. All the training data have to be augmented to improve the accuracy of age estimation methods. For example, we employ the three augmentation methods: (i) flipping, (ii) scaling, and (iii) shifting. Note that the augmentation methods should add only realistic variation to T1-weighted images. We empirically confirmed that the estimation accuracy is decreased by other augmentation methods such as rotation, noise, deformation, etc. The method (i) flips input images horizontally with a probability of 0.5. Method (ii) randomly crops input

images within [0, 10] voxel around and resizes them to the original size. Method (iii) randomly shifts input images within [−8, 8] voxels. The input for 2D CNN is a set of slice images extracted from T1-weighted images. We empirically confirmed that all the slices are not always effective for estimating age. For example, the adjacent slices have almost the same information, and upper and lower slices do not have any effective information. Therefore, the best accuracy of 2D CNN-based method is obtained by selecting the optimal slices for age estimation.

19.5 Age Estimation Using 3D CNN

The 2D CNN-based method exhibited a higher prediction accuracy than conventional methods using manually defined local features. 3D features may not be utilized in age estimation since 2D convolution was used to extract features from slice images of volume data. This section describes the recent approach using 3D CNN proposed in [12]. The 3D CNN-based method consists of two steps: (i) preprocessing and (ii) age estimation using CNN. Preprocessing has to be applied to T1-weighted images before age estimation. We empirically confirmed that preprocessing improves the accuracy of age estimation as well as 2D CNN [10]. Preprocessing used in the proposed method consists of the following three steps. First, all the T1-weighted images are aligned with the standard template using the procedure proposed by Good et al. [18] as well as the 2D CNN-based method. Next, the aligned images are resized to 95 × 79 × 7 voxels to reduce the computation time and memory usage. Finally, the voxel values are normalized to have zero mean and unit variance. Figure 19.3 illustrates a 3D-CNN architecture used for age estimation from brain T1-weighted images. This architecture has 4 convolution blocks consisting of a 3D convolutional layer (kernel size: 3 × 3 × 3, stride: (1), a 3D batch-normalization layer, a rectified linear unit (ReLU) activation layer and a max-pooling layer (kernel size: 2 × 2 × 2, stride: (2). The number of feature channels is 8, 16, 32, and 64 for each block, respectively. The last three layers are fully con-

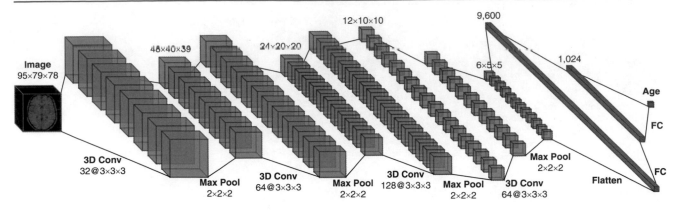

Fig. 19.3 Network architecture used in 3D CNN for age estimation [12]

nected layers to combine the feature vectors. The output is a scalar value of the predicted age. This architecture is designed to be simple, taking into account the large computation time and the large memory usage in 3D-CNN. The loss function for 3D CNN is the same as that for 2D CNN in Eq. (19.1). The three data argumentation methods are also applied to T1-weighted images.

19.6 Performance Evaluation Using a Large-Scale Dataset

This section describes performance evaluation using a large-scale T1-weighted image dataset. T1-weighted MR images collected by the Aoba Brain Imaging Project (Aoba 1) in Sendai, Japan [21] and the Tsurugaya Project (Tsurugaya 1) in Sendai, Japan are used in the experiment. T1-weighted images were taken by the same 0.5 T MR scanner (Signa contour, GE-Yokogawa Medical Systems, Tokyo) in both projects, where the image size is $256 \times 256 \times 124$ voxels. The image size is changed to $189 \times 157 \times 156$ voxels after alignment with the ICBM 152 template and then is resized to $95 \times 79 \times 78$ voxels. The subjects of both projects were all healthy and had neither present illness nor a history of neurological disease, psychiatric disease, brain tumor, or head injury. 1101 subjects aged from 20 to 80 years from the dataset are used in the experiment. We randomly select 768 subjects for the training data and also select the remaining 333 subjects for test data. We evaluate the accuracy of age estimation using the mean absolute error (MAE), the root mean square error (RMSE), and the correlation coefficient (Corr.). We evaluate the accuracy of age estimation for methods: PCA [7], BLF [9], 2D CNN [10], and 3D CNN [11, 12]. The hyperparameters for 2D and 3D CNNs are shown in Table 19.2. Table 19.3 shows a summary of experimental results. The CNN-based methods exhibit higher estimation accuracy than the handcrafted feature-based methods [7, 9].

Table 19.2 Hyperparameters for 2D CNN and 3D CNN used in training

Hyperparameter	2D CNN [10]	3D CNN [12]
Learning rate	0.0001	0.00005
Learning rate decay	0.0001	0.0001
Weight decay	0.001	0.0001
Momentum	0.9	0.9
Batch size	28	16

Table 19.3 Summary of experimental results of age estimation from healthy subjects for each method

Method	MAE [y/o]	RMSE [y/o]	Corr.
Franke et al. [7] PCA	5.56	6.95	0.92
Kondo et al. [9] BLF	4.39	5.57	0.94
Huang et al. [10] 2D CNN	4.18	5.31	0.94
Cole et al. [11] 3D CNN	3.73	4.73	0.95
Ueda et al. [12] 3D CNN	3.23	4.18	0.96

The accuracy of 3D CNN methods is much higher than that of 2D CNN [10] since the 3D CNN-based method can fully utilize volume data in age estimation.

19.7 Analysis

We analyze effective local regions in the age estimation in order to discuss the medical implication of the experimental result. We add the random noise to one local region of T1-weighted images and then estimate the age using BLF [9], 2D CNN [10], and 3D CNN [12]. If the MAE increases, the effectiveness of the masked local region is high. If the MAE decreases, its effectiveness is low. The masked region is determined by the 90 AAL atlas [14] or the 1024 AAL atlas [15]. The use of the above procedure makes it possible to evaluate the effective local regions in the CNN-based approach although, in general, it is difficult for CNN-based

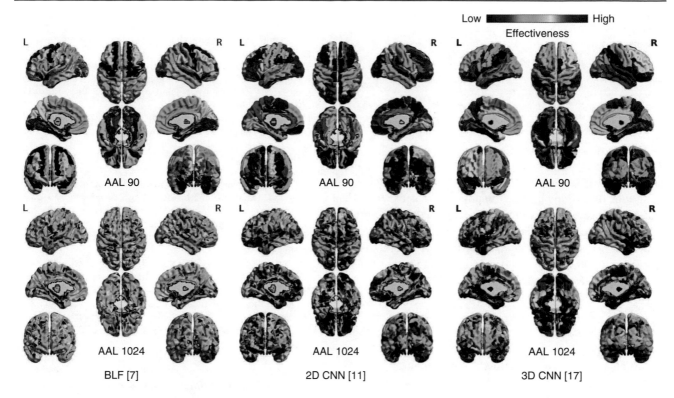

Fig. 19.4 Effective local regions in age estimation for each method

approaches to explain which local features are effective. Figure 19.4 shows effective local regions for each method. Effective local regions are concentrated in the frontal association area, the Wernicke's area, the angular gyrus and the primary motor cortex. The frontal association area takes on the function of behavioral decisions and working memory. The functions of Wernicke's area include language comprehension, semantic processing, language recognition, and language interpretation. The angular gyrus takes on the functions related to language, spatial cognition, memory retrieval, attention, and theory of mind. The above regions take on the high-order function compared with other regions and hence are impaired with aging. The primary motor cortex exhibits high effectiveness in age estimation, although this region takes on the low-order function. The location of the primary motor cortex is close to that of the central sulcus. The central sulcus becomes dilated by atrophying the frontal area, and the primary motor cortex also becomes dilated, which looks like atrophy. On the other hand, ineffective local regions are concentrated in the parietal lobe and the occipital lobe. These regions take on the low-order function and hence are robust against aging. The above result corresponds to the statistical analysis of age-related morphological changes [4]. The proposed method shows a clearer trend than other methods, although all the methods indicate almost the same trend in the statistical analysis.

19.8   Application to AD Identification

In this section, we explore the performance of age estimation methods in identifying AD using the database released by the Alzheimer's Disease Neuroimaging Initiative (ADNI),[3] which includes patients with AD. We evaluate the effectiveness of age estimation methods in supporting the diagnosis of Alzheimer's disease by estimating the ages of Alzheimer's patients and healthy individuals and comparing estimation errors. MRI images in the ADNI datasets are divided into healthy subjects (Normal or Control: CN), mild cognitive impairment (MCI), and Alzheimer's disease (AD) based on the cognitive assessment. We use T1-weighted images with $192 \times 192 \times 160$ voxels in ADNI1, which are acquired by 1.5 T MR scanners. Randomly selected 462 CNs are used as training data, and the remaining CN, MCI, and AD are used as test data. We evaluate the average of MAE and standard deviation (SD) for actual and estimated age. Table 19.4 shows the summary of experimental results for ADNI. In all the methods, the estimation errors of MCI and AD were larger than those of CN. The age of CN is estimated by an error of ±10 years, while MCI and AD have a large error. In particular, there is a large margin of error for MCI and AD subjects with respect to the actual age of subjects under 70 years of age.

---

[3]http://adni.loni.usc.edu/

Table 19.4 Summary of experimental results for ADNI, where the value indicates MAE SD for each method

Method	CN	MCI	AD
Franke et al. [7]	1.90 ± 2.37	5.57 ± 6.66	5.67 ± 6.99
Kondo et al. [9]	1.25 ± 1.76	4.96 ± 6.26	5.32 ± 6.43
Huang et al. [10]	1.68 ± 2.24	5.14 ± 6.21	5.50 ± 6.30
Cole et al. [11]	1.65 ± 2.23	5.02 ± 6.04	5.68 ± 6.13
Ueda et al. [12]	2.16 ± 2.77	5.10 ± 6.21	5.77 ± 6.30

This indicates that the changes in brain morphology in MCI and AD are not caused by normal aging. On the other hand, there was no difference in estimation error between MCI and AD subjects and CN in the data over the 80s. The brain atrophy of the elderly is more pronounced with normal aging, and it is difficult to distinguish MCI and AD from CN in the brain. Therefore, it may lead to the early detection of brain diseases at a relatively young age.

19.9   Conclusion

In this chapter, we have described the potential applications of age estimation and AD identification from T1-weighted images as examples of brain MR image analysis, which is a part of the temporal and pathological axes in MCA model. With the rapid development of deep learning, MR image analysis is shifting from handcrafted features to features automatically extracted by CNN. Since the availability of enormous amounts of training data is a critical factor in the performance of CNNs, the most important issue is how to collect such data. In addition, how to collect data on a small number of diseases is also a major challenge for medical imaging applications. In the field of machine learning, several methods have been investigated, such as a learning method using only normal data, a learning method using a small amount of disease data, and a data augmentation method to increase the number of data, and these methods are also expected to be useful in medical image analysis. In the future, the interdisciplinary integration of machine learning and medical image analysis will be essential, and great progress can be expected.

References

1. Jernigan T, Archibald S, Fennema-Notestine C, Gamst A, Stout J, Bonner J, Hesselink J. Effects of age on tissues and regions of the cerebrum and cerebellum. Neurobiol Aging. 2001;22: 581–94.
2. Allen J, Bruss J, Brown C, Damasio H. Normal neuroanatomical variation due to age: the major lobes and a parcellation of the temporal region. Neurobiol Aging. 2005;26:1245–60.
3. Terribilli D, Schaufelberger M, Duran F, Zanetti M, Curiati P, Menezes P, Scazufca M, Amaro E Jr, Leite C, Busatto G. Age-related gray matter volume changes in the brain during non-elderly adulthood. Neurobiol Aging. 2011;32:354–68.
4. Taki Y, Thyreau B, Kinomura S, Sato K, Goto R, Kawashima R, Fukuda H. Corre- lations among brain gray matter volumes, age, gender, and hemisphere in healthy individuals. PLoS One. 2011;6(7):e22734–1–e22734–13.
5. Lao Z, Shen D, Xue Z, Karacali B, Resnick SM, Davatzikos C. Morphological classi- fication of brains via high-dimensional shape transformations and machine learning methods. NeuroImage. 2004;21(1):46–57.
6. Neeb H, Zilles K, Shah NJ. Fully-automated detection of cerebral water content changes: study of age- and gender-related H2O patterns with quantitative MRI. NeuroImage. 2006;29(3):910–22.
7. Franke K, Ziegler G, Kloppel S, Gaser C. Estimating the age of healthy subjects from T1-weighted MRI scans using kernel methods: exploring the influence of various parameters. NeuroImage. 2010;50(3):883–92.
8. Wang J, Li W, Miao W, Dai D, Hua J, He H. Age estimation using cortical surface pattern combining thickness with curvatures. Med Biol Eng Comput. 2014;52(4):331–41.
9. Kondo C, Ito K, Wu K, Sato K, Taki Y, Fukuda H, Aoki T. An age estimation method using brain local features for T1-weighted images. In: Proc. Int'l Conf. IEEE Eng. Med. Biol. Soc; 2015. p. 666–9.
10. Huang TW, Chen HT, Fujimoto R, Ito K, Wu K, Sato K, Taki Y, Fukuda H, Aoki T. Age estimation from brain MRI images using deep learning. In: Proc. Int'l Symp. Biomed. Imaging; 2017. p. 849–52.
11. Cole JH, Poudel RPK, Tsagkrasoulis D, Caan MWA, Steves C, Spector TD, Mon-tana G. Predicting brain age with deep learning from raw imaging data results in a reliable and heritable biomarker. NeuroImage. 2017;163:115–24.
12. Ueda M, Ito K, Wu K, Sato K, Taki Y, Fukuda H, Aoki T. An age estimation method using 3D-CNN from brain MRI images. In: Proc. Int'l Symp. Biomed. Imaging; 2019. p. 380–3.
13. Tipping ME. Sparse Bayesian learning and the relevance vector machine. J Machine Learn- ing Research. 2001;1:211–44.
14. Tzourio-Mazoyer N, Landeau B, Papathanassiou D, Crivello F, Etard O, Delcroix N, Mazoyer B, Joliot M. Automated anatomical labeling of activations in SPM using a macro- scopic anatomical parcellation of the MNI single-subject brain. NeuroImage. 2002;15(1):273–89.
15. Zalesky A, Fornitoa A, Hardinga I, Cocchia L, Yucela M, Pantelisa C, Bullmorect E. Whole-brain anatomical networks: does the choice of nodes matter? NeuroImage. 2010;50(3):970–83.
16. Goodfellow I, Bengio Y, Courville A. Deep learning. The MIT Press; 2016.
17. Ji S, Xu W, Yang M, Yu K. 3D convolutional neural networks for human action recognition. IEEE Trans Pattern Anal Mech Intell. 2013;35(1):221–3.
18. Good CD, Johnsrude IS, Ashburner J, Henson RNA, Friston KJ. A voxel-based morphometric study of ageing in 465 normal adult human brains. NeuroImage. 2001;14(1):21–36.
19. Talairach J, Tournoux P. Co-planar stereotaxic atlas of the human brain. George Thieme Verlag. 1988;
20. Simonyan, K., Zisserman, A.: Very deep convolutional networks for large-scale image recog-nition. CoRR abs/1409.1556 (2014).
21. Sato K, Fukuda H, Kawashima R. Neuroanatomical database of normal Japanese brains. Neural Netw. 2003;16(9):1301–10.

A Computer-Aided Support System for Deep Brain Stimulation by Multidisciplinary Brain Atlas Database

Ken'ichi Morooka, Shoko Miyauchi, and Yasushi Miyagi

Abstract

Our research group has been constructing a 3D digital atlas of a Japanese brain (Fukuda et al, Neurosci Res. 67:260–265, 2010, Miyauchi et al. Proc. Computer Assisted Radiology and Surgery. 232, 2017). The purpose of our project is to develop a support system for Deep Brain Stimulation using the brain atlas. Practically, the support system estimates a patient brain atlas by using a multidisciplinary brain atlas database based on our 3D brain atlas. To achieve this, we have been developing two fundamental techniques. The first is to deform our brain atlas nonlinearly to estimate a reliable atlas of a patient with acceptable accuracy for practical medical cases. The second is to determine the correspondences among the atlases of different individuals to construct the brain atlas database.

Keywords

Brain atlas · Deep brain stimulation

20.1 Introduction

Stereotactic neurosurgery is a minimally invasive surgery in which an electrode is inserted deeply into the brain from an arbitrary position on the skull of a patient, and an electrical stimulus was placed to a target neural structure through the electrode. Recent researches have reported the high therapeutic benefit of stereotactic neurosurgery on intractable movement disorder, epilepsy, and central pain. Therefore, the number of stereotactic neurosurgery cases has been increasing every year. Moreover, electrical neurostimulation therapies, most of which are performed in stereotactic neurosurgery, have been applied to various kinds of neural circuit disorders, such as anorexia or bulimia nervosa, major depression and obsessive-compulsive disorder.

In order to guide the electrode to the target neural structure of a patient, classical human brain atlases are used to identify the patient brain structure. There are two famous classical brain atlases widely used for stereotactic neurosurgery: the atlases made by Schaltenbrand and Wahren [1], and Talairach and Tournoux [2]. Each of the atlases is derived from the brain preparations of some Westerners. In the atlas construction, the preparations are divided into left and right hemispheres, and each hemisphere is serially sliced into one of three planes: coronal, horizontal, or sagittal planes. For each slice, the contours of the brain surface and neural structures are manually extracted from the slice. Therefore, the brain atlas is the set of the extracted contours from all the slices.

Each person has a different brain structure from the others. Therefore, when the brain atlas is applied to estimate the patient brain structure, the atlas needs to be deformed to fit the brain structure of the individual patient to be treated. The atlas deformation is made by simple scaling along each axis to make the anatomical feature points of the brain atlas be fitted to their corresponding points of the patient's brain.

However, the classical atlases contain the following problems when applying to real clinical cases. Coronal, horizontal, and sagittal slices of the classical atlases are obtained from different persons. As mentioned above, there are differences in the brain structures of individuals. Therefore, when the 3D brain atlas is constructed from all its slices, the coordinates of the same structure are often in conflict with those from the different slices. Moreover, the brain morphology varies significantly in individual, age, gender, and probably

K. Morooka (✉)
Graduate School of Natural Science and Technology, Okayama University, Okayama, Japan
e-mail: morooka@okayama-u.ac.jp

S. Miyauchi
Graduate School of Information Science and Electrical Engineering, Kyushu University, Fukuoka, Japan

Y. Miyagi
Department of Functional Neurosurgery, Fukuoka Mirai Hospital, Fukuoka, Japan

race [3]. Owing to these factors, the atlas deformation is regarded as the complex and nonlinear mapping of the atlas onto the patient brain make. However, since the patient brain structure is estimated by changing simply the scale of the atlas, the estimation accuracy is not acceptable for real stereotactic neurosurgery.

On the other hand, the recent development of medical imaging devices, such as PET scan and fMRI, provides the structure and activity of a human brain. However, our knowledge about morphological details of the human brain has not been advanced so much from observations by Brodmann one hundred years ago, who discriminated some forty-five discrete areas in the human cerebral cortex and established a complete map of the cortex. One of the reasons for this situation is that human brains are too large in size to be subjected to sophisticated studies as have been done in rodent brains. Moreover, few clinical equipments can deal with a whole brain of human. Therefore, most of the previous basic researches has focused on animal brains, whereas clinical researches for revealing anatomo-functional interrelation in human brains still rely on classical knowledge on Brodmann areas.

In the last few decades, there are some projects to create digital databases for human anatomy, including the visible human project [4] and the Korean project [5]. These projects use 1-meter-large microtome and a freezer in which the box with length 2 [m] can be stored. However, preparing such expensive equipment and research environments are not realistic in general laboratories. As the more serious problem, the research using the whole body of the cadaver for the above purpose is very difficult in Japan because of the legal, social, and ethical limitations specific to Japanese culture.

We have been studying the construction of a digital Japanese brain atlas [6, 7]. This is the first attempt to create the complete brain atlas of Japanese. Figure 20.1 shows our digital brain atlas constructed from a brain of a cadaver, 89-year-old male Japanese. The whole surface of the brain was scanned by a 3D digitizer. Using the specially designed brain-cutting machine, the brain embedded in agar was divided into serial blocks, each 1 cm in thickness. Moreover, each block is partitioned into three sub-blocks. The subblocks were further sliced into serial 100[μm]-thick sections with a vibrating microslicer. After staining the sections with Nissl-and myelin-stainings, experienced neurosurgeons traced manually the contours of the section and the internal neural structures from all the section images. Each contour model includes its structure name, the set of the 3D points on the contour, and the connection of the points. Our digital brain atlas is generated by collecting all the contours in the sub-blocks as shown in Fig. 20.1.

The purpose of our project is to construct a support system using our digital brain atlas for Deep Brain Stimulation

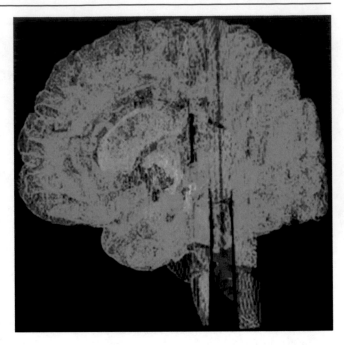

Fig. 20.1 Our digital brain atlas generated from a Japanese cadaver

(DBS), one of the stereotactic neurosurgeries. To achieve this, we have been developing two fundamental techniques. The first is to deform our brain atlas nonlinearly to estimate a reliable atlas of a patient with acceptable accuracy for practical medical cases. The second is to determine the correspondences among the atlases of different individuals to construct the brain atlas database.

20.2 Patient Brain Atlas Estimation by Deep Neural Networks

We have been developing a new method for deforming the brain atlas using multiple deep neural networks (DNNs) [8] (Fig. 20.2). Each DNN, called a nodal behavior estimator (NBE), estimates the behavior of one node in the mesh model of the atlas. By connecting many DNNs, our method estimates a whole deformation of the atlas.

The NBE of a target node trains by observing many patterns of the node behavior. Practically, the input of the NBE is the displacement and stress of the node and its neighbor nodes and the relative velocities and accelerations between the node and its neighbor nodes. When the data of the node at time t is given, the NBE outputs the displacement and stress of the node at time (t-1). In the NBE construction, stacked autoencoder is applied to obtain the initial weights between two sequential layers. Moreover, the NBE is finetuned by using backpropagation. Finally, the system for estimating the whole deformation of the atlas is generated by integrating all the NBE of the nodes in the mesh model of the atlas.

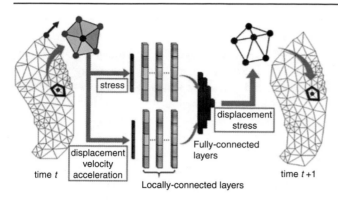

Fig. 20.2 Concept of the brain atlas deformation by deep neural networks

Similar to our method, there are methods for simulating the object deformation. Among them, the finite element method (FEM) is one of well-known techniques for simulating the behavior of real objects accurately. To estimate more accurate deformations of an object, FEM needs the object mesh model with many nodes. However, the increase in the number of nodes leads to a heavy computational burden. Several studies tackled the problem in FEM by replacing the nonlinear constitutive equations in FEM with their linear approximated equations. However, the use of the approximated computations sacrificed the accuracy of estimating the deformation. Hence, in conventional techniques for object deformation simulation, there is a trade-off between the estimating accuracy and the computational cost.

Unlike the conventional methods for speed up FEM, the computations in DNNs are the weighted sum of the simple nonlinear functions. Moreover, since the behavior of each node is estimated independent of other NBEs, our method enables to efficiently estimate the atlas deformation by simulating the behaviors of all the NBEs simultaneously while keeping the accuracy of estimating the tissue deformation.

To validate the applicability of our proposed method, we made experiments using a triangular mesh model of putamen that is a neural structure of the brain atlas. The putamen model (29×59 [mm]) consists 59 internal nodes and 42 nodes on the contour of the model. We simulate the deformations of the putamen when 48 types of external forces act on each active node. Therefore, for each node, 33,744 (= $37 \times 48 \times (N - 1)$) nodal behavior patterns are obtained. Moreover, we use as test data another 7030 (= $37 \times 10 \times (N - 1)$) patterns generated by using 10 force vectors. Here, the forces used in the test data generation are chosen randomly from the range of the external forces used in the generation of the behavior patterns of the putamen.

The proposed NBE is evaluated by the difference between the deformation results estimated by the proposed NBE and the original FEM. Here, for each NBE, the difference to the original FEM is measured by an L2 distance between the displacements (or the stresses) estimated by the NBE and the FEM when each input data is given. The differences for the displacement and stress are 3.2 ± 1.4 [mm $\times 10^{-2}$] and 2.8 ± 1.2 [N/mm^2], respectively. From these experimental results, the proposed method can estimate the node behavior accurately compared with FEM.

20.3 Brain Atlas Database Construction

One of the main problems for constructing the database of brain atlases is to determine the correspondence among the atlas models. Here, a human brain has complex internal neural structures, including lateral ventricle and thalamus. Therefore, we need to establish the two factors: (1) the correspondence among both the surfaces and internal structures of brains of different individuals; (2) the spatial relationships among the internal structures.

To solve these problems, we have proposed a volumetric Self-organizing Deformable Model (vSDM) [9]. vSDM maps the brain volume model onto its target volume so that the surface of the brain is fitted to that of the target volume while each internal structure of the organ is deformed to fit its corresponding local target. When the vSDM is applied to all the brain models, all the brain volume models can be represented with a unified structure of the common target volume. The use of such brain models makes it easy to find the correspondences among the brain models. Moreover, vSDM preserves geometrical properties of the original brain volume model before and after the mapping. These characteristics allow to obtain the reliable correspondence among the brain volume models easily. Also, vSDM includes a process for correcting inverted tetrahedral by moving vertices without changing the mesh structure of the brain volume model. From this process, when shape deformation caused by the control of the mapping positions is small, the mapped model by the vSDM has no inverted tetrahedra. Owing to these factors, the use of our vSDM mapping results enables us to recover models with acceptable accuracy for determining the correspondences among the models.

20.3.1 Volumetric Self-Organizing Deformable Mode (vSDM)

In vSDM, a tetrahedral volume model M_v of a human brain is used as an initial vSDM. The external surface of M_v is regarded as the outer model surface (OMS) of the vSDM. vSDM contains the volume models of the internal structures inside the OMS. When some of the internal structures are selected to analyze their shapes, the surfaces of the selected internal structures are used as the inner model surfaces (IMSs) of the vSDM. As an example of the initial vSDM, a brain volume model is shown in Fig. 20.3 (a-c). The brain volume model consists of brain surface (the pink part in Fig. 20.3 c) and two internal structures: the set of lateral ventricle and third ventricle (the blue part), and fourth ventricle (the green part). In this example, the brain surface is used as the OMS while the surfaces of the set of lateral ventricle and third ventricle, and the fourth ventricle are used as the IMSs.

The vertices in M_v are classified into three types. The vertices on the OMS and IMSs, respectively, are named as the OMS and IMS vertices, while the rest vertices are regarded as the inner vertices. For each vertex except the OMS vertices, its 1-ball region is the set of the tetrahedra containing the vertex.

When vSDM is mapped onto a target volume, the external surface of the target volume, called the outer target surface (OTS), is the mapping destination of the OMS. The target volume includes inner targets within the OTS. Each IMS is mapped onto its corresponding inner target surface (ITS). Here, the position of the initial vSDM is determined so that the OMS and the OTS overlap each other as large as possible. As an example, a target volume used in our experiment is a volume model with an average shape of brain surfaces (the light blue region in Fig. 20.3 (d and e)). The target volume contains two inner targets: a volume model with an average shape of the sets of lateral ventricle and third ventricle (the yellow regions in Fig. 20.3 e) and an ellipsoid (the orange regions in Fig. 20.3 e). In this case, the OTS is the average surface of the brain surfaces, while the ITSs are the ellipsoidal surface and the average surface of the sets of lateral ventricle and third ventricle.

The algorithm of vSDM deformation is as follows.

1. Map the OMS vertices of the initial vSDM onto the OTS by step S1, S2, and S4 of mSDM deformation [11].

2. For each vertex except the OMS vertices, move the vertex toward the centroid of its polyhedron. Here, the polyhedron of a vertex v is generated by removing from its 1-ball region the vertex v and the edges connecting with v. This movement process is repeated until no vertex is moved.
3. Correct inverted tetrahedra.
4. Perform an angle- and/or volume-preserving mapping by moving the vertices except for the OMS vertices.
5. For each IMS,
   • Determine the position and pose of the corresponding ITS by using the mapped IMS vertices.
   • Map the IMS vertices onto the ITS by step S1, S2, and S4 of mSDM deformation.
6. Move each inner vertex toward the centroid of its polyhedron. This movement process is repeated until all inner vertices are not moved.
7. Correct inverted tetrahedra.
8. Perform an angle- and/or volume-preserving mapping by moving only the inner vertices while fixing the OMS and IMS vertices.

We made mapping experiments by using the volume models of six patient brains (average age: 42.2) shown in Fig. 20.3 (a-c). To construct each brain model, an experienced neurosurgeon manually extracts the contours of OMS and IMSs from medical images. A common target volume is shown in Fig. 20.3 (d-e) has 15,000 points on the OTS and ITSs averagely. The position and pose of each ITS are determined manually based on the positions of its corresponding IMS vertices. The final mapping results are shown in Fig. 20.3 f, show the mapped brain volume models onto the target volume. Here, in this figure, the mapping results are represented with the shape of the target volume.

The mapping results are evaluated by two criteria. The first is the number of inverted tetrahedra in the final resulting

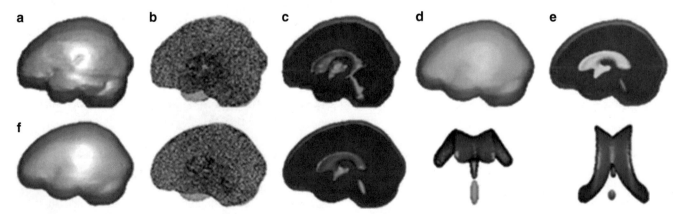

Fig. 20.3 (a) The surface of the brain volume model; (b) The brain volume model cut by a virtual plane for the interior visualization; (c) A cross-section of the OMS (pink), IMS1 (blue), and LMS2 (green); (d) The surface of the target volume; (e) A cross-section of the OTS (light blue), ITS1 (yellow), and ITS2 (orange); (f) Final mapping result of a brain volume model: from left, its OMS, its cross-section, a cross-section of its OMS and IMSs, and IMSs viewed from front and top (published from [10])

models. From our mapping results, the number of the inverted tetrahedra is 27 at most, although the mapping positions of lateral ventricles with complicated shapes are controlled. Moreover, one mapping result has no inverted tetrahedra.

The second criterion is the mapping accuracy of OMS and IMSs. The accuracy is measured by the distance between the surfaces and internal structures of the brain model and their target surfaces. When the distance is closed to zero, the surface fits its target surface completely. The average value for IMS1s and IMS2s are 0.3 and 0.2 [mm], respectively. Here, the mean width, height, and length of ITS2s are 39.9 [mm] $\times$ 14.4 [mm] $\times$ 22 [mm]. Compared with the sizes of the ITS2s, the values for IMS1s and IMS2s are small enough. Therefore, vSDM mapping enables to map of brain volume models to their target volumes while fitting the OMS and IMSs to their corresponding OTS and ITSs.

20.4   Conclusion

We proposed two fundamental techniques for constructing our multidisciplinary brain atlas database. The first is to deform our brain atlas nonlinearly to estimate a reliable atlas of a patient with acceptable accuracy for practical medical cases. The second is to determine the correspondences among the atlases of different individuals to construct the brain atlas database. Now we have been developing the two techniques, and our research project can be contributed to the development of human brain analysis and the establishment of new neurosurgeries.

Now, we have been developing our multidisciplinary brain atlas database. The use of the database helps neurosurgeons to understand the 3D structure of a patient brain. Therefore, our brain atlas database is a powerful and useful tool for safe and accurate neurosurgery, including DBS. Our current brain atlas consists of the 3D surface anatomy and internal structures of a human brain. Moreover, other information about the human brain, such as its functional connectivity and genomic information, can be incorporated into our brain atlas. This means the brain atlas database is used as a platform for various researches of the human brain. Accordingly, the brain atlas database can be contributed to the development of human brain science.

References

1. Schaltenbrand G, Wahren W. Atlas for Stereotaxy of the human brain. Stuttgart: Thieme; 1977.
2. Talairach J, Tournoux P. Co-planar stereotactic atlas of the human brain. Stuttgart: Thieme; 1988.
3. Kouchi M. Secular changes in the Japanese head form viewed from somatometric data. Anthropol Sci. 2004;112:41–52. https://doi.org/10.1537/ase.00071.
4. Spitzer VM, Ackerman MJ, Scherzinger AL, Whitlock D. The visible human male: a technical report. J AmericanMed Inform Assoc. 1996;3:118–30. https://doi.org/10.1136/jamia.1996.96236280.
5. Park JS, Chung MS, Hwang SB, Lee YS, Har DH, Park HS. Visible Korean human: improved serially sectioned images of the entire body. IEEE Trans Medical Imaging. 2005;24:352–60. https://doi.org/10.1109/TMI.2004.842454.
6. Fukuda T, Morooka K, Miyagi Y. A simple but accurate method for histological reconstruction of the large-sized brain tissue of the human that is applicable to construction of digitized brain database. Neurosci Res. 2010;67:260–5. https://doi.org/10.1016/j.neures.2010.03.005.
7. Miyauchi S, Morooka K, Sasaki S, Tsuji T, Miyagi Y, Fukuda T, Samura K, Kurazume R. Brain atlas construction by non-rigid registration of brain block models. Proc Computer Assisted Radiol Surg 2017.232.
8. Kobayashi K, Morooka K, Miyagi Y, Fukuda T, Tsuji T, Kurazume R, Samura K. Simulation of deforming human tissue by multiple deep neural networks. Proc the Int Forum Med Imag Asia. 2017:187–98.
9. Miyauchi S, Morooka K, Tsuji T, Miyagi Y, Fukuda T, Kurazume R. Volume representation of Parenchymatous organs by volumetric self-organizing deformable model. Int Workshop Spectral Shape Analy Med Imag. 2016:39–50.
10. Miyauchi S, Morooka K, Miyagi Y, Fukuda T, Kurazume R. Volumetric brain volume mapping for constructing volumetric statistical shape model. Int Forum Med Imag Asia 2019, 2019. https://doi.org/10.1117/12.2519819.
11. Miyauchi S, Morooka K, Tsuji T, Miyagi Y, Fukuda T, Kurazume R. Fast modified self-organizing deformable model: geometrical feature-preserving mapping of organ models onto target surfaces with various shapes and topologies. Comput Methods Prog Biomed. 2018;157:237–50. https://doi.org/10.1016/j.cmpb.2018.01.028.

Integrating Bio-metabolism and Structural Changes for the Diagnosis of Dementia

21

Yuichi Kimura

Abstract

We introduce some aspects to develop an algorithm to diagnose dementia diseases, such as Alzheimer's disease. First, we present specific points of dementia. Dementia follows a long period of more than 20 years and is irreversible. Therefore, we need to detect the onset of dementia before any occurrence of neurological symptoms. To satisfy this aspect, functional information derived from PET and anatomical or structural information derived from MRI are incorporated for feature values to diagnose dementia. Recently, Alzheimer's disease, which is the most common dementia disease, has been the target, and PET with an amyloid β probe is useful for diagnosing Alzheimer's disease. We focus on PET as a source of the features. MRI is also an attractive modality, and we describe some data preprocessing procedures for MRI. A statistical modeling or AI-based algorithm is the key to realizing CAD for dementia, but the issue is how to collect an adequate number of training images. We discuss this point in which the generative adversarial network algorithm is applied.

Keywords

Dementia · Alzheimer's disease · PET · MRI
CycleGAN

21.1 Introduction

Dementia is a degenerative disease, and it causes loss of cognitive function. Because of improvements in medical services, particularly in developing countries, patients and potential people who have dementia are increasing. WHO estimated the number of patients to be 150 hundred million worldwide before 2050 [1]. Dementia has an extremely negative impact on human society. Hence, we need to develop a method to diagnose dementia. Major diseases presenting with dementia are Alzheimer's disease; more than 50% of patients with dementia have Alzheimer's disease. The disease-modifying drug aducanumab [2] is under development and is being put on the market. I introduce some aspects to diagnose dementia diseases, including Alzheimer's disease.

21.2 Specific Points to Diagnose Dementia

Dementia follows more than 20 years and is irreversible. In the case of Alzheimer's disease, deposition of amyloid β causes Alzheimer's disease [3], as presented in Fig. 21.1; amyloid β demolishes nerve cells. In typical cases, the deposition starts in the 50s in potential patients with Alzheimer's disease. During the first 10 years, no significant neurological symptoms were observed. In the next 10 years, a non-negligible number of nerve cells lost their functionality because of continued amyloid β deposition, and they suffered from mild cognitive infarction, followed by Alzheimer's disease. It is not adequate to diagnose dementia after presenting any neurological symptoms, and we should detect functional and structural changes in the brain; they may be slightly before the occurrence of symptoms.

Various modalities can be used to diagnose dementia. MRI provides an image of the anatomical structure with fine resolution, such as 1-mm^3 regions. Dementia causes cerebral atrophy, and MRI can be visualized clearly owing to its resolution. PET is a powerful tool for diagnosis because it can visualize various brain functionalities quantitatively. Two radiopharma-

Y. Kimura (✉)
Department of Computational Systems Biology,
Faculty of Biology-Oriented Science and Technology,
and Cyber Informatics Research Institute, Kindai University,
Osaka, Japan
e-mail: ukimura@ieee.org

M. Hashizume (ed.), *Multidisciplinary Computational Anatomy*, https://doi.org/10.1007/978-981-16-4325-5_21

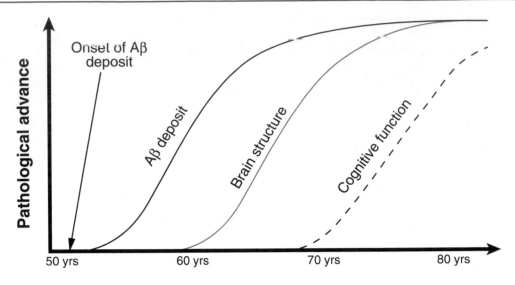

Fig. 21.1 Degenerative dementia disease, such as Alzheimer's disease (AD), follows a long time history. The horizontal axis denotes the age, and vertical axis represents the advancement of the pathological condition. Amyloid β is the agent of AD, and around 10 years after the onset of the deposition, the brain structure begins to deteriorate. After the next 10 years, dementia symptoms appear. Dementia is irreversible, and therefore, we need to detect the onset when no significant dementia symptoms are observed. The figure is the modified version of Fig. 2 [3]

ceuticals are applied with PET for dementia: ^{18}F-FDG and amyloid probes. FDG is a fluorine-labeled deoxyglucose that is an analog of glucose. It is trapped in a brain cell in its phosphorylated form [4], and the voxel where high glucose metabolism is observed presents high radioactivity. Therefore, we acquired an image of glucose metabolism that reflects neuronal activity via FDG-PET. Amyloid probes can bind to amyloid β molecules specifically, and its spatial distribution can be visualized quantitatively or semi-quantitatively [5]. Because amyloid β causes Alzheimer's disease, amyloid PET is a powerful tool to diagnose it. If high radioactivity is observed in some cerebral regions, amyloid β deposition begins and the patient may suffers from Alzheimer's disease.

To diagnose dementia, we need to freely use these modalities to detect the onset of dementia in very early phases.

21.3 Feature Values in a Functional Image

We can derive functional features from PET that can be utilized to diagnose dementia, such as neuronal activity and the amount of depositing amyloid β. We should consider the stability and noise in the PET voxel value.

Dementia follows a long period of more than 20 years, and we need to monitor its progression periodically, e.g., every year. Stability along the time axis is important for its application in diagnosing dementia. This issue was discussed in [6], in which the authors concluded that a binding potential [7] is better than SUVR. The binding potential is a quantitative measure of the density of amyloid β, which is usually used in PET neuroreceptor imaging [8]; however, we should conduct a dynamic PET scan that involves multiple scans, 20 or 30 scans for 60 or 90 min after the administration of PET radiopharmaceutical. On the other hand, SUVR is a semi-quantitative version of a binding potential derived from a static PET scan. This is a single PET scan conducted 60 min after the administration. Hence, SUVR is favorable for an exact clinical situation. Although the binding potential is computed by considering a regional blood flow [8], SUVR is affected by a fluctuation in regional blood flow that is unavoidable. Moreover, we used the cerebellar gray matter as a reference region to compute the binding potential and SUVR, and systematic algorithms to define the region for the binding potential were proposed [9, 10].

Noise in PET images is another issue. If we develop a diagnostic algorithm for dementia with an adequate clinical impact, the algorithm should detect the onset of dementia before dementia symptoms occur. Therefore, we should detect small and faint changes in the brain. In particular, the noise in PET images is problematic. If we can administer high radioactivity, noise statistics in the image can be developed; however, it is clinically impractical because of radiation exposure to the patient and a medical staff attending a PET scan; a PET image is noisier than MRI and CT images. Consequently, the noise reduction algorithm should be applied to a PET image. An ordinary noise reduction algorithm smooths an original PET image spatially, averaging neighboring voxels. This causes a loss of spatial resolution and makes the small and faint amyloid β deposit undetectable. Another approach was proposed based on the kinetics of the administered amyloid probe. If we measure the time history of radioactivity concentration in the brain that can be acquired to conduct multiple PET scans after the administration of the radiopharmaceutical, which is a dynamic scan, we can obtain some information on the kinetics of the administered radiopharmaceutical. If the brain tissue is rich in specific binding sites or amyloid β, the time history decreases slowly. The kinetic-based idea was proposed in [11] and applied to amyloid imaging [12]. Another approach was presented [13] in which an unsupervised clustering algorithm was invoked.

In general, first, a functional disorder occurs, and subsequently, an anatomical structure is impaired. In this context, PET is better than MRI because MRI is used to image anatomical structures. However, MRI has a higher resolution than PET (1 mm^3 for MRI, and 5 or 6 mm^3 for PET). Moreover, we can carry out an MRI scan easily and more inexpensively than PET because PET scans require radiopharmaceutical. Accordingly, MRI can be used as a feature for dementia.

Another issue is the individual differences in the shape of the brain. Overall, the shape is common, but details are different individually. Moreover, brain function is localized; every local region in the brain has a different function. One approach to handle the shape difference is "template." A representative template is the AAL Atlas [14], in which the brain is parcellated to 120 regions. More fine versions are available, with around 1500 regions [15]. An individual MRI or PET scan is superimposed onto a standard brain on which the template is defined, and the averaged voxel values in each ROI are then applied as feature values.

21.4 Statistical Modeling and Training Images

As aforementioned, diagnosing dementia means monitoring the deterioration of cognitive functions in its very early phase. It may be useful to apply a statistical model to incorporate the information derived from various modalities. We can account for the time history of the progression of dementia. However, the elapsed time after the onset of dementia is difficult because there are no significant symptoms in the early stage of dementia.

Another drawback of investigating the statistical model for dementia is the difficulty in collecting an adequate number of clinical images. We can expect a statistical inference algorithm or an AI-based algorithm to be a powerful tool to diagnose dementia. However, thousands of images are required to train the algorithm [16]. It is challenging to consider the time history because it is difficult to gather an adequate number of clinical images for various stages of dementia. We can use a large dataset of amyloid imaging from the ADNI project,

Alzheimer's Disease Neuroimaging Initiative, but we do not have such datasets in the public domain that contain other dementia diseases, such as dementia with Lewy bodies or frontotemporal lobar degeneration. An AI-based algorithm for Alzheimer's disease with good performance has been published. For example, the algorithm of [15] can diagnose such cases of mild cognitive impairment that progress to Alzheimer's disease with 83% accuracy. We cannot ignore other dementia diseases than Alzheimer's disease to realize CAD for dementia because the incidence rates for the two dementia diseases are around 20% and 10%, respectively. Here, an AI-based algorithm to synthesize images has been tried to tackle this issue. It is a generative adversarial network (GAN) [17] applied to the CycleGAN algorithm [18] to synthesize CT images using MRI for radiotherapy. Kimura et al. [19] was a preliminary work to synthesize dementia PET images from healthy images. Frid-Adar et al. [20] applied GAN to train the CNN for liver classification.

21.5 Conclusion

Research on the use of CAD algorithms for dementia is increasing [21] owing to the coming aging society and improvement in medical services. PET and MRI are useful for deriving features for the diagnosis. We can utilize various algorithms, and more investigations are required to realize a huge image set to train the algorithm.

References

1. Abbott A. A problem for our age. Nature. 2011;475:S2–4.
2. Servigny J, Bussière PCT, Weinreb PH, Williams L, Maier M, Dunstan R, Salloway S, Chen T, Ling Y, O'Gorman J, Qian F, Arastu M, Li M, Chollate S, Brennan MS, Quintero-Monzon O, Scannevin RH, Arnold HM, Engber T, Rhodes K, Brennan MS, Quintero-Monzon O, Scannevin RH, Arnold HM, Engber T, Rodes K, Ferrero J, Hang Y, Milulskis A, Grimm J, Hock C, Mitsch RM, Sandrock A. The antibody aducanumab reduces Aβ plaques in Alzheimer's disease. Nature. 2016;537:50–6.
3. Jack CR, Knopman DS, Jagust WJ, Shaw LM, Alsen PS, Weiner MW, Petersen RC, Trojanowski JQ. Hypothetical model of dynamic biomarkers of the Alzheimer's pathological cascade. Lancet Neurol. 2010;9:119–28.
4. Sokoloff L, Reivichl M, Kennedy C, Des Rosiers MH, Patlak CS, Pettigrew KD, Sakurada O, Shinohara M. The [14C] deoxyglucose method for the measurement of local cerebral glucose utilization: theory, procedures, and normal values in the conscious and anesthetized albino rat. J Neurochem. 1977;28:897–916.
5. Price JC, Klunk WE, Lopresti BJ, Lu X, Hoge JA, Ziolko SK, Holt DP, Meltzer CC, DeKosky ST, Mathis CA. Kinetic modeling of amyloid binding in humans using PET imaging and pittsburgh compound-B. J Cereb Blood Flow Metab. 2005;25:1528–47.
6. van Berckel BNM, Ossenkoppele R, Tolboom N, Yaqub M, Foster-Dingley JC, Windhorst AD, Scheltens P, Lammertsma AA, Boellaard R. Longitudinal amyloid imaging using 11C–PiB: methodogic considerations. J Nucl Med. 2013;54:1570–6.

7. Innis RB, Cunningham VJ, Delforge J, Fujita M, Gunn RN, Holden J, Houle S, Huang S-C, Ichise M, Ito H, Kimura Y, Koeppe RA, Knudsen GM, Knuuti J, Lammertsma AA, Laruelle M, Maguire RP, Mintun M, Morris ED, Parsey R, Slifstein M, Sossi V, Suhara T, Votaw J, Wong DF, Carson RE. Consensus nomenclature for in vivo imaging of reversibly-binding radioligands. J Cereb Blood Flow Metab. 2007;27:1533–9.

8. Watabe H, Ikoma Y, Kimura Y, Naganawa M, Shidahara M. PET kinetic analysis—compartmental model. Ann Nucl Med. 2006;20:583–8.

9. Razifar P, Engler H, Ringheim A, Estrada S, Wall A, Långström B, Engler H. An automated method for delineating a reference region using masked volumewise principal-component analysis in ^{11}C–PIB PET. J Nucl Med Tech. 2009;37:38–44.

10. Yamada T, Watanabe S, Nagaoka T, Nemoto M, Hanaoka K, Kaida H, Ishii K, Kimura Y. Automatic delineation algorithm of reference region for amyloid imaging based on kinetics. Ann Nucl Med. 2019;34:102–7.

11. Kimura Y, Hsu H, Toyama H, Senda M, Alpert NM. Improved signal-to-noise ratio in parametric images by cluster analysis. NeuroImgae. 1999;9:554–61.

12. Yamada T, Kimura Y, Sakata M, Nagaoka T, Nemoto M, Hanaoka K, Kaida H, Ishii K. Clustering-based data reduction algorithm with simplified reference tissue model to generate parametric images in amyloid imaging. The 14th International Conference on Quantification of BrainFunction with PET, PP02–L07. 2019.

13. Turkheimer FR, Edison P, Pavese N, Roncaroli F, Anderson AN, Hammers A, Erhard A, Hinz R, Tai YF, Brooks D. Reference and target region modeling of [^{11}C]–(R)-PK11195 brain studies. J Cereb Blood Flow Metab. 2007;48:158–67.

14. Tzourio-Mazoyer N, Landeau B, Papathanassiou D, Crivello F, Etard O, Delcroix N, Mazoyer B, Joliot M. Automated anatomical labeling of activations in SPM using a macroscopic anatomical parcellation of the MNI MRI single-subject brain. NeuroImage. 2002;15:273–89.

15. Lu D, Popuri K, Ding GW, Balachandar R, Beg MF. Alzheimer's disease neuroimaging initiative. Multiscale deep neural network based analysis of FDG-PET images for the early diagnosis of Alzheimer's disease. Med Image Anal. 2018;46:26–34.

16. Van Ginneken B, Schaefer-Prokop CM, Prokop M. Computeraideddiagnosis: how to move from the laboratory to the clinic. Radiology. 2011;261:719–32.

17. Hiasa Y, Otake Y, Takao M, Matsuoka T, Takashima K, CarassA PJL, Sugano N, Sato Y. Cross-modality image synthesisfrom unpaired data using CycleGAN. In: Int Workshop SimulSynthesis Med Imag, Lecture Notes in Computer Science, vol. 11037. Cham: Springer; 2018. p. 31–41.

18. Zhu JY, Park T, Isola P, Efros AA. Unpaired image-to-image translation using cycle-consistent adversarial networks. In Proceedings of the IEEE International Conference on Computer Vision. 2017, p. 2223–2232. arXiv: 1703.10593v6.

19. Kimura Y, Watanabe A, Yamada T, Watanabe S, Nagaoka T, Nemoto M, Miyazaki K, Hanaoka K, Kaida H, Ishii K. AI approach of cycle-consistent generative adversarial networks to synthesize PET images to train computer-aided diagnosis algorithm for dementia. Ann Nucl Med. 2020;34:521–15.

20. Frid-Adar M, Diamant I, Klang E, Amitai M, Goldberger J, Greenspan H. GAN-based synthetic medical image augmentation for increased CNN performance in liver lesion classification. arXiv. 2018; https://doi.org/10.1016/j.neucom.2018.09.013.

21. Rathore S, Habes M, Iftikhar MA, Shacklett A. A review on neuroimaging-based classification studies and associated feature extraction methods for Alzheimer's disease and its prodromal stages. NeuroImage. 2017;155:530–48.

Normalized Brain Datasets with Functional Information Predict the Glioma Surgery

22

Manabu Tamura, Ikuma Sato, and Yoshihiro Muragaki

Abstract

The goal of this study is to transform to the digitized intraoperative imaging and the compiled brain-function database for predicting glioma surgery that is based on the patient's future perspective depending on the tumor resection rate as well as the postoperative complication rate. Firstly, we successfully acquired log data with the location of medical device integrated into intraoperative MR image and digitized brain function was converted to a normalized brain data format in 20 cases with acceptable accuracy. There were totally 22 speech arrest (SA), 10 speech impairment (SI), 12 motor, and 7 sensory responses (51 responses). Secondly, we simulated the projection of the normalized brain data to the individual pre- and intraoperative MR image. These image integration and transformation methods for brain normalization should facilitate practical intraoperative brain mapping. In the future, these methods may be helpful for preoperatively and/or intraoperatively predicting brain function.

Keywords

Brain mapping · Digitization · Transformation
Normalization · Predictive surgery

M. Tamura (✉) · Y. Muragaki
Faculty of Advanced Techno-Surgery, Institute of Advanced Biomedical Engineering and Science, Tokyo Women's Medical University, Tokyo, Japan

Department of Neurosurgery, Neurological Institute, Tokyo Women's Medical University, Tokyo, Japan
e-mail: tamura.manabu@twmu.ac.jp

I. Sato
Faculty of System Information Science Engineering, Future University Hakodate, Hakodate, Japan

22.1 Introduction

In recent papers on glioma surgery, the extent of resection has shown positive correlations with patients' survival [1–4]. On the other hand, aggressive tumor resection carries a risk of causing postoperative complications, so various methods of intraoperative monitoring have been developed to reduce the risk of neurological deterioration in motor, sensory, and language functions [5, 6]. After having made an appropriate evaluation of current surgical conditions, it became evident that MRI was needed to choose the optimal treatment method and procedure, as well as maximize the preservation of neurological function, in glioma surgery. We have developed the optimal surgery system in two major areas in which this method can provide with assistance. First, in 2000, an MRI unit was installed in the operating theatre, enabling surgeons to use intraoperative MRI to evaluate the extent of tumor resection in the operating theatre. A navigation system that promptly updates these MR images and can accurately localize the surgeon's procedure for tumor removal and provide an understanding of the position of the surgical area has been implemented, providing groundbreaking precision-guided surgery in over 2000 operations [7]. Second, starting in 2004, a dedicated device, IEMAS™ (Intraoperative Examination Monitor for Awake Surgery), has been used when performing brain mapping in awake craniotomy, providing a time-synchronized recording display with electric stimulation, updated navigation system, and patient conditions while examining linguistic function [8]. This system has now been used in over 508 brain tumor removals for eloquent lesions during awake craniotomy without interruption of intra-operative surgical manipulations [9, 10]. However, this visualization using IEMAS is limited to analog information that has been recorded on video and seen with the naked eye. In addition, because the position of the probes used for electrical stimulation on the brain surface is unclear, this information cannot be integrated with intraoperative MRI during brain mapping.

With the goal of developing a new approach to solve this problem, the electrical stimulation probe results are recorded as log (digitized) data, and together with task information, electrical stimulus conditions, and patient response during mapping, they are stored in an integrated dataset as digital information [11]. The purpose of this study was to transform brain mapping analog data into a digitized intraoperative MRI and integrated brain-function dataset for predictive glioma surgery considering tumor resection volume, as well as the intraoperative and postoperative complication rates.

22.2 Methods

The first steps of our system include performing MR acquisition with a strong magnetic field (3T) scanner (ACHIEVA, Philips Medical Systems) for patients scheduled in awake surgery due to tumor localization, accumulate 3D volumetric axial T1-weighted image (WI) images (TE: 4.6 ms, TR: 6.8 ms, voxel size 1 × 1 mm) of thin slices (2 mm), and prepare for tumor resection preoperatively. The acquired MRI

images are analyzed based on object-based morphometry, which can provide sulci analysis that gets to the core of brain development and cortical dysplasia. For the analysis, the BrainVISA software developed by Jean François Mangin, a cooperative researcher in the Computer-Assisted Neuroimaging Laboratory, Neurospin, was used. This software firstly extracts elementary cortical folds using geometric and topographic features from preoperative T1-weighted MRI. The folds are then pattern-recognized comparing the models and labeled automatically [12]. In daily clinical practice, preoperative a 3D brain image produced by automatically labeled sulcus and gyrus analysis is always used and because this accurately displays the position of the elementary sulci (e.g., central sulcus, anterior/ascending lateral fissure), the position of motor and language areas can be estimated, which is crucial prior preparation for brain mapping during tumor removal (e.g., Fig. 22.1, left lower image of brain surface).

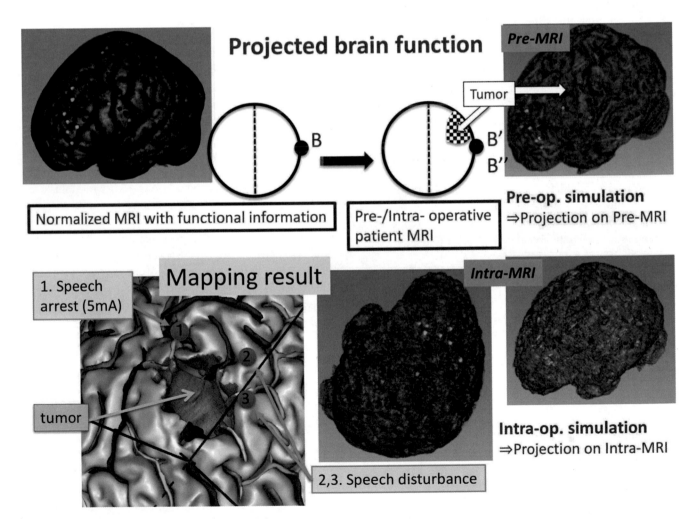

Fig. 22.1 Illustrative case (Case 20 in Table 22.1): Projection of point (B) on Japanese type normalized-MRI to intra-operative MRI via pre-operative MRI (ref. Fig. 22.4). After pre-processing, (B) is converted to (B′) on pre-operative MRI and then converted to (B″) on intra-operative MRI. This pre- and intra-operative simulated case with the left pre-central glioma reveals the MRI is firstly converted to the modified

3D-surface map with functional plots on the pre-operative brain (pre-operative simulation). Then, this integrated dataset is converted to the intra-operative MRI (intra-operative simulation), confirming the mapping result retrospectively. Practical speech function was recorded near the projected imaginary speech functional dataset on the 3D-surface map of intra-operative MRI

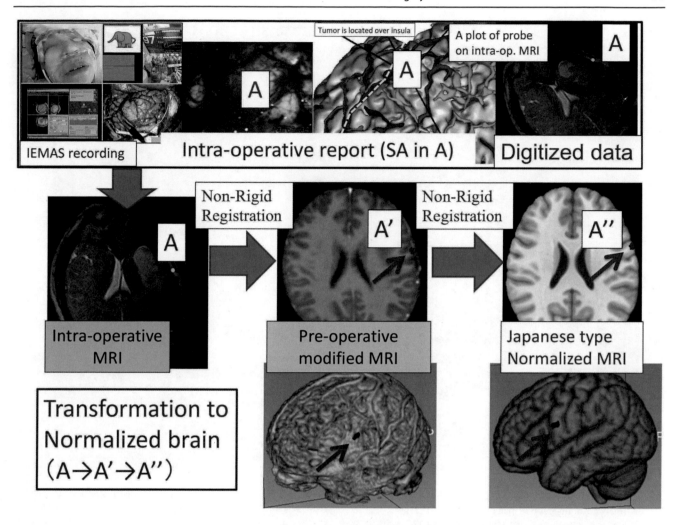

Fig. 22.2 The study protocol for a transformation to normalized brain is showed. *Step 1*: The brain-function database was stored using preparatory mapping process analysis by referring to IEMAS (intra-operative examination monitor for awake surgery). *Step 2*: Brain mapping log data of the electrical stimulus probe is acquired using neuronavigation system (BRAINLAB Curve™) that provides updated intra-operative MRI was installed on the electrical stimulus probe with interpolar distance of 5 mm and tips diameter of 1 mm (Unique Medical Corporation). *Step 3*: Reading in images and position information log in a digital for-

mat was performed in the 3D-Slicer image analysis software. Illustrative case (Case 3 in Table 22.1) of digitization of a sterilized electrical stimulation probe to localize the mapping point (Plot A). From an intra-operative analog mapping report, in which a plot A stimulus included a speech arrest response, the plot A point was tagged digitally on intra-operative MRI. Secondly, after pre-processing, plot A is converted with the physical-based non-rigid registration to plot A' on pre-operative modified MRI. Plot A' is then converted with the physical-based non-rigid registration to plot A'' on Japanese-type normalized MRI

Brain-function database using preparatory mapping and IEMAS™, is transformed to normalized brain digitized brain mapping localization with the modified electric stimulus probe [13] (Fig. 22.2). Normalized brain with functional information (Fig. 22.3) is projected to individual patient's brain and predicted brain function (Fig. 22.4). Then, structural and functional digitized data predict an intraoperative brain information. Finally, the normalized brain with functional information is projected to the individual patient's brain for prediction of brain function (Fig. 22.5). The ethics committee of Tokyo Women's Medical University approved brain mapping during tumor removal, and each patient provided informed consent before surgery.

Generally, video analysis using IEMAS™ for visualization of mapping processes during awake craniotomy was

conducted. The area where the electrical stimulus is applied by the surgeon and position where a response occurred is visualized alongside testing tasks conducted by the examiner (mainly number counting, picture naming (n), verb generation (v), and reading (r) kanji characters and hiragana, calculation) and electrical current values from 2 to 8 mA. The electrical stimulus conditions were as follows: biphasic stimulus using an Ojemann cortical stimulator (OCS-1™; Integra Radionics, Inc., Burlington, MA) and bipolar electrode probe with interpolar distance of 5 mm and tips diameter of 1 mm (Unique Medical Corporation), frequency: 50 Hz, pulse width: 0.5 ms. The evaluation confirmed whether there was any speech arrest (SA) and/or speech impairment (SI) following electrical stimulation, and the area stimulated, task presented, and stimulus cur-

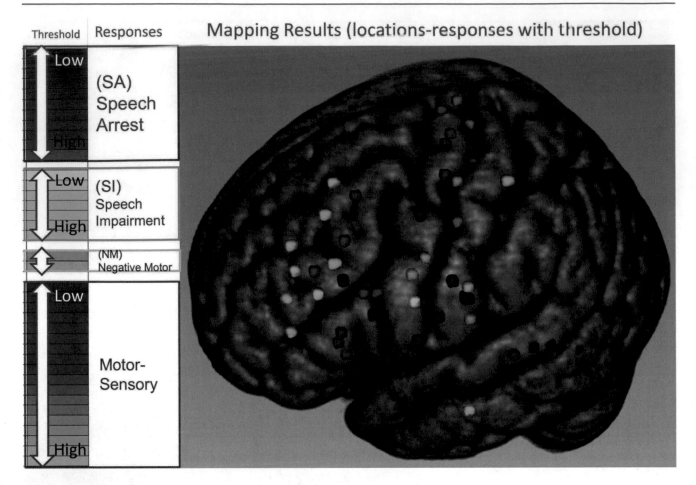

Fig. 22.3 SA, SI and NM language, motor and sensory response plots onto normalized brain in color map. Speech arrest (SA, red), speech impairment (SI, green), negative motor language (NM, yellow), and motor-sensory (blue) responses in 20 cases are plotted onto the Japanese type normalized 3D-brain map including color-labelled information of the mapping threshold (from low [2 mA] to high [6 mA] stimulus intensity). Language function (SA and SI) is distributed widely from the ALA near Broca to PLA near Wernicke and crosses the motor/sensory function area near the NM area. The ALA is located not only in Broca on the inferior frontal gyrus, but also in the middle frontal gyrus. Motor-sensory function is clearly distributed along the leg, hand, oro-facial and tongue response areas from the vertex

rent were determined through the video records [14]. The visualized data is entered into the electrical stimulus probe's database and is also essential for map creation on a normalized brain.

More specifically, a sterilized instrument that could be read into the navigation system (BRAINLAB Corporation) that provides updated intraoperative MRI (0.4 Tesla magnetic field, AIRIS II, Hitachi Medical Corporation) was installed on the bipolar electrode probe. Reading in images and position information was performed using the 3D-Slicer [15] image analysis software, and the position information log of the electrical stimulus probe was acquired in a digital format using a method we developed. A coordinate transformation to convert the digital function information into the MNI-icbm normalized brain data for-

mat (voxel size $1 \times 1 \times 1$ mm) [16], and plot it on a 3-dimensional brain image for visualization was carried out (Fig. 22.2). Do not convert brain function response points from the intraoperative MRI (T1-WI, TE: 10 ms, TR: 27 ms, voxel size 0.9×0.9 mm) [13] during surgery directly into normalized brain data, and first convert using the nonrigid registration [17] for each patient to the preoperative MRI. Each preoperative MRI is converted to our Japanese-type normalized format using the nonrigid registration [13, 17]. Preoperative and normalized brain data 3-dimensional images are created using BrainVISA software [12]. In the final step, the Japanese-type normalized brain with functional information is projected to the individual patient's brain for prediction of brain practical function (Fig. 22.4).

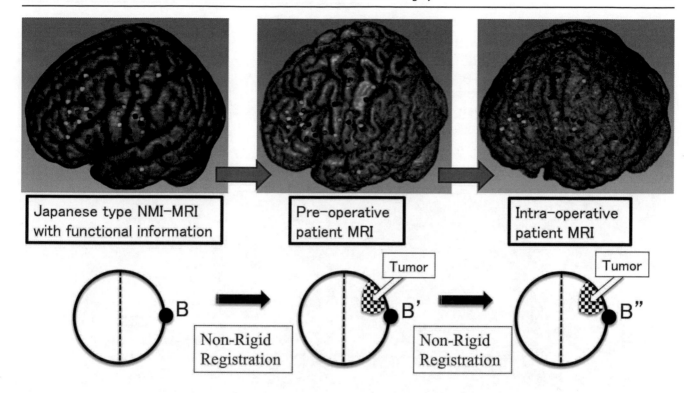

Fig. 22.4 Transformation (Projection) of point (B) on Japanese type NMI-MRI to intra-operative MRI via pre-operative MRI. After pre-processing, (B) is converted with the physical based non-rigid registration to (B′) on pre-operative MRI. (B′) is then converted with the physical based non-rigid registration to (B″) on intra-operative MRI. Japanese type NMI-MRI is converted backwards to individual MRI and projected onto the patient's brain to predict the patient's intra-operative brain function

22.3   Result

For all 20 cases with 32 SA/SI and 19 motor/sensory plots, the intraoperative electro-stimulator position data were used in combination with preoperative MRI for conversion to the normalized brain data format (Table 22.1). All functional responses in the 20 cases were integrated into digitized functional datasets on a Japanese-type normalized 3D brain map (Fig. 22.3). There was variation in the distribution of the mapping response (SA, SI, NM, Motor-Sensory) including color-labelled information of mapping threshold (from low [2 mA] to high [8 mA] electrical stimulus intensity). Language function (SA and SI) was distributed from the ALA (anterior language area) near Broca to the PLA (posterior language area) near Wernicke and crossed motor/sensory function in the NM (negative motor) area [10]. The interesting thing is that the ALA was located not only in so-called Broca on the inferior frontal gyrus, but also in the middle frontal gyrus. Motor or sensory function was clearly distributed with leg, hand, orofacial, and tongue responses from the vertex direction.

According to the evaluation of transformation accuracy for the three subjects, referring to (1) each two neighboring sulci on the electro-stimulator position and (2) the cortex surface near each tumor, the first transformation from intra- to preoperative MRI with nonrigid registration was calculated at (1) 2.6 ± 1.5 mm and (2) 2.1 ± 0.9 mm in average, while the second transformation from preoperative to normalized brain with nonrigid registration was calculated at (1) 1.7 ± 0.8 mm, (2) 1.4 ± 0.5 mm in average.

Finally, Japanese-type normalized MRI was converted backward to individual MRI and projected to the patient's brain to predict the patient's intraoperative brain function. For example, in Fig. 22.1, a left precentral glioma patient's MRI was converted to the modified 3D-surface map with functional plots in the preoperative brain. Then, the integrated datasets were converted to the intraoperative MRI, and the mapping record was confirmed retrospectively. Practical motor function was recorded near the projected imaginary hand-motor functional datasets on the 3D-surface map of intraoperative MRI.

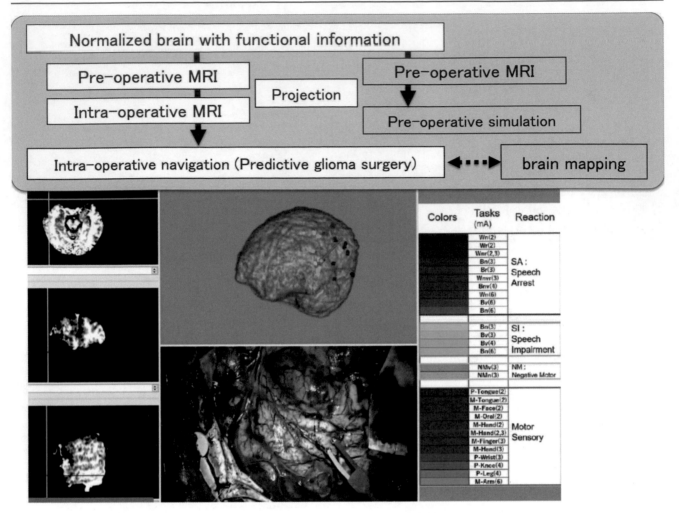

Fig. 22.5 Normalized brain datasets with functional information project the predictive glioma surgery. With the collaboration concerning image registration regarding to normalization of shifted brain and electrical distribution simulation for cortical brain mapping regarding to physics, we aim at the intra-operative brain mapping assistance using projected patient's MRI. A case simulation of predictive glioma surgery (Case 6, in Table 22.1) is demonstrated; Left column: navigation of intra-operative MRI fused the electrical stimulation probe. Middle upper: functional regions are plotted on the individual patient's intra-operative 3D surface image brain. Middle lower: time-synchronized representation of electrical probe movement as well as an imitation probe in middle upper

22.4 Discussion

22.4.1 Cortical Brain Mapping

With the positive-response 20 cases of cortical brain mapping, the area and precision of SA event observation could vary depending on the type of tasks presented during the language examination (n, v, r). The required current strength for the cortical stimuli varied from 2 to 6 mA for the ALA, PLA, and NM areas. These values were determined using the mapping stimulus conditions implemented at our facilities (f: 50 kHz, D: 0.5 ms, biphasic stimulation), and they differ from the values reported at other facilities [18, 19].

In our institution (unpublished data), out of the 85 patients for which cortical language mapping was performed, SA in the ALA was examined in 54 subjects, and observed near Broca's area in 34 subjects. The testing tasks during which SA occurred were confirmed as naming in 31 cases and verb generation in 11 cases (including multiple events), and the average stimulus current threshold was 4.6 ± 1.3 mA. In the NM area, the average stimulus current threshold was confirmed as 5.1 ± 1.3 mA for 16 subjects (30%), and in 7 cases (13%), there were no clear SA events confirmed on the brain surface, even at the tumor surface. In the frontal language area, SA was observed in 47 cases (87%), differentiated from 40 cases (74%) in the NM and Broca's area locations.

Because there is no fixed standard for mapping stimulation intensity, further investigation is required in this area. The method of confirming SA with direct cortical electrical stimulation is the gold standard technology and the only method by which language function can be confirmed [20].

Table 22.1 Patient characteristics, cortical mapping results

						Cortical mapping results		
Case	Age/sex	Pathology, WHO grade	Initial or additional	Side/location	Object for mapping	SA	SI	Motor/ sensory
1	38, M	Anaplastic astrocytoma, 3	Initial	Lt-Insula-Temporal	B, M	No	Bn(3)	No
2	28, F	Anaplastic astrocytoma, 3	Initial	Lt-Temporal	W, M	Wnvr(3)	No	No
3	24, M	Diffuse astrocytoma, 2	Additional	Lt-Insula-Temporal	B, M	Bn(3)	No	No
4	42, F	Oligodendroglioma, 2	Initial	Lt-Temporal	W	Wnr(2,3), Wn(2), Wr(2)	No	No
5	37, F	Anaplastic oligo-astrocytoma, 3	Initial	Lt-Frontal	B. M	NMn(3)	No	M(6)
6	56, M	Anaplastic oligodendroglioma, 3	Initial	Lt-Frontal	B, M	Bn(6), Bv(6)	Bv(3)	No
7	60, M	Oligodendroglioma, 2	Initial	Lt-Frontal	B, M	Bn(3)	Bn(3,6), Bv(4)	No
8	19, M	Anaplastic oligodendroglioma, 3	Initial	Lt-Temporal	W	No	Wnr(3,4)	No
8	37, M	Diffuse astrocytoma, 2	Initial	Lt-Frontal	B, M	Bnv(4), NMv(3)	No	P(2)
9	44, F	Anaplastic astrocytoma, 3	Initial	Lt-Frontal	M	No	No	P(3–4)
10	35, F	Oligodendroglioma, 2	Initial	Lt-Frontal	M	No	No	M(2–3)
11	36, M	Oligodendroglioma, 2	Additional	Lt-Frontal	B, M	Bn(3)	Bn(6), Bn(6)	No
12	37, M	Oligodendroglioma, 2	Initial	Lt-SMA	B, M	No	No	M(2)
13	31, F	Oligoastrocytoma, 2	Initial	Lt-Frontal	B, M	Bn(6)	No	M(2)
15	22, M	Anaplastic astrocytoma, 3	Additional	Lt-Front-Parietal	B, M	No	Bnv(3)	No
16	37, F	Anaplastic astrocytoma, 3	Recurrent	Lt-Parietal	W	Wn(6)	No	P(3)
17	42, F	Oligodendroglioma, 2	Initial	Lt-Frontal	B, M	Bnv(6,6), Bv(6)	No	No
18	45, M	Anaplastic oligodendroglioma, 3	Initial	Lt-Frontal	M	No	No	M(3)
19	44, M	Anaplastic oligodendroglioma, 3	Recurrent	Lt-Frontal	B, M	Bn(6,6,6)	Bc(3)	No
20	47, M	Anaplastic astrocytoma, 3	Initial	Lt-Frontal	B, M	Bnv(5,6), Bn(6)	No	No

Lt Left, *SMA* supplementary motor area, Object for Mapping (B: frontal language, W: posterior language, M: motor-sensory), Tasks for language examination (n: naming, v: verb generation, r: reading) and cortical mapping results (SA: speech arrest, SI: speech impairment, NM: negative motor, P: Sensory) with mapping threshold in (mA) are shown as a same item. Some cases include multi-response lesions

However, further efforts are required to consider false positives, as well as safety and standardization of the results, using this testing method [21].

22.4.2 Conversion of Digitized Information to Normalized Brain Data Format

The response points are first extrapolated from the intraoperative MRI and applied to the preoperative MRI, then conversion is applied with the normalized brain data precision retained, creating normalized brain data based on the complete brain function database. The use of this data is not limited to individual tumor patients. Through the accumulation of cases, a standard brain format map for functional areas such as the motor area and language area can be built. This data would allow an explanation for standard function positions to be developed and converting backward from this normalized brain to individual patient's MRI would probably enable the creation of functional maps for individual patients' brain function pre- and intraoperatively (Fig. 22.1).

22.4.3 Predictive Glioma Surgery Based on Database

As above mentioned, this backward conversion was tried and now we evaluate the accuracy of the projection comparing the practical brain function. Our goal, predictive surgery, is based on the database (including brain function, tumor location, and mapping threshold), and would allow explanations of each expected resection rate linking survival rate and postoperative neurological complication and high-level brain function damages that can hinder reintegration into society.

Functional region on normalized brain was plotted on a patient's intraoperative 3D surface image and the patient of left frontal cortical mapping is simulated as the practical mapping (Case 6 in Table 22.1, Fig. 22.5). Now, this is a simulated case, however, it has a strong potential to drive this concept in the real-time operating theatre when we will prepare an accurate and rapid intraoperative system to realize the predictive glioma surgery.

We aim for clinical application while deepening the localization of spatial structure and brain function by using

MRI, and the computational anatomy based on the multidimensional axis including the pathological axis of brain tumor. To predict complications associated with individual tumor removal from the normalized brain based on database and to make full use of highly intelligent preoperative diagnosis, we develop applications for predictive surgery. We should also consider sharing a database in which all researchers involved in brain tumor to solve clinical multifaceted problems such as brain shift during craniotomy, movement / plasticity of functional sites in tumor recurrence, and temporal change of pathological image information due to genetic change.

22.5 Conclusion

Within the limited stimulation time in awake craniotomy, brain mapping accuracy is always evaluated to clarify brain function for glioma surgery. Digitized intraoperative datasets for cortical brain mapping should be acquired precisely, and integrated image fusion and transformation on normalized datasets that compiled brain function can lead to predictive glioma surgery based on our new concept to the projection of the normalized brain.

Acknowledgment This study is supported by Grant-in-Aid for Scientific Research (B-22300093, C-12007086, C-19K12845), CREST (JST), JSPS Grant-in-Aid for Scientific Research on Innovative Areas (Multidisciplinary Computational Anatomy, JSPS KAKENHI Grant-15H01128, 17H05306), and the MIC Grant-162101001 (SCOPE). The authors would like to thank Drs. Takashi Maruyama, Kazuma Ohshima, Jean-Francois Mangin, Masayuki Nitta, Taiichi Saito, Ken Masamune, Takakazu Kawamata, and Hiroshi Iseki for advisory assistance with this text.

References

1. Grabowski MM, Recinos PF, Nowacki AS, Schroeder JL, Angelov L, Barnett GH, Vogelbaum MA. Residual tumor volume versus extent of resection: predictors of survival after surgery for glioblastoma. J Neurosurg. 2014;121(5):1115–23. https://doi.org/10.3171/2014.7.JNS132449.
2. Chaichana KL, Jusue-Torres I, Navarro-Ramirez R, Raza SM, Pascual-Gallego M, Ibrahim A, Hernandez-Hermann M, Gomez L, Ye X, Weingart JD, Olivi A, Blakeley J, Gallia GL, Lim M, Brem H, Quinones-Hinojosa A. Establishing percent resection and residual volume thresholds affecting survival and recurrence for patients with newly diagnosed intracranial glioblastoma. Neuro-Oncology. 2014;16(1):113–22. https://doi.org/10.1093/neuonc/not137.
3. Nitta M, Muragaki Y, Maruyama T, Ikuta S, Komori T, Maebayashi K, Iseki H, Tamura M, Saito T, Okamoto S, Chernov M, Hayashi M, Okada Y. Proposed therapeutic strategy for adult low-grade glioma based on aggressive tumor resection. Neurosurg Focus. 2015;38(1):E7. https://doi.org/10.3171/2014.10.FOCUS14651.
4. Hervey-Jumper SL, Li J, Lau D, Molinaro AM, Perry DW, Meng L, Berger MS. Awake craniotomy to maximize glioma resection: methods and technical nuances over a 27-year period. J Neurosurg. 2015;123(2):325–39. https://doi.org/10.3171/2014.10.jns141520.
5. Saito T, Muragaki Y, Maruyama T, Tamura M, Nitta M, Okada Y. Intraoperative functional mapping and monitoring during glioma surgery. Neurol Med Chir (Tokyo). 2015;55(1):1–13. https://doi.org/10.2176/nmc.ra.2014-0215.
6. Mandonnet E, Sarubbo S, Duffau H. Proposal of an optimized strategy for intraoperative testing of speech and language during awake mapping. Neurosurg Rev. 2017;40(1):29–35. https://doi.org/10.1007/s10143-016-0723-x.
7. Muragaki Y, Iseki H, Maruyama T, Tanaka M, Shinohara C, Suzuki T, Yoshimitsu K, Ikuta S, Hayashi M, Chernov M, Hori T, Okada Y, Takakura K. Information-guided surgical management of gliomas using low-field-strength intraoperative MRI. Acta Neurochir Suppl. 2011;109:67–72. https://doi.org/10.1007/978-3-211-99651-5_11.
8. Fukutomi Y, Yoshimitsu K, Tamura M, Masamune K, Muragaki Y. Quantitative evaluation of efficacy of Intraoperative Examination Monitor for Awake Surgery (IEMAS). World Neurosurg. 2019; https://doi.org/10.1016/j.wneu.2019.02.069.
9. Yoshimitsu K, Maruyama T, Muragaki Y, Suzuki T, Saito T, Nitta M, Tanaka M, Chernov M, Tamura M, Ikuta S, Okamoto J, Okada Y, Iseki H. Wireless modification of the intraoperative examination monitor for awake surgery. Neurol Med Chir (Tokyo). 2011;51(6):472–6.
10. Tamura M, Muragaki Y, Saito T, Maruyama T, Nitta M, Tsuzuki S, Iseki H, Okada Y. Strategy of surgical resection for glioma based on intraoperative functional mapping and monitoring. Neurol Med Chir (Tokyo). 2015;55(5):383–98.
11. Yamada H, Maruyama T, Konishi Y, Masamune K, Muragaki Y. Reliability of residual tumor estimation based on navigation log. Neurol Med Chir. 2020;60(9):458–67. https://doi.org/10.2176/nmc.oa.2020-0042.
12. BrainVISA. http://brainvisa.info/
13. Tamura M, Sato I, Maruyama T, Ohshima K, Mangin JF, Nitta M, Saito T, Yamada H, Minami S, Masamune K, Kawamata T, Iseki H, Muragaki Y. Integrated datasets of normalized brain with functional localization using intra-operative electrical stimulation. Int J Comput Assist Radiol Surg. 2019;14(12):2109–22. https://doi.org/10.1007/s11548-019-01957-7.
14. Saito T, Tamura M, Chernov MF, Ikuta S, Muragaki Y, Maruyama T. Neurophysiological monitoring and awake craniotomy for resection of intracranial gliomas. Prog Neurol Surg. 2018;30:117–58. https://doi.org/10.1159/000464387.
15. 3D Slicer. https://www.slicer.org
16. The McConnell Brain Imaging Centre. http://www.bic.mni.mcgill.ca/ServicesAtlases/ICBM152NLin2009
17. Liu Y, Kot A, Drakopoulos F, Yao C, Fedorov A, Enquobahrie A, Clatz O, Chrisochoides NP. An ITK implementation of a physics-based non-rigid registration method for brain deformation in image-guided neurosurgery. Front Neuroinform. 2014;8:33. https://doi.org/10.3389/fninf.2014.00033.
18. Sanai N, Mirzadeh Z, Berger MS. Functional outcome after language mapping for glioma resection. N Engl J Med. 2008;358(1):18–27. https://doi.org/10.1056/NEJMoa067819.
19. Boetto J, Bertram L, Moulinie G, Herbet G, Moritz-Gasser S, Duffau H. Low rate of intraoperative seizures during awake craniotomy in a prospective cohort with 374 supratentorial brain lesions: electrocorticography is not mandatory. World Neurosurg. 2015;84(6):1838–44. https://doi.org/10.1016/j.wneu.2015.07.075.
20. Kayama T. The guidelines for awake craniotomy guidelines committee of the Japan awake surgery conference. Neurol Med Chir (Tokyo). 2012;52(3):119–41.
21. Mandonnet E, Winkler PA, Duffau H. Direct electrical stimulation as an input gate into brain functional networks: principles, advantages and limitations. Acta Neurochir. 2010;152(2):185–93. https://doi.org/10.1007/s00701-009-0469-0.

New Frontier of Technology in Clinical Applications Based on MCA Models: Cardio-respiratory System

MCA Analysis for the Change in the Cardiac Fiber Orientation Under Congestive Heart Failure

23

Toshiaki Akita, Hirohisa Oda, Noriko Usami, and Kensaku Mori

Abstract

We developed the structure tensor analysis methods to investigate the cardiac fiber orientation from Micro-focus X-ray CT (μCT) imaging (Oda et al., J Med Imaging (Bellingham) 7(2):026001, 2020). Using this technique, we analyzed the cardiac fibers and sheets orientation of canine hearts under normal and congestive heart failure (HF) conditions. The average cardiac fibers orientation of the normal and HF canine model was not different, but the standard deviation of the angle was more diverse in HF. The sheet angle to the horizontal plane was more vertical in HF than in normal.

Keywords

Micro CT · Cardiac fiber orientation · Cardiac sheet Heart failure · Dilated cardiomyopathy

23.1 Introduction

The spiral orientation of cardiac fibers relates directly to systolic & diastolic function [1]. Heart failure (HF) may cause the changes of fibers orientation of the left ventricle (LV)

from spiral (±60°) to more horizontal angel [2, 3]. Diffusion tensor magnetic resonance imaging (DT-MRI) is commonly used to investigate the cardiac fibers orientation [4, 5]. However, long acquisition time and technical difficulty for DT-MRI make it for limited use.

We are focusing on using micro-focus X-ray CT (μCT), which has much higher spatial resolution and faster acquisition time than DT-MRI. We developed the structure tensor analysis to track the cardiac fibers from μCT images [6]. Using this technique, we planned to compare the cardiac fiber and sheet orientation of the canine heart under normal and HF conditions.

23.2 Materials and Methods

(a) Heart Preparation: Animal experimental protocol was approved by the IRB of Nagoya University, Graduate School of Medicine. The canine heart was harvested from normal or HF model. The HF canine model was created by rapid atrial pacing at 230 bpm for 8 weeks, which resulted in the left ventricular (LV) diastolic diameter 25 mm and LV ejection fraction 19%. 20 ml of 10% KCl solution was injected into the aortic root for euthanasia. 10% formaldehyde solution was injected into the aortic root after excision of the heart. We kept the heart in 7.5% I_2KI solution for 1 day to enhance cardiac fibers contrast in μCT volumes [7]. The heart was put into a plastic cage and scanned by a μCT scanner, inspeXio SMX-90CT Plus (Shimadzu).

(b) Analysis of Cardiac Fiber Orientation by μCT

For each CT volume, fiber orientation at a point is estimated by using the structure tensor analysis [6]. From each estimated fibers orientation vector, the inclination angle that commonly used in anatomical studies [4] is computed. On comparison-target points that are defined as grid patterns in the LV, we compare the mean

T. Akita (✉) · N. Usami
Department of Cardiac Surgery, Nagoya University Graduate School of Medicine, Nagoya, Aichi, Japan
e-mail: takita@med.nagoya-u.ac.jp

H. Oda
Graduate School of Information Science, Nagoya University, Nagoya, Aichi, Japan

K. Mori
Graduate School of Informatics, Nagoya University, Nagoya, Aichi, Japan

and standard deviation of angles between normal and HF models. The trajectories are colored by the inclination angle. This allows us to observe how cardiac fibers are structured visually.

23.3 Results

Fiber tracking results of normal and HF canine heart are shown as 3D images in Fig. 23.1 (3D). Fiber inclination angle versus the depth from the center of the wall at the equator point of these hearts was shown in Fig. 23.2. The average inclination angles from inner to outer layer were similar in normal and HF hearts (Fig. 23.3). However, the inclination angles of HF hearts were much diverse than the normal one. Cardiac sheet inclination angle to horizontal plane at equator position was also analyzed as the patch angles constructed by the stacks of extracellular matrix region, which is expressed as low CT number (Fig. 23.4). Histograms of sheet inclination angles at equator point shows that more vertical angle sheet existed in HF than in normal (Fig. 23.5).

23.4 Discussion

To our knowledge, this is the first report that cardiac fibers orientation of HF heart becomes divergent under similar average fibers orientation in normal and HF hearts. These findings were only accomplished by the high spatial resolution of μCT images, not by DT-MRI. Divergency of cardiac myofibers should result in weaker force generation than convergent fibers orientation, which exacerbates cardiac function with reduced cardiomyocytes. These results also match the disarrangement of cardiac fibers observed in the histologic specimen of dilated cardiomyopathy (DCM) patients.

The cardiac sheet inclination angle of the HF heart had a more vertical position than that of the normal heart. Helm et al. reported [8] that the orientation of laminar sheets become more vertical in the early-activated septum in dyssynchronous failing canine hearts using DT-MRI. These changes are accompanied by LV chamber dilation and LV wall thinning. Therefore, the vertical shift of cardiac sheet angles may not be the cause but the result from LV chamber dilatation and wall thinning.

We could not confirm the latter part of Buckberg's thesis that the cardiac shape changes from the natural elliptical form into a spherical contour, causing the systolic dysfunction of DCM, as global ventricular stretch transforms the normal 60° helical fiber orientation into a more transverse geometric configuration [6, 8]. Although tachycardia-induced HF canine model is well-established method to investigate the dilated cardiomyopathy [9–11], a longer pacing time may be required to establish the structural change.

This work mainly extends over the Multidisciplinary Computational Anatomy (MCA) axes: space and pathology. For the space axis, we explored high-resolution cardiac imaging using μCT, which allowed us to observe with the micro-meter-level spatial resolution. Using computational image processing by structure tensor analysis [6] from these data, it is now possible to analyze cardiac fiber structures in 3D. Our discoveries offered the methodologies for investigating the microscopic structure changes under pathological processes such as heart failure, which contributes to the pathology axis of MCA.

This is the preliminary experiment of a small sample size. We are planning to analyze additional normal and HF canine hearts to validate our preliminary findings.

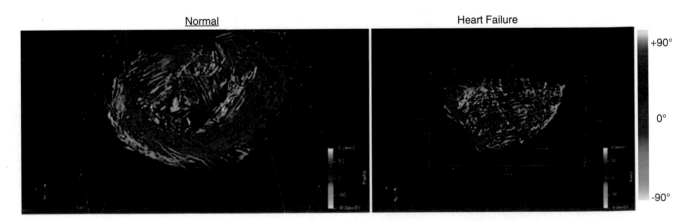

Fig. 23.1 3D view of cardiac fibers orientation in normal and HF canine heart Fiber angles are expressed by color

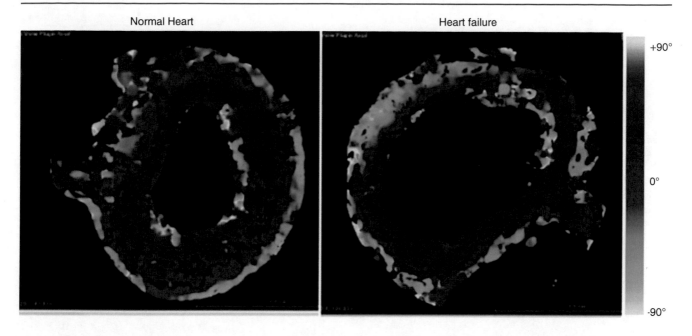

Fig. 23.2 2D view of cardiac fibers angles at equator line of normal and HF canine hearts

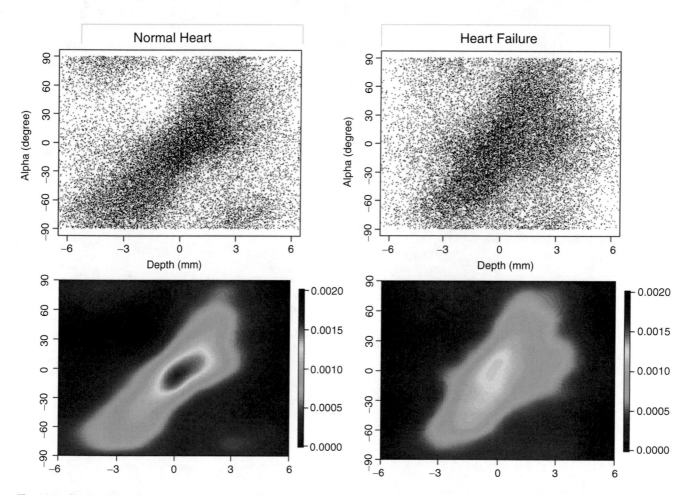

Fig. 23.3 Cardiac Fiber angles at various points on the equator plane in normal and HF canine hearts. The average cardiac fiber angles of both samples shift from around −70°at outer side to around +50°at inner side. Angles in HF hearts are more diverse than normal heart

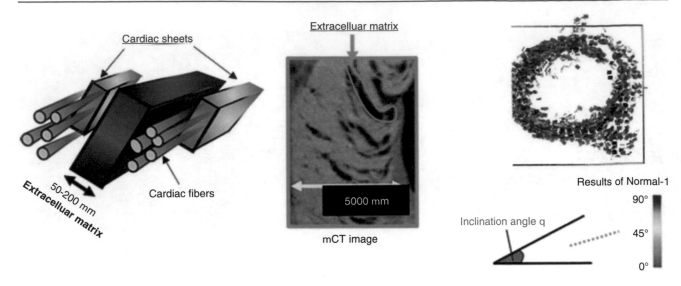

Fig. 23.4 Shema of cardiac sheets, fibers, and extracellular matrix(left). Cardiac sheet angles are determined by stacks of extracellular matrix images and are expressed in color

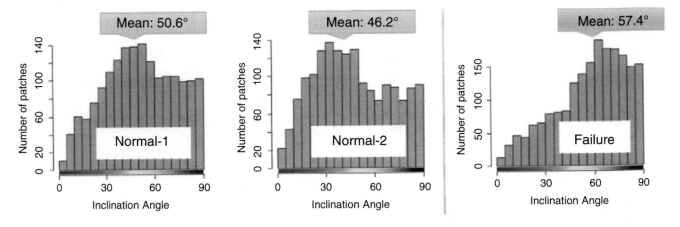

Fig. 23.5 Count of the angles of extracellular patches, which indicate sheet inclination angles in normal and HF canine hearts. HF heart has a more vertical sheet inclination angle than the normal one

23.5   Conclusion

The cardiac fibers orientation analysis by μCT will be a powerful tool to investigate the three-dimensional microscopic structural changes under congestive HF.

Acknowledgments This work is supported by MEXT/JSPS KAKENHI (Grant Number 17H05288).

References

1. Buckberg G, Mahajan A, Saleh S, Hoffman JI, Coghlan C. Structure and function relationships of the helical ventricular myocardial band. J Thorac Cardiovasc Surg. 2008;136(3):578–89, 89 e1–11.

2. Buckberg GD, Hoffman JI, Coghlan HC, Nanda NC. Ventricular structure-function relations in health and disease: part I. The normal heart. Eur J Cardiothorac Surg. 2014;47(4):587–601.

3. Buckberg GD, Hoffman JI, Coghlan HC, Nanda NC. Ventricular structure-function relations in health and disease: part II. Clinical considerations. Eur J Cardiothorac Surg. 2014;47(5):778–87.

4. Helm P, Beg MF, Miller MI, Winslow RL. Measuring and mapping cardiac fiber and laminar architecture using diffusion tensor MR imaging. Ann N Y Acad Sci. 2005;1047:296–307.

5. Rohmer D, Sitek A, Gullberg GT. Reconstruction and visualization of fiber and laminar structure in the normal human heart from ex vivo diffusion tensor magnetic resonance imaging (DTMRI) data. Investig Radiol. 2007;42(11):777–89.

6. Oda H, Roth HR, Sugino T, Sunaguchi N, Usami N, Oda M, et al. Cardiac fiber tracking on super high-resolution CT images: a comparative study. J Med Imaging (Bellingham). 2020;7(2):026001.

7. Aslanidi OV, Nikolaidou T, Zhao J, Smaill BH, Gilbert SH, Holden AV, et al. Application of micro-computed tomography with iodine staining to cardiac imaging, segmentation, and computational model development. IEEE Trans Med Imaging. 2013;32(1):8–17.

8. Helm PA, Younes L, Beg MF, Ennis DB, Leclercq C, Faris OP, et al. Evidence of structural remodeling in the dyssynchronous failing heart. Circ Res. 2006;98(1):125–32.

9. Houser SR, Margulies KB, Murphy AM, Spinale FG, Francis GS, Prabhu SD, et al. Animal models of heart failure: a scientific statement from the American Heart Association. Circ Res. 2012;111(1):131–50.

10. Shinbane JS, Wood MA, Jensen DN, Ellenbogen KA, Fitzpatrick AP, Scheinman MM. Tachycardia-induced cardiomyopathy: a review of animal models and clinical studies. J Am Coll Cardiol. 1997;29(4):709–15.

11. Monnet E, Chachques JC. Animal models of heart failure: what is new? Ann Thorac Surg. 2005;79(4):1445–53.

Computerized Evaluation of Pulmonary Function Based on the Rib and Diaphragm Motion by Dynamic Chest Radiography

Rie Tanaka

Abstract

The respiratory system is essential to maintain life but can be affected by many types of diseases. Thus, a better understanding of pulmonary structure, function, and physiology is critical to solving the universal problems associated with peoples' health in the respiratory system. In this chapter, we focus on the rib and diaphragm motion, as well as the lung density, closely related to the pulmonary function. As an effective tool for comprehensively understanding pulmonary function, dynamic chest radiography (DCR) and its findings in animal and clinical studies are introduced, followed by a proposal of cross-disciplinary approaches based on DCR in the respiratory system.

Keywords

Pulmonary function · Rib · Diaphragm · Lung density Dynamic chest radiography

24.1 Introduction

The respiratory system is essential to maintain life but can be affected by many types of diseases. Therefore, a better understanding of pulmonary structure, function, and physiology is critical for dealing with the universal problems associated with peoples' health in the respiratory system. The knowledge of respiratory physiology was established in the mid-nineteenth century mainly based on the findings in pulmonary function test (PFT) and lung scintigraphy [1]. After that, medical imaging technology has significantly improved and is now providing multidimensional information regarding the respiratory system, ranging from the cell level to the organ level. For example, micro-computed tomography (μCT) provides extremely high-resolution images of samples and can be employed as a non-destructive inspection tool [2]. Recent advancements in computed tomography (CT) and magnetic resonance imaging (MRI) technologies have facilitated functional lung imaging. Dynamic 4-dimensional CT and dual-energy CT allow the estimation of lung ventilation based on Hounsfield unit values [3]. Conversely, cine MRI provides regional ventilation differences based on lung signal [4], as well as lung volume and diaphragm motion [5]. On the other hand, dynamic chest radiography (DCR) provides sequential images with a large field of view (FOV) during respiration at a high imaging rate of 15–30 frames/s [6, 7]. Based on this information, multidisciplinary computational anatomy (MCA) might innovate respiratory medicine and surgery. In this chapter, we focus on the rib and diaphragm motion, as well as the lung density, closely related to pulmonary function. As an effective tool for comprehensively understanding the pulmonary function, DCR, and its findings in animal and clinical studies are introduced, followed by a proposal of cross-disciplinary approaches based on DCR in the respiratory system.

24.2 Dynamic Chest Radiography (DCR)

Recent digital radiography technology allows DCR with a total exposure dose comparable to that of conventional chest radiography. DCR is a flat-panel detector (FPD) based on functional lung imaging that can be performed in a general X-ray room [6, 7]. The imaging system has been commercialized, receiving FDA approval in 2019.

Sequential chest radiographs are obtained during forced respiration with use of a dynamic FPD system and X-ray generator capable of pulsed irradiation (15 frames/s). The large FOV of FPDs permits real-time observation of the

R. Tanaka (✉)
School of Health Sciences, College of Medical,
Pharmaceutical and Health Sciences, Kanazawa University,
Kanazawa, Ishikawa, Japan
e-mail: rie44@mhs.mp.kanazawa-u.ac.jp

entire lungs and simultaneous right-and-left evaluation of diaphragm kinetics. Except for breathing manner, imaging is performed in the same way as is the conventional chest examination, i.e., the standing position and the PA (postero-anterior) direction. The total patient dose is adjustable by changes in the imaging time, imaging rate, and source to image distance (SID) and can be less than the dose limit for two projections (PA + LA) recommended by the International Atomic Energy Agency (IAEA) (1.9 mGy). Although over-lapping lung structures have been disturbing quantitative analysis of lung dynamics on 2D projection images, the bone suppression (BS) image-processing technique has allowed us to separate bones from soft tissue in a chest radiograph [8, 9]. DCR combined with the BS technique has the potential to allow a rapid and better understanding of respiratory changes/motions of each lung structure [10].

24.3 What Is Reflected on Dynamic Chest Radiographs?

Dynamic chest radiographs contain a wealth of functional information, such as rib and diaphragm motion, cardiac motion, pulmonary ventilation, and circulation. In particu-lar, respiratory muscle function has a direct association with pulmonary function, and one of the functional disor-ders appear the other one. For example, pulmonary impair-ments alter lung behavior, resulting in a phase shift or/and extension of the respiratory period [11–14]. On the other hand, bone fracture and diaphragm disorders limit lung mobility during respiration, resulting in a decrease in vital capacity (VC) [15–17]. In addition, contraction of the dia-phragm alone accounts for about 75% of inspiration. Muscles of the thoracic wall (intercostal muscles) and selected muscles of the neck and abdomen also can partici-pate in inspiration and assist the diaphragm, especially dur-ing active breathing (e.g., during exercise) [1, 17]. Therefore, an understanding of rib and diaphragm kinemat-ics is crucial to evaluate pulmonary function and treatment effects. The other important information is a change in radiographic lung density during respiration and cardiac beating. The changes are caused by relative increases and decreases in the volume of lung vessels and bronchi per unit of lung volume [18, 19]. Therefore, pulmonary impair-ments can be detected as reduced changes in radiographic lung density, even without the use of contrast media or radioactive medicine.

However, their interpretation is challenging for radiolo-gists; therefore, computerized methods have been developed for the evaluation of pulmonary function on dynamic chest radiographs [6, 7]. The following findings have been revealed by DCR through animal and clinical studies.

24.3.1 Diaphragm Motion

Pulmonary function is generally assessed by PFT as a total lung capacity, such as forced expiratory volume and forced vital capacity. While, analysis of diaphragm motion allows for the evaluation of pulmonary function in each lung because pulmonary function directly reflects on diaphragm motion. In normal subjects, there is a high correlation between diaphragm excursion and tidal volume and no sig-nificant difference in diaphragm excursion between both lungs [20, 21]. In contrast, lungs with atelectasis indicate a significant reduction in diaphragm excursion ($P < 0.05$). Temporal cross-sectional image is useful for a better under-standing of such a diaphragm motion [22]. Some clinical studies report that pulmonary impairments could be detected based on abnormal diaphragm motion, such as no/reduced motion, time-lagged motion, and paradoxical motion [22]. In addition, unilateral pulmonary impairment can be detected based on reduced diaphragm excursion in a patient with nor-mal findings in PFT. Dynamic analysis of diaphragm motion is helpful for following the progress of a patient with pulmo-nary diseases and assessing the results of treatment.

24.3.2 Rib Motion

Abnormal rib motion can be detected as asymmetrical distri-bution of local velocity in the lungs. For this purpose, dynamic bone images created by BS technique are useful to quantify and distinguish movements of ribs from those of other lung structures [10]. Previous studies indicated that normal controls show the symmetrical distribution of rib movement in both the velocity vector maps and velocity magnitude maps. On the other hand, in many patients with respiratory disease and/or lung cancer, limited rib mobility appears as a reduced velocity field, resulting in an asymmet-rical distribution of rib motion in both maps. Therefore, abnormal cases are more likely to show large variations of vector sum in the horizontal direction throughout all frames [10]. Furthermore, paradoxical rib movements can be observed in some frames, resulting in less synchronous between the vector sum in the vertical direction with the respiratory phase, and no correlation with the diaphragm movements. Dynamic analysis of rib motion is expected to be an effective index for predicting pulmonary function, especially in a patient with scoliosis with limited rib motion.

24.3.3 Pulmonary Ventilation and Circulation

Changes in lung volume and circulation dynamics are observed as temporal changes in radiographic lung density on dynamic chest radiographs [18, 19]. Therefore, the pulmonary function can be evaluated by time-series analysis of radiographic lung density [23–30]. Previous animal and clinical studies indicated that trapped air, limited airflow, and pulmonary embolism could be detected as reduced changes in pixel value in the lung regions on dynamic chest radiographs [31, 32]. In addition, there is high linearity between changes in radiographic lung density and tidal volume, as well as blood volume in the lungs ($r = 0.99$). Areas of atelectasis or pulmonary embolism displayed significantly reduced changes in pixel values ($P < 0.05$). Furthermore, the color-mapping visualization technique provides new functional images, which is useful for understanding slight changes in radiographic lung density during respiration and cardiac pumping. DCR is capable of ventilation- and perfusion-related parameters based on temporal changes in radiographic lung density, even without the use of radioactive agents or contrast media. DCR allows real-time observations of pulmonary function so that it might be applied for emergency medicine and supporting operations in the future.

24.4 Pulmonary Function Evaluation Based on MAC

In Fig. 24.1, we illustrate the concept of the pulmonary function evaluation based on MCA, which is realized by integrating a variety of information. The 2D/3D image registration technique is essential for MCA in the respiratory system, for example, to recover the time-series information from 3D static data. In a preliminary study, it was indicated that the rib motion could be reproduced using combination of DCR and one-time-phase CT; accordingly, the rib motion could be comprehensively understood [33]. MCA enables the further utilization of the information existing in clinical settings, thereby introducing new findings into the respiratory system. Cross-disciplinary approaches play a key role in the evaluation of pulmonary function based on MCA.

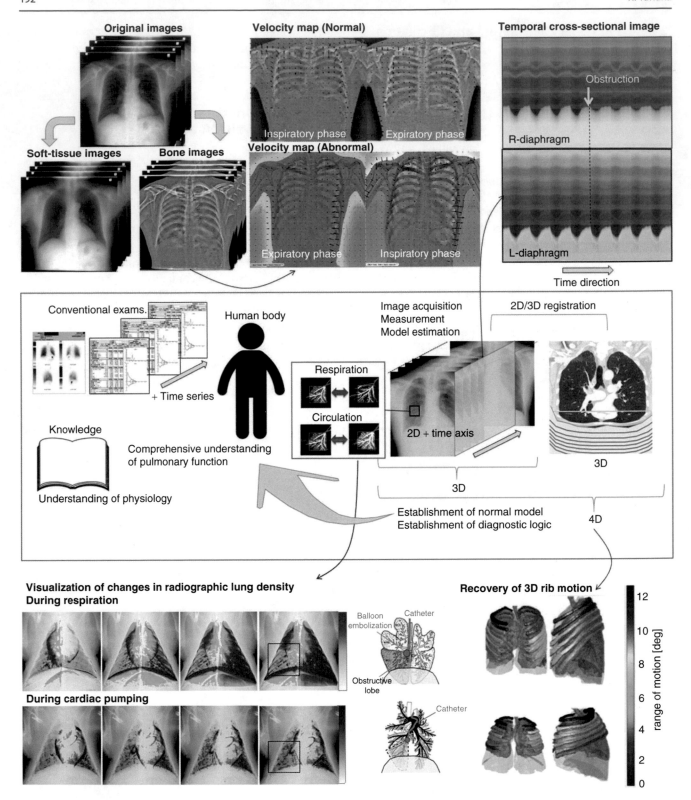

Fig. 24.1 A concept of the pulmonary function evaluation with dynamic chest radiography (DCR) based on multidisciplinary computational anatomy (MCA) approach

Acknowledgments This research was supported in part by a grant-in-Aid for Scientific Research (c) (16K10271) and a grant-in-aid for Scientific Research on Innovative Areas (Multidisciplinary Computational Anatomy) (17H05286) from the Ministry of Education, Culture, Sports, Science, and Technology (MEXT), Tokyo, Japan, a grant-in-aid program for revitalization in Fukushima, Japan, and Konica Minolta, Inc., Tokyo, Japan. The author sincerely thanks staff in Dept. of respiratory medicine, thoracic surgery, and radiology at Kanazawa University Hospital, who gave me variable clinical advice, as well as staff in the biomedical innovation center, Dept. of radiology and Dept. of emergency, and I.C.U., at the Shiga University of Medical Science and Konica Minolta Inc. for the technical support in animal studies.

References

1. West JB. Respiratory physiology. The essentials. 1st ed. Philadelphia: Lippincott Williams & Wilkinss; 1974.
2. Nakamura S, Mori K, Iwano S, Kawaguchi K, Fukui T, Hakiri S, Ozeki N, Oda M, Yokoi K. Micro-computed tomography images of lung adenocarcinoma: detection of lepidic growth patterns. Nagoya J Med Sci. 2020;82(1):25–31.
3. Yamashiro T, Moriya H, Tsubakimoto M, et al. Continuous quantitative measurement of the proximal airway dimensions and lung density on four-dimensional dynamic-ventilation CT in smokers. Int J Chron Obstruct Pulmon Dis. 2016;11:755–64.
4. Ohno Y, Hatabu H. Basics concepts and clinical applications of oxygen-enhanced MR imaging. Eur J Radiol. 2007;64:320–8.
5. Bhave S, Lingala SG, Newell JD Jr, et al. Blind compressed sensing enables 3-dimensional dynamic free breathing magnetic resonance imaging of lung volumes and diaphragm motion. Investig Radiol. 2016;51:387–99.
6. Tanaka R. Dynamic chest radiography: flat-panel detector (FPD) based functional X-ray imaging. Radiol Phys Technol. 2006;9:139–53.
7. Tanaka R, Sanada S. 12. Respiratory and cardiac function analysis on the basis of dynamic chest radiography. In: Suzuki K, editor. Part III Image Processing and Analysis, Computational Intelligence in Biomedical Imaging. Berlin: Springer; 2013. p. 317–45.
8. Suzuki K, Abe H, MacMahon H, Doi K. Image-processing technique for suppressing ribs in chest radiographs by means of massive training artificial neural network (MTANN). IEEE Trans Med Imaging. 2006;25:406–16.
9. Knapp J, et al. Feature based neural network regression for feature suppression. U.S. Patent Number, 8,204,292 B2, June 12, 2012.
10. Tanaka R, Sanada S, Sakuta K, Kawashima H. Quantitative analysis of rib kinematics based on dynamic chest bone images: preliminary results. J Med Imaging. 2015;2(2):024002. https://doi.org/10.1117/1.JMI.2.2.024002.
11. Gilmartin JJ, Gibson GJ. Abnormalities of chest wall motion in patients with chronic airflow obstruction. Thorax. 1984;39:264–71.
12. Gilmartin JJ, Gibson GJ. Mechanisms of paradoxical rib cage motion in patients with chronic obstructive pulmonary disease. Am Rev Respir Dis. 1986;134:683–7.
13. Yamada Y, Ueyama M, Abe T, Araki T, Abe T, Nihino M, et al. Time-resolved quantitative analysis of the diaphragms during tidal breathing in a standing position using dynamic chest radiography with a flat panel detector system ("dynamic X-ray phrenicography"): initial experience in 172 volunteers. Acad Radiol. 2017;24(4):393–400.
14. Culham EG, Jimenez HA, King CE. Thoracic kyphosis, rib mobility, and lung volumes in normal women and women with osteoporosis. Spine. 1994;19:1250–5.
15. Leong JC, Lu WW, Luk KD, Karlberg EM. Kinematics of the chest cage and spine during breathing in healthy individuals and in patients with adolescent idiopathic scoliosis. Spine. 1999;24:1310–5.
16. Smyth RJ, Chapman KR, Wright TA, Crawford JS, Rebuck AS. Pulmonary function in adolescents with mild idiopathic scoliosis. Thorax. 1984;39:901–4.
17. Hansen JT, Koeppen BM. Netter's atlas of human physiology. New Jersy: Icon Learning Systems LLC; 2003. p. 94–5.
18. Squire LF, Novelline RA. Overexpansion and collapse of the lung: causes of mediastinal shift. In: Fundamentals of radiology. 4th ed. Cambridge, MA: Harvard University Press; 1988. p. 88–103.
19. Fraser RS, Muller NL, Colman NC. Part III: Radiologic signs of chest disease. In: Fraser and Pare's diagnosis of diseases of the chest. 4th ed. Philadelphia: W.B. Saunders Company; 1999. p. 431–594.
20. Tanaka R, Tani T, Nitta N, Tabata T, Matsutani N, Muraoka S, et al. Pulmonary function diagnosis based on diaphragm movement using dynamic flat-panel detector imaging: An animal-based study. Proc SPIE 10578, Medical Imaging 2018: Biomedical Applications in Molecular, Structural, and Functional Imaging, 10578V-1-6, 2018.
21. Ohkura N, Kasahara K, Watanabe S, Hara J, Abo N, Sone T, et al. Dynamic-ventilatory digital radiography in air flow limitation: change in lung area reflects air trapping. Respiration. 2020;99:382–8.
22. Tanaka R, Kasahara K, Matsumoto I, Sanada S. Computerized evaluation of the rib kinetics and pulmonary function based on the rib and diaphragm motion by dynamic chest radiography. In Proceedings of the 5th international symposium on the project "Multidisciplinary computational anatomy". 2019. p. 141–5
23. Tanaka R, Sanada S, Okazaki N, Kobayashi T, Fujimura M, Yasui M, Matsui T, Nakayama K, Nanbu Y, Matsui O. Evaluation of pulmonary function using breathing chest radiography with a dynamic flat-panel detector (FPD): primary results in pulmonary diseases. Investig Radiol. 2006;41:735–45.
24. Tanaka R, Sanada S, Okazaki N, Kobayashi T, Nakayama K, Matsui T, Hayashi N, Matsui O. Quantification and visualization of relative local ventilation on dynamic chest radiographs. The international society for optical engineering. Medical imaging 2006. Proc SPIE. 2006;6143(2):62432Y1–8.
25. Tanaka R, Sanada S, Fujimura M, Yasui M, Nakayama K, Matsui T, Hayashi N, Matsui O. Development of functional chest imaging with a dynamic flat-panel detector (FPD). Radiol Phys Technol. 2008;1:137–43.
26. Tanaka R, Sanada S, Fujimura M, Yasui M, Tsuji S, Hayashi N, Okamoto H, Nanbu Y, Matsui O. Ventilatory impairment detection based on distribution of respiratory-induced changes in pixel values in dynamic chest radiography: a feasibility study. IJCARS. 2011;6:103–10.
27. Tanaka R, Sanada S, Tsujioka K, Matsui T, Takata T, Matsui O. Development of a cardiac evaluation method using a dynamic flat-panel detector (FPD) system: a feasibility study using a cardiac motion phantom. Radiol Phys Technol. 2008;1:27–32.
28. Tanaka R, Sanada S, Fujimura M, Yasui M, Tsuji S, Hayashi N, Okamoto H, Nanbu Y, Matsui O. Pulmonary blood flow evaluation using a dynamic flat-panel detector: feasibility study with pulmonary diseases. IJCARS. 2009;4:449–54.
29. Tanaka R, Sanada S, Fujimura M, Yasui M, Tsuji S, Hayashi N, Okamoto H, Nanbu Y, Matsui O. Development of pulmonary blood flow evaluation method with a dynamic flat-panel detector (FPD): quantitative correlation analysis with findings on perfusion scan. Radiol Phys Technol. 2010;3:40–5.
30. Tanaka R, Sanada S, Fujimura M, Yasui M, Tsuji S, Hayashi N, Okamoto H, Nanbu Y, Matsui O. Ventilation-perfusion study with-

out contrast media in dynamic chest radiography. The international society for optical engineering. Medical imaging 2011. Proc SPIE. 2011;7965:79651Y1 17.

31. Tanaka R, Tani T, Nitta N, Tabata T, Matsutani N, Muraoka S, et al. Detection of pulmonary embolism based on reduced changes in radiographic lung density during cardiac beating using dynamic flat-panel detector: an animal-based study. Acad Radiol. 2019;26(10):1301–8.

32. Tanaka R, Tani T, Nitta N, Tabata T, Matsutani N, Muraoka S, et al. Pulmonary function diagnosis based on respiratory changes in lung density with dynamic flat-panel detector imaging: an animal-based study. Investig Radiol. 2018;53(7):417–23.

33. Hiasa Y, Otake Y, Tanaka R, Sanada S, Sato Y. Recovery of 3D rib motion from dynamic chest radiography and CT data using local contrast normalization and articular motion model. Med Image Anal. 2019;51:144–56.

Computer-Aided Diagnosis of Interstitial Lung Disease on High-Resolution CT Imaging Parallel to the Chest

25

Shingo Iwano, Hiroyasu Umakoshi, and Shinji Naganawa

Abstract

Interstitial lung diseases (ILDs) are conditions in which inflammation and fibrosis diffusely affect the pulmonary interstitium and parenchyma and include a variety of subsets, such as idiopathic interstitial pneumonia (IIP), collagen vascular disease-related ILD and chronic hypersensitivity pneumonitis. High-resolution computed tomography (HRCT) is essential for visual assessment. Additionally, high-speed multi-detector row CT can provide high-resolution images of the whole lung. However, evaluation of the entire lung field on a lot of axial HRCT images can be complicated even for expert radiologists. To resolve these difficulties, we developed a novel 3D imaging system of curved HRCT (3D-cHRCT) that only targets the peripheral lung field where is the predominant location of ILDs, especially usual interstitial pneumonia (UIP). The 3D-cHRCT can evaluates visually an image depicting the zone of the peripheral lung where there are little pulmonary vessels or bronchi. Therefore, we attempted to visually and quantitatively assess ILDs using the 3D-cHRCT at a constant depth from the chest wall.

Keywords

Interstitial lung disease · Interstitial pneumonia · COPD High-resolution CT · Computer-aided diagnosis

25.1 Introduction

Interstitial lung diseases (ILDs) are conditions in which inflammation and fibrosis diffusely affect the pulmonary interstitium and parenchyma and include a variety of subsets,

such as idiopathic interstitial pneumonia (IIP), collagen vascular disease-related ILD and chronic hypersensitivity pneumonitis [1, 2]. High-resolution computed tomography (HRCT) is essential for visual assessment. Additionally, high-speed multi-detector row CT can provide high-resolution images of the whole lung. However, evaluation of the entire lung field on a lot of axial HRCT images can be complicated even for expert radiologists. To resolve these difficulties, we developed a novel 3D imaging system of curved HRCT (3D-cHRCT) that only targets the peripheral lung field where is the predominant location of ILDs, especially usual interstitial pneumonia (UIP). In this study, we adapted the MCA model to the combination of three-dimensional space and pathological findings.

25.2 3D-cHRCT

All preoperative CT scans were performed using a 64-multidetector row CT scanner in the craniocaudal direction during inspiratory apnea. Axial thin-section CT images of the whole lung were reconstructed with a slice thickness of 0.5-mm or 1-mm at the same increment using a high-spatial frequency algorithm. The axial HRCT imaging data of the study patients were transferred to the 3D-workstation that automatically used our original software to reconstruct each 3D-cHRCT image of the lung at a constant 1-cm depth from the chest wall [3]. Figure 25.1 shows the overall scheme of the 3D-cHRCT reconstruction procedure and examples of 3D-cHRCT images.

25.3 Visual Assessment of ILDs

In September 2018, a guideline for the clinical recommendations for the diagnosis of idiopathic pulmonary fibrosis (IPF) was provided [4, 5]. Therefore, we reviewed the HRCT images of 27 patients with or without ILDs (Figs. 25.2, 25.3, 25.4, and 25.5). The concordance rate between the conven-

S. Iwano (✉) · H. Umakoshi · S. Naganawa
Department of Radiology, Nagoya University Graduate School of Medicine, Nagoya, Japan
e-mail: iwano45@med.nagoya-u.ac.jp

Fig. 25.1 Overall schematic diagram of the three-dimensional, curved high-resolution CT (3D-cHRCT) image at a constant depth from the chest wall [3]. Axial HRCT image and dashed line indicating 1-cm depth from the chest wall; multiple-view 3D-cHRCT images showing the 3D distribution of interstitial pneumonia infiltrates; left lateral view, posterior view, right lateral view

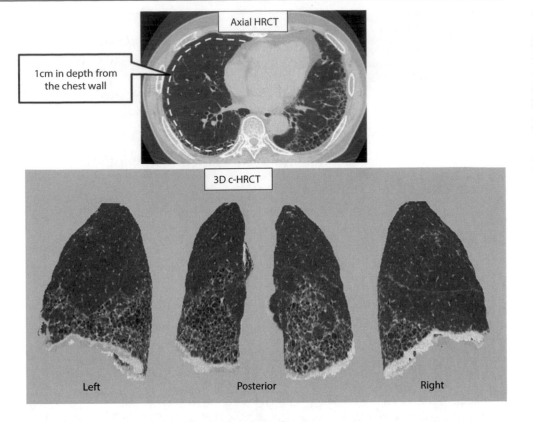

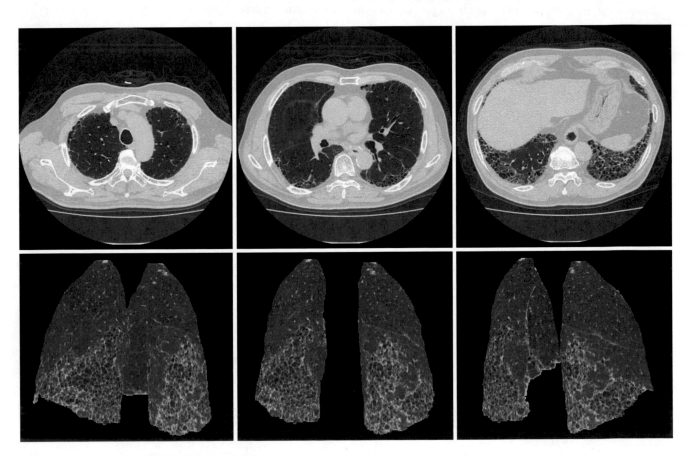

Fig. 25.2 A case of 3D-cHRCT of definite UIP pattern

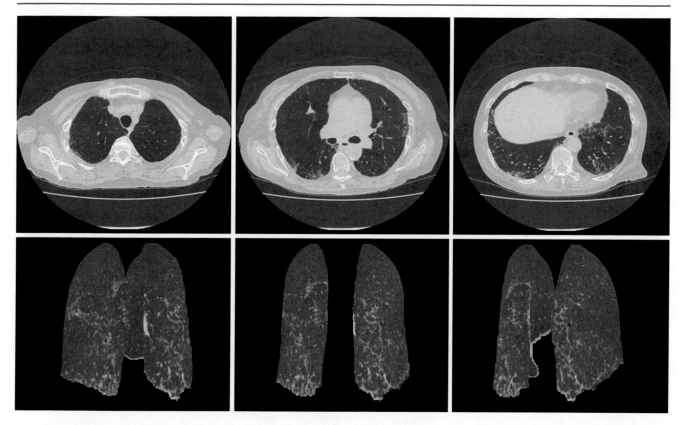

Fig. 25.3   A case of 3D-cHRCT of probable UIP pattern

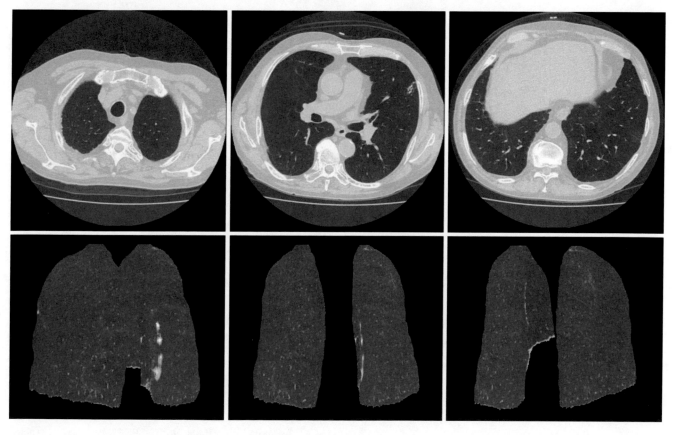

Fig. 25.4   A case of 3D-cHRCT of indeterminate for UIP pattern

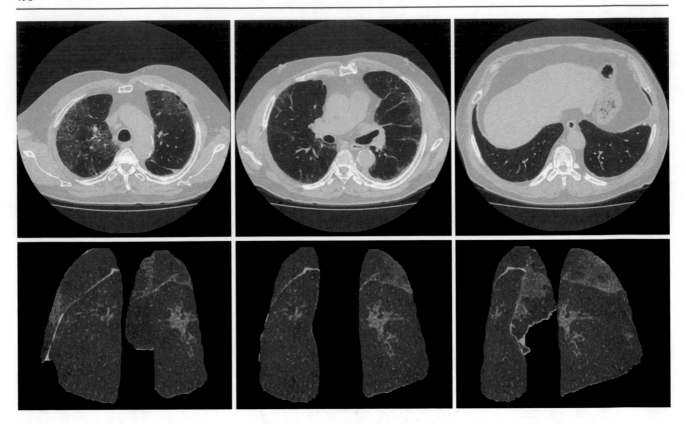

Fig. 25.5 A case of 3D-cHRCT of alternative diagnosis pattern. Upper or mid-lung distribution is shown in 3D-cHRCT

Table 25.1 Cross-reference Table for axial HRCT and 3D-cHRCT in ILDs pattern diagnosis

		3D-cHRCT				
		Definite UIP	Probable UIP	Indeterminate for UIP	Alternative diagnosis	Normal
Axial HRCT	Definite UIP	2				
	Probable UIP		8			
	Indeterminate for UIP	1		5	1	
	Alternative diagnosis				2	
	Normal				1	7

tional axial HRCT and 3D-cHRCT was 89% (24/27) with excellent reproducibility ($k = 0.855$, $p = 0.001$, Table 25.1).

25.4 Visual Assessment of Lung Cancer

Frequently, patients with IPF occur primary lung cancer as complications because of their smoking history. Although most of the lung cancers develop spherically, subsets of peripheral lung cancers show flattened shape because the progression is prevented by the chest wall. However, even in such a case, a tumor may show round shape on 3D-cHRCT images of the lung at a constant 1-cm depth from the chest wall. Therefore, we tried to diagnose peripheral lung cancer (Fig. 25.6).

25.5 Quantification of Emphysema and IP by CAD

Smoking-related diffuse pulmonary diseases such as emphysema and interstitial pneumonia (IP) often occur in combination with primary lung cancer [1]. Patients with combined pulmonary fibrosis and emphysema (CPFE), which is a unique disorder of the lungs that comprises upper lung emphysema and lower lung fibrosis, are especially at high risk for lung cancer and have a poor prognosis [6]. Pulmonary function testing (PFT) is usually performed for the evaluation of preoperative respiratory function. However, the percent vital capacity (%VC) and the ratio of forced expiratory volume in 1 s to forced vital capacity (FEV1/FVC) are often normal in patients with CPFE, whereas diffusion capacity for carbon monoxide (DLCO) is low [6]. Then we attempted to quantify the extent of comorbid emphysema and IP in patients with lung cancer using 3D-cHRCT and

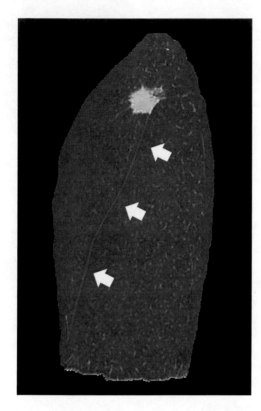

Fig. 25.6 A case of peripheral lung cancer in the left upper lobe. The upper row is conventional axial HRCT and the lower row is 3D-cHRCT. The tumor shows a flattened shape on the axial image, although it shows a round-like shape on the 3D-cHRCT image. Additionally, the correlation with a major fissure line (arrow) is obvious

compare the results to visual assessment of IP and the results of PFT [3].

The total area (TA) of the lung and low-attenuation area (LAA) (<−950 HU), which denotes emphysema, on the 3D-cHRCT images were calculated by the CAD program on the workstation. The percentage of low-attenuation area (%LAA) was defined as follows: %LAA = LAA/TA * 100 (%). In a similar fashion, the percentage of the high-attenuation area (%HAA) (higher than the threshold value [>−500 HU]), which denotes IP, was defined as follows: %HAA = HAA/TA * 100 (%). Fifty-one patients with primary lung cancer and IP were selected, and their DLCO was reviewed.

Figure 25.7 shows the 3D scatter diagrams of %LAA, %HAA, and %DLco. Both %LAA and %HAA were significantly and negatively correlated with %DLco. That is, the %HAA and %LAA values computed using 3D-cHRCT images at a 1-cm depth from the chest wall were significantly correlated with DLco, and may be important quantitative parameters for IP, emphysema, and CPFE. Furthermore, TA, HAA, and %HAA ratios at 2-cm showed significant correlations with physiologically progressive ILD [7].

25.6   Conclusion

We have proved that the 3D-cHRCT image can evaluate disease condition and progression of ILDs quantitatively. Additionally, it was also useful for the morphological classification of ILDs. It was recognized that the 3D-cHRCT could add new findings to diagnosis by conventional axial HRCT images. In the near future, 3D-cHRCT can be applied to the diagnosis of peripheral lung cancer.

Fig. 25.7 3D scatter diagrams show the correlation between %DLCO and quantitative 3D c-HRCT parameters. Both %HAA and% LAA are indicators of low diffusion capacity ($p < 0.001$ and $p < 0.001$, respectively)

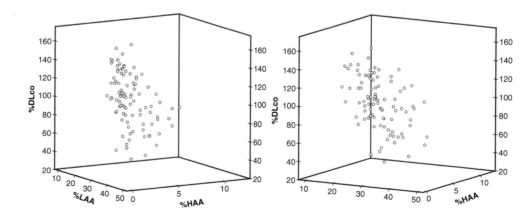

References

1. American Thoracic Society/European Respiratory Society. International Multidisciplinary Consensus Classification of the Idiopathic Interstitial Pneumonias. This joint statement of the American Thoracic Society (ATS), and the European Respiratory Society (ERS) was adopted by the ATS board of directors, June 2001 and by the ERS Executive Committee, June 2001. Am J Respir Crit Care Med. 2002;165(2):277–304.

2. Travis WD, Costabel U, Hansell DM, et al. An official American Thoracic Society/European Respiratory Society statement: update of the international multidisciplinary classification of the idiopathic interstitial pneumonias. Am J Respir Crit Care Med. 2013;188(6):733–48.

3. Umakoshi H, Iwano S, Inoue T, Li Y, Naganawa S. Quantitative evaluation of interstitial pneumonia using 3D-curved high-resolution CT imaging parallel to the chest wall: a pilot study. PLoS One. 2017;12(9):e0185532.

4. Raghu G, Collard HR, Egan JJ, et al. An official ATS/ERS/JRS/ALAT statement: idiopathic pulmonary fibrosis: evidence-based guidelines for diagnosis and management. Am J Respir Crit Care Med. 2011;183(6):788–824.

5. Raghu G, Remy-Jardin M, Myers JL, et al. Diagnosis of idiopathic pulmonary fibrosis. An official ATS/ERS/JRS/ALAT clinical practice guideline. Am J Respir Crit Care Med. 2018;198(5):e44–68.

6. Jankowich MD, Rounds SI. Combined pulmonary fibrosis and emphysema syndrome: a review. Chest. 2012;141(1):222–31.

7. Umakoshi H, Iwano S, Inoue T, Li Y, Nakamura K, Naganawa S. Quantitative follow-up assessment of patients with interstitial lung disease by 3D-curved high-resolution CT imaging parallel to the chest wall. Nagoya J Med Sci. 2019;81(1):41–53.

Postoperative Prediction of Pulmonary Resection Based on MCA Model by Integrating the Temporal Responses and Biomechanical Functions

26

Fei Jiang, Xian Chen, Kazuhiro Ueda, and Junji Ohgi

Abstract

The goal of this research is to predict postoperative compensatory growth of the residual lung based on preoperative anatomical information by integrating biomechanical functions into the lung anatomical model along the time axis in the framework of multidisciplinary computational anatomy. We first develop a biomechanical simulation system for the human respiratory system. After constructing a finite element model of the respiratory system including lung, trachea, rib cage, intercostal muscles and diaphragm, the behavior of muscle contractions was represented by introducing a Hill-type transversely isotropic hyperelastic continuum material model, while the lung, including the airflow, was characterized as porous hyperelastic materials based on a multiphasic model using mixture theory. The developed numerical respiratory system is able to reproduce the data in terms of thorax displacement, diaphragm movement as well as lung deformation by introducing contraction of the respiratory muscle. The results were validated by comparing the thorax deformation with the four-dimensional computed tomography (4D-CT) images during normal quiet breathing. We also proposed a voxel-based flow simulation approach directly based on the medical CT images by using the lattice Boltzmann method. This approach enables us to provide detailed airflow information in the lung bronchus. With this simulation platform of the human respiratory system, the respiratory function recovery and compensatory lung growth can be evaluated from the information of stress and strain distribution by the virtual lobectomy simulation.

Keywords

Virtual lobectomy simulation · Finite element method Biomechanics

26.1 Introduction

Lobectomy and stereotactic radiotherapy are performed as treatments for lung cancer. In the former case, the problem is the reduction of respiratory function after treatment. In the latter case, deformation of the lung due to respiration causes tumor migration and results in difficulty in focusing irradiation only on tumor cells. In recent years, simulation approaches of lung deformation due to respiration have been proposed. However, in previous researches, the lung deformation was represented by shape reconstruction using medical images or by prescribing boundary conditions rather than represented based on physiological phenomena. To predict the respiratory function after lobectomy and postoperative compensatory growth of the residual lung, we aim to develop a biomechanical simulation method for the whole respiratory system.

The prediction of residual lung dilation may help to determine the minimum amount of lung resection necessary for lung cancer treatment, because lung function lost in pulmonary resection for lung cancer treatment is recognized to be compensated by expansion of the residual lung after surgery [1]. Recent research has revealed that the biomechanical state of the residual lung, which depends on the intrathoracic mechanical environment, affects the expansion of the residual lungs [2]. Therefore, the residual lung expansion may be predicted by combining the biomechanical state in the residual lung with dynamic anatomical information.

In this study, we first developed a biomechanical simulation approach for the human respiratory system. Then, we

F. Jiang · X. Chen (✉) · J. Ohgi
Department of Mechanical Engineering, Graduate School of Sciences and Technology for Innovation, Yamaguchi University, Ube, Japan
e-mail: xchen@yamaguchi-u.ac.jp

K. Ueda
Department of General Thoracic Surgery, Graduate School of Medical and Dental Sciences, Kagoshima University, Kagoshima, Japan

carried out a biomechanical simulation based on time series medical images before and after lung resection surgery to investigate the biomechanical state in the lung and pulmonary function change. The final stage is to achieve the goal of predicting postoperative residual lung expansion based on preoperative anatomical information by integrating the biomechanical function into the lung anatomical information along the time axis. In this work, the biomechanical and anatomical information are integrated not only along the time axis but also the cross function axis. The approach proposed in this research makes it possible to extract the physical information of the living human body from anatomical information based on medical images. Therefore, this study contributes to the aim of a comprehensive understanding of the living human body based on medical images.

26.2 Methods

To simulate the postoperative compensatory growth of the residual lung, a platform establishing an integrated computational mechanics model of the human respiratory system has to be developed first. This platform includes a finite element model of respiration system including sternal, rib, vertebral bones, intercostal muscle, diaphragm, heart and lung and a voxel model for the bronchus. A voxel dataset of the chest was first segmented from CT slices of a male volunteer. Based on this dataset, we constructed a 3D model of the normal respiratory system, including lung, bronchi, rib cage, intercostal muscles, and diaphragm (Fig. 26.1). Since it is difficult to segment the intercostal muscles and diaphragm from CT slices, the 3D models of intercostal muscles were created by connecting the muscle attachments between upper and lower rib bones, and the diaphragm was generated to attach the lung bottom by referencing the anatomy textbook [3]. The finite element meshes (424,304 tetrahedra elements and 92,841 nodes) were then generated from the 3D models by using mesh generation software ANSYS ICEM CFD (ANSYS, Inc.). To reproduce reasonable chest movement, the fiber direction of muscles has to be carefully determined for contraction. The fiber direction of intercostal muscles was assigned based on the experimental data from Loring [4]. For the diaphragm, reference to the anatomy textbook [3], the fiber directions were determined by assuming that the fiber distribution has a radial pattern from the center of the top to the bottom edge.

The transversely isotropic hyperelastic material model proposed by Martins et al. [5] was adopted as the constitutive

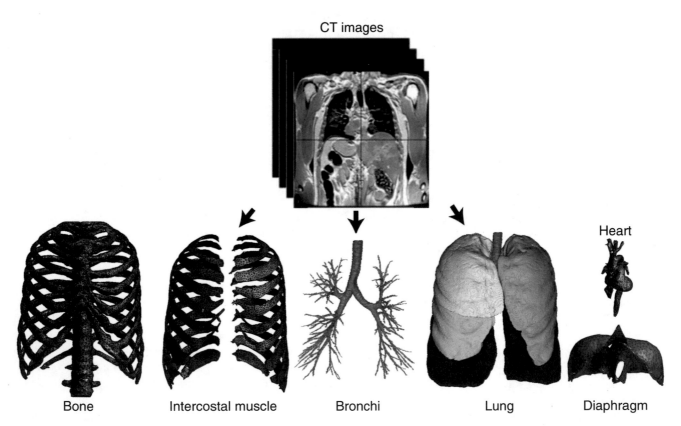

Fig. 26.1 A 3D model of the normal respiratory system including lung, bronchi, rib cage, intercostal muscles and diaphragm

model for reproducing the mechanical behavior of intercostal muscles and diaphragm. In this model, the traditional Hill's three-element muscle model is extended to three-dimensional cases. In this three-element muscle model, the parallel element PE and series element SE is nonlinear springs representing passive behavior. The third contractile element, CE produces contractile force when the muscle is excited. The muscles were modeled as incompressible transversely isotropic hyperelastic materials, such that the strain energy function can be written as follows:

$$U = U_I\left(\overline{I}_1^C\right) + U_f\left(\overline{\lambda}_f, \alpha\right) \tag{26.1}$$

$$U_I\left(\overline{I}_1^C\right) = c\left[\mathrm{EXP}\left(b\left(\overline{I}_1^C - 3\right) - 1\right)\right] \tag{26.2}$$

$$U_f\left(\overline{\lambda}_f, \alpha\right) = U_{\mathrm{PE}}\left(\overline{\lambda}_f\right) + U_{\mathrm{CE}}\left(\overline{\lambda}_f, \alpha\right) \tag{26.3}$$

$$U_{\mathrm{PE}}\left(\overline{\lambda}_f\right) = T_0^{\mathrm{M}} \int_1^{\overline{\lambda}_f} f_{\mathrm{PE}}\left(\lambda\right) \mathrm{d}\lambda \tag{26.4}$$

$$U_{\mathrm{CE}}\left(\overline{\lambda}_f, \alpha\right) = T_0^{\mathrm{M}} \int_1^{\overline{\lambda}_f} f_{\mathrm{CE}}\left(\lambda\right) \alpha \, \mathrm{d}\lambda \tag{26.5}$$

where U_I and U_f stand for the strain energies stored in the isotropic hyperelastic matrix and muscle fiber, respectively. $\overline{\lambda}_f$ is the stretch ratio in the fiber direction, T_0^{M} denotes the maximum muscle stress determining the maximum muscle contraction force, α represents the muscle's activation level that ranges from 0 to 1. The 2nd Piola-Kirchhoff stress tensor can be derived as:

$$S = -\mathrm{p}I + \frac{\partial U}{\partial E} = -\mathrm{p}I + \frac{\partial U_I}{\partial E} + \frac{\partial U_f}{\partial E}$$

$$= -\mathrm{p}I + \frac{\partial U_I}{\partial \overline{I}_1^C} \frac{\partial \overline{I}_1^C}{\partial E} + \frac{\partial U_f}{\partial \overline{\lambda}_f} \frac{\partial \overline{\lambda}_f}{\partial E} \tag{26.6}$$

All the degrees of freedom at the bottom of the vertebra were fixed as displacement boundary conditions, and the simulations were performed with a nonlinear finite element analysis program developed in-house. The compliance of the costovertebral joint was set at relatively small in order to eliminate the influence from the compliance of the costovertebral and interchondral joints and model a neutral position of the rib cage during breathing. All the material parameters used for the intercostal muscles, the diaphragm, bone, and other tissues can be found in the reference paper [6].

To evaluate the pulmonary function, acquiring detailed information of the airflow inside bronchi is also important for the reason that pulmonary function reduction is highly related to the local flow velocity distribution, flow pattern,

etc. To simulate the airflow, a voxel airway model was directly constructed from the total data set of the chest. We built a 10-generation airway model consisting of a fluid mesh with 500,573 cubes measuring 0.4 mm³. The term "generation" indicates the number of branches from the trachea to the outermost peripheral airways. As the CT data consist of voxels aligned in Cartesian coordinates, a computational mesh can be directly generated from the CT data using a Cartesian mesh. Simulation with many generations involves thousands of peripheral airways. It becomes impossible to mark out all the peripheral outlet boundaries manually. An efficient detection algorithm [7] is adopted to automatically select the outermost peripheral airways voxels for applying the boundary conditions of outlets. With this airway model, the pulmonary airflow was simulated by using the lattice Boltzmann method (LBM), which is a fully explicit incompressible flow solver. The fluid velocity and pressure are set to zero for the initial condition. No-slip boundary conditions are set at the 3D airway walls. To drive the flow, we used an inflow boundary condition at the inlet and specified pressure outlet boundary conditions at all terminal bronchi.

With this simulation platform of the human respiratory system, we performed biomechanical simulation of the respiratory cycle by contracting intercostal muscles and diaphragm based on physiological phenomena. The lobectomy operation was numerically reproduced by virtually removing a lobe of the lung from the constructed respiratory system (Fig. 26.2). By simulating the behavior of the respiratory system after the virtual operation, reduction of respiratory function can be evaluated. The distribution of the residual lung surface curvature and the equivalent nodal concentrated area obtained from the surface nodal information of the residual lung mesh model and the distribution of the equivalent nodal concentrated volume calculated for all the remaining nodal points can be used to represent the anatomy characteristics of postoperative lung. In the simulation after lung resection, the surface area and the volume change rates can be evaluated from the deformation of the lungs due to the contraction of the intercostal muscles and the diaphragm. By comparing such biomechanical information with the information from the medical image, it becomes possible to associate the physical state with the anatomy characteristics. In addition, the effect of resection on the airflow in the lung can be evaluated by performing airflow analysis after reproducing changes in the airway due to resection of the lung. Furthermore, by reproducing the deformation of the residual lung due to the contraction of the intercostal muscles and the diaphragm, it is possible to make the mechanical factors clear by investigating the distribution of strain and stress in the residual lung.

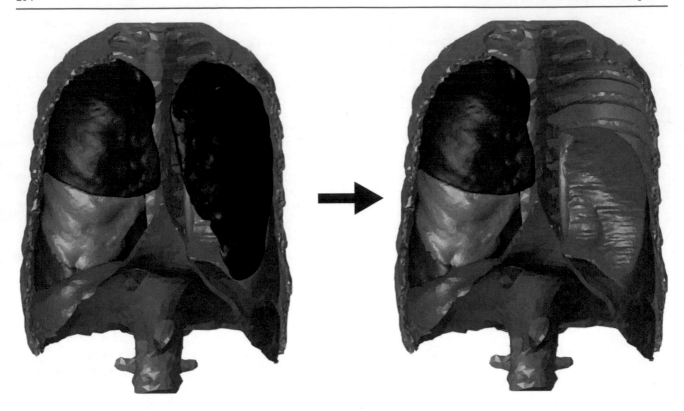

Fig. 26.2 Virtual lobectomy operation by removing a lobe of the lung

26.3　Results

The simulation was performed by assuming a normal quiet breathing. During inspiration, by the contraction of external intercostal muscle and diaphragm, the thoracic cavity and lung were expanded. Therefore, the low pressure in the lung was generated to allow the air to move into the lung. Conversely, the intercostal muscle and diaphragm were relaxed during expiration to produce a pressure in the lung higher than atmospheric pressure thus, the air was expelled. Our proposed numerical respiratory system is able to reproduce the data in terms of thorax displacement, diaphragm movement, airway flow as well as lung deformation by introducing contraction of the respiratory muscle. Our results show that the diaphragm was not only descended depending on contraction but also deformed by the rising of the thorax. This is validated by comparing the thorax deformation with the four-dimensional computed tomography (4D-CT) images during normal quiet breathing [8]. Furthermore, simulation results for the variations of alveolar, pleural pressures and lung volume during normal breathing are compared with the reference data [9] for the validation of the proposed multiphasic model for lung parts.

To estimate the respiratory function decrease, we performed a virtual lobectomy on the developed respiratory system. The lobectomy operation was numerically reproduced by virtually removing a lobe of the lung from the constructed respiratory system (Fig. 26.2). By simulating the behavior of the respiratory system after the virtual operation, we first evaluated the reduction of respiratory function by comparing the airway flow rate before and after the operation. The flow rate in the trachea dropped about 30% immediately after operation. To maintain the same respiratory function, a more negative alveolar pressure is necessary. From our airway flow simulation, a 50% lower value of alveolar pressure is required to produce the same flow rate before operation. The change of airway flow patterns between pre-operation and post-operation is illustrated in Fig. 26.3. The airway flow velocities are much higher after lobectomy operation due to the cut of the airway tree in the affected lobe. Furthermore, increasing the airflow in the bronchus in the remaining lung was obtained. It is considered that the decreased airflow due to the loss of bronchi in the resected lung lobe was compensated by the bronchi in the remaining lung lobe.

The respiratory function recovery and lung compensatory growth have been investigated from the information of stress and strain distribution obtained by the virtual lobectomy simulation. After the virtual operation, the lung expands in shape post-upper lobectomy to conform to the apical part of the chest wall. It was confirmed that the volume change rate due to the deformation was also large in the place where the residual lung surface curvature was high, which shows the possibility that there is an anatomical relationship between the feature and the deformation. Regards to the stress distri-

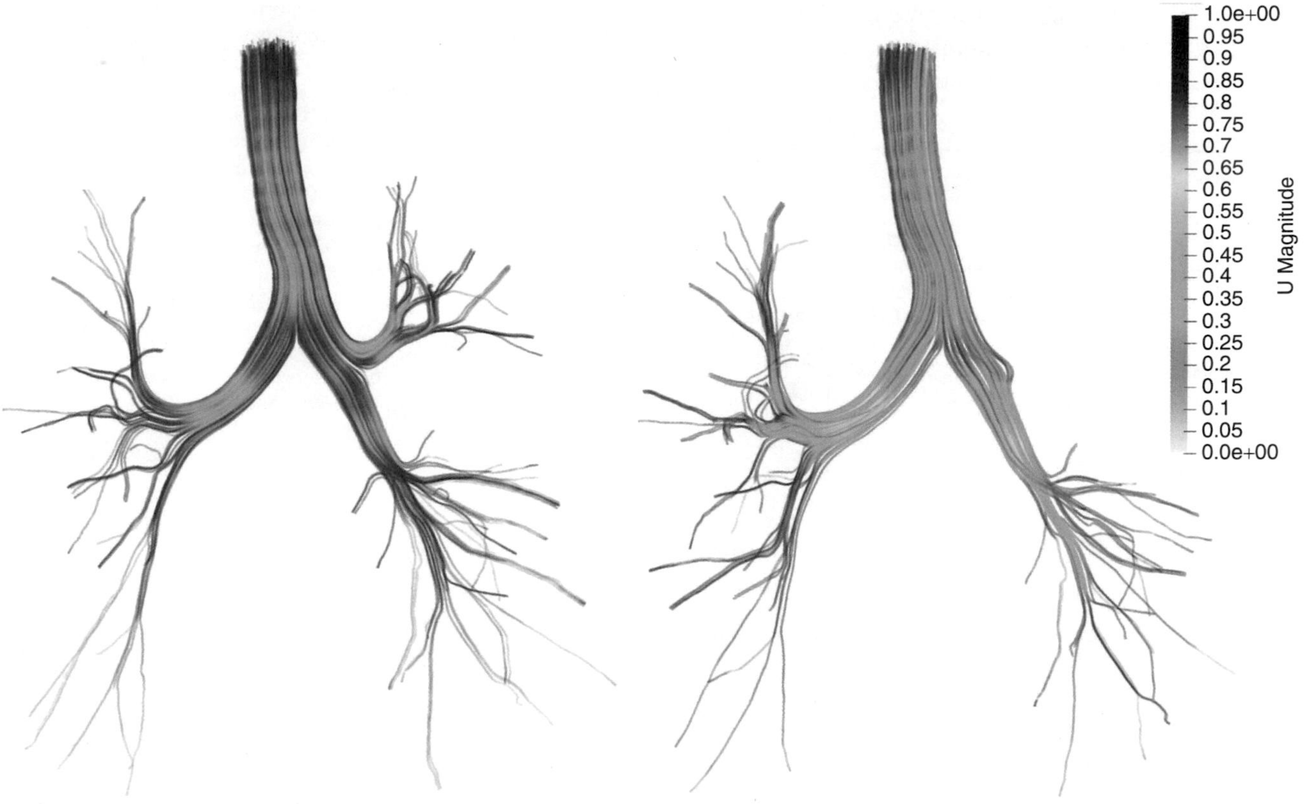

Fig. 26.3 Airway flow patterns at steady inspiration; left: pre-operation; right: post-operation. Color legend shows the magnitude of normalized flow velocities

bution after the operation, the top of the expanded lung showed a high-stress state. The pleural pressures post-operation become more negative at the apex than at the pulmonary base. Since it is assumed that the space of the removed lung lobe is filled with air, it is considered that the resistance to the deformation of the diaphragm was reduced and thus resulted in the reduction of the pleural pressure. The more negative pleural pressures help recovering the respiratory function according to the airway flow simulation. On the other hand, the tortuosity of the bronchi becomes higher due to the deformation of the lung lobe. High tortuosity of the bronchi results in low airflow velocities. Therefore, the respiratory function deteriorated after the lobectomy operation.

26.4 Conclusion

A platform for establishing an integrated computational mechanics model of the human respiratory system has been developed. This platform includes a finite element model of respiration system including sternal, rib, vertebral bones, intercostal muscle, diaphragm, heart and lung and a voxel model for the bronchus.

The physiologically based simulation of respiration was carried out by achieving the chest movement with the con-

traction of the intercostal muscle and diaphragm. The chest deformation modes were obtained compatible with the conventional inference. The intrathoracic and intrapulmonary pressures were also obtained consistent with clinical observation. The effectiveness of the proposed computational model was demonstrated. In addition, the respiratory function was evaluated by airflow simulation performed before and after the virtual lobectomy. The mechanism of respiratory function recovery and lung compensatory growth were investigated from the relationship between biomechanical information (e.g., stress and strain distribution) and morphological change of the lung.

Our developed numerical respiratory system has great potential for not only providing useful information in terms of predicting accurate lung tumor position for the radiation therapy but also estimating the respiratory function for postoperative period of lobectomy. On the other hand, establishing clinically useful postoperative prediction of pulmonary resection requires simulations for many cases to make clear the relationship between physiological and biomechanical indices. Since creating simulation models takes much time and long period is necessary for accumulating postoperative clinical data, the postoperative prediction has not reached a clinically practical level in this study. However, by establishing the methodology of postoperative prediction of lung resection, we believe that the final purpose of this study can be achieved by performing simulations based on as many clinical cases as possible in the future.

References

1. Ueda K, Hayashi M, Tanaka N, Tanaka T, Hamano K. Long-term pulmonary function after major lung resection. Gen Thorac Cardiovasc Surg. 2014;62(1):24–30.
2. Dane DM, Yilmaz C, Estrera AS, Hsia CCW. Separating in vivo mechanical stimuli for postpneumonectomy compensation: physiological assessment. J Appl Physiol. 2013;114(1):99–106.
3. Netter FH. Atlas of human anatomy. Philadelphia: Saunders; 2006.
4. Loring SH. Action of human respiratory muscles inferred from finite element analysis of rib cage. J Appl Physiol. 1992;72(4):1461–5.
5. Martins JAC, Pato MPM, Pires EB. A finite element model of skeletal muscles. Virtual Phys Prototyp. 2006;1(3):159–70.
6. Zhang G, Chen X, Ohgi J, et al. Biomechanical simulation of thorax deformation using finite element approach. Biomed Eng Online. 2016;15:18. https://doi.org/10.1186/s12938-016-0132-y.
7. Jiang F, Hirano T, Ohgi J, Chen X. A voxel image-based pulmonary airflow simulation method with an automatic detection algorithm for airway outlets. Int J Numer Meth Biomed Engng. 2020;36:e3305. https://doi.org/10.1002/cnm.3305.
8. Zhang G, Chen X, Ohgi J, Jiang F, Sugiura S, Hisada T. Effect of intercostal muscle contraction on rib motion in humans studied by finite element analysis. J Appl Physiol. 2018;125:1165–70.
9. Guyton AC, Hall JE. Textbook of medical physiology, Guyton physiology series. Philadelphia: Elsevier Saunders; 2006.

Part IX

New Frontier of Technology in Clinical Applications Based on MCA Models: Abdominal Organs

Analysis in Three-Dimensional Morphologies of Hepatic Microstructures in Hepatic Disease

Hiroto Shoji

Abstract

Microstructures in the liver primarily composed of hepatocytes, hepatic blood, and biliary vessels. Each hepatocyte comes in contact with both vessels; hence, these vessels form three-dimensional (3D) periodic network patterns. In this chapter, we present an estimation of 3D gaps by using a reaction-diffusion algorithm. The proposed method realizes a reliable tool for image segmentation for 3D periodic network patterns. We also applied this approach to examine the 3D sinusoidal network patterns of rats fed a high-fat/high-cholesterol diet; these rats exhibited pathological features similar to those of human patients with nonalcoholic steatohepatitis related to metabolic syndrome. Significant difference was found in diffusion scaling parameter among the experimental groups. Moreover, extending the RD mechanism, we have developed the method to segmentation the sinusoidal network and bile canaliculi at the same time. Therefore, this approach may have the power not only for image segmentation of 3D network patterns but also for pattern recognition problems in diseased animals.

Keywords

Sinusoid · Bile canaliculi · Turing · Reaction-diffusion

27.1 Morphology of 3D Microstructures in Hepatic Lobules

The liver handles several chemical reactions in the body such as protein synthesis, nutrition storage, detoxification action, and bile synthesis necessary for digestion of food [1]. The liver is constructed of basic units called hepatic lobule, which stack in hexagonal columns with a diameter of one to two mm to form a polygonal prism [2]. Hepatic lobules are primarily composed of hepatocytes, hepatic blood vessels, sinusoids (SDs), and the biliary system (bile canaliculus (BC)). Hepatocytes have abundant eosinophilic cytoplasm and nuclei. They are arranged in cords that are one or two cells thick. These cords are separated by SDs, which are the capillary networks that supply the nutrients to hepatocytes. In addition, each hepatocyte is in contact with BC, which are thin tubes that collect bile secreted by hepatocytes. These are formed by modification of the contact surfaces of liver cells. Therefore, hepatocytes partly face SDs, and partly face BC, with the adjacent hepatocytes. As each hepatocyte is in contact with both the networks as described above, these microstructures constitute a 3D conformation [2, 3]. These structures are arranged in such a way that they cannot be written in a plan two-dimensional (2D) image plan. We have elucidated 3D morphology of SDs and BC of liver in diseased rats based on mathematical view, and have developed a pattern recognition method of a 3D structures of microstructures to apply to each process of Multidisciplinary Computational Anatomy, MCA. In this chapter, we introduce the reaction–diffusion (RD) algorithms to extract a 3D SD and BC network efficiently and mathematically.

27.2 Reaction–Diffusion Algorithm for Segmentation of 3D Sinusoidal Networks

Confocal microscopy has been used to analyze 3D structures of cells and tissues after immunofluorescence staining that allows for the examination of the relationship between cell arrangement and metabolic function [1, 2]. Animal experiments were performed using 6-week-old male Wistar rats. All rats were sacrificed, and their livers removed. An immunofluorescence technique was applied to 50-μm-thick frozen sections of liver. Confocal Z-stack images were obtained using an Olympus FV 1000 confocal microscope running

H. Shoji (✉)
Department of Science and Technology, Kwansei Gakuin University, Sanda, Hyogo, Japan
e-mail: shoji@kwansei.ac.jp

Fluo View version 2.0 c software (Olympus, Tokyo, Japan). For each 3D fluorescence image, 50 frames (640 × 640 pixels) were obtained with a length of 0.50 μm between pixels and frames. Sinusoids are formed by sinusoidal endothelial cells, and these cells can be imaged by immunostaining using confocal microscopy. Therefore, to analyze the 3D sinusoidal network structures, explicitly segmenting the sinusoidal networks is necessary.

The segmentation process can be very time consuming; thus, it is fundamental to choose the right techniques for properly filtering images [4]. Furthermore, it is necessary to perform appropriate noise removal suitable for the image processing that follows [4]. On the other hand, several examples of signal processing algorithms employing biological pattern formation mechanisms, such as the RD model, have been proposed [5]. In relation to these, Alan Turing, in 1952, demonstrated that spatially heterogeneous patterns can be formed out of a completely homogeneous field, in which two kinds of diffusive chemical substances, called *morphogens*, react with one another and diffuse through fields, if certain conditions are met [6, 7].

3D sinusoidal segmentation have carried out based on a Turing RD model [8]. In this system, the fascinating phenomenon of Turing pattern formation has been reported [8]. The method used for image segmentation was based on the RD model (FitzHugh-Nagumo equation), with modifications as follows:

$$\frac{\partial u(\vec{r},t)}{\partial t} = D_u \delta \nabla^2 u(\vec{r},t) + u(\vec{r},t) - u(\vec{r},t)^3 - v(\vec{r},t) + \varepsilon U(\vec{r}),$$

$$\partial v(\vec{r},t)/\partial t = D_v \delta \nabla^2 v(\vec{r},t) + \gamma \left(u(\vec{r},t) - \alpha v(\vec{r},t) - \beta \right),$$
(27.1)

where D_u, D_v, α, β, and ε are positive constants; δ is the control parameter; the variable $u(\vec{r},t)$ and $v(\vec{r},t)$ are local concentrations of the activator and inhibitor, respectively; and $U(\vec{r})$ indicates the intensity of 3D fluorescence images of sinusoid endothelial cells. We first scaled the [0, 255] scale image into the [−0.5, 0.5] range linearly. The initial conditions of $u(\vec{r},t)$ and $v(\vec{r},t)$ were given the equilibrium value of reaction terms with white noise without any spatial correlations. Parameters were selected as satisfying conditions for self-organizing periodic patterns.

In the case where $\varepsilon = 0$, self-organized patterns were generated. Figure 27.1a–c shows the time evolution of the distributions of u and v in one dimension. Previous studies have shown that static periodic patterns are self-organized. Furthermore, 3D Turing patterns have previously been studied. In cases in where $\varepsilon > 0$, self-organized patterns were entrained to the distribution of $U(\vec{r})$. Figure 27.1d–f shows the time evolution of the distributions of u and v with external data $U(\vec{r})$ in one-dimension. Considering the situations in which local differences in intensities of fluorescence

images exist and the distributions are kink type, with dents and different periodicities, the prepared distribution of U was utilized, as shown in Fig. 27.1d–f. Figure 27.1f shows the obtained distribution. Although the prepared distribution U was bumpy and exhibited spatially different amplitudes, the amplitude of the obtained distribution u was identical after the numerical calculation.

To extend this method to 3D spaces, numerical simulations were carried out in 3D space. The chosen space size was the same size as the image size of 3D images obtained by confocal microscopy (640 × 640 × 50, 320 μm × 320 μm × 25 μm). We performed calculations of Eq. (27.1).

The parameter choice of δ introduced in the diffusion coefficient is one of the crucial problems. Changing the parameter δ changes the spatial period of the self-organized pattern. In order to select the most suitable δ for fluorescence imaging, the spatial–autocorrelation function between u and U were calculated. The simulation was repeated by changing the parameter δ. The δ takes the largest correlation extracted as the chosen parameter δ^* and adopted it as a parameter for segmentation. Figure 27.1g–k is an example of segmentations performed.

This method is a new one that differs from the conventional linear image processing utilizing filters. Additionally, we have also developed an application that uses the parameter for generating an index for detecting the pathological characters described below.

27.3 Developing an Index of Liver Disease Progression

Utilizing the parameter δ^* described above, we performed morphological comparison of sinusoidal networks in fatty livers of rats fed a high-fat/high-cholesterol (HFC) diet. These rats exhibited pathological features similar to those of human patients with nonalcoholic steatohepatitis related to metabolic syndrome [2, 3, 9]. The parameter δ^* captured the variations in feeding patterns for rats fed HFC diets.

The HFC groups were fed an HFC diet for 3, 6, 9, 12 weeks (HFC 3, 6, 9, 12 w), and the control groups were fed the control diet for 3, 6, 9, or 12 weeks (Cont 3, 6, 9, 12 w). The livers of each of the three animals were removed, and confocal microscopy images were obtained, and their δ^* values were examined.

Figure 27.2a–h shows representative results for 3D segmentation of the SD network from fluorescence pixel information utilizing the RD algorithms. Changing the parameter δ in Eq. (27.1), we calculated $\delta^* = 1.00, 1.05, 1.10$, and 1.10 for Cont at 3, 6, 9, and 12 weeks, and $\delta^* = 1.10, 1.45, 1.65$, and 1.70 for HFC at 3, 6, 9, and 12 weeks, respectively. We independently calculated δ^* for four segmentations of SD networks for each three individuals. We calculated 12 δ^* for

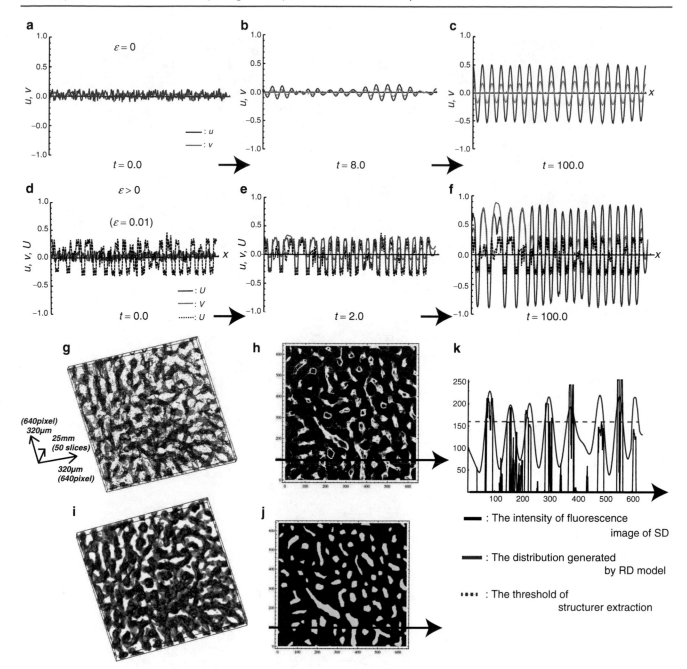

Fig. 27.1 The 3D reconstructed confocal image of liver from rats utilizing the RD algorithm. (**a–f**) Formation of spatial patterns in one dimension obtained from Eq. (27.1). The thick red line, the thick blue line, and the dotted line indicate $u(\vec{r},t)$, $v(\vec{r},t)$, and $U(\vec{r})$, respectively. Since the patterns in (**d–f**) were generated much faster than that for (**a–c**). The time of the patterns is different between (**b**) and (**e**), and between (**c**) and (**f**). (**g–k**) An example of 3D segmentation of sinusoidal network patterns of HFC6w. (**g**), (**h**): 3D segmentation patterns using raw pixel data of the fluorescence image of sinusoidal endothelial cells, and (**i**), (**j**): 3D segmentation patterns obtained by RD processing of Eq. (27.1) with $\delta^* = 1.35$. (**g**) and (**i**) show the 3D segmentation patterns of sinusoidal network (red tubes), and (**h**) and (**j**) indicate the slices at the middle position of the z-axis from (**g**) and (**i**), where the white area indicates the positions inside the sinusoids. (**k**) indicates the spatial variations of the distributions of pixel data (black line) and the scaled distribution (red line) after RD processing of Eq. (27.1) along the black arrow in (**h**) and (**j**). The dotted line shows the threshold of segmentations in the image processing, where the values above the red line were considered to be inside the SD

each experimental group. Significant differences were observed between Cont and HFC at 3, 6, 9, and 12 weeks, and δ^* were increased depending on the number of weeks of HFC diet, as shown in Fig. 27.2i.

Here, we mention the proposed index described above. Since the parameter δ^* is related to the period of the patterns obtained using Eq. (27.1), it is possible that the periodicity of the 3D network pattern is essential to detecting differences

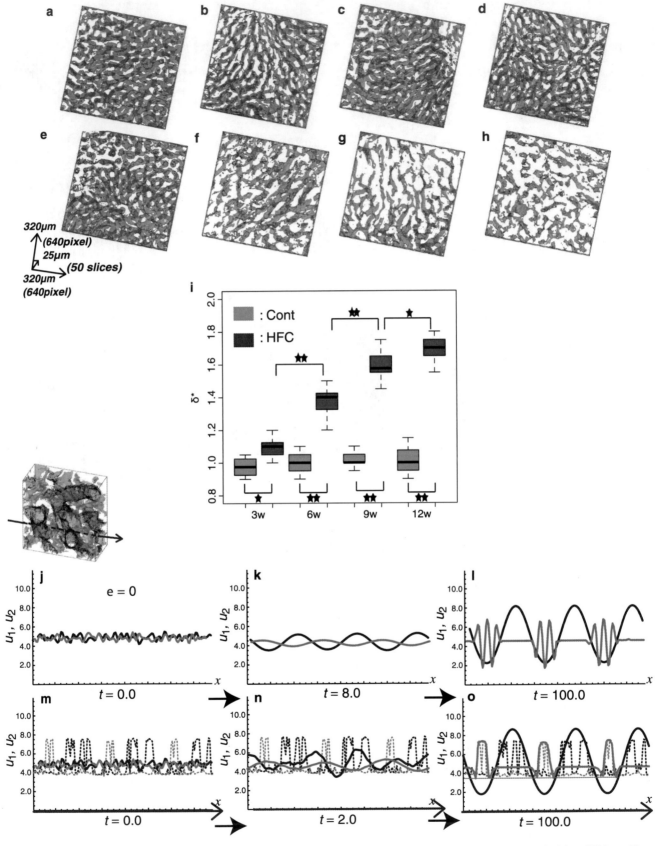

Fig. 27.2 Application and expansion of RD algorithm. (**a–h**) Example of 3D segmentation of sinusoidal networks utilizing Eq. (27.1). (**a–d**): Cont, and (**e–h**): HFC, (**a**) and (**d**): 3 weeks, (**b**) and (**f**): 6 weeks, (**c**) and (**g**): 9 weeks, (**d**) and (**h**): 12 weeks. The red area indicates the sinusoidal veins. (**i**) Change in δ^* with the largest correlation function. The number of stars shows the statistical level of significance (filled star: $p < 0.05$, double filled star: $p < 0.01$) in the Mann-Whitney U-test test between HFC and Cont at each week. (**j–o**) Formation of spatial patterns in one dimension obtained from Eq. (27.2). The thick red line, the thick blue line, the dotted red line, and the dotted green line indicate $u_1(\vec{r},t)$, $u_2(\vec{r},t)$, $U(\vec{r})$, and $V(\vec{r})$, respectively.

among the obtained patterns. However, strong localities in periodicities of 3D sinusoidal network patterns were observed. Therefore, it was not possible to detect clear periodicities in 3D SD patterns using calculations such as the Fourier analysis or spatial correlation analysis.

27.4 Two-Layer Reaction–Diffusion Model for an Evaluation of Two Types of Networks

In the previous sections, we have considered the reaction–diffusion model that captures the morphology of sinusoidal networks. However, as described above, the hepatocytes partly face SDs, and partly face biliary canaliculi, with adjacent hepatocytes. They also have a portion adjacent to the canaliculi that attach to the neighboring hepatocytes [2]. Namely, two types of 3D networks with different thicknesses, SD and BC, are self-organized, and do not cross each other.

Bile salt export pump of the canalicular membrane transporter in bile canaliculi can be also imaged by immunostaining using confocal microscopy. Therefore, in this section, we study the following type of RD model for self-organizing two types of 3D networks with different thicknesses.

3D SD and BC network segmentation was based on the following a Turing RD model. The method for image segmentation was based on the RD model with modifications as follows:

$$\partial u_1(\vec{r},t)/\partial t = D_{u1}\delta_1\nabla^2 u_1(\vec{r},t) + a - (1+b)u_1(\vec{r},t) + u_1^2(\vec{r},t)v_1(\vec{r},t) + \varepsilon U((\vec{r})),$$

$$\partial v_1(\vec{r},t)/\partial t = D_{v1}\delta_1\nabla^2 v_1(\vec{r},t) + bu_1(\vec{r},t) - u_1^2(\vec{r},t)v_1(\vec{r},t),$$

$$\partial u_2(\vec{r},t)/\partial t = D_u\delta_2\nabla^2 u_2(\vec{r},t) + \alpha(u_1(\vec{r},t) - u_2(\vec{r},t)) + a - (1+b)u_2(\vec{r},t) + u_2^2(\vec{r},t)v_2(\vec{r},t) + \varepsilon V((\vec{r})),$$

$$\partial v_2(\vec{r},t)/\partial t = D_{v2}\delta_2\nabla^2 v_2(\vec{r},t) + \alpha(v_1(\vec{r},t) - v_2(\vec{r},t)) + bu_2(\vec{r},t) - u_2^2(\vec{r},t)v_2(\vec{r},t), \tag{27.2}$$

where $D_{u1}, D_{u2}, D_{v1}, D_{v2}, a, b, \alpha$, and ε are positive constants; δ_1 and δ_2 are the control parameter; the variable $u_1(\vec{r},t), u_2(\vec{r},t), v_1(\vec{r},t)$, and $v_2(\vec{r},t)$ are local concentrations of the activator for the sinusoid layer, inhibitor for the sinusoid layer, activator for the bile canaliculus layer and inhibitor for the bile canaliculus layer, respectively; and $U(\vec{r})$ and $V(\vec{r})$ indicate the intensities of 3D fluorescence images of sinusoid endothelial cells and bile salt export pump. Following the same manner described above, we scaled the fluorescence image scale into the range of the RD distributions, and the initial distributions of $u_1(\vec{r},t), u_2(\vec{r},t), v_1(\vec{r},t)$, and $v_2(\vec{r},t)$ were given.

In the cases where $\varepsilon = 0$, self-organized patterns were generated. Figure 27.2j–l shows the time evolution of the distributions of u and v in one dimension. As shown in the previous section, in cases where $\varepsilon > 0$, self-organized patterns were entrained to the distribution of $U(\vec{r})$ and $V(\vec{r})$. Figure 27.2m–o shows the time evolution of the distribution of $u_1(\vec{r},t), u_2(\vec{r},t)$ with external data $U(\vec{r})$ and $V(\vec{r})$ in one dimension.

Following the manner described in the previous section, numerical simulations were carried out in 3D space. In the case where $\varepsilon = 0$, self-organized patterns were formed so that they do not intersect and were orthogonal to each other. In cases where $\varepsilon > 0$, self-organized patterns were entrained to the distribution of $U(\vec{r})$ and $V(\vec{r})$. Therefore, as we choose appropriate thresholds for segmentation of SD and BC in the image processing, where the values above the red line and green line were considered to be inside the SD and BC, 3D SD and BC segmentation can be carried out.

27.5 Future Outlook

In this chapter, we have introduced a Turing reaction–diffusion algorithm for extracting complicated 3D SD and BC network patterns. Using this mechanism, we proposed a method for segmentation of 3D sinusoidal networks using a Turing RD model and information interpolation for complicated SD and BC network patterns.

On the other hand, it is known that the physiology of hepatocytes differs slightly depending on their position within the hepatic lobule [2, 3]. While developing these techniques and gaining a resolution of the sub-micrometer order size that is equal to the structure of the BC network, we would like to advance the detailed analysis using the expanded area information about 3D conformations of SDs, BC, and hepatocytes.

References

1. Kierszenbaum AL, Tres LL. Histology and cell biology. 3rd ed. Philadelphia: Elsevier; 2012.
2. Shelila S, Dooley J. Diseases of the liver and bilary systems. 12th ed. Oxford: Wiley-Blackwell; 2002.
3. Sanyal SAJ, Boyerm TD, Lindor KD, Terrault NA. Zakim and Boyer's hepatology: a textbook of liver disease. 7th ed. Philadelphia: Elsevier; 2018.
4. Dawant DM, Zijdenbos AP. Image segmentation. In: Fitzpatrick JM, Sonka M, editors. Handbook of medical imaging. Bellingham: SPIE Press; 2000.
5. Kuhner L, Agladze KI, Krinsky VI. Image processing using light-sensitive chemical waves. Nature. 1989;337:224.
6. Turing AM. The chemical basis of morphogenesis. Philos Trans R Soc London. 1952;237:37–72.
7. Murray JD. Mathematical biology. 3rd ed. Cham: Springer; 2003.
8. Shoji H, Yamada K, Ueyama D, Ohta T. Turing patterns in three dimensions. Phys Rev. 2007;E75:46212.
9. Yatti H, Naito H, Jiax X, Shindo M, Taki H, Tamada H, Kiitamori K, Hayashi Y, Ikeda K, Yamori Y, Nakajima T. High-Far-Cholesterol Diet Mainly Induced Necrosis in Fibrotic Steatophepatitis Rat by Suppressing Caspase Activity. Life Sci. 2013;93:673–80.

Quantitative Evaluation of Fatty Metamorphosis and Fibrosis of Liver Based on Models of Ultrasound and Light Propagation and Its Application to Hepatic Disease Diagnosis

28

Tsuyoshi Shiina, Makoto Yamakawa, Kengo Kondou, and Masatoshi Kudo

Abstract

With the aim of early diagnosis of chronic hepatitis, we investigated a noninvasive method to quantify hepatic fibrosis and fat deposition with high sensitivity. In order to quantitatively evaluate hepatic fibrosis, we examined the effect of changes in the liver fibrous structure on shear wave dispersion by modeling the mechanical properties of hepatic tissues using hepatitis specimen sections. Simulation analysis indicated that the influence of liver fibrous structure cannot be ignored in the evaluation of liver viscosity, and that combining the dispersion slopes estimated at different frequency ranges is suitable for viscoelasticity analysis to evaluate fibrosis progression in chronic hepatitis diagnosis.

Fat deposition was evaluated using photoacoustic imaging technology by modeling the optical properties of different lipid concentrations in liver tissues. The fatty liver models were composed of mixing different lipid percentages and homogenized livers. We proposed the "spectral ratio" as an index of fatty deposition. We demonstrated that the spectral ratio reflected lipid percentage and verified the feasibility of an ex vivo experiment using a nonalcoholic steatohepatitis (NASH) mouse model. The results confirmed the usefulness of this system in the evaluation of fatty liver.

Keywords

Tissue characterization · Ultrasound elastography Photoacoustic imaging · Chronic hepatitis · Liver fibrosis Fatty metamorphosis · Nonalcoholic steatohepatitis

28.1 Introduction

Diagnosing chronic liver disease in the early stage is necessary for the treatment of liver disease because chronic hepatitis C virus disease progresses towards cirrhosis, which often leads to liver cancer [1]. In addition to chronic viral hepatitis, recently, the incidence of nonalcoholic steatohepatitis (NASH), a malignant fatty liver disease is increasing. NASH can also develop into progressive liver fibrosis, leading to cirrhosis and an increased risk of cancer [2, 3]. At present, the gold standard for the assessment of liver fibrosis is liver biopsy, which is invasive, causes patient discomfort, and a concomitant disease risk. In addition, the accuracy of liver biopsy is limited because of significant intra- and inter-observer variability and sampling errors [4, 5]. Thus, there is a need for a noninvasive method to quantify hepatic fibrosis and fat, which is a biomarker for hepatic disease and metabolic syndrome.

A positive correlation has been demonstrated between liver stiffness and hepatic fibrosis stage in chronic hepatitis [6–8]. Therefore, ultrasound elastography has been developed to evaluate fibrosis progression from shear wave speeds. Recently, it has been hypothesized that viscosity analysis, in addition to elasticity measurement, could improve the accuracy of fibrosis staging. Recent studies have used shear wave elastography to evaluate viscosity by the dispersion slope of shear wave phase velocity [9–10]. However, the shear wave cannot propagate properly when the thickness of the subject is less than the shear wavelength. Therefore, the shear wave

T. Shiina (✉) · M. Yamakawa · K. Kondou
Graduate School of Medicine, Kyoto University, Kyoto, Japan
e-mail: shiina.tsuyoshi.6w@kyoto-u.ac.jp

M. Kudo
Department of Gastroenterology and Hepatology, Faculty of Medicine, Kindai University, Osaka, Japan
e-mail: m-kudo@med.kindai.ac.jp

dispersion is expected to be affected not only by viscosity, but also by fibrous structure.

Photoacoustic imaging has attracted attention in recent years as an in vivo imaging method for visualizing details of the neovascularization structure of tumors and the distribution of oxygen saturation, which is related to the tumor grade. It presents the benefits of deep imaging of ultrasound with high contrast and resolution, which are merits of optical imaging [11]. Photoacoustic imaging is also applicable for diagnosing properties of various tissues, such as fat related to arteriosclerosis, fatty liver, and fibrous tissues related to hepatitis [12–16].

Therefore, with the aim of early diagnosis of chronic hepatitis, we investigated a noninvasive method to quantify hepatic fibrosis and fat deposition with high sensitivity as shown Fig. 28.1. In order to quantitatively evaluate hepatic fibrosis, we examined the effect of changes in the liver fibrous structure on shear wave dispersion by modeling the mechanical properties of hepatic tissues using hepatitis specimen sections derived from patients with chronic hepatitis C. Fat deposition was evaluated by photoacoustic imaging

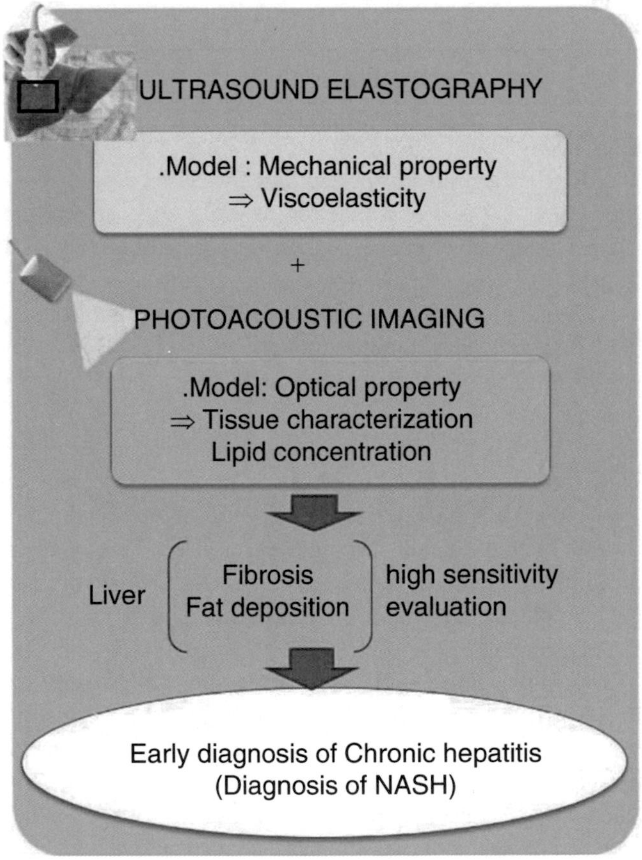

Fig. 28.1 Quantitative evaluation of hepatic fibrosis and fat deposition by modeling the mechanical properties of liver tissues using ultrasound elastography and the optical properties with photoacoustic technology

technology by modeling the optical properties of different lipid concentrations of liver tissues.

28.2 Evaluation of Viscoelasticity Related to Fibrosis Structure

Based on the positive correlations between liver stiffness and hepatic fibrosis stage, ultrasound elastography is now applied to noninvasively differentiate mild-to-moderate fibrosis from advanced fibrosis and cirrhosis [17, 18]. However, soft tissues are viscoelastic; therefore, viscosity is also necessary to characterize tissue properties. In particular, early diagnosis of NASH is expected using both the properties, elasticity, and viscosity, of tissues [9, 10]. Several studies have examined the frequency-dependent shear wave properties for viscoelasticity evaluation. However, it is not clear how the frequency-dependent behavior of shear wave propagation or dispersion affects the diagnostic utility of shear wave elastography.

Therefore, for quantitative evaluation of hepatic fibrosis, we investigated the mechanism by which fibrosis progression affects the estimation of viscoelastic properties by using a mechanical model based on hepatitis histological specimen sections [19].

28.2.1 Evaluation of Shear Wave Dispersion

In shear wave elastography, shear waves are generated by acoustic radiation force (ARF) near the measurement points and propagate along the lateral direction. First, the shear wave phase velocity is estimated from the phase difference ($\Delta\theta$) of the shear waves and the distance (Δd) between two reference pixels [A, B in Fig. 28.2a, b] located at the same depth. The phase difference is measured by Fourier transforming the shear wave particle velocity in time.

The shear wave phase velocity, $c_s(f)$ is obtained by Eq. (28.1).

$$c_s(f) = \frac{\Delta d}{\Delta\theta / 2\pi f} \tag{28.1}$$

This procedure is repeated for the entire pixel area of the particle velocity data to estimate the distribution of the shear wave phase velocity. Second, the averaged shear wave phase velocity at each frequency is plotted on a graph as represented in Fig. 28.2c, and parameterized for a linear dispersion model defined as:

$$c_s(f) = c_0 + \frac{dc_s}{df} f \tag{28.2}$$

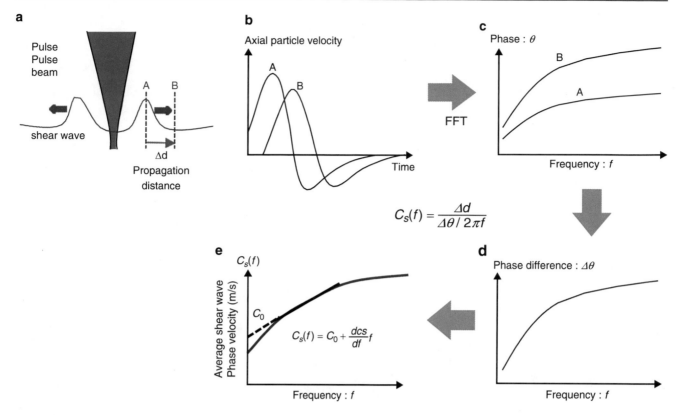

Fig. 28.2 Estimation of shear wave phase velocities and their dispersion slope in shear wave elastography. (**a**) Shear wave generated by acoustic radiation force of push pulse beam and its propagation. (**b**) Particle velocity detected at two points A and B. (**c**) Phase as a function of frequency, which is calculated by Fourier transforming waveform in (**b**). (**d**) The phase difference of shear waves obtained from (**c**). (**e**) Shear wave phase velocity versus frequency calculated by Eq. (28.1) and dispersion slope estimated by a linear regression of phase velocity based on Eq. (28.2)

where $c_s(f)$ denotes the shear wave phase velocity at each frequency, f expresses the frequency, c_0 signifies the intercept, and dc/df stands for the dispersion slope. The intercept and dispersion slope were estimated by performing a linear regression of the phase velocity as a function of frequency. The slope value dc/df is evaluated as the dispersion slope. In the linear regression, a frequency range needs to be selected; in this case, it ranges over 30–450 Hz [20].

28.2.2 Mechanical Model Analysis of Liver Fibrosis

First, the structural model is formed in accordance with the fibrosis binary image in the 40 mm × 40 mm 2-dimensional area derived from the chronic hepatitis C specimen section, as shown in Fig. 28.3. Next, the elasticity values are assigned to models based on elasticity values measured by transient elastography, for example, parenchyma as 3 kPa and fiber as 75 kPa. Histological models were created from multiple specimens, that is, three specimens for fibrosis stages F0, F1,

F2, F3, and two specimens for F4. Each model was shifted horizontally, vertically, or transposed to increase the variety of models. Consequently, 12 models for F0, F1, F2, F3, and 8 models for F4 were generated for the simulation. The averaged elasticity values for each stage are shown in Table 28.1.

28.2.3 Analysis of Shear Wave Dispersion by Fibrous Structure

We investigated how the shear wave dispersion is affected by fibrous structure. Shear wave propagation within this model was simulated using the open-source k-Wave MATLAB toolbox. The simulation area was 40 mm × 40 mm, and the ARF excitation was set at 7.5 mm from the left side of the model. Plane wave excitation was performed by a 1-mm wide Gaussian function. The simulation time was 20 ms, the sampling frequency was 50 MHz, and the tracking PRF was 50 kHz. The shear wave velocity is determined for each model based on the elasticity value at every pixel in each model, using $E = 3\rho c_s^2$, where E is Young's modulus, ρ is density, and

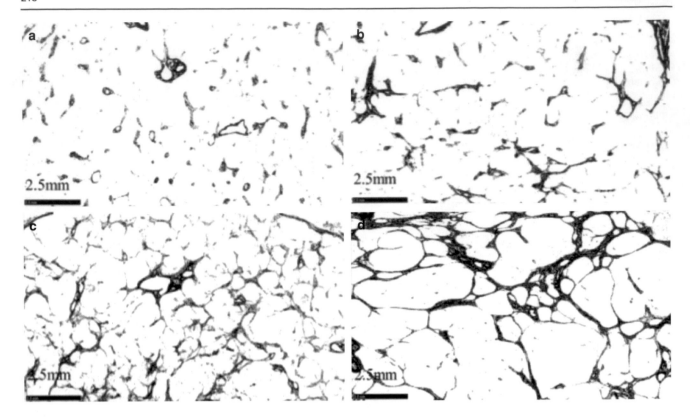

Fig. 28.3 An example of histological binary fiber image from hepatic histological specimens derived from patients of chronic hepatitis C. (**a**) Fibrosis stage F1, (**b**) F2, (**c**) F3, (**d**) F4

Table 28.1 Averaged elasticity in fibrosis model

Fibrous stage	F0	F1	F2	F3	F4
Histological model (kPa) ($n = 12$ or 8)	4.56 ± 0.24	6.60 ± 0.97	10.63 ± 0.61	18.42 ± 1.01	25.55 ± 7.94

c_s is shear wave velocity. Tissue viscosity was set to zero to examine the effect of the liver fibrous structure alone. The shear wave dispersion slope in the range of 100–600 Hz is calculated from the simulated shear wave particle velocity. The calculation area is 2-dimension Fourier transformed after performing the Turkey window function. Then, the data are directionally filtered and inverse Fourier transformed. The shear wave phase velocity is calculated by the phase spectroscopy method in the ROI picked up from the calculation area. The phase velocity is calculated from the phase difference and interval between two points parted by 0.3 mm.

Dispersion analysis requires the averaged phase velocity from every point in the ROI at each frequency. At each frequency, the data with a standard deviation over 50 is excluded from the analysis; a linear regression of the remaining data is performed to determine the dispersion slope using the least-square method. The dispersion slope is derived from the approximation straight line, frequency vs. averaged shear wave phase velocity.

The relation between shear wave phase velocity at 200 Hz and the dispersion slope at 100–600 Hz frequency range are summarized at each stage in Fig. 28.4. The average values of the phase velocity at 200 Hz and dispersion slope at each fibrosis stage are tabulated in Table 28.2. These results show that both the dispersion slope and shear wave phase velocity at 200 Hz tend to increase with hepatic fibrosis stage progression.

28.2.4 Tissues Viscoelasticity Evaluation by Shear Wave Propagation

Shear wave propagation with viscoelastic materials was simulated based on a viscoelastic model that is similar to the Voight model as shown below [19].

$$c(\omega) = \sqrt{\frac{2(A^2 + B^2)}{\rho(A + \sqrt{A^2 + B^2})}} \qquad (28.3)$$

$$A = \mu_1 + \frac{\mu_2 \omega^2 \eta^2}{\mu_2^2 + \omega^2 \eta^2}, \qquad (28.4)$$

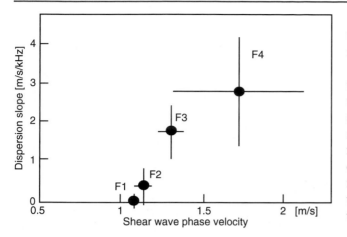

Fig. 28.4 Increment in dispersion slope and average shear wave phase velocity at 200 Hz as fibrous stage progresses

Table 28.2 Dispersion slope and phase velocity

Fibrous stage	Phase velocity (m/s)	Dispersion slope (m/s/kHz)
F1 ($n = 12$)	1.060 ± 0.015	-0.007 ± 0.136
F2 ($n = 12$)	1.111 ± 0.018	0.338 ± 0.395
F3 ($n = 12$)	1.295 ± 0.096	1.671 ± 0.785
F4 ($n = 8$)	1.720 ± 0.414	2.795 ± 1.423

$$B = \frac{\mu_2^2 \omega \eta}{\mu_2^2 + \omega^2 \eta^2} \quad (28.5)$$

where μ_1 and μ_2 denote the shear moduli and η represents the viscosity. A liver fibrosis progression model was developed based on reference studies [21]. The average Young's modulus of these models is given in Table 28.1. Viscosity η was set homogeneously and was varied over 0.1, 0.5, 1.0, 1.5, and 2.0 Pa.s to conduct simulations in various viscoelastic materials.

The two-dimensional model, which was 40×40 mm^2, included 400×400 elements. The element size was 0.1×0.1 mm^2. Gaussian envelope excitation with 1 mm FWHM is applied to the model for 0.5 ms to initiate plane shear wave propagation. The shear wave phase velocities were measured, and the dispersion slope was estimated from the average shear wave phase velocity in the ROI, as shown in Fig. 28.5.

Dispersion slope estimation requires a certain frequency range to calculate the dispersion slope. Some studies have used different frequency ranges. We used frequency ranges of two types: low frequency and high frequency. The low-frequency range is 25–100 Hz; the high-frequency range extends from 200 Hz to a selected frequency because the shear wave phase velocity has different linearity in the two frequency ranges with frequency changes between 100 and 200 Hz, as shown in Fig. 28.5a.

Figure 28.5b, c shows the averaged dispersion slopes in the low-frequency range and high-frequency range, respectively.

These results show that the dispersion slope depends particularly on viscosity in the high-frequency range. In addition, the results of the comparison between the homogeneous and fibrosis models show that the effect of the liver fibrous structure is almost not in the dispersion slope in the high-frequency range. These results indicate that estimating the dispersion at low frequency is expected to be suitable for fibrosis staging considering the degree of the fibrous structure, while the dispersion at high frequency can be used for the evaluation of viscosity.

28.3 Tissue Characterization of Fatty Liver Using Photoacoustic Imaging

To evaluate the feasibility of using photoacoustic methods for tissue characterization of fatty liver, we investigated the relationship between photoacoustic spectra and lipid percentage using the fatty liver model. The fatty liver models were composed of mixing lipid and homogenized chicken liver. Models with different lipid percentages, 0–30%, were developed.

Figure 28.6a shows the experimental setup. A sample of each model embedded in the 2% agar base material was irradiated with nanosecond pulses of laser light (800–1300 nm wavelength, 5 nm step, 30 Hz repetition rate, 1–4 mJ/pulse). Photoacoustic signals of respective wavelengths were obtained using a linear array ultrasound probe (4–15 MHz).

The photoacoustic spectra of liver and lipid in the near-infrared region were measured. The liver has a peak around 900 nm in the photoacoustic spectrum, while lipid has a peak around 930 and 1200 nm. Figure 28.6b shows the photoacoustic spectra of the fatty liver models for different percentages of lipids.

The results show that all models have a peak around 900 and 1200 nm in the spectra, which reflect the properties of their components. To extract the feature regarding lipid percentage, a parameter, spectral ratio κ, was defined as shown in Eq. (28.6):

$$\kappa = \frac{P(1200)}{P(900)} \quad (28.6)$$

where $P(1200)$ and $P(900)$ represent the amplitude of the photoacoustic spectra at 1200 nm and 900 nm, respectively.

Figure 28.6c shows the relation between the spectral ratio κ and lipid percentage, which are calculated using three sets of data. The results indicate that the spectral ratio strongly correlates with lipid percentage and suggest the possibility of estimating lipid percentage by spectral ratio.

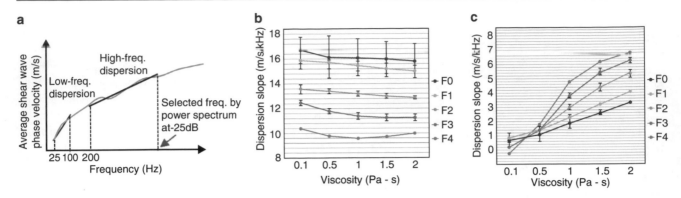

Fig. 28.5 Averaged dispersion slope at all fibrosis stages estimated at the two frequency ranges in the fibrosis model ($n = 3$). (**a**) Obtained shear wave phase velocity and two frequency ranges for estimation of the dispersion. (**b**) Averaged dispersion slope for different fibrosis stages at low-frequency range: 25–100 Hz (**c**) Averaged dispersion slope for different fibrosis stages at high-frequency range: 200–400, 600 Hz

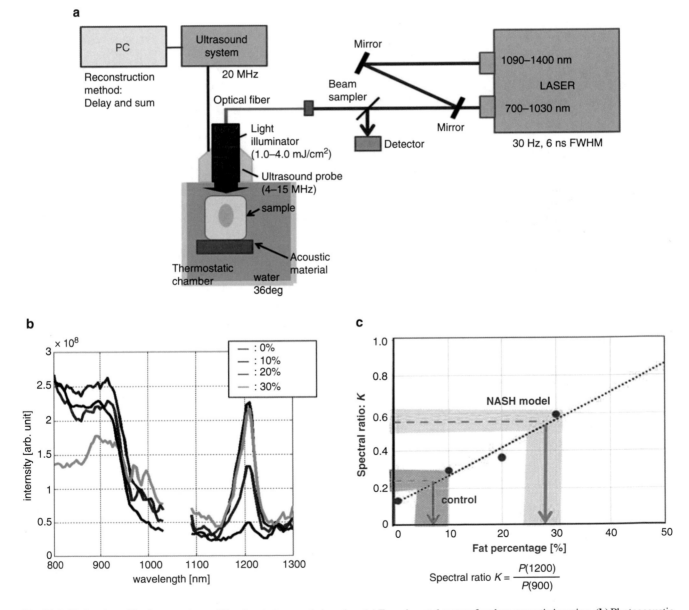

Fig. 28.6 Estimation of the fat percentage of liver by photoacoustic imaging. (**a**) Experimental system for photoacoustic imaging. (**b**) Photoacoustic spectra of the fatty liver models for different lipid content ratios. (**c**) Relation between the spectral ratio and lipid percentage

Next, we conducted ex vivo experiments using control and two fatty liver model mice, as shown in Fig. 28.7.

Figure 28.7b shows tissue sections of mouse liver, which indicate that the livers of NASH model mice include many spots of fatty deposition compared with the control. B-mode images of the liver of normal or control mouse and NASH model mouse are shown in Fig. 28.7c, while the photoacoustic images for the control and NASH mouse models are repre-

sented in Fig. 28.7d. It can be observed that the photoacoustic signal from the liver boundary of the NASH model mouse is stronger than that of the control mouse, which reflects the photoacoustic spectrum of the fatty tissue, as shown in Fig. 28.5b. The photoacoustic spectra of the control and NASH mouse models are shown in Fig. 28.7d, e, respectively.

Next, the spectral ratios [defined in Eq. (28.6)] for the control and NASH mouse models were obtained from

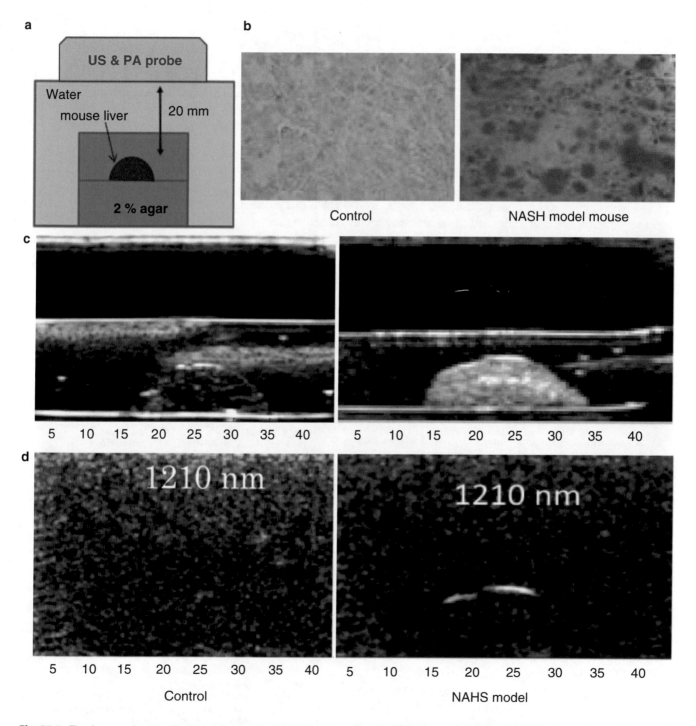

Fig. 28.7 Ex vivo experiment with livers of control and NASH mice models, with ultrasound B-mode and photoacoustic (PA) imaging. (**a**) Experimental setup (**b**) Tissue section of mouse liver (**c**) Ultrasound B-mode (**d**) PA image (at 1210 nm) (**e**) PA spectra of control mouse (**f**) PA spectra of NASH model mouse

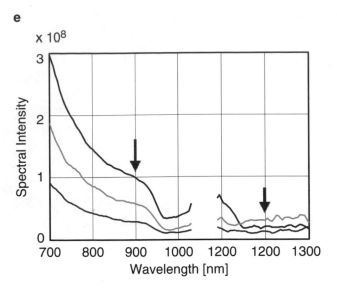

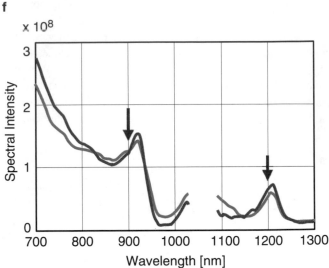

Fig. 28.7 (continued)

Fig. 28.7d, e. Finally, by applying these values to the relationship between spectral ratio and lipid percentage in Fig. 28.6c, the fat concentration of liver was estimated: 23–32% for the NASH model vs. 6–10% for the control.

28.4 Conclusion

We performed tissue viscoelasticity evaluation by modeling the shear wave propagation; according to in vivo measurements in NAFLD patients [3], dispersion slope values range from 0 to 10 m/s/kHz, from F0 to F4 stage. Our results shown in Table 28.2 are about 30% of the dispersion slope in in vivo measurements. Therefore, these results suggest that the influence of liver fibrous structure cannot be ignored while evaluating liver viscosity. The results of the simulation analysis of shear wave propagation with a viscoelastic tissue model indicate that the dispersion slope in the high-frequency range reflects the viscosity change, while dispersion at low frequencies reflects the progress of the fibrosis stage. Therefore, we infer that combining the dispersion slopes estimated at different frequency ranges is suitable for viscoelasticity analysis to evaluate fibrosis progression in the diagnosis of chronic hepatitis.

To evaluate the characteristics and functional properties of liver tissues by photoacoustic imaging, we demonstrated that the spectral ratio reflects lipid percentage and verified the feasibility of an ex vivo experiment using a NASH mouse model. The results confirmed the usefulness of this system in the evaluation of fatty liver. Experiments using more realistic models featuring subcutaneous tissues, optimization of irradiated light intensity, selection of appropriate wavelength,

and signal processing to improve SNR must be conducted in future studies.

References

1. National Institutes of Health. National Institutes of Health Consensus Development Conference Statement. Management of Hepatitis C. Hepatology. 2002;36:S3.
2. Wieckowska A, Feldstein AE. Diagnosis of nonalcoholic fatty liver disease: invasive versus noninvasive. Semin Liver Dis. 2008;28:386–95.
3. Vuppalanchi R, Chalasani N. Nonalcoholic fatty liver disease and nonalcoholic steatohepatitis: selected practical issues in their evaluation and management. Hepatology. 2009;49:306–17.
4. Bedossa P, Dargere D, Paradis V. Sampling variability of liver fibrosis in chronic hepatitis C. Hepatology. 2003;38(6):1449–57.
5. Ziol M, Handra-Luca A, Kettaneh A, et al. Noninvasive assessment of liver fibrosis by measurement of stiffness in patients with chronic hepatitis C. Hepatology. 2005;41(1):48–54.
6. Arena U, Vizzutti F, Corti G, et al. Acute viral hepatitis increases liver stiffness values measured by transient elastography. Hepatology. 2008;47(2):380–4.
7. Shiina T, et al. WFUMB guidelines and recommendations for clinical use of ultrasound elastography: part 1: basic principles and terminology. Ultrasound Med Biol. 2015;41(5):1126–47.
8. Ferraioli G, Tinelli C, Bello BD, Zicchetti M, Filice G, Filice C. Accuracy of real-time shear wave elastography for assessing liver fibrosis in chronic hepatitis C: a pilot study. Hepatology. 2012;56(6):2125–33.
9. Nightingale KR, Rouze NC, Rosenzweig SJ, Wang MH, Abdelmalek MF, Guy CD, Palmeri ML. Derivation and analysis of viscoelastic properties in human liver: impact of frequency on fibrosis and steatosis staging. IEEE Trans Ultrason Ferroelectr Freq Control. 2015;62(1):165–75.
10. Barry CT, Mills B, Hah Z, Mooney RA, Ryan CK, Rubens DJ, Parker KJ. Shear wave dispersion measures liver steatosis. Ultrasound Med Biol. 2012;38(2):175–82.

11. Toi M, et al. Visualization of tumor-related blood vessels in human breast by photoacoustic imaging system with a hemispherical detector array. Sci Rep. 2017;7:41970-1-11.

12. Wang LV, Yao J. A practical guide to photoacoustic tomography in the life sciences. Nat Methods. 2016;13(8):627–38.

13. Hirano S, et al. Aortic atherosclerotic plaque detection using a multiwavelength handheld photoacoustic imaging system. Proc SPIE. 2016;97084 https://doi.org/10.1117/12.2229272.

14. Shiina T, Toi M. Photoacoustic imaging technology to support vascular health science. Optoronics; 2017.

15. Uchimoto Y, Namita T, Kondo K, Yamakawa M, Shiina T. Feasibility evaluation of 3D photoacoustic imaging of blood vessel structure using multiple wavelengths with a handheld probe. Proc SPIE. 2018.

16. Namita T, Uchimoto Y, Murata Y, Asada K, Nakao Y, Kondo K, Yamakawa M, Shiina T. Tissue characterization using multi-wavelength photoacoustic imaging system. 38th Annual Meeting of the Laser Society of Japan, Jan, 2018.

17. Ferraioli G, et al. WFUMB guidelines and recommendations for clinical use of ultrasound elastography: part 3: liver. Ultrasound Med Biol. 2015;41(5):1161–79.

18. Poynard T, Munteanu M, Luckina E, et al. Liver fibrosis evaluation using real-time shear wave elastography: applicability and diagnostic performance using methods without a gold standard. J Hepatol. 2013;58(5):928–35.

19. Fujii S, Yamakawa M, Kondo K, Namita T, Shiina T. Evaluation of shear wave dispersion in hepatic viscoelastic models including fibrous structure. Jpn J Appl Physics. 2019;58:SGGE07.

20. Parker KJ, Partin A, Rubens DJ. What do we know about shear wave dispersion in normal and steatotic livers? Ultrasound Med Biol. 2015;41:1481–87.

21. Shiina T, Maki T, Yamakawa M, Mitake T, Kudo M, Fujimoto K. Mechanical model analysis for quantitative evaluation of liver fibrosis based on ultrasound tissue elasticity imaging. Jpn J Appl Phys. 2012;51:07GF11.

MCA Analysis for Hepatology: Establishment of the In Situ Visualization System for Liver Sinusoid Analysis

29

Keiichi Akahoshi, Takeshi Ishii, and Atsushi Kudo

Abstract

The liver sinusoid has important role to maintain the function of the liver. By combining a microscope, a high-speed camera and a new Particle Image Velocimetry software "Flownizer2D," we established the in-situ visualization system for liver sinusoid analysis. This system enabled us to measure small changes in the motion of red blood cells in the liver sinusoids of rats. This system has multiple potentials to evaluate the effects of selective hepatic artery or portal vein clamping on the sinusoidal flow and is expected to contribute to clarify the function of liver sinusoids.

Keywords

In situ visualization · Sinusoidal flow · Flownizer2D

29.1 Introduction

The liver is a vital organ that performs multiple critical functions in vertebrates. It produces bile. It detoxifies various drugs. It converts nutrients into forms that can be used by the body and stores the nutrients. These amazing multiple functions are maintained by the liver sinusoid. The liver sinusoid is a type of capillary that has the fenestrated endothelium that permits blood plasma to touch with hepatocytes. Thus, to analyze the change of the liver sinusoid is very important to investigate the functions of the liver. We

have started to try to observe sinusoidal flow 15 years ago. First, we visualized the sinusoidal flow of rats by using a microscope. However, the quality to visualize the blood flow in the liver sinusoid was not adequate, as red blood cells are tiny and the width of the liver sinusoid is only 5 μm. At that time, we evaluate the change of sinusoidal flow in the post-warm ischemic rat liver by counting the number of perfused midzonal sinusoids [1]. Some researchers proposed similar intra-viral microscopic systems [2]. Recently, we utilized a new Particle Image Velocimetry software, "Flownizer2D" (DITECT, Japan). Flownizer2D can instantly process the sequential image provided from the intravital videomicroscopic system into digital data with a new computer architecture and accurately can trace the motion of red blood cells in the liver sinusoids [3]. By using this system, we became to detect the precise and tiny change of the sinusoidal flow in the rat model.

29.2 Animal Preparation

Male Wister rats (weight 270–320 g) were housed in cages under a 12 h light/dark cycle with access to food and water ad libitum. Under subcutaneous urethane anesthesia (2 g/kg), laparotomy was performed. Ligamentous attachments from the liver to the diaphragm and abdominal wall were dissected. The lateral lobe of this rat was placed on the table of the microscope.

29.3 Structure of the In Situ Video Microscopic System

An inverted system microscope (IX71, OLYMPUS, Japan) and a high-speed camera (HAS-L1, DETECT, Japan) were prepared. The high-speed camera was connected with the eyepiece lens of the microscope. And, the real-

K. Akahoshi (✉) · T. Ishii · A. Kudo
Department of Hepatobiliary and Pancreatic Surgery, Graduate School of Medicine, Tokyo Medical and Dental University, Tokyo, Japan
e-mail: akahmsrg@tmd.ac.jp

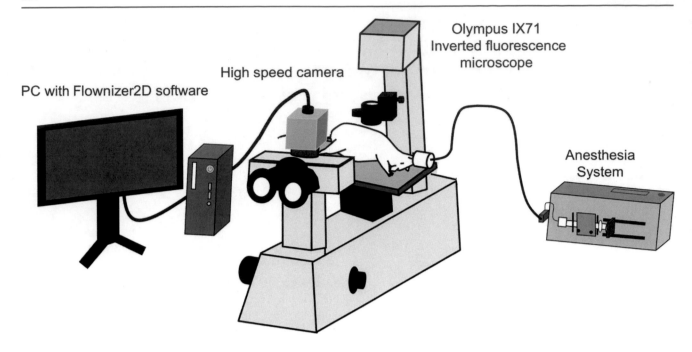

Fig. 29.1 Establishment of the in situ visualization microscopic system

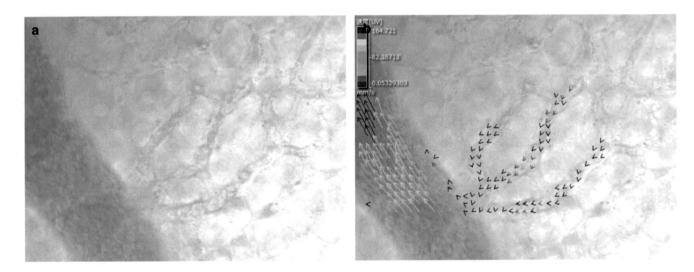

Fig. 29.2 Image capture of liver sinusoids of rats. (**a**) Raw data (×200), (**b**) The data analyzed by Flownizer2D. The motion and its speed of each red blood cell are visualized by allows

time imaging view of the camera was sent to the computer unit. This system enabled us to watch and record real-time sinusoidal flow (Fig. 29.1). After recording the sinusoidal flow, the data was analyzed by using a Particle Image Velocimetry software "Flownizer2D" (DITECT, Japan).

29.4 Observation of the Sinusoidal Flow

As shown in Fig. 29.2, the sinusoidal flow of rats was clearly visualized by this microscopic system (Fig. 29.2a). Then, the Flownizer2D can trace the motion of each red blood cells in

the sinusoids and quantify the speed of red blood cells sequentially (Fig. 29.2b).

29.5 Examples of Sinusoidal Flow Analysis

When we measured the sequential change of sinusoidal flow of rats, the average maximum velocity was 26.1 mm/s, and the flow was a pulsatile flow of 97 times/min. The respiratory rate of the rats was about 100, suggesting that they were affected by the respiration of the rat (Fig. 29.3).

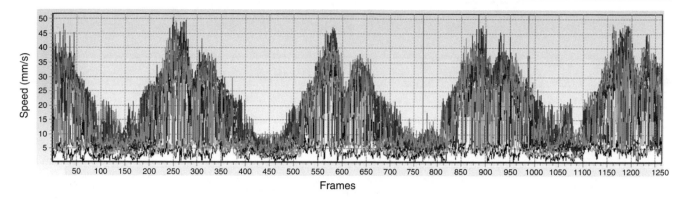

Fig. 29.3 Sequential changes of sinusoidal flow. Sequential changes of the speed of each red blood cells are shown. The vertical axis shows the speed, and the horizontal axis shows the number of frames. Five hundred frames are equivalent to 1 s

29.6   Conclusion

The novel in situ visualization system for liver sinusoid analysis was established by combining a current microscope, a high-speed camera and particle image velocimetry software. This system enabled us to perform a multidisciplinary computational anatomical approach to investigate the function of liver sinusoids, physically, physiologically, pathologically, and sequentially.

References

1. Ban D, Kudo A, Sui S, Tanaka S, Nakamura N, Ito K, Suematsu M, Arii S. Decreased Mrp2-dependent bile flow in the post-warm ischemic rat liver. J Surg Res. 2009;153:310–6.
2. Uhlmann S, Uhlmann D, Spiegel HU. Evaluation of hepatic microcirculation by in vivo microscopy. J Investig Surg. 1999;12(4):179–93. https://doi.org/10.1080/089419399272458.
3. Sakai R. In: Wilson SP, Verschure PFMJ, Mura A, Prescott TJ, editors. Biomimetic and Biohybrid Systems: 4th International Conference, Living Machines 2015, Barcelona, Spain, July 28–31, 2015, Proceedings. Heidelberg: Springer; 2015.

Simulation Surgery for Hepatobiliary-Pancreatic Surgery

Yukio Oshiro

Abstract

Purpose: In April 2012, "Image-assisted navigation in liver resection" was covered by the Japanese medical insurance and was widely used in many institutions in Japan. In recent years, three-dimensional (3D) image-assisted surgery has been expanded to not only liver surgery but also biliary and pancreatic surgery. We have developed computer-aided surgery (CAS) systems in hepatobiliary-pancreatic surgery by utilizing knowledge of multidisciplinary computational anatomy.

Methods: We have independently developed various computer-aided surgery (CAS) systems for hepatobiliary and pancreatic surgery: (1) a surgical simulation using 3D printing in hepatectomy, (2) a 3D virtual hepatectomy simulation combined with real-time deformation, (3) development of our original 3D captured hepatectomy navigation system using 3D camera, and (4) development of our original simulation system for biliary-pancreatic surgery.

Results: The 3D liver print model, the surgical simulation system that enables organ deformation, the 3D captured hepatectomy navigation system using 3D camera, and the biliary pancreatic surgery simulation system that fuses CT and MRCP, which we have originally developed, have greatly contributed to our safe surgery and the improvement of surgical results.

Conclusion: Based on the multidisciplinary computational anatomy, we have developed and introduced various CAS systems for hepatobiliary-pancreatic surgery. Our novel CAS systems have contributed to safe and secure surgeries. It is expected that the possibilities of CAS will continue to expand, as multidisciplinary computational anatomy develops.

Keywords

3D · Computer-aided surgery (CAS) · Simulation surgery Navigation surgery · Hepatobiliary-pancreatic surgery

30.1 Introduction

Surgical imaging support, such as three-dimensional (3D) simulation and navigation surgery based on the patient's independent medical image, has made rapid progress with the development of computer-aided surgery (CAS) in the last 10 years [1–6]. In Japan, image-assisted hepatic surgery has been covered by medical insurance since 2012. Since then, 3D simulation of liver resection before surgery has been widely used throughout Japan. Currently, in many institutions in Japan, it is a standard process for surgeons to share visual information using 3D images reconstructed from patient multidetector computed tomography (MDCT) datasets. In recent years, 3D image-assisted surgery has been expanded to not only liver surgery but also biliary and pancreatic surgery [1, 7, 8]. In recent years, a real-time navigation system similar to the neuro-navigator (Mizuho Medical Industry Co., Ltd., Tokyo), which has been developed and used in head and neck region surgery, and navigates surgery like car navigation, is expected to be used in abdominal surgery. In addition, indocyanine green (ICG) fluorescent navigation surgery using an infrared light observation image system of fluorescence imaging technology has recently become popular in gastroenterological surgery [9, 10]. In this report, our team reviewed the history of hepatobiliary-pancreatic surgery using CAS and reported our research to date.

Y. Oshiro (✉)
Department of Gastroenterological Surgery,
Ibaraki Medical Center, Tokyo Medical University, Ibaraki, Japan
e-mail: oshiro@tokyo-med.ac.jp

M. Hashizume (ed.), *Multidisciplinary Computational Anatomy*, https://doi.org/10.1007/978-981-16-4325-5_30

30.2   Progress of 3D Medical Image Processing and 3D Surgical Simulation

In April 2012, "Image-assisted navigation in liver resection" was covered by the Japanese medical insurance and was rapidly used in many institutions in Japan. The development and popularization of 3D medical image processing workstations such as SYNAPSE VINCENT (Fuji Film Medical Co., Ltd.), Ziostation (Ziosoft Co., Ltd.), and AZE VirtualPlace (AZE Co., Ltd.), have contributed greatly to image-assisted navigation in liver resection. "Image-assisted navigation in liver resection" does not mean a navigation for the operator during surgery like a car navigation; rather, a 3D model is constructed from the computed tomography (CT) scan of the patient before surgery. Virtual liver resection volume is measured using volumetry, surgical planning is performed, the simulated cut surface is shared by the surgical team, and the surgery is performed by comparing the 3D simulation image with the surgical field. There are few studies that have reported on the utilization of 3D surgical simulations; however, many studies have reported new knowledge of the unique liver anatomy using 3D simulation [1–6]. In biliary and pancreatic surgery, simulation methods, such as integrated images of magnetic resonance cholangiopancreatography (MRCP) and CT [1, 7, 8], integrated images of the bile duct, and CT visualized with CO_2 [11], have been reported. In esophageal and gastric surgery, simulations using 3D CT angiography have been performed [12, 13], and in colon surgery, 3D CT angiography and virtual endoscopic images have been integrated [14–16]. There have also been reports on methods using virtual reality (VR) and augmented reality (AR) technology that superimposes preoperative 3D images on the abdomen and organs, and navigation surgery that uses a head-mounted display [17, 18]. In Germany, MeVisLab (MeVis, Bremen) and in France, 3DVSP (IRCAD, Strasbourg) have been used to perform surgical simulations [19, 20]. Although the organs of the digestive system have the property of deformation, AR navigation for laparoscopic surgery using magnetic resonance imaging (MRI) images taken during laparoscopic surgery in the open MRI surgery room have been developed [21]. In addition, for the purpose of real-time navigation for liver resection, such as car navigation, a method of synchronizing intraoperative ultrasound image and preoperative CT using a magnetic sensor has been reported [22]. Furthermore, a method for synchronizing the display image of an actual laparoscopic hepatectomy with a 3D liver model using an infrared sensor has been performed [23, 24]. There have been reports on real-time navigation using infrared sensors for laparoscopic hepatectomy and da Vinci robotic hepatectomy [25].

30.3   3D Surgical Simulation for Liver Surgery

30.3.1   Liver Resection Simulation Using SYNAPSE VINCENT

SYNAPSE VINCENT (Fuji Film Medical, Tokyo, Japan) is a 3D image processing software. This software helps in understanding the complex structure of intrahepatic vessels and allows the assessment of the liver volume supplied by the blood vessels. We used the SYNAPSE VINCENT to perform 3D visualization and virtual resection of the liver. Based on the patient's CT, the liver parenchyma, inferior vena cava, portal vein, hepatic vein, and tumor were traced by the function of automatic and manual segmentation of the SYNAPSE VINCENT. Then, the 3D reconstructed liver was completed (Fig. 30.1). Regarding the benefits of liver resection using 3D surgical simulation with the SYNAPSE VINCENT, several reports have suggested that it is useful for performing subsegmentectomy of the liver and [3] that the operation time was shortened [26].

30.3.2   Utilization of Our Original 3D Printing for Surgery

Recent advances in sophisticated 3D printing technology have made it possible to create 3D models as accurate 3D prints. The 3D model of the liver including the liver parenchyma, portal vein, hepatic vein, and inferior vena cava reconstructed using 3D analysis software such as the SYNAPSE VINCENT has been output as a stereolithography

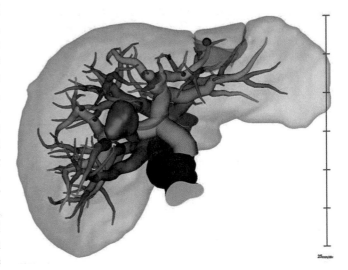

Fig. 30.1 3D reconstructed liver model using the SYNAPSE VINCENT: The 3D reconstructed liver is completed using the SYNAPSE VINCENT

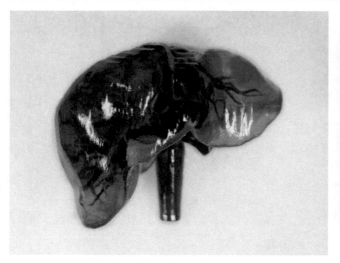

Fig. 30.2 The conventional 3D-printed liver model: As the shape of the liver is not flat and the transparent resin is affected by the refraction of light, the blood vessels inside the liver are distorted and difficult to observe

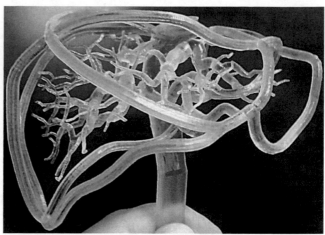

Fig. 30.3 3D-printed frame model of the liver: It is easy to see the complicated vascular structure and the spatial relationship between the blood vessels and the tumor

file, and 3D printed as a real model using a 3D printer. Preoperative simulations, rehearsals, and surgical planning are performed using these 3D prints [27]. In addition, a surgical training system has been developed and put into practical use; it uses the chest and abdominal contours, and the organ model that represents the texture of the living body created with 3D printing based on the patient's CT data [28]. For methods of 3D surgical simulation, 3D images are generally displayed and observed on a two-dimensional (2D) screen; however, many of the advantages of 3D images on a 2D screen are lost. The 3D-printed model has a great advantage in that we can grasp the spatial structure on our hands [27].

In the conventional 3D-printed liver model, structures such as intrahepatic vessels and tumors are made of opaque or colored resin. Additionally, the liver parenchyma that fills the inside of the liver is created using transparent acrylic resin as the loading material (Fig. 30.2). In the conventional model because the shape of the liver is not flat and the transparent resin is affected by the refraction of light, the blood vessels inside the liver are distorted and difficult to observe. Moreover, because acrylic resin is very expensive, the high cost of the material is a problem. In 2015, the University of Tsukuba developed a frame model in which the inside of the liver was hollowed out without using loading materials and the surface of the liver was surrounded by a nylon frame (Fig. 30.3). In this frame model, the amount of resin has been greatly reduced, and low cost has been achieved. In this frame model because we can directly observe the blood vessels inside the liver, it is easy to see the complicated vascular structure and the spatial relationship between the blood vessels and the tumor can be understood only in our hands. In addition, this frame model has the property of enhancing the effects of image sharing within the surgical team, preoperative simulation, and intraoperative navigation [29].

30.4 Development of Our Original Hepatectomy Simulation Software

30.4.1 3D Virtual Hepatectomy Simulation Combined with Real-Time Deformation, "Liversim"

As mentioned above, although image analysis software, such as SYNAPSE VINCENT, has become popular, the 3D liver model is a rigid model and does not deform. The liver is generally soft and deforms during surgical procedures. With conventional software, it is possible to visualize the shape of the cut surface of the liver and the appearance of blood vessels; however, it is impossible to visualize the process of resection of the liver. Therefore, we have developed a novel surgical simulation software that allows virtual surgery of the liver on a computer screen [30].

In addition to the functions of conventional software, the main functions of Liversim are the deformation and resection of a 3D liver model and the visualization and cutting of intrahepatic blood vessels that appear on the liver's cut surface. We can cut the liver parenchyma linearly at any set depth by placing a resection line on a 3D liver model using a PC mouse. Similar to the actual surgery, pulling the liver parenchyma on both sides of the resection line outward makes it easier to see the cut surface of the liver (Fig. 30.4). The 3D liver model is reconstructed using the SYNAPSE VINCENT based on the patient's CT data. The STL data of the 3D liver model reconstructed using the SYNAPSE VINCENT is the input for Liversim, and subsequently, a virtual hepatectomy deforming the liver can be quickly performed. With Liversim, it is easier to understand the depth of blood vessels and the timing of their appearance than with conventional software. Liversim is also useful for young surgeons and medical students to understand the spatial recog-

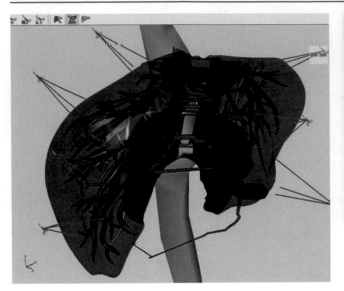

Fig. 30.5 Indocyanine green (ICG) fluorescence navigation surgery using an infrared light observation image system: The fluorescence-guided surgical navigation is performed in liver resection

Fig. 30.4 3D virtual hepatectomy simulation combined with real-time deformation, "Liversim": We can cut the liver parenchyma linearly at any set depth by placing a resection line on a 3D liver model using a PC mouse

nition of liver anatomy and the resection processes of various surgical procedures before surgery.

30.4.2 Intraoperative Navigation Using Liversim

We can perform virtual surgery on a PC using a patient's independent 3D liver model before surgery, which is exactly a rehearsal of the surgery. Furthermore, it is a useful navigation method to continuously replay a rehearsal video recorded before surgery on a screen during surgery and refer to it during surgery. Regarding the accuracy of Liversim, there were no discrepancies in the depth, direction, and positions of the portal and hepatic veins during the liver resection process compared to the actual liver resection. Liversim is also very useful in subsegmentectomy because the volume of the subsegment area can be measured using the volumetric function and the demarcation line of the measurement area, which facilitates the preoperative simulation [30].

30.5 Movement to Navigation Surgery

Preoperative simulation of hepatectomy has been performed in many institutions, and clinical application of real-time navigation technology is desired in the future. Recently, ICG fluorescence navigation surgery using an infrared light observation image system has become popular in gastroenterological surgery [31, 32]. Since liver tumors with accumulated ICG fluoresce when exposed to infrared light, fluorescence-guided surgical navigation is performed in liver

resection (Fig. 30.5). Fluorescence imaging has the advantage of clearly distinguishing surgical anatomy. The ICG fluorescent navigation system is already on the market and has been approved as a medical device. In the fiscal year 2018, intraoperative blood vessel imaging was covered by the Japanese medical insurance, hence, its use is expected to spread rapidly for gastrointestinal surgery, particularly for the esophagus, stomach, and large intestine surgery.

Recently, as a research and development of real-time navigation surgery such as car navigation, trials of navigation surgery in laparoscopic hepatectomy and laparoscopic gastrectomy using infrared sensors have been performed [33]. Real-time navigation surgery requires integration and accurate registration between the real world and virtual objects. Although NeuroNavigator (Applied Neuroscience Inc.) has already been commercialized in the field of head and neck surgery, it is difficult to realize the integration of virtual objects into the real world in gastrointestinal surgery that deals with moving and deformable nonrigid organs. Research on nonrigid registration has been reported, but breakthroughs based on new ideas such as deep learning are expected in the future.

30.6 Development of Our Original Hepatectomy Navigation System

30.6.1 Original 3D Reconstruction of Surgical Field and Development of Surgical Navigation System

Liversim is a simulation software that enables liver deformation, where surgeons can simulate the operation plan on a computer system. Even though Liversim can display a deformed liver image, it is far different from a navigation system such as car navigation [30]. A car navigation system uses GPS to measure the location and provide feedback to the map system to show the current locations on the map. However, the Liversim simulation system does not have such a sensing system. Therefore, we aim to develop a surgical navigation system for hepatectomy similar to a car navigation system based on Liversim. An overview of this system is as follows. First, the real-time capture of the surgical field by the 3D camera system will be performed to reconstruct a 3D model of the liver and calculate the resection state. Second, the resection and orientation information of the 3D liver model will be sent to the LiverNavi system. Lastly, the liver resection status is reflected in LiverNavi showing the progress of the surgery and the deformed liver to be displayed during the surgery.

30.6.2 Configuration of the Novel 3D Captured Liver Resection Navigation System

This system consists of two 3D camera sensors for real-time 3D capture of the liver in the surgical field: a G-Capture software that analyzes, captures data, and extracts the resection line; and a LiverNavi software, which is an extended version of Liversim for navigation use. LiverNavi can show a deformed liver by resection using resection and orientation information from the G-Capture software (Fig. 30.6).

- A G-Capture software: The G-Capture software analyzes the 3D shape data of the liver captured by the 3D camera and extracts the information about the resection line. To cooperate with LiverNavi, calibration between the real space and the simulation space is required. Therefore, G-Capture has the function of calibrating the space and orientation of the simulation 3D liver model (LiverNavi space) with the reconstructed liver 3D model space (real space). Along with the progress of the surgery, G-Capture can extract the information of the resection and orienta-

tion from the captured 3D model and send the data to the LiverNavi system. With this data, LiverNavi shows the resection process and the deformed liver model as a surgical navigation (Fig. 30.7).

- A LiverNavi software: The planned resection line of the liver needs to be set in advance. This planned resection line can be set by anatomical analysis of the patient's liver stereolithography file and preset the three marks required to calibrate with G-Capture's 3D model space. The three marks consist of the origin (the intersection of the resection line and the edge of the liver), the arbitrary point on the resection line from the origin, and the arbitrary point on the resected liver edge from the origin. By using these three marks on both the simulation liver model and the reconstructed 3D liver model, the calibration process will mostly be done automatically. Subsequently, LiverNavi displays the progress of resection of the liver based on the resection line information from G-Capture. In addition, the orientation information from G-Capture enables LiverNavi to match the viewpoint of the simulation liver to the surgeon's view (Fig. 30.7).

30.6.3 Clinical Application of the Novel 3D Captured Liver Resection Navigation System

As a clinical application, it was used as a trial in 10 cases of liver resection: two right hepatectomies, three left hepatectomies, two posterior segmentectomies, two anterior segmentectomies, and one lateral segmentectomy. As described above, with this navigation system, the point cloud of the liver captured by the 3D camera was made into a polygon, and the surface and normal lines were calculated, from which the resection line could be extracted. The calibration of the real space and simulation space was simplified by setting

Fig. 30.6 A novel original 3D captured liver resection navigation system configuration: This system consists of two 3D camera sensors, a G-Capture software and a LiverNavi software

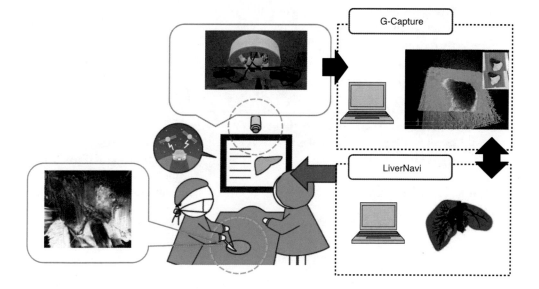

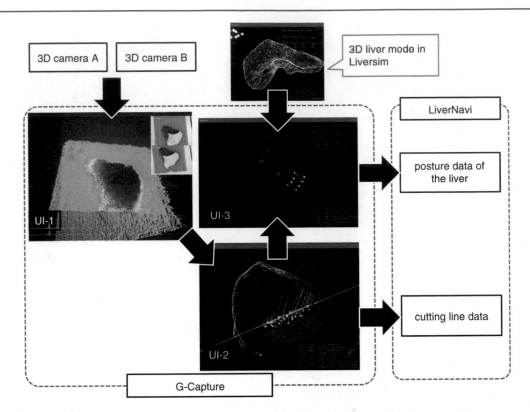

Fig. 30.7 A data flow of 3D captured liver resection navigation system. 1. Registration of the images of two 3D cameras and 3D image creation (UI-1): A measuring gauze is used to automatically register the captured images of the two cameras. The 3D images from the two registered 3D cameras are composited in real time. 2. 3D image analysis (UI-2): The data of the cut surface and the resection line of the liver are extracted. The extracted resection line data are sent to UI-3. 3. Fusion screen (registration of real space and virtual space) (UI-3): The registration of the real space and the virtual space of Liversim is performed. The organ posture data and the resection line data are sent to LiverNavi, and the navigation image that reflects the resection process is displayed on the LiverNavi screen

three marks. Therefore, real-time automatic resection of the liver could be performed with LiverNavi, based on the posture data of the liver and resection line data obtained by the abovementioned system (Fig. 30.8). However, some problems with this system have become apparent. It was found that the depth of the resection line cannot be measured with the current method when the resection line becomes deep and the resection surface becomes large, frequent calibrations arise for surgical procedures that require liver mobilization. In the future, we aim to solve these problems and make more practical improvements.

30.7 Development of Our Original Simulation for Biliary-Pancreatic Surgery

30.7.1 Surgical Simulation for Biliary-Pancreatic Surgery

In hepatobiliary and pancreatic surgery with complicated anatomical variations, an accurate understanding of surgical anatomy is essential for performing safe surgery. Therefore, preoperative 2D images such as MDCT and MRCP are examined in detail to grasp the positional relationship between the lesion and surrounding organs. However, constructing a 3D anatomical image from the 2D modality often depends on many surgical experiences. It is difficult for a young hepatobiliary-pancreatic surgeon with little experience in surgical anatomy to accurately understand anatomy with preoperative images constructed from a 2D MDCT. We usually use 3D MDCT for a detailed evaluation prior to surgery and also project it as a guide during surgery on a large screen in the operating room.

In addition, we evaluated the usefulness of pancreatic resection using a 3D surgical simulation in which MDCT and MRCP are fused [34, 35]. The SYNAPSE VINCENT was used to create a 3D model from 2D CT data. We have enabled the fusion technology of CT and MRCP by aligning the normalized mutual information between two different modalities [1]. This 3D fusion image contains blood vessels as well as bile ducts and pancreatic ducts, and accurately represents the surgical anatomy of individual cases [36]. Therefore, at the preoperative conference, it became possible to determine the

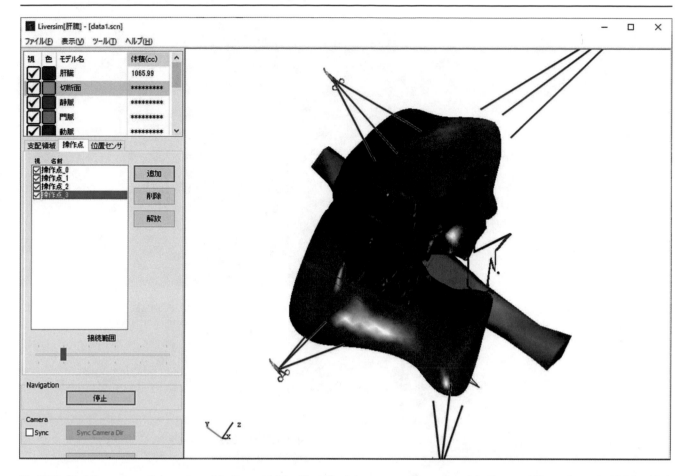

Fig. 30.8 Real-time automatic resection of the liver with LiverNavi: Real-time automatic resection of the liver could be performed with LiverNavi based on the posture data of the liver and resection line data

appropriate surgical procedures by estimating the bile duct and pancreatic duct anatomy using these 3D images. In addition, we can confirm the 3D anatomy displayed in the operating room, and navigation surgery can be performed while confirming the anatomical critical points by the surgical team.

The fusion method involved four steps. First, the Heavily T2-weighted MRCP and the thin slice 3D-T1 TFE cross-sectional image were fused, and then the thin slice 3D-T1 TFE cross-sectional image and the CT cross-sectional image were fused. Therefore, the fusion of the MRCP image and CT image using as an intermediary, the thin-slice 3D-T1 TFE cross-sectional image was completed. Third, we focused on the hilar region and made a rough correction of the position of the bile duct in the fusion image. Fourth, we set three checkpoints and checked that there was no gap in the positional relationship between the portal vein and the bile duct. The three checkpoints were as follows: (1) the main portal vein, (2) the umbilical portion of the left portal vein (whether the lateral segment bile duct runs in front of the umbilical portion or not), and (3) the right portal vein (whether the posterior segment bile duct runs in front of the right portal vein) [1].

30.7.2 Surgical Simulation for Biliary-Pancreatic Surgery

The patient was a 72-year-old man who was referred from a local hospital complaining of jaundice. Abdominal CT showed a low-concentration tumor measuring 18 mm in the middle bile duct, which invaded the right hepatic artery and the right wall of the portal vein. MRCP showed severe stenosis in the middle bile duct and dilation of the intrahepatic bile duct. Therefore, he was diagnosed with middle bile duct cancer, T4, N1, M0, Stage IIIa. We created a 3D fusion image using CT and MRCP and performed the surgical simulation (Fig. 30.9a). In the 3D fusion image, it was easy to grasp the positional relationship of the tumor, bile duct, portal vein, and hepatic artery spatially, and to share the image with the surgical team. The volume of the remaining liver after right hepatectomy was approximately 30%. Hence, after right portal vein embolization, right hepatic lobectomy with extrahepatic bile duct resection, portal vein resection, and reconstruction were planned and performed (Fig. 30.9b).

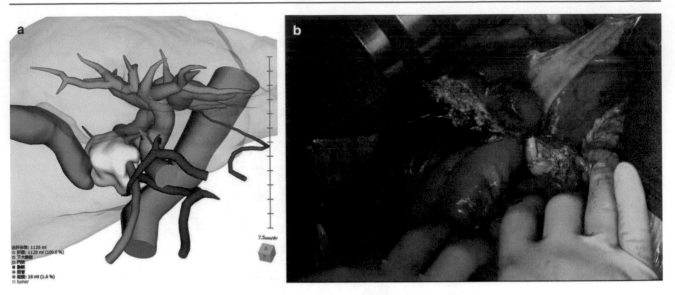

Fig. 30.9 Surgical simulation for biliary-pancreatic surgery. (**a**) 3D fusion image using CT and MRCP. It is easy to grasp the positional relationship of the tumor, bile duct, portal vein, and hepatic artery spa-

tially. (**b**) A surgical photograph. The right hepatic lobectomy with extrahepatic bile duct resection, portal vein resection, and reconstruction are planned and performed

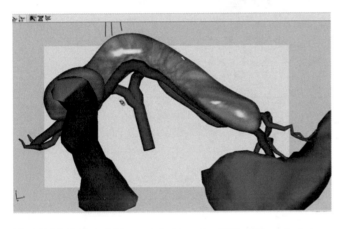

Fig. 30.10 Deformable pancreatectomy simulation: 3D surgical simulation of laparoscopic distal pancreatectomy is performed

30.7.3 Usefulness of 3D Simulation in Pancreatic Resection

For 117 patients who underwent pancreaticoduodenectomy (PD), operation time (min), blood loss (g), complications, pancreatic fistula, and postoperative hospital stay (day) were compared before and after the introduction of the 3D surgical simulation. Statistical analyses were performed using the chi-squared test or unpaired t-test. Therefore, in the group after the introduction of 3D simulation, a significant decrease in blood loss was observed (group before introduction of 3D simulation, 1174 ± 862 g; group after introduction of 3D simulation, 810 ± 668 g, $p = 0.012$). However, no significant

difference was observed between other perioperative factors before and after the introduction of 3D simulation [35].

30.7.4 Deformable Pancreatectomy Simulation

We are using the "Liversim" simulation software that allows deformation of organs to visualize the surgical process, not only for the liver but also for pancreatic surgery [30]. Figure 30.10 shows a 3D surgical simulation of laparoscopic distal pancreatectomy and a surgical photograph. It was confirmed that the spatial relationship between the splenic artery and splenic vein when the pancreas was elevated was similar to that in the simulation. This is not possible with existing software that cannot deform the pancreas.

30.8 Conclusion

Based on the knowledge of multidisciplinary computational anatomy, we have developed and introduced various CAS for hepatobiliary and pancreatic surgery. Our newly developed CAS has contributed to safe and secure surgeries. It is expected that the possibilities of CAS will continue to expand as multidisciplinary computational anatomy develops.

Acknowledgment I thank Mr. Sakamoto of nextEDGE Technology, Inc., Mr. Seitoku, Mr. Itoh of LEXI, Inc., Mr. Kotera of Fujifilm medical, Inc, Dr. Mitani, Dr. Kitahara, and Dr. Yano of University of Tsukuba.

References

1. Oshiro Y, Sasaki R, Nasu K, et al. A novel preoperative fusion analysis using three-dimensional MDCT combined with three-dimensional MRI for patients with hilar cholangiocarcinoma. Clin Imaging. 2013;37:772–4.
2. Sasaki R, Kondo T, Oda T, et al. Impact of three-dimensional analysis of multidetector row computed tomography cholangiography in operative planning for hilar cholangiocarcinoma. Am J Surg. 2011;202:441–8.
3. Takamoto T, Hashimoto T, Ogata S, et al. Planning of anatomical liver segmentectomy and subsegmentectomy with 3-dimensional simulation software. Am J Surg. 2013;206:530–8.
4. Saito S, Yamanaka J, Miura K, et al. A novel 3D hepatectomy simulation based on liver circulation: application to liver resection and transplantation. Hepatology. 2005;41:1297–304.
5. Ariizumi S, Takahashi Y, Kotera Y, et al. Novel virtual hepatectomy is useful for evaluation of the portal territory for anatomical sectionectomy, segmentectomy, and hemihepatectomy. J Hepatobiliary Sci. 2013;20:396–402.
6. Oshiro Y, Sasaki R, Takeguchi T, et al. Analysis of the caudate artery with three-dimensional imaging. J Hepatobiliary Pancreat Sci. 2013;20:639–46.
7. Oshiro Y, Gen R, Hashimoto S, et al. Neuroendocrine carcinoma of the extrahepatic bile duct: a case report. World J Gastroenterol. 2016;22:6960–4.
8. Miyamoto R, Oshiro Y, Nakayama K, et al. Three-dimensional simulation of pancreatic surgery showing the size and location of the main pancreatic duct. Surg Today. 2017;47:357–64.
9. Aoki T, Murakami M, Yasuda D, et al. Intraoperative fluorescent imaging using indocyanine green for liver mapping and cholangiography. J Hepatobiliary Pancreat Sci. 2010;17:590–4.
10. Ishizawa T, Fukushima N, Shibahara J, et al. Real-time identification of liver cancers by using indocyanine green fluorescent imaging. Cancer. 2009;115:2491–504.
11. Okuda Y, Taura K, Seo S, et al. Usefulness of operative planning based on 3-dimensional CT cholangiography for biliary malignancies. Surgery. 2015;158:1261–71.
12. Wada T, Takeuchi H, Kawakubo H, et al. Clinical utility of preoperative evaluation of bronchial arteries by three-dimensional computed tomographic angiography for esophageal cancer surgery. Dis Esophagus. 2013;26:616–22.
13. Natsume T, Shuto K, Yanagawa N, et al. The classification of anatomic variations in the perigastric vessels by dual-phase CT to reduce intraoperative bleeding during laparoscopic gastrectomy. Surg Endosc. 2011;25:1420–4.
14. Okuda J, Matsuki M, Narabayashi I, Tanigawa N. Minimally invasive tailor-made surgery for colorectal cancer navigated by integrated 3D-CT imaging. J JSCAS. 2004;6:91–5.
15. Miyamoto R, Tadano S, Sano N, et al. The impact of three-dimensional reconstruction on laparoscopic-assisted surgery for right-sided colon cancer. Videosurgery. 2017;12(3):251.
16. Miyamoto R, Nagai K, Kemmochi A, et al. Three-dimensional reconstruction of the vascular arrangement including the inferior mesenteric artery and left colic artery in laparoscope-assisted colorectal surgery. Surg Endosc. 2016;30:4400–4.
17. Sugimoto M, Yasuda H, Koda K, et al. Image overlay navigation by markerless surface registration in gastrointestinal, hepatobiliary and pancreatic surgery. J Hepatobiliary Pancreat Sci. 2010;17:629–36.
18. Yu S, Sugimoto M, Imura S, et al. Intraoperative 3D hologram support with mixed reality techniques in liver surgery. Ann Surg. 2020;271:e4–7. https://doi.org/10.1097/SLA.0000000000003552.
19. Wang Y, Zhang Y, Peitgen H-O, et al. Precise local resection for hepatocellular carcinoma based on tumor-surrounding vascular anatomy revealed by 3D analysis. Dig Surg. 2012;29:99–106.
20. Bégin A, Martel G, Lapointe R, et al. Accuracy of preoperative automatic measurement of the liver volume by CT-scan combined to a 3D virtual surgical planning software (3DVSP). Surg Endosc. 2014;28:3408–12.
21. Tsutsumi N, Tomikawa M, Uemura M, et al. Image-guided laparoscopic surgery in an open MRI operating theater. Surg Endosc. 2013;27:2178–84.
22. Satou S, Aoki T, Kaneko J, et al. Initial experience of intraoperative three-dimensional navigation for liver resection using real-time virtual sonography. Surgery. 2014;155:255–62.
23. Igami T, Tanaka H, Nojiri M, et al. Navigation in open and laparoscopic hepatectomy. Shokakigeka. 2016;39:37–44. (in Japanese)
24. Kingham TP, Jayaraman S, Clements LW, et al. Evolution of image-guided liver surgery: transition from open to laparoscopic procedures. J Gastrointest Surg. 2013;17:1274–82.
25. Buchs NC, Volonté F, Pugin F, et al. Augmented environments for the targeting of hepatic lesions during image-guided robotic liver surgery. J Surg Res. 2013;184:825–31.
26. Nakayama K, Oshiro Y, Ohkohchi N. The effect of three-dimensional preoperative simulation on liver surgery. World J Surg. 2017;41:1840–7.
27. Igami T, Nakamura Y, Hirose T, et al. Application of a three-dimensional print of a liver in hepatectomy for small tumors invisible by intraoperative ultrasonography: preliminary experience. World J Surg. 2014;38:3163–6.
28. FASOTEC. http://www.fasotec.co.jp/
29. Oshiro Y, Mitani J, Okada T, et al. A novel three-dimensional print of liver vessels and tumors in hepatectomy. Surg Today. 2017;47:521–4.
30. Oshiro Y, Yano H, Mitani J, et al. A novel 3-dimensional virtual hepatectomy simulation combined with real-time deformation. World J Gastroenterol. 2015;21:9982–92.
31. Nishino H, Hatano E, Seo S, et al. Real-time navigation for liver surgery using projection mapping with indocyanine green fluorescence: development of the novel medical imaging projection system. Ann Surg. 2017;267(6):1134–40.
32. Aoki T, Yasuda D, Shimizu Y, et al. Image-guided liver mapping using fluorescence navigation system with indocyanine green for anatomical hepatic resection. World J Surg. 2008;32:1763–7.
33. Hayashi Y, Misawa K, Hawkes DJ, et al. Progressive internal landmark registration for surgical navigation in laparoscopic gastrectomy for gastric cancer. Int J CARS. 2016;11:837–45.
34. Miyamoto R, Oshiro Y, Nakayama K, et al. Impact of three-dimensional surgical simulation on pancreatic surgery. Gastrointest Tumors. 2018;4:84–9.
35. Miyamoto R, Oshiro Y, Sano N, Inagawa S, Ohkohchi N. Three-dimensional surgical simulation of the bile duct and vascular arrangement in pancreatoduodenectomy: a retrospective cohort study. Ann Med Surg (Lond). 2018;36:17–22. https://doi.org/10.1016/j.amsu.2018.09.043.
36. Miyamoto R, Oshiro Y, Hashimoto S, Kohno K, Fukunaga K, Oda T, Ohkohchi N. Three-dimensional imaging identified the accessory bile duct in a patient with cholangiocarcinoma. World J Gastroenterol. 2014;20:11451–5.

New Frontier of Technology in Clinical Applications Based on MCA Models: Musculoskeletal System

Development of Multiple Skeletal Muscle Recognition Technique in the Thoracoabdominal Region for Respiratory Muscle Function Analysis

Naoki Kamiya

Abstract

This chapter describes the purpose and summary of our achievement in the research work about respiratory muscle segmentation as respiratory function analysis. Our goal is to develop a complex segmentation technique of the skeletal muscle in the thoracoabdominal region for respiratory function analysis using skeletal muscle. Here, we aim to develop complex segmentation and analysis technology of thoracoabdominal respiratory muscle by improving automatic segmentation method of deep muscle and surface muscle region that we already established. Specifically, we construct direction and shape models of the respiratory muscle based on muscle fibers, and realize automatic site-specific segmentation of respiratory muscle. This makes it possible to clarify the information of respiratory muscle of each part and the relationship between skeletal muscles necessary for elucidating the relationship between respiratory function and skeletal muscle. Then, we generate a respiratory muscle mathematical model applicable to three axes (space, function, and pathology). First, we constructed a sternocleidomastoid muscle and intercostal muscle model. In addition, we performed two-dimensional segmentation of the spinal column erector, which is closely related to COPD, which is one of respiratory diseases. Then, we proposed a three-dimensional segmentation method of spinal column erector muscles as the main target. In particular, we propose a method based on machine learning and compared it with a method based on deep learning. These results were shared with the group "Function integrated diagnostic assistance based on MCA Models" and contributed to the fusion of deep learning and model-based methods.

Keywords

Skeletal muscle · Musculoskeletal analysis Musculoskeletal segmentation · Respiratory muscle Random forest · FCN · 2D U-Net · 3D U-Net

31.1 Introduction

Skeletal muscles are present throughout the human body. It applies not only to the surface part but also to the deep parts. For this reason, skeletal muscles are often drawn on images in images of various modalities photographed for diagnosing organs. However, automatic segmentation of the skeletal muscle is one of the difficult tasks, as their structure (shape and distribution) is complex, as well as differences among individuals are large. In particular, since skeletal muscle and other organ regions are similar in gray value distribution, skeletal muscle segmentation is an important task as with organ segmentation. For skeletal muscle segmentation, various methods such as methods focusing on the shape of skeletal muscle and statistical features have been proposed, but there is no established method yet. In order to develop the advanced systems, our goals in this project are to develop anatomical and functional muscle segmentation methods for diagnosis of musculoskeletal functions in thoracoabdominal regions.

In the previous project, we developed computational anatomy models of various muscles in torso CT images [1]. In this project, we not only continue to improve model construction and application but also focus on establishment of muscular models for functional imaging. These muscle models can be effectively combined to process multidisciplinary information. There are also several studies on skeletal muscle segmentation in the MCA project. In particular, Fujita et al. are working on functional analysis of surface muscles for whole-body CT and torso CT, and Sato et al. are working on segmentation in the lower limb. Our roles in this project are to investigate image muscle segmentation methods in thoracoabdominal region and to establish methodologies of computer-aided diagnostic (CAD) systems for respiratory

N. Kamiya (✉)

School of Information Science and Technology, Aichi Prefectural University, Nagakute, Aichi, Japan
e-mail: n-kamiya@ist.aichi-pu.ac.jp

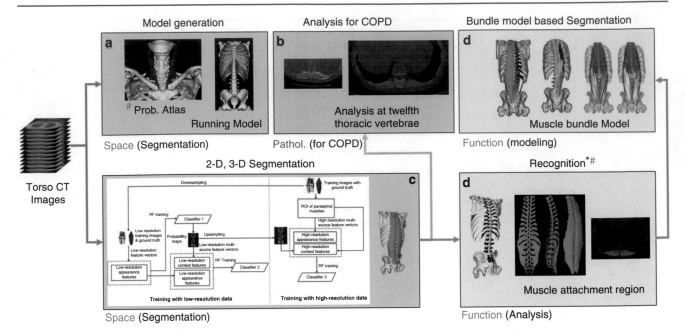

Fig. 31.1 Schematic diagram of thoracoabdominal muscle recognition for analysis of respiratory muscle function. (**a**) Model-based recognition of muscle, (**b**) Muscle analysis for COPD, (**c**) 2D, 3-D segmenta- tion for spinal erector muscle based on machine learning based method (Random forest), and (**d**) spinal erector muscle segmentation using muscle bundle model

muscle function analysis from the aspect of the musculoskel- etal segmentation. Figure 31.1 shows the schematic diagram of our research achievement, thoracoabdominal muscle seg- mentation for analysis of respiratory muscle function.

In this study, to target the respiratory muscles. Respiratory diseases are the 4th and 5th largest causes of death in the world [2]. The respiratory muscles consist of a number of muscles, many are present in the chest and abdomen region. Then, the relationship between chronic obstructive pulmo- nary disease (COPD), which is one of respiratory diseases, and muscles have been clarified in part, and the correlation between the cross-sectional area of the spinal column erector muscle and the prognosis of COPD is shown [3]. In this research, in order to elucidate not only the respiratory mus- cle but also the relationship between respiratory disease and skeletal muscle, we propose a segmentation method of skel- etal muscle of the thoracoabdominal part. The major topics will be reported in the following sections.

31.2 Model Generation for Respiratory Muscle Function Analysis

1. Purpose: Model-based segmentation has also been shown to be effective in skeletal muscle segmentation. In this project, a model of sternocleidomastoid muscle and inter- costal muscle was constructed.

   The sternocleidomastoid muscle is one of the respira- tory muscles. In our group, muscle analysis targeting

ALS (amyotrophic lateral sclerosis), which is a difficult- to-treat disease using whole-body CT images, is per- formed. In addition, the sternocleidomastoid muscle is also a part susceptible to ALS. Here, we use automatic probabilistic atlas to automatically recognize sternoclei- domastoid muscle.

Intercostal muscles attached between the upper and lower ribs, perform vertical motion of the ribs. In other words, the intercostal muscle is a muscle, which expands and contracts the thorax and performs exhalation/inspira- tion movement. The intercostal muscles consist of the interior and exterior intercostal muscles and are distrib- uted in layers. Not only related to the respiratory func- tion, it can also occur damage to the muscle itself. In this study, in order to analyze the motor function of the inter- costal muscle, we considered not only the accurate seg- mentation of the intercostal muscle region but also the running of the muscles.

2. Method: Our study was approved by the institutional review board. Twenty normal cases of non-contrast CT images were collected. We generated sternocleido- mastoid muscle model based on probabilistic atlas and proposed an automatic segmentation method using this atlas on torso CT images (Fig. 31.1a left).

   Next, our scheme of the intercostal muscle model con- sisted of the following steps: (1) costal cartilage segmen- tation, (2) landmarks detection and boundary detection, and (3) muscle fiber modeling. In the first step, costal car- tilage is recognized using CT value. Costal cartilage recognized here is for use as a boundary of the internal and external intercostal muscles, precision extraction of costal cartilage was not carried out. The boundary of this costal cartilage is the boundary of each intercostal muscle. Next, anatomical landmarks are recognized on each rib. Anatomical landmarks are detected on each rib and costal cartilage and become the intercostal muscle boundary and attachment point. Finally, connect the upper and lower ribs, which are origin and insertion, to obtain a muscle running model simulating running of the muscle fibers (Fig. 31.1a right).

3. Results: We successfully segmented the major area of the sternocleidomastoid muscle. This is because the atlas of sternocleidomastoid muscle deformed using the information of bone anatomical location and edge of the sternocleidomastoid muscle is fitted in the shape of the individual muscle [4].

Since it is difficult to exactly observe the running of the intercostal muscle fibers on the CT image, we visually evaluated whether the position and running of the intercostal muscle were correctly obtained. By 2-D slice and 3-D volume rendering, the adhesion position and the running of the muscle fibers at the origin and insertion were confirmed. We did not address precise segmentation of bone features such as rib nodules, so we could express muscle running in the intercostal muscle region necessary for segmentation of the intercostal muscle although it was not strictly distinguishable between the internal and external intercostal muscles [5, 6].

4. Conclusion: An automated scheme was proposed to generate the muscle model of sternocleidomastoid and intercostal muscle. Muscle model can detect initial region of each muscle. The results suggest the efficiency of the muscle running model-based image interpretation. Muscle running as muscle fiber direction is needed for muscle segmentation. In the next step, it is necessary to generate stratified muscle fiber running model based on the detection of the precise feature point on the skeleton.

31.3 Automated Segmentation of the Erector Spinae Muscle in Torso CT Images and Its Segmentation Based on Muscle Bundle Model

1. Purpose: Although the erector spinae muscle is not a respiratory muscle, the cross-sectional area of the erector spinae muscle in the 12th thoracic cross section is correlated with the prognosis of respiratory disease COPD [3]. However, its automatic segmentation has not been realized. In addition, since the erector spinae muscle is composed of a large number of muscle groups, it is considered

difficult to apply the proposed skeletal muscle segmentation procedure. Therefore, in this research we aim to automatically recognize the erector spinae muscle in 2-D cross-sectional area (Fig. 31.1b) and 3-D volume (Fig. 31.1c).

2. Methods: We performed automatic segmentation of the erector spinae muscle using machine learning. In the method based on machine learning, random forest-based method is used. We learned using the original image and the correct image of the bone region and manually segmented the spinal erector region. In the A02-3 group is also try to segment this muscle using deep CNN. However, the region of the spinal erector muscle is large, and it takes a huge amount of time to prepare a correct answer image necessary for learning.

In addition, we proposed a method to integrate deep learning and model-based methods. In particular, in large muscles such as the spinal column erector muscle, it is often composed of multiple anatomical muscles. As already mentioned, it is a very difficult task to prepare a correct image of a plurality of muscles constituting the spinal column standing muscle as well as preparing a correct image of the spinal column standing muscle itself. Therefore, firstly, the muscle bundle of the muscle group constituting the spinal column erector is modeled. Then, by applying a model of the muscle bundle to the spinal column erector muscle recognized by machine learning, it is a method of classifying the spinal column erector muscles into muscles of different parts (Fig. 31.1d).

3. Results: In the automatic segmentation of the erector spinae muscle based on the running of the muscles, the concordance, recall, and relevance rate in the 12th thoracic vertebra were 78.5%, 93.8%, and 83.1%, respectively [7]. Next, in the method using machine learning, the concordance, recall, and relevance rate in the 12th thoracic vertebra were 87.0%, 96.4%, and 89.9%, respectively [8]. Also, with the deep CNN based method collaborate with A02-3, the accuracy of coincidence rate, recall rate and relevance rate by 3-D volume were 80.9%, 93.0% and 86.6%, respectively [9].

We automatically recognized attachment areas on the bone of the muscle from the area of the spinal column standing muscle recognized by machine learning. Then, a model of the muscle bundle was fitted to the attachment region on the bone, and classification of the skeletal muscle obtained by machine learning was realized (Fig. 31.1d) [10, 11].

4. Conclusion: In this study, automatic segmentation of erector spinae muscle correlated with COPD prognosis of respiratory disease was performed. The results suggest the possibility of applying a quantitative image analysis. In the method using machine learning, it is possible to

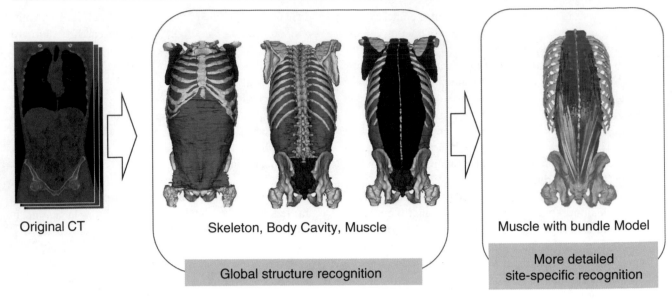

Original CT   Skeleton, Body Cavity, Muscle   Muscle with bundle Model

Global structure recognition   More detailed site-specific recognition

Fig. 31.2 A schematic diagram of detailed musculoskeletal recognition in combination with model-based methods from the global structure recognition using deep learning

recognize with 3-D volume, and high segmentation rate was obtained even in the region other than the 12th thoracic cross section. On the other hand, in the method using muscle fiber modeling, clinically useful information such as each muscle group segmentation and muscle attachment region can be used.

31.4 Current Status and Future Issues in Thoracoabdominal Skeletal Muscle Recognition

Though skeletal muscle segmentation of thoracoabdominal region for the respiratory muscle function analysis was described by the previous section, it is considered that the compartmentation with deep learning is necessary in future. Figure 31.2 shows an example of our current efforts. In deep learning, it is important to make an annotation image, but it is very time consuming and labor consuming to paint a part of skeletal muscle in detail. In order to realize a more detailed recognition of site-specific skeletal muscles, it is considered that muscles and global structures which can be painted by manual work as shown in the previous section are recognized by deep learning and used in combination with a model-based method [12]. In addition, we have proposed a method to recognize surface muscles in the whole-body by learning voxel patches composed of a limited number of slices which obtained by selectively [13], and are tackling skeletal muscle recognition by an approach from the whole to the detail.

31.5 Conclusion

Our progress in developing the methods for constructing the multidisciplinary muscle models and applying the models for muscle image segmentation and its analysis were described. The results obtained so far are promising, which convince us the success of our research project.

In the next step, it is to promote muscle analysis by fusion of DCNN and model-based method which is our recent research topic (Figs. 31.1d and 31.2). This is an indispensable technology not only for respiratory muscles but also for muscle function analysis and relationship analysis with disease where it is difficult to prepare a large number of correct answer images.

Acknowledgments The authors would like to thank all the members of the Fujita Laboratory in the Graduate School of Medicine, Gifu University, for their collaboration. This work was supported in part by a JSPS Grant-in-Aid for Scientific Research on Innovative Areas (Multidisciplinary Computational Anatomy, #26108005 and # 17H05301), JAPAN.

References

1. Fujita H, Hara T, Zhou X, et al. Model construction of computational anatomy: progress overview FY2009-2013. In Proceedings of Fifth International Symposium on the Project "Computational Anatomy"; 2014. pp. 25–35.
2. World Health Organization. Top 10 cause of death. Global Health Observatory (GHO) data. http://www.who.int/gho/mortality_burden_disease/causes_death/top_10/en/

3. Tanimura K, Sato S, Fuseya Y, et al. Quantitative assessment of erector spinae muscles in patients with chronic obstructive pulmonary disease. Novel chest computed tomography–derived index for prognosis. Ann Am Thorac Soc. 2016;13(3):334–41.

4. Kamiya N, Ieda K, Zhou X, et al. Automated segmentation of sternocleidomastoid muscle using atlas-based method in X-ray CT images: Preliminary study. Med Imaging Inf Sci. 2017;34(2):87–91.

5. Yamada S, Kamiya N, Sato S, et al. Construction of an intercostal muscle running model based on segmentation of the intercostal muscle origin and insertion in torso CT images. In Proceedings of the 3rd International Conference on Radiological Science and Technology, IO-04; 2017. p. 32.

6. Kamiya N, Zheng G, Zhou X, et al. Skeletal feature segmentation and skeletal muscle modeling for skeletal muscle analysis in whole-body CT images. In: Proceeding of 40th annual international conference of the IEEE engineering in medicine and biological society, ThBT19.4; 2018. p. 50.

7. Kume M, Kamiya N, Zhou X, et al. Automated segmentation of the erector spinae muscle based on deep CNN at the level of the twelfth thoracic vertebrae in torso CT images. In Proceedings of the 36th JAMIT Annual Meeting; 2017. p. 74–76 (in Japanese)

8. Kamiya N, Li J, Kume M, et al. Fully automatic segmentation of paraspinal muscles from 3D torso CT images via multi-scale iterative random forest classifications. Int J Comput Assist Radiol Surg. 2018;13(11):1697–706. https://doi.org/10.1007/s11548-018-1852-1.

9. Kume M, Zheng G, Kamiya N, et al. Automatic segmentation of spinal erector muscle region using deep CNN in torso CT image and early investigation of attachment site segmentation. IEICE Tech Rep. 2018;117:33–4. (in Japanese)

10. Kamiya N. Muscle segmentation for orthopedic interventions, intelligent orthopaedics. Advances in experimental medicine and biology, vol. 1093. Singapore: Springer; 2018. p. 81–91. https://doi.org/10.1007/978-981-13-1396-7_7.

11. Kamiya N, Kume M, Zheng G, et al. Automated segmentation of erector spinae muscles and their skeletal attachment region via deep learning in torso CT images, computational methods and clinical applications in musculoskeletal imaging. Cham: Springer; 2019. p. 1–10. https://doi.org/10.1007/978-3-030-11166-3_1.

12. Kamiya N. Deep learning technique for musculoskeletal analysis, deep learning in medical image analysis. Advances in experimental medicine and biology, vol. 1213. Cham: Springer; 2020. p. 165–76. https://doi.org/10.1007/978-3-030-33128-3_11.

13. Kamiya N, Oshima A, Zhou X, et al. Surface muscle segmentation using 3D U-net based on selective voxel patch generation in whole-body CT images. Appl Sci. 2020;10(13):4477. https://doi.org/10.3390/app10134477.

Morphometric Analysis for the Morphogenesis of the Craniofacial Structures and the Evolution of the Nasal Protrusion in Humans

32

Motoki Katsube

Abstract

The facial morphology of humans, which is very complicated in three dimensions, is closely related to various functions, such as support of the brain and the eyeball and formation of the mastication apparatus and respiratory system. Such a complicated shape is acquired by mid-fetal life. The early stage of this developmental process, the embryonic stage, is closely related to many congenital anomalies. Various studies have been conducted on the morphogenesis of the face during this period. However, specimens in the late stage of facial morphogenesis, the early fetal period, are too large to be sectioned for the histological analysis and have been difficult to study using traditional medical imaging modalities, such as X-rays, because the facial skeleton has not been calcified. The development of medical imaging modalities has allowed us to obtain high-definition images of these early fetal facial skeletons. In addition, geometric morphometrics have enabled us to perform three-dimensional morphological analyses and visualize the growth trait in the early fetal life.

Keywords

Geometric morphometrics · Facial morphogenesis · Fetus Nasal septum · Mid-face

32.1 Introduction

In the process of human ontogenesis, the shape of the organs drastically changes until 10 weeks of gestation, which is the embryonic period, and acquires almost the normal morphology. The maturation process of organs occurs during the fetal period following the embryonic period. As for the facial morphogenesis in the embryonic period, the medial nasal process, lateral nasal process, maxillary process, and mandibular process are fused precisely. If these fusions are impaired, congenital anomalies, such as cleft lip and palate, develop. Most congenital anomalies arise from dysplasia during the embryonic period. However, as shown in Fig. 32.1, the facial morphology is still very immature, even at the end of the embryonic stage. In the mid-fetal period, the fetuses acquire neonatal-like facial features.

32.1.1 Nasal Protrusion in Humans

The facial morphology of humans has unique features compared to primates, such as the anteriorly protruding forehead and protruding chin and nose. These appearances have changed with various functional adaptations over the course of human evolution of approximately 7 million years. In early developmental stages, such as the embryonic stage, the facial morphology is similar to that of other primates, and the human-specific morphology is acquired with growth. The greatest contribution to this is presumed to occur during the early fetal period [1–3].

32.1.2 Nasal Septum Development in Human Ontogeny

As a human-specific facial appearance characteristic, nasal protrusion is supported by the nasal septum. Disturbances in the development of the nasal septum in the early prenatal period can cause severe deformity of the nose, called the flat nose, which is a main characteristic of the Binder phenotype [4, 5]. The development of the nasal septum is widely known to be a major factor for the growth of the nose in childhood and for the thrusting force of the midface in the forward and downward directions with growth [6, 7]. Epidemiological studies have reported that maternal exposure to warfarin dur-

M. Katsube (✉)
Department of Plastic and Reconstructive Surgery, Kyoto University Graduate School of Medicine, Kyoto, Japan
e-mail: katsube@kuhp.kyoto-u.ac.jp

© The Author(s), under exclusive license to Springer Nature Singapore Pte Ltd. 2022
M. Hashizume (ed.), *Multidisciplinary Computational Anatomy*, https://doi.org/10.1007/978-981-16-4325-5_32

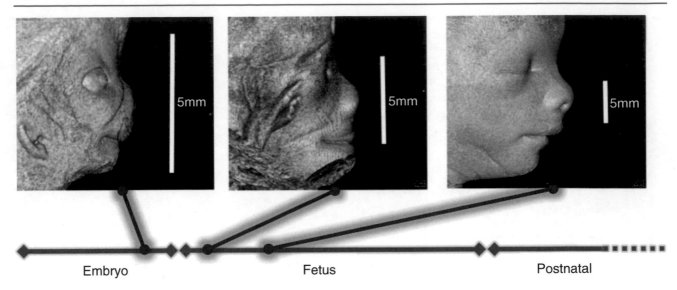

Fig. 32.1 The lateral view of three-dimensional images reconstructed from the magnetic resonance (MR) imaging data during the end of the embryo and the early fetal period. MR images were obtained by using a 7-T MR system (Biospec 70/20 USR, Bruker Biospin MRI Gmbh, Ettlingen, Germany)

ing the first trimester of pregnancy could cause the Binder phenotype, called warfarin embryopathy [8–12]. Therefore, the development of the nasal septum during the first trimester of pregnancy could be considered a major factor affecting nasal protrusion; however, this has not been well analyzed embryologically.

32.1.3 Two-Dimensional Morphometric Analysis for Nasal Septum Growth Allometry

Only a few studies have been carried out using human fetuses and have shown that the nasal septum maintains an almost isometric shape with growth during the second trimester of pregnancy [13, 14]. The morphometric analysis involved a linear measurement called traditional morphometrics, which was the application of multivariate statistical analyses to sets of quantitative variables, such as length, width, and height. While multivariate morphometrics combined multivariate statistics and quantitative morphology, it is difficult to find the shape difference because of the loss of the geometric relationships among the variables. For instance, the oval shape and rectangular shape objects could have the same maximum length and maximum width, although they are clearly different in shape. In the late 1980s, a morphometric analysis that captured the geometry of the morphological structures of interest and preserved this information throughout the analyses was developed [15]. This approach is called geometric morphometrics (GM).

Katsube et al. [1] quantified the shape of the nasal septum of human embryos and fetuses using magnetic resonance (MR) images (Fig. 32.2) by applying GM. They carried out

a generalized Procrustes analysis, followed by the principal component analysis. The allometric shape vector indicating the "growth vector" was calculated from the multivariate regression of PCs 1–3 on the centroid size (CS) [16]. According to the transformation of the deformation grids and MR images along the allometric shape vector, the nasal septum developed mainly in the anteroposterior direction and was not isometric (Fig. 32.3). Furthermore, the plots of all specimens in the PC space were projected onto an allometric shape vector, and their scores were plotted as allometric shape scores (ASS) against CS. The scatter plot showed that the ASS rapidly increased as CS increased in the smaller specimens, while the change was almost constant in the larger specimens (Fig. 32.4). The switch point was found to occur at 14.2 weeks of gestation according to Sahota's equation [17]. Thus, the nasal septum expanded in the anteroposterior direction until around 14 weeks of gestation, developing with an almost constant profile after this period.

32.1.4 Nasal Protrusion

Anteroposterior expansion, or longitudinal growth, of the nasal septum plays an important role in enabling protrusion of the nose and development of the anterior nasal spine (ANS); disturbances in this process could cause a low nasal profile, including the Binder phenotype. Two mechanical forces have been reported to contribute to the development of the ANS. The anterior growth of the nasal septum beyond the anterior edge of the premaxilla provides a growing force to the premaxilla, which contributes to the morphogenesis of the ANS via the septo–premaxillary ligament [18]. In addition to the anteroposterior development of the nasal septum,

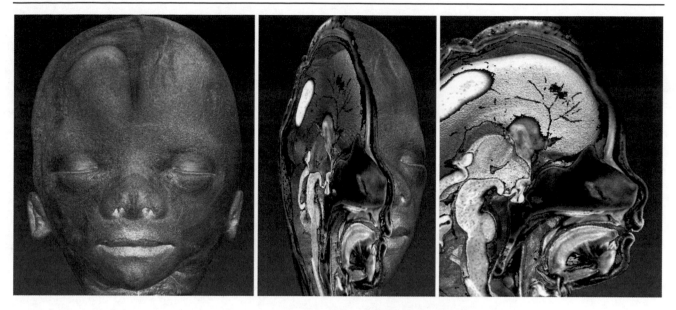

Fig. 32.2 Three-dimensional and mid-sagittal images are reconstructed from the MRI data of a human fetus

Fig. 32.3 The deformation grids and MR images from (**a**) to (**c**) are located on -2 SD, origin, and +2 SD (indicated by red stars) on the allometric shape vector (black arrow), respectively. These deformation grids and MR images show that the shape of the nasal septum expands mainly in the anteroposterior direction along with the allometric shape vector

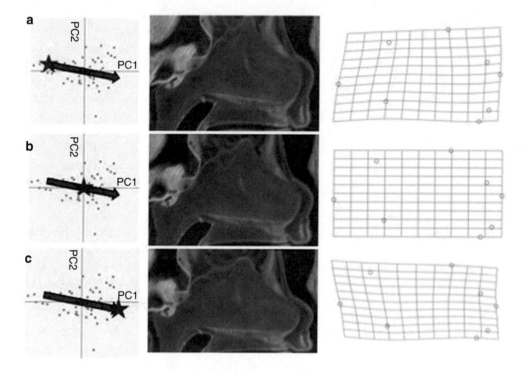

the stability of the premaxilla, which results from the fusion of the premaxillary–maxillary suture, has been reported to contribute to the development of the ANS [19]. Interestingly, the timing of fusion of the premaxillary–maxillary suture varies among humans, contributing to the mid-facial profile [19, 20]. A population with a low nose generally has a smaller ANS, and a population with a high nose generally has a larger ANS [21]. Mooney and Siegel [19] reported that the timing of fusion of the suture is earlier in people of European origin than in people of African origin. The nasal septum and

the premaxilla-maxilla suture may play a key role in the protruding human nose.

32.1.5 Morphogenesis of the Mid-Face

Congenital anomalies in the midface usually represent the conspicuous disharmony, often require multiple treatment procedures, and remain a challenging condition for craniofacial surgeons [22]. The skeleton of the midface is a complex

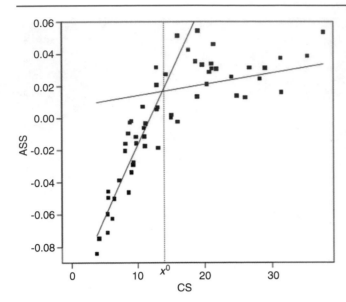

Fig. 32.4 The scatter plot of the allometric shape scores (ASS) against the centroid size (CS) shows that the ASS rapidly increased as CS increased in the smaller specimens, while the change was almost constant in the larger specimens. The switch point (X^0) between these groups was calculated using a regression equation

of various facial bones. In addition, several types of essential and functional organs are packed in small spaces. Accordingly, the shape of the midfacial skeleton is a very complicated three-dimensional structure. The shape of the midfacial skeleton has been reported to change drastically; it is presumably established during the early fetal period [1, 13, 23, 24]. Most of congenital craniofacial anomalies are induced by impairment of the normal growth trait, such as that in the cleft lip and palate or midfacial hypoplasia. Therefore, understanding the normal growth trait during the early fetal period could elucidate the pathogenesis of congenital craniofacial anomalies. However, the mechanisms underlying these phenomena remain unclear.

32.1.6 Traditional Morphometric Studies for the Mid-Facial Development

Diewert [25] investigated the craniofacial growth during 7–10 weeks of the gestational age using mid-sagittal histological sections in the morphometric analysis, such as cephalometry, and found that the mandibular growth was faster than the nasomaxillary complex and the cranial base angulation and maxillary position to the cranial base develop during the late embryonic period. Burdi [13, 14] reported the nasomaxillary growth during 12–24 weeks of gestation using the mid-sagittally sectioned head of human fetuses in the morphometric analysis, such as cephalometry, and found that almost isometric expansion occurred and the upper face grew in downward and forward directions. Trenouth [26] investi-

gated the craniofacial growth during 10–22 weeks of gestation using photographs and radiographs in the linear analysis and found that relative growth rates were greater in cranial length, height, and width in the order, and greater in facial length, width, and height in the order.

32.1.7 Three-Dimensional Morphometric Analysis for the Mid-Facial Development

Recent developments in medical imaging modalities, e.g., MR imaging and/or computed tomography (CT), allowed us to study these small specimens in three dimensions. Furthermore, GM enables us to grasp three-dimensional shape changes. Katsube et al. [2] investigated the allometry of the mid-facial skeleton during the early fetal period using 7 and 3 T MR images. The sample size was 60, including one embryo and 59 fetuses. They carried out the GM analysis in 3D and calculated the allometric shape vector and visualized the warped surface model along the vector. The transformation of the mid-face along the AS vector showed a shape change as follows: In the anterior view, the width of the alveolar arch changed a little while the middle to upper part of the maxilla was reduced in width, accompanied by a reduction in the relative size of the nasal cavity. In the lateral view, the zygoma drastically expanded in the anterolateral dimension; a structure such as the malar prominence was formed from the greater development in the superolateral portion of the zygoma compared to the inferior portion, the lateral part of the maxilla developed forward while the central part of the maxilla and the nasal developed little. The shape of the cranial base changed as follows: From a superior view, the anterior cranial base expanded in the anteroposterior direction, and the central part of the middle cranial fossa was reduced in width and length of the clivus (Fig. 32.5). Their results suggested that the shape of the facial structure related to the primary organs, such as the orbit, alveolar arch, and nasal cavity, has already been established before the fetal period; on the other hand, the zygoma drastically expanded in the anterolateral direction in the fetal period. Furthermore, they found a close correlation between the growth centers, including the nasal septum and spheno-ethmoidal synchondrosis with mid-facial growth.

32.1.8 Ontogenetic Allometry of the Face

The ontogenetic allometry of the facial shape results from the complex interaction between the genetic and epigenetic actions [27]. In the very early embryonic stage, specific genomes were presumed to regulate the morphogenesis of the central nervous system and the facial shape. The impair-

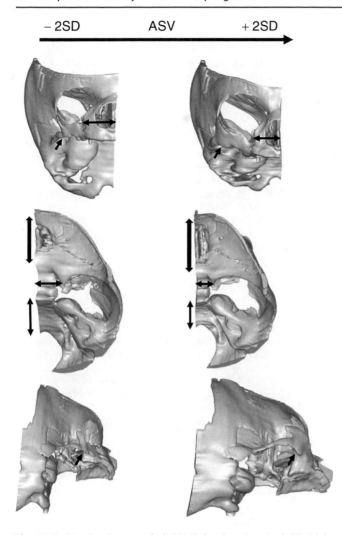

-2SD   ASV   +2SD

Fig. 32.5 Warping images of -2 SD (left column) and +2 SD (right column) of the allometric shape vector (ASV). These warping images show that the maxilla had a reduced width with the reduction of the relative size of the nasal cavity (double-headed arrow), the zygoma drastically expanded in the anterolateral dimension (arrow), the anterior cranial base expanded in the anteroposterior direction (double-headed arrow), and the central part of the middle cranial fossa had reduced width and length in the clivus (double-headed arrow)

ment of such gene expression causes severe congenital anomalies, such as holoprosencephaly [28]. Moss suggested the influence of the epigenetics on facial growth as a functional matrix hypothesis and stated that "Bones do not grow; bones are grown" [29–31]. He divided the matrix into periosteal and capsular matrices. According to the study of Katsube et al. [2], in the ontogenetic allometry of the human facial skeleton, the capsular matrix, such as the cranial base, orbit, or nasal cavity, develop first, followed by the periosteal matrix, such as zygoma, was developed later. The mastication muscle activity, observed as the mouth opening from around 11 weeks of gestation [32], probably influences on the zygoma as the periosteal matrix. Thus, epigenetic factors play an important role in the morphogenesis of the face.

The complexity of the facial skeleton is closely related to its various functions, such as support of the brain and eyeball and formation of the mastication apparatus and respiratory system. Therefore, elucidating its morphogenesis remains challenging. Application of GM will probably enable us to grasp the complicated facial morphogenesis in three dimensions.

The results of our study demonstrated the morphological changes along the time axis and its integration with functional axis in early fetal life. The results will also contribute to understand the pathology of the congenital facial anomalies.

Acknowledgments I acknowledge the contribution of collaborating obstetricians and the members of the Congenital Anomaly Research Center, Kyoto University Graduate School of Medicine. I would like to express our deepest gratitude to Akinobu Shimizu and Atsushi Saito from Tokyo University of Agriculture and Technology for their suggestions and technical support in the preparation of warping the images.

References

1. Katsube M, Yamada S, Miyazaki R, Yamaguchi Y, Makishima H, Takakuwa T, et al. Quantitation of nasal development in the early prenatal period using geometric morphometrics and MRI: a new insight into the critical period of Binder phenotype. Prenat Diagn. 2017;37(9):907–15. https://doi.org/10.1002/pd.5106.

2. Katsube M, Yamada S, Yamaguchi Y, Takakuwa T, Yamamoto A, Imai H, et al. Critical growth processes for the midfacial morphogenesis in the early prenatal period. Cleft Palate Craniofac J. 2019;56(8):1026–37. https://doi.org/10.1177/1055665619827189.

3. Katsube M, Rolfe SM, Bortolussi SR, Yamaguchi Y, Richman JM, Yamada S, et al. Analysis of facial skeletal asymmetry during foetal development using μCT imaging. Orthod Craniofac Res. 2019;22(S1):199–206. https://doi.org/10.1111/ocr.12304.

4. Goh RCW, Chen YR. Surgical management of Binder's syndrome: lessons learned. Aesthetic Plast Surg. 2010;34(6):722–30. https://doi.org/10.1007/s00266-010-9533-7.

5. Holmstrom H. Clinical and pathologic features of maxillonasal dysplasia (Binder's syndrome): significance of the prenasal fossa on etiology. Plast Reconstr Surg. 1986;78(5):559–67.

6. Delaire J, Precious D. Interaction of the development of the nasal septum, the nasal pyramid and the face. Int J Pediatr Otorhinolaryngol. 1987;12(3):311–26. https://doi.org/10.1016/S0165-5876(87)80007-1.

7. Grymer LF, Bosch C. The nasal septum and the development of the midface. A longitudinal study of a pair of monozygotic twins. Rhinology. 1997;35(1):6–10.

8. Keppler-Noreuil KM, Wenzel TJ. Binder phenotype: associated findings and etiologic mechanisms. J Craniofac Surg. 2010;21(5):1339–45. https://doi.org/10.1097/SCS.0b013e3181ef2b71.

9. Howe AM, Lipson AH, de Silva M, Ouvrier R, Webster WS. Severe cervical dysplasia and nasal cartilage calcification following prenatal warfarin exposure. Am J Med Genet. 1997;71(4):391–6.

10. Howe AM, Hawkins JK, Webster WS. The growth of the nasal septum in the 6-9 week period of foetal development—Warfarin embryopathy offers a new insight into prenatal facial development. Aust Dental J. 2004;49(4):171–6.

11. Hall JG, Pauli RM, Wilson KM. Maternal and fetal sequelae of anticoagulation during pregnancy. Am J Med. 1980;68(1):122–40. https://doi.org/10.1016/0002-9343(80)90181-3.

12. Van Driel D, Wesseling J, Sauer PJJ, Touwen BCL, Van Veer ED, Heymans HSA. Teratogen update: Fetal effects after in utero exposure to coumarins overview of cases, follow-up findings, and pathogenesis. Teratology. 2002;66(3):127–40. https://doi.org/10.1002/tera.10054.

13. Burdi AR. Cephhalometric growth analyses of the human upper face region during the last two trimesters of gestation. Am J Anat. 1969;125(1):113–22. https://doi.org/10.1002/aja.1001250106.

14. Burdi AR. Sagittal growth of the nasomaxillary complex during the second trimester of human prenatal development. J Dent Res. 1965;44:112–25. https://doi.org/10.1177/00220345650440010401.

15. Adams DC, Rohlf FJ, Slice DE. Geometric morphometrics: ten years of progress following the 'revolution'. Ital J Zool. 2004;71(1):5–16. https://doi.org/10.1080/11250000409356545.

16. Zollikofer CPE, Ponce de León MS. Neanderthals and modern humans—chimps and bonobos: similarities and differences. New York: Springer; 2008.

17. Sahota DS, Leung TY, Leung TN, Chan OK, Lau TK. Fetal crown-rump length and estimation of gestational age in an ethnic Chinese population. Ultrasound Obstet Gynecol. 2009;33(2):157–60. https://doi.org/10.1002/uog.6252.

18. Latham RA. Maxillary development and growth: the septopremaxillary ligament. J Anat. 1970;107(Pt 3):471–8.

19. Mooney MP, Siegel MI. Developmental relationship between premaxillary-maxillary suture patency and anterior nasal spine morphology. Cleft Palate J. 1986;23(2):101–7.

20. Schultz AH. Relation of the external nose to the bony nose and nasal cartilages in whites and negroes. Am J Phys Anthropol. 1918;1(3):329–38. https://doi.org/10.1002/ajpa.1330010304.

21. Park HS. Lambda-shaped implant for augmentation of anterior nasal spine in Asian rhinoplasty as an ancillary procedure. Aesthetic Plast Surg. 2001;25(1):8–14. https://doi.org/10.1007/s002660010086.

22. Renier D, Lajeunie E, Arnaud E, Marchac D. Management of craniosynostoses. Childs Nerv Syst. 2000;16(10–11):645–58. https://doi.org/10.1007/s003810000320.

23. Diewert VM. Development of human craniofacial morphology during the late embryonic and early fetal periods. Am J Orthod. 1985;88(1):64–76.

24. Johnston LE. A cephalometric investigation of the sagittal growth of the second-trimester fetal face. Anat Rec. 1974;178(3):623–30. https://doi.org/10.1002/ar.1091780309.

25. Diewert VM. A morphometric analysis of craniofacial growth and changes in spatial relations during secondary palatal development in human embryos and fetuses. Am J Anat. 1983;167(4):495–522.

26. Trenouth MJ. Relative growth of the human fetal skull in width, length and height. Archiv Oral Biol. 1991;36(6):451–6. https://doi.org/10.1016/0003-9969(91)90136-I.

27. Carlson DS. Theories of craniofacial growth in the postgenomic era. Semin Orthod. 2005;11(4):172–83. https://doi.org/10.1053/j.sodo.2005.07.002.

28. DeMyer W, Zeman W, Palmer C. The face predicts the brain: diagnostic significance of median facial anomalies for holoprosencephaly (arhinencephaly). Pediatrics. 1964;34:256.

29. Moss ML, Salentijn L. The primary role of functional matrices in facial growth. Am J Orthod. 1969;55(6):566–77. https://doi.org/10.1016/0002-9416(69)90034-7.

30. Moss ML. The differential roles of periosteal and capsular functional matrices in orofacial growth. Eur J Orthod. 2007;29(Supplement 1):i96–i101. https://doi.org/10.1093/ejo/cjl097.

31. Moss ML. The functional matrix hypothesis revisited. 1. The role of mechanotransduction. Am J Orthod Dentofacial Orthop. 1997;112(1):8–11.

32. Humphrey T. Development of oral and facial motor mechanisms in human fetuses and their relation to craniofacial growth. J Dent Res. 1971;50(6):1428–41. https://doi.org/10.1177/00220345710500061301.

Development of Bone Strength Prediction Method by Using MCA with Damage Mechanics

33

Mitsugu Todo

Abstract

The goal of this research project was to establish an effective prediction method of bone strength using CT-FEM. In the present study, three sub-projects were conducted, namely, deformation analysis of micro-CT image-based cancellous bone model, comparison of CT-FEA with cadaveric experiment, and analysis of the correlation between vertebra strength and lumbar YAM.

Keywords

Computer tomography · Finite element method Bone strength

33.1 Introduction

Osteoporosis has been one of the most important diseases in the current aged society of Japan. It has been known that osteoporosis easily causes bone fractures such as vertebral compression fracture (VCF), and such spontaneous bone fractures have become one of the most important issues in the field of recent orthopedics [1, 2]. In the current clinical situations, the degree of osteoporosis is examined on the basis of bone mineral density (BMD) measured by X-ray, such as the dual-energy X-ray absorptiometry (DEXA). Doctors then try to use a medical index YAM to diagnose the patient with the osteoporotic symptom. YAM can be recognized as the percentage ratio of the measured BMD to the average of young BMD (20–44 y.o.). In general, low YAM values are well correlated with a high risk of bone fracture; however, such averaged material property is not enough to predict bone strength which is a mechanical property and strongly related to the micro- and macro-structures of bone.

In recent years, as a field of multidisciplinary computational anatomy (MCA), CT-image based finite element method (CT-FEM) has widely been utilized to develop complicated 3D bone models using CT images and to analyze their mechanical performance. Furthermore, BMD distribution within the bone models can be estimated from CT values of the CT images by assuming a linear relationship between them. Once distributed BMD values are obtained, the elastic modulus (Young's modulus) can be estimated using Keyak's or Keller's method [3–5]. In their methods, the relationship between BMD and modulus is expressed by empirical equations in which the modulus is expressed as an exponential function. Those equations were obtained from cadaveric experiments. Damage mechanics-based analysis method can be incorporated with CT-FEM code to analyze microscopic damage formation behavior and therefore, bone fracture problems [6, 7]. Our research group has actively been working on the application of CT-FEM with damage mechanics to clinical biomechanical problems of the femur [8, 9] and spine [10, 11].

The objective of this study was to develop a computational method for bone strength analysis using MCA, such as CT-FEM combined with damage mechanics. There were three primary researches involved in this project. The first one was the deformation analysis of micro-CT image-based cancellous bone model that was conducted to understand the micro-deformation mechanism of porous cancellous bone. The second was the comparison of CT-FEA with the cadaveric experiment that was performed to assess the applicability of the standard Keyak's method to cadaveric experimental results and to develop a more accurate method to predict elastic modulus. The third research was the analysis of the correlation between vertebra strength and lumbar YAM using patient's data. Total 244 vertebrae were analyzed by CT-FEM to estimate their compressive strength and correlated with BMD data of the patients.

M. Todo (✉)

Research Institute for Applied Mechanics, Kyushu University, Fukuoka, Japan

e-mail: todo@riam.kyushu-u.ac.jp

© The Author(s), under exclusive license to Springer Nature Singapore Pte Ltd. 2022

M. Hashizume (ed.), *Multidisciplinary Computational Anatomy*, https://doi.org/10.1007/978-981-16-4325-5_33

33.2 Deformation Analysis of μ-CT Image-Based Cancellous Bone Model

Micro-CT images of a femoral head of osteoarthritis (OA) patient was prepared for the finite element (FE) modeling of cancellous bone. A cubic 3D-FE model of cancellous bone was then constructed from these images using Mechanical Finder Software (RCCM, Inc). The size of the cubic model is 350 μm³. The Young's modulus and Poisson's ratio of the model were set to 18 GPa and 0.3, respectively. The model was then sandwiched between two plates, and distributed compressive load of 10 N was applied to the top surface. A damage model was also introduced to predict microscopic damage accumulation in the cancellous bone structure.

Distribution patterns of strain energy density (SED) and damaged elements are shown in Fig. 33.1. It is well known that SED is thought to be related to bone remodeling and has been used as a control parameter in bone remodeling analysis. Localized SED concentrations are seen in the figure. Most of the damage modes are related to compressive failures, yielding, and fracture.

33.3 Comparison of CT-FEA with Cadaveric Experiment

In this research, CT-FEA results of mechanical tests of femur were compared with the corresponding experimental results of cadaveric femurs to assess the prediction method of the relationship between bone mineral density and Young's modulus. A modified prediction method was also proposed in this study. Four different femurs retrieved from cadavers were tested in Chiba University Hospital with the approval of the ethics committee. CT images of the femurs were also obtained prior to the tests. Then, 3D femoral models were developed from the CT images using Mechanical Finder Software. The boundary conditions were set to imitate the experimental testing conditions.

It was found that the load-displacement curves obtained from the experimental testing were very different from each other. These load-displacement curves were tried to be predicted by CT-FEM, and the results showed that the experimental load-displacement curve cannot be predicted by using Keyak or Keller equations very well, although their equations have widely been used in bone analysis worldwide. A new prediction method was then proposed by considering the distribution of bone mineral density within the femurs. This equation to predict the relationship between BMD and Young's modulus was based on the Keller equation for vertebrae. The FE results are shown in Fig. 33.2. It is clearly seen that the initial slope of the load-displacement curve corresponding to the stiffness was well predicted by the newly proposed method.

33.4 Analysis of Correlation Between Vertebra Strength and Lumbar YAM

The degree of osteoporosis of an elderly patient has clinically been determined on the basis of DEXA result, which corresponds to the averaged bone mineral density, although osteoporosis has been defined as a disease characterized by a reduction of bone strength. In this study, strength values of 244 vertebrae of 84 patients were examined by CT-FEM, and the correlation between the strength and YAM was assessed. CT data of the total 103 patients were examined, and 84 data were chosen for CT-FEA of vertebra strength. 3D-models of total of 244 vertebrae, including T11, T12, L1, L2, and L3, were constructed from the CT images using Mechanical Finder CLINIC software (RCCM, Inc). The 3D models were

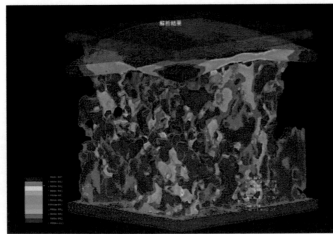

Strain energy density

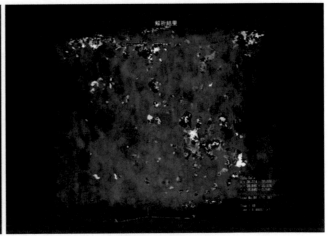

Damaged elements

Fig. 33.1 Distributions of strain energy density and damaged elements

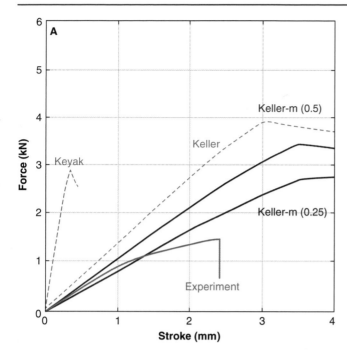

Fig. 33.2 Prediction results using CT-FEM with modified Keller method

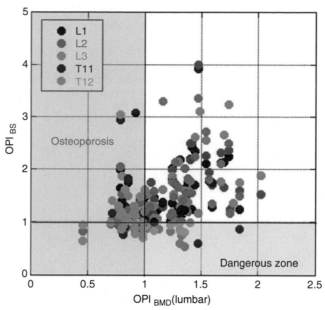

Fig. 33.3 Relationship between OPI$_{BS}$ and OPI$_{BMD}$

then analyzed under compressive loading conditions, and the strength was evaluated from FEA with damage mechanics. In general, a low BMD value corresponds well to osteoporosis, and therefore, YAM can be recognized as a useful parameter in clinical situations. Two kinds of osteoporosis index were newly defined by:

$$OPI_{BMD} = YAM/70 \qquad (33.1)$$

$$OPI_{BS} = \text{Vertebra strength}/2500\,\text{N} \qquad (33.2)$$

Thus, OPI$_{BMD}$ < 1 corresponds to the clinical determination of osteoporosis and OPI$_{BS}$ < 1 corresponds to high risk of vertebra compression fracture (VCF). The relationship between OPI$_{BMD}$ and OPI$_{BS}$ is shown in Fig. 33.3. It is seen that strong correlation does not exist between the two indexes. It is very important to note that some vertebrae exhibited low OPI$_{BS}$ (<1) with high OPI$_{BMD}$ (>1), suggesting the following misleading: the patients having these vertebrae will be examined as no osteoporosis because of their high YAM values, although their vertebrae have a high risk of VCF.

33.5 Conclusions

In this 2-year project, three different researches were conducted mainly, and the results were obtained as follows:

1. 3D model of cancellous bone was developed using micro-CT images of an OA femur, and FEA was performed to understand the deformation mechanism of the porous structure. Localized SED distribution was clearly observed, suggesting localized failure of microstructures. Such localized damage accumulation was also analyzed by introducing the damage mechanics for bone.

2. CT-FEA results of three cadaveric femurs were compared with their experimental results. It was clearly shown that the standard Keyak or Keller methods cannot be used to predict the macroscopic load-displacement behavior. A modified Keller method was proposed, and the prediction of stiffness was dramatically improved.

3. Compressive strength values of 244 vertebrae of 84 patients were evaluated by CT-FEM, and their correlations with YAM values were examined. It was worth noting that some vertebrae exhibited high risk of vertebra fracture, although the corresponding YAM values were considered as normal conditions. These results suggest that we need to introduce the strength-based diagnostic method along with the standard BMD-based methods.

References

1. Johnell O, Kanis JA. An estimate of the worldwide prevalence and disability associated with osteoporotic fractures. Osteoporos Int. 2006;17:1726–33.
2. Liang D, Ye LQ, Jiang XB, Yang P, Zhou GQ, Zhang SC, et al. Biomechanical effects of cement distribution in the fractured area on osteoporotic vertebral compression fractures: a three-dimensional finite element analysis. J Sur Res. 2015;195:246–56.
3. Keyak JH, Lee IY, Skinner HB. Correlations between orthogonal mechanical properties and density of trabecular bone: use of different densitometric measures. J Biomed Mater Res. 1994;28:1329–36.

4. Keyak JH, Rossi SA, Jones KA, Skinner HB. Prediction of femoral fracture load using automated finite element modeling. J Biomech. 1998;31:125–33.
5. Keller TS. Predicting the compressive mechanical behavior of bone. J Biomech. 1994;27:1159–68.
6. Bessho M, Ohnishi I, Matsuyama J, et al. Prediction of strength and strain of the proximal femur by a CT-based finite element method. J Biomech. 2007;40:1745–53.
7. Bessho M, Ohnishi I, et al. Prediction of proximal femur strength using a CT-based nonlinear finite element method: differences in predicted fracture load and site with changing load and boundary conditions. Bone. 2009;45:226–31.
8. Abdullah AH, Todo M. Effects of hip arthroplasties on bone adaptation in lower limbs: a computational study. J Biosci Med. 2015,3.1–7.
9. Abdullah AH, Todo M. Stress evaluation of lower limbs with hip osteoarthritis and hip arthroplasty. J Med Bioeng. 2015;4:100–4.
10. Takano H, Yonezawa I, Todo M, et al. Biomechanical study of the effects of balloon kyphoplasty on the adjacent vertebrae. J Biomed Sci Eng. 2016;9:478–87.
11. Mazlan MH, Todo M, et al. Effect of cage insertion orientation on stress profiles and subsidence phenomenon in posterior lumbar innerbody fusion. J Med Bioeng. 2016;5:93–7.

Part XI

New Frontier of Technology in Clinical Applications Based on MCA Models: Emerging Innovasive Imaging Technology

Development of a Generation Method for Local Appearance Models of Normal Organs by DCNN

34

Shouhei Hanaoka

Abstract

The objective of this chapter is to describe how to develop generation methods for statistically modeling the local appearances of normal organs. Mainly we focus upon abnormality detection tasks, which is the main application of such local appearance models. Firstly, we developed an unsupervised anomaly detection system with an autoencoder in emergency head CT volumes. In the system, an autoencoder encodes the normal local appearances of the brain. Using this autoencoder, various abnormalities in the given emergency CT datasets were detected. Secondly, we developed abnormality detection system for chest FDG-PET-CT images and residual network-based unsupervised temporal image subtraction for highlighting bone metastases.

Keywords

Deep convolutional neural network · Emergency head CT · FDG PET-CT · Bone metastasis · Unsupervised deep learning

34.1 Introduction

Manual labeling of each abnormal lesion for supervised machine learning is a time-consuming work. Therefore, manual labeling is a factor hindering the improvement of CAD performance. However, that time-consuming work can be avoided if the method can be trained in an unsupervised manner.

Therefore, we prefer a general anomaly detection method to a combination of lesion detection methods designed to detect individual disorders separately.

The purpose of this chapter is to show several methods/tasks to detect various abnormal findings in medical images. As you will see later, building a normal appearance model is critically important in such tasks. We used deep learning-based appearance models to solve this problem.

34.2 Detection of Abnormalities in Head CT Volumes

Abnormal findings of head emergency CT may be overlooked in busy emergency situations. Therefore, people working in an emergency department desire a computer-assisted detection system for emergency head CT.

The target disorders of emergency head CT are wide-ranging, such as infarction, intracerebral hemorrhage, subarachnoid hemorrhage, extradural hemorrhage, subdural hemorrhage, and tumors. Moreover, multiple disorders may be found at the same time.

Many studies have investigated the lesion detection methods in the head CT. Some studies have focused on acute ischemic stroke [1–3], and other studies have focused on cerebral hemorrhage [4, 5]. Merkowa et al. [6] used deep convolutional neural networks to recognize 30 traits in CT head images. A few studies have focused on anomaly detection methods in emergency head CT [7–9].

An autoencoder is a neural network to train the features of a dataset in an unsupervised manner. An autoencoder is trained by minimizing the reconstruction error between the input data and its reconstruction. After training by normal cases, the reconstruction error calculated by the autoencoder can interpret as an abnormality of input data. Therefore, we can use the reconstruction error for anomaly detection. Some studies used reconstruction error as an abnormality.

The purpose of this study was to propose an unsupervised anomaly detection method in emergency head CT using an autoencoder and to evaluate the anomaly detection performance of our method in emergency head CT. To the best of

S. Hanaoka (✉)
Department of Radiology, The University of Tokyo, Tokyo, Japan
e-mail: hanaoka-tky@umin.ac.jp

© The Author(s), under exclusive license to Springer Nature Singapore Pte Ltd. 2022
M. Hashizume (ed.), *Multidisciplinary Computational Anatomy*, https://doi.org/10.1007/978-981-16-4325-5_34

our knowledge, this [10] is the first study of the potentialities of an autoencoder in anomaly detection in CT.

34.2.1 Methods

We used a 3D convolutional autoencoder (3D-CAE) to detect anomaly lesions of emergency head CT. In the training phase, we trained the 3D-CAE using 3D patches extracted from normal cases. In the test phase, we calculated abnor-

malities of each voxel in the emergency head CT (normal or abnormal) and evaluate the likelihood of anomaly existence.

34.2.1.1 Construction of Our Autoencoder

The autoencoder used in this study as 3D convolutional neural network (3D-CNN) block and 3D deconvolutional neural network (3D-deCNN) block. The 3D-CNN block contains 11 convolutional layers, and the 3D-deCNN block has six deconvolutional layers. A schematic diagram is shown in Fig. 34.1.

Fig. 34.1 A schematic diagram *of 3D-CAE*

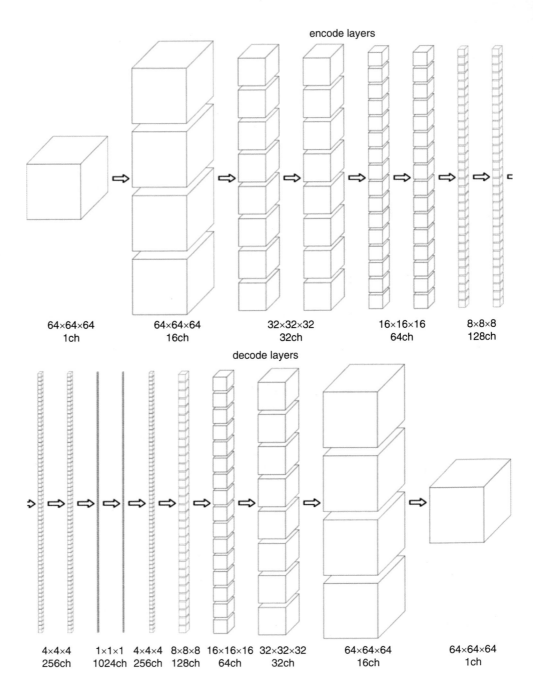

34.2.1.2 Patch Extraction

Each CT volume was rescaled to $1 \times 1 \times 1$ mm³/voxel. We extracted the intracranial region automatically by using global thresholding and morphological operations. We extracted 3D partial volume patches of $64 \times 64 \times 64$ voxels whose center point was each voxel in the intracranial region.

34.2.1.3 Training Process

We utilized 10,000 3D patches from 50 cases of normal head CT and trained the 3D-convAE to minimize reconstruction errors, which were calculated by mean squared errors (MSE). The training was stopped when the reconstruction error stopped to improve.

34.2.1.4 Evaluation Process

Using the trained 3D-CAE, we calculated the MSE of each 3D patch extracted from our test CT cases (normal and abnormal). We considered that MSE as the patch abnormality. Then we defined an abnormality of a case as the maximum of all patch abnormalities in the case. Receiver operating characteristic (ROC) analysis was performed to determine the threshold to differentiate between normal CT cases and abnormal ones.

34.2.1.5 Dataset

We utilized 50 emergency head CT volumes of normal cases for training 3D-convAE and other 38 emergency head CT volumes for evaluation. Twenty-two of 38 evaluation cases were abnormal ones. We determined the ground truth label of each CT case (i.e., as normal or abnormal) by the corre-

sponding diagnostic report written by radiologists. Abnormal cases contained two acute infarctions, six old infarctions, seven intracerebral hemorrhages, two brain injuries, six subarachnoid hemorrhages, two subdural hemorrhages, five intraventricular hemorrhages. The original voxel size of each CT volume was $0.43 \times 0.43 \times 1.0$ mm.

34.2.2 Results

Figure 34.2 shows that the 2D exploded views of input and output 3D patches. Images with noise removed and relatively clear white matter, gray matter, cerebrospinal fluid, cerebral sickle, and bones were output.

Figure 34.3 shows examples of the abnormality map generated by the 3D-convAE corresponding to a single slice of CT images. Figure 34.4 shows a ROC curve of abnormalities among the 38 evaluation cases. This method achieved a sensitivity of 68% and specificity of 88%, with the area under the curve (AUC) of the ROC of 0.87.

34.2.3 Discussion

The results show that our method has a moderate accuracy to distinguish normal CT cases from abnormal ones. This means that 3D-CAE has potentialities for anomaly detection in emergency head CT. According to the results, a part of the structure not in the normal brain was calculated by this method with a high degree of abnormality.

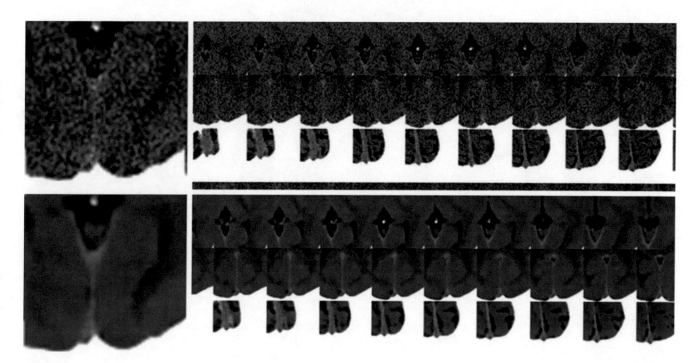

Fig. 34.2 An example of exploded views of input 3D patch (top row) and output (bottom row). The left column shows enlarged views

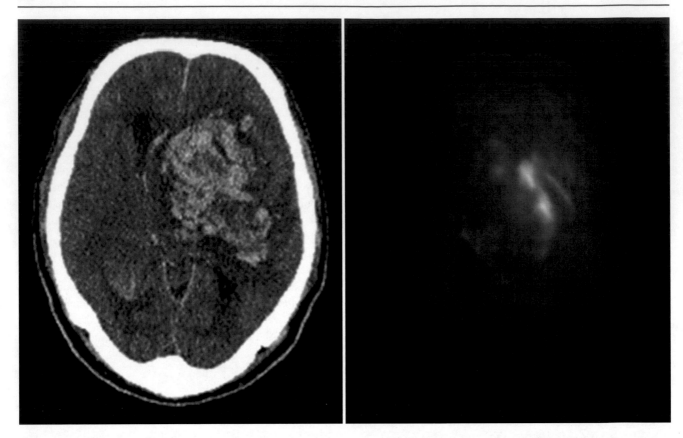

Fig. 34.3 Examples of the abnormality map generated by the 3D-CAE corresponding to a single slice of CT images. Arrows indicate hemorrhage lesions

Fig. 34.4 The ROC curve of the 16 normal and 22 abnormal cases. AUC = 0.87

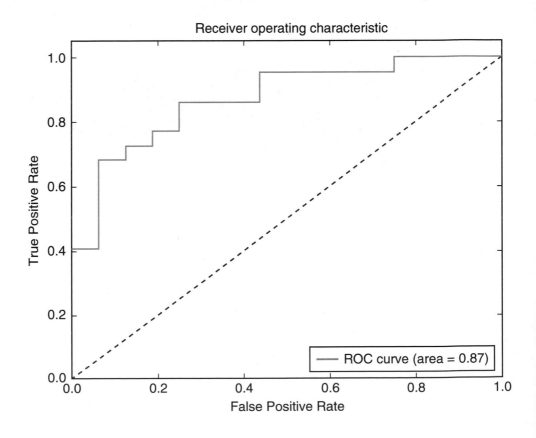

The small lesions make them difficult to detect by our method. The small old infarct lesion is partially similar in shape to structures such as the ventricles and the cerebral vasculature.

34.2.4 Conclusion

We proposed an unsupervised anomaly detection in emergency head CT using a 3D convolutional autoencoder. According to the evaluation, the AUC of our method achieved 0.87. This shows that autoencoder has potentialities for anomaly detection in emergency head CT.

34.3 Abnormality Detection System for Chest FDG-PET-CT Images

In this study [11], we developed an abnormality detection system for PET-CT. Because only images from normal cases were used in the training process, the proposed method requires no manual input of abnormal lesions.

The proposed method is based on 3-D ResNet [12] without any pooling layer. The network is illustrated in Fig. 34.5. The training and detection were performed using a sliding window method in which each voxel in the chest region was evaluated one by one. The input of the network is a partial VOI (volume of interest) around the target voxel extracted from the CT volume. The outputs of the network were an estimated mean and standard deviation of the SUV (standardized uptake value) of the target voxel in the corresponding PET volume. In other words, the network estimates the probability distribution of the SUV value of the center voxel of the VOI as a normal distribution, using the CT values of the voxels within the VOI. Using this estimated distribution, a z-score of the SUV value at each voxel is calculated and used as an abnormality score.

The proposed method was trained using 498 normal FDG PET-CT cases from our health-check program in our hospital. In each dataset, the lung regions were extracted automatically. Then, the network was trained using a sliding window method, in which a VOI is placed and moved across all the chest voxel. In evaluation, 28 PET-CT datasets with abnormal FDG uptake lesion(s) were used. A radiological

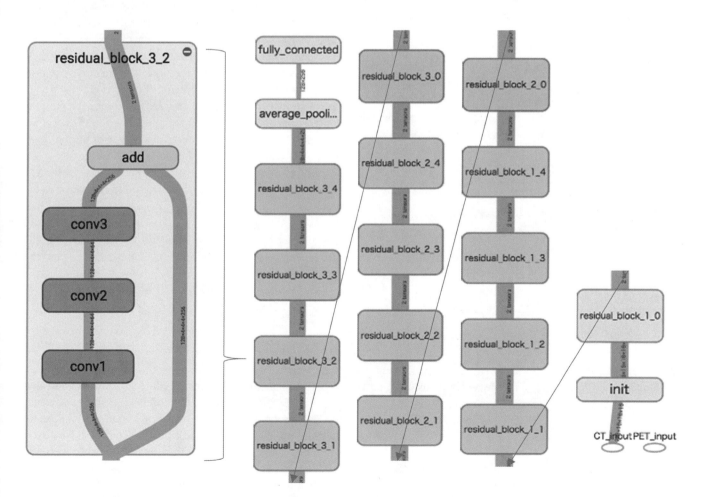

Fig. 34.5 The network diagram

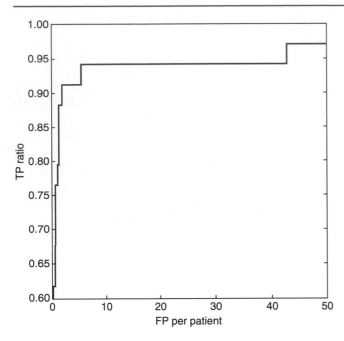

Fig. 34.6 The FROC curve

expert manually segmented the lesions (these label images were used for the evaluation purpose only). The Z-score of each voxel in the lung fields was calculated and binarized using a threshold of 3.0. The final output is connected to components of the binarized z-score. The likelihood of each connected component was determined as the maximum value of z-scores within the connected component. Using these likelihoods, an FROC (free-response receiver operating characteristic) analysis was performed (Fig. 34.6).

The number of false positives was three per case. Figure 34.7 shows an example of the z-score maps. Note that the z-score image clearly visualized the lesion when it is compared to the SUV image.

In conclusion, we developed a novel system for an abnormality detection system for FDG-PET-CT of lung fields. It showed a fair sensitivity without any abnormal training datasets.

34.4 Residual Network–Based Unsupervised Temporal Image Subtraction For Highlighting Bone Metastases

34.4.1 Purpose

The purpose of this study [13] is to develop a system to highlight bone metastasis lesions in a couple of time-series CT volumes. Using two time-series of CT datasets, our algorithm maximizes the information of the intertemporal differences and highlight the metastases.

34.4.2 Methods

In our previous study [14], bony landmark detection and landmark-based demons algorithm, which can register two bony images, had been established. Furthermore, the bony regions (the spine and the pelvis) are automatically segmented. Using these methods, the previous and current CT volumes are registered and segmented. Then, the subtraction is performed by the Residual network-based method. Firstly, from every voxel x in the given previous CT volume, a $15 \times 15 \times 15$ adjacent voxel set $N_{\text{previous}}(x)$ is extracted. This input is processed by a ResNet-based network illustrated in Fig. 34.8 right. Finally, this network outputs not only the estimated mean of the current CT image $\mu(x)$, but also the estimated standard deviation of the current CT image $\sigma(x)$.

In the training phase, this network only uses normal temporal pair cases (without bone metastasis)—the proposed network estimates $N_{\text{previous}}(x) \rightarrow \mu(x)$, $\sigma(x)$. Here, let the CT value of current CT volume (which has been registered to the previous volume) at x be $I_{\text{new}}(x)$. It is assumed that $I_{\text{new}}(x) \sim \mathcal{N}(\mu(x), \sigma(x))$. Then, the proposed ResNet-based [11] network estimates the best couple of μ and σ by minimizing the loss function

$$\mathcal{L} = -\ln \Pr\left(I_{\text{new}}(x) \mid \mu, \sigma\right) = -\left[\frac{1}{2}\ln 2\pi - \ln \sigma\left(N_{\text{previous}}(x)\right) - \frac{1}{2}\left(\frac{I_{\text{new}}(x) - \mu\left(N_{\text{previous}}(x)\right)}{\sigma\left(N_{\text{previous}}(x)\right)}\right)^2\right].$$

In other words, our ResNet estimates two functions $N_{\text{previous}}(x) \rightarrow \mu(x)$ and $N_{\text{previous}}(x) \rightarrow \sigma(x)$ simultaneously so as to minimize L. Again, note that this loss function minimization is performed only via normal temporal CT pairs in which no metastasis occurred. Thus, this study is a kind of unsupervised learning.

In the test phase, if there is a metastasis at the position x in the current volume, the estimated z score $z(x) = \dfrac{I_{\text{new}}(x) - \mu\left(N_{\text{previous}}(x)\right)}{\sigma\left(N_{\text{previous}}(x)\right)}$ should be a large positive

(e.g., osteoblastic) or negative (e.g., osteolytic) value. Thus, in our implementation, after a sliding window method that makes a 3D z-score volume, both of these positive and negative value are highlighted by using maximum and minimum intensity projection (MIP and MinIP) techniques. Finally, a color map that illustrates both osteoblastic and osteolytic lesions is generated.

The training phase was performed with 50 normal CT temporal pairs. The test was performed with 40 CT temporal pairs in which bone metastases have occurred between the

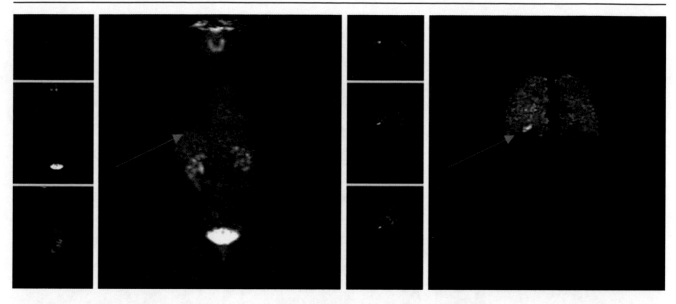

Fig. 34.7 An example of detection. Left: the cross-sections and a maximum intensity projection (MIP) image of the SUV values. Right: the cross-sections and a MIP image of the z-scores outputted by the proposed method. The arrow indicates an abnormal legion

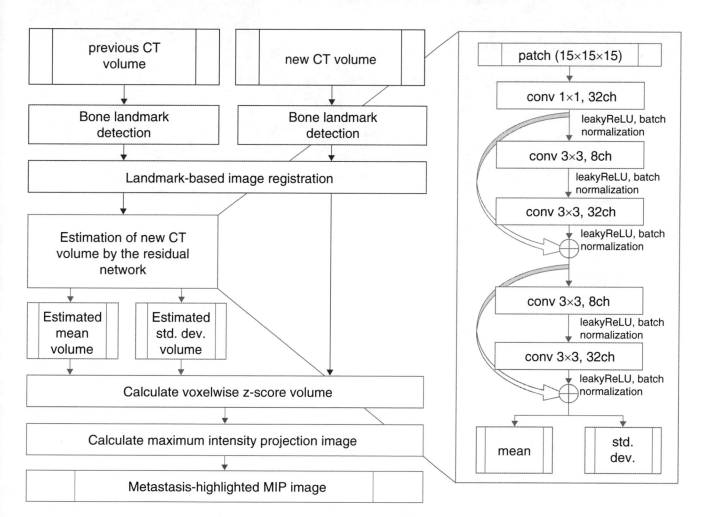

Fig. 34.8 The outline of the proposed method

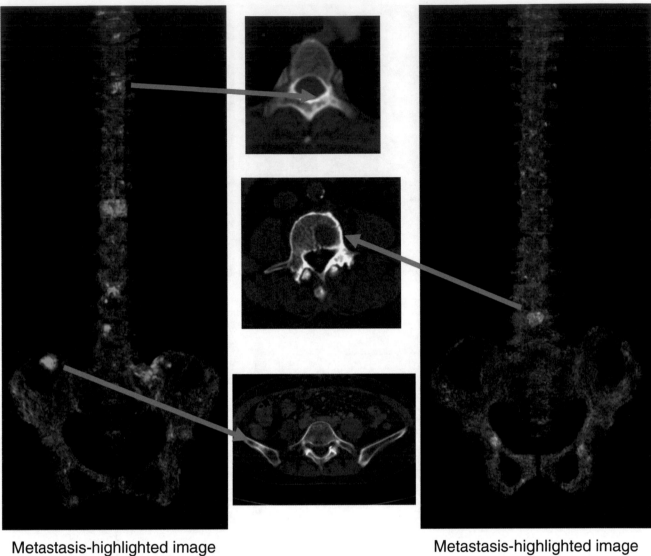

Metastasis-highlighted image
(osteoblastic case)

Metastasis-highlighted image
(osteolytic case)

Fig. 34.9 An exemplar result of the proposed method

previous and current CT acquisitions. For each test dataset, colored MIP and MinIP images are created to grasp the entire bony lesion in one glance.

34.4.3 Results

The exemplar results of the proposed method are illustrated in Fig. 34.9. As illustrated, the proposed method could highlight both osteolytic and osteoblastic lesions. Among 40 test cases, 12 cases had image distortion/false positive due to large registration errors. However, most of the false positives were suppressed due to the introduction of the standard deviation σ. All of the 40 cases, the metastases were clearly highlighted.

34.4.4 Conclusion

A method to emphasize bony metastatic lesions is presented. Because the proposed method is unsupervised, it can be readily applied to the daily clinical environment. Therefore, we consider that the proposed method is beneficial in routine CT examinations for patients with cancer.

References

1. Nagashima H, et al. Computerized detection of acute ischemic stroke in brain computed tomography images. Med Imaging Technol. 2009;27(1):30–8. (in Japanese)

2. Takahashi N, Lee Y, Tsai DY, Matsuyama E, Kinoshita T, Ishii K. An automated detection method for the MCA dot sign of acute stroke in unenhanced CT. Radiol Phys Technol. 2014;7(1):79–88.

3. Herweh C, Ringleb PA, Rauch G, Gerry S, Behrens L, Möhlenbruch M, Gottorf R, Richter D, Schieber S, Nagel S. Performance of e-ASPECTS software in comparison to that of stroke physicians on assessing CT scans of acute ischemic stroke patients. Int J Stroke. 2016;11(4):438–45.

4. Chan T. Computer aided detection of small acute intracranial hemorrhage on computer tomography of brain. Comput Med Imaging Graph. 2007;31:285–98.

5. Li Y, Wu J, Li H, Li D, Du X, Chen Z, Jia F, Hu Q. Automatic detection of the existence of subarachnoid hemorrhage from clinical CT images. J Med Syst. 2012;36:1259–70.

6. Jameson M, Robert L, Kim N, Stefano S, Zhuowen T, Andrea V. DeepRadiologyNet: radiologist level pathology detection in CT head images. arXiv preprint arXiv:1711.09313; 2017.

7. Chawla M, Sharma S, Sivaswamy J, Kishore L. A method for automatic detection and classification of stroke from brain CT images. Conf Proc IEEE Eng Med Biol Soc. 2009;2009:3581–4.

8. Zaki WMDW, Fauzi MFA, Besar R, Ahmad WSHMunirah W. Abnormalities detection in serial computed tomography brain images using multi-level segmentation approach. Multimed Tools Appl. 2011;54:321–40.

9. Lee Y, Takahashi N, Tsai DY. Computer-aided diagnosis for acute stroke in CT images. In: Saba L, editor. Computed tomography—clinical applications. InTech; 2012.

10. Sato D, Hanaoka S, et al. A primitive study on unsupervised anomaly detection with an autoencoder in emergency head CT volumes. SPIE Medical Imaging 2018.

11. Kamesawa R, Sato I, Hanaoka S, Nomura Y, Nemoto M, Hayashi N, Sugiyama M. Lung lesion detection in FDG-PET/CT with Gaussian process regression. SPIE Medical Imaging 2017, Florida, USA, February; 2017. p. 11–6,

12. He, K, et al. Deep residual learning for image recognition. In Proceedings of the IEEE conference on computer vision and pattern recognition. 2016.

13. Hanaoka S, Masumoto T, Hoshiai S, Nomura Y, Takenaga T, Murata M, Miki S, Yoshikawa T, Hayashi N, Abe O. Residual network–based unsupervised temporal image subtraction for highlighting bone metastases. CARS 2018 the 31st International Congress and Exhibition of Computer Assisted Radiology and Surgery, Berlin, Germany, June 20-23, 2018. Int J Comput Assist Radio Surg. 2018;13(suppl 1):S173–4.

14. Hanaoka S, et al. Landmark-guided diffeomorphic demons algorithm and its application to automatic segmentation of the whole spine and pelvis in CT images. Int J Comput Assist Radiol Surg. 2017;12(3):413–30.

Comprehensive Modeling of Neonatal Brain Image Generation for Disorder Development Onset Prediction Based on Generative Adversarial Networks

Saadia Binte Alam and Syoji Kobashi

Abstract

Medical images with a variety of modalities are helpful for better understanding the changes of the human anatomy. Multidisciplinary computational anatomy (MCA) provides a platform of mathematical analysis for a comprehensive and useful understanding of both static anatomies of human organs and its dynamic properties. In this chapter, we explored a method to generate synthetic brain images using generative adversarial networks (GANs). In medical image analysis, the exchange of clinical image data is a crucial issue for the implementation of diagnostic support systems. This relic is a demanding and vital task to produce realistic medical images that are entirely different from the original ones. At the same time, it is difficult for researchers to obtain medical image data due to the personalized information contained in it. Recent studies opened a scope to distribute training images without seeing actual image data using GAN models where the artificially recreated images have anonymized personal information completely. These synthetically created images can be used as training images for the classification of medical image, encouraging medical image analysis as a feasible choice. Instead of collecting a large amount of MR image data, an approach to image generation has been implemented in this chapter. In order to generate neonatal brain image, we exploit progressive growing GAN (PGGAN), a method that can be used for brain MR image classification and brain development disease prediction tasks. The PGGAN slowly discovers the related features in MR images by adding new layers during the training phase. The method of synthetic image generation shows that it can produce brain MR images avoiding artificial artifacts and have clinical characteristics of the target symptom.

Keywords

Generative adversarial network · Synthetic medical image generation · PGGAN · Neonates · MR image

35.1 Introduction

In research works related to medical imaging, a set of multiple images is usually required to find a desired output. For instance, for accurate diagnosis and segmentation of cerebral disease detection, multiple MR contrast such as T1-weighted, post-contrast T1 weighted, T2 weighted, and T2-FLAIR are required. Tactlessly, the complete set of input data are often difficult to obtain due to the different acquisition protocol at each organization, lengthy acquisition time, operative errors, or patient movement during the data acquisitions. Furthermore, it is often impossible to use contrast agents for neonates. Without the complete contrast, the subsequent analysis can be susceptible to significant biases and errors that can decrease the numerical efficiency, and the accurate segmentation of the neonatal brain may not be feasible.

Furthermore, in some circumstances, although multiple contrast images are available, some of the images suffer from systematic errors. For example, a synthetic MR imaging technique called Magnetic Resonance Image Compilation (MAGiC, GE Healthcare) [1] allows the generation of the various contrasts MR images using a Multi-Dynamic Multi-Echo (MDME) scan which can provide clinically useful synthetic MR images various contrasts. Unfortunately, it is often stated that sometimes synthetic contrasts have readily recognizable with artifacts [2]. Particularly, the characteristic granulated hyperintense artifacts apparent in the margins

S. B. Alam (✉)
Department of Electrical and Electronic Engineering, IUBAT-International University of Business, Agriculture and Technology, Dhaka, Bangladesh
e-mail: saadiabinte@ieee.org

S. Kobashi
Graduate School of Engineering, University of Hyogo, Kobe, Japan
e-mail: kobashi@amec-hyogo.org

M. Hashizume (ed.), *Multidisciplinary Computational Anatomy*, https://doi.org/10.1007/978-981-16-4325-5_35

along the cerebrospinal fluid (CSF)–tissue boundaries on MAGiC FLAIR can be mistaken for true pathologic conditions. Also, flow and/or noise artifacts are more common on MAGiC FLAIR than conventional FLAIR. This leads to further MR acquisition to confirm the diagnosis, which requires massive amount of cost and patient inconvenience, especially for neonates.

Hence, rather than re-acquiring all data as a complete set in this unforeseen situation, it is often vital to fill the missing data with substituted data. In statistical writings, this process is often mentioned as missing data imputation. Once all data have been credited, the dataset can be used as an input for standard analysis. Recently, the field of image imputation has been expressively progressive due to the massive success of deep neural networks [3–5].

For medical image generation methods that allow the generation of meaningful synthetic information, researchers are being able to develop and validate more sophisticated techniques for the recognition of images. Typically, the scenario can be formulated as an image translation problem from one domain to another domain [6, 7], whose performance has been greatly improved with [8]. The core purpose of GAN architecture is to generate realistic samples/images. The use of synthetic data would lead to image recognition tasks for minor inspections that include sharing and updating clinical data. In this study, we targeted neonatal brain MRI images for the diagnosis of brain disorder disease onset prediction.

In this chapter, we exploit PGGAN, a progressively growing Generative Adversarial Network. This featured method studies how to distribute the target data and how to produce the image after distribution from a latent space. Hence, created synthetic images are not linked to individual patient image information and can easily be used by researchers to construct support systems.

We addressed two problems; which GAN architecture is well adjusted for practical medical image generation and how can we treat MR images with distinctive intra- sequence variability. So our contribution in this chapter is to explore PGGAN and see if it can produce realistic brain MR images, which can potentially lead to effective clinical applications for data augmentation for machine learning and medical imaging tasks. This work observes the way to use medical images with an underlying intra-sequence variation to maximize GAN-based synthetic data generation for medical imaging.

35.2 Preliminary

35.2.1 Generated Adversarial Network

Characteristically, GAN consists of two neural networks: a generator and a discriminator. The discriminator tries to find the features to differentiate false image from real images, while the generator learns to synthesize images so that the discriminator face difficulty to judge as real or fake. After training both neural networks, the generator produces realistic outputs, which cannot be distinguished as fake samples by the discriminator. Both simulators are trained at the same time as a stochastic gradient descent (SGD) algorithm. The training actions can be seen as a two-player min–max step with the following objective function:

$$\min_{X} \max_{Y} V(X,Y) = \mathbb{E}_{a \sim p_{\text{data1}}(a)} \Big[\log X(a) \Big] + \mathbb{E}_{z\text{-p.(s)}} \Big[\log\big(1 - X(Y(z))\big) \Big] \tag{35.1}$$

In the above Eq. (35.1), the discriminator X attempts to maximize $V(X, Y)$ and the generator Y attempts to minimize it. In other words, the discriminator X separates the images in a p_{data} from those of $Y(z)$, while the generator Y produces samples to fool the discriminator X. Since the introduction of the original, many inventive additions have appeared. For example, for the translation between two domains A and B, CycleGAN constructs two generators, GA → B and GB → A, and two discriminators, DA and DB, so that the images between two domains can be fruitfully translated by cycle consistency loss [9]. In another variation, to handle the multiple domains more than two, StarGAN utilized the shared feature learning by means of a single generator and a single discriminator. Using the concatenated input image with target domain vector, the generator produces the fake image, which is classified as the target domain by the discriminator.

In the sense of conditional image creation, the principle of GANs has also been applied to supervised and unsupervised image domain transformations. For example, pix-to-pix attains image-to-image translation using paired data samples. In order to address the problem of collecting paired data samples, UNIT [10], CoGAN [11], CycleGAN [12], and DiscoGAN [13] have suggested unpaired image-to-image translation. Yet, the two-player objective function leads to difficult training associated with inauthenticity and mode collapse [14], especially in high resolution.

Due to the stable training results generated by Deep Convolutional GAN (DCGAN) [15], several multistage generative training methods have been proposed: Composite GAN feats multiple generators to separately generate different parts of an image [16]. The PGGAN implements multiple training procedures from low to high resolution in order to incrementally generate a realistic image [17]. In addition, current domain classification-based GANs that monitor the characteristics of the generated images by consecutively creating a latent distribution have shown encouraging results.

35.3   Method

35.3.1   Data Preprocessing

In this study, we used 12 neonatal brain data which are manually segmented from raw MR images. The images are of 320×320 pixels and each of which has approximately 200 slices. We pick the slices from number 30 to number 120 from all the slices to omit the initial or final slices, since they relay a marginal amount of useful information and can negatively impact the training of PGGAN.

35.3.2   PGGAN Implementation

In this study, the Wasserstein loss PGGAN architecture with gradient penalty has been used. In general, the discriminator X belongs to the set of 1-Lipschitz functions, P_r is the distribution of the data by the true data sample y, and P_g is the distribution of the model by the synthetic sample produced by the conditioning of the image noise samples y using a uniform distribution in $[-1, 1]$.

Mathematically, we can represent it as follows:

$$\mathbb{B}_{\tilde{y} \sim \mathbb{P}_g}\left[X\left(\tilde{y}\right)\right] - \mathbb{B}_{y \sim \mathbb{P}_r}\left[X\left(y\right)\right] + \lambda \mathbb{B}_{\hat{y} \sim \mathbb{P}_{\hat{y}}}\left[\left(\left\|\nabla_{\hat{y}} Y\left(\hat{y}\right)\right\|_2 - 1\right)^2\right] \quad (35.2)$$

The last word of the term is the gradient penalty for the random sample $\hat{y} \sim P_{\hat{y}}$. The training lasts 20,000 steps with a batch size of 4 and 2.0×10^{-4} learning rates for the Adam Optimizer. Once in three times, we flip the real/synthetic labels of the discriminator for robustness.

It is important to categorize between irregular and normal characteristics when trying to generate synthetic images for the task of classification. Though, abnormal brain MR images frequently vary marginally from normal brain MR images and are difficult to understand; to distinguish the subtle differences between abnormal and normal images, we use progressive growing adversarial network (PGGAN) architecture. PGGAN training starts with low-resolution images. After that, it gradually increases resolution by adding new layers to the generator and the discriminator. Our networks with a low resolution of 4×4 pixels.

Network architecture is shown in Fig. 35.1. PGGAN learns the outlines of the training images at the low-resolution level. In the high-resolution phase, our PGGAN learns the detailed regions of the training images through progressive training methods, the generator will learn the features from training images. Conditional information is also added to generate synthetic images in a high-resolution phase.

35.4   Experiment

35.4.1   Experimental Settings

For experimental purpose, 12 neonate MR images were used. All images were in grayscale and 320×320 pixels, separated into several patches of 256×256 pixels. The scale of the patches has been determined through experimental observations. For the image generation procedure, these patches have been resized during training session. We constructed

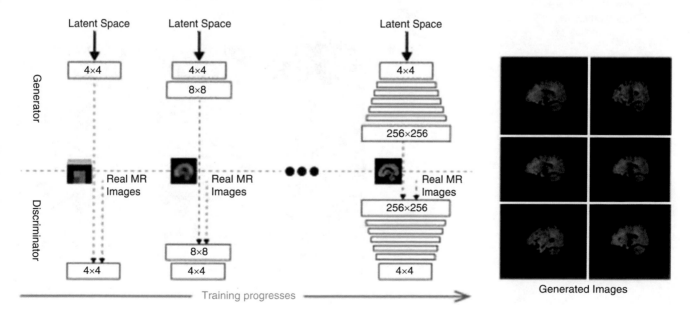

Fig. 35.1   Network architecture to generate synthetic brain MR images

our image generation training data set through random selection of images from original data. Evaluation metrics, such as Inception Score [18], Fréchet Inception Distance (FID) [19], and Sliced Wasserstein Distance (SWD)] [20], have been proposed for evaluation of the quality of the images created. But, these benchmarks are not suitable for assessing images for problems with classification and prediction. GAN-train measures the classification performance of a classifier trained on generated synthetic images and tests the output of a set of real images.

35.4.2 Result

In this research, the support vector machine (SVM) was used as an estimator for the MR image classification in the GAN train. A deep learning-based estimator is the first choice in terms of precision. However, such an estimator has many parameters, and the efficiency of the classification heavily count on parameters settings. As a result, we implied the simplest SVM as our estimator to correctly evaluate the effectiveness of the images generated. Variation in features also affect the performance of the classification. Manually crafted features are an old-fashioned approach. In this inves-

tigation, we extracted high-level semantic features from the pre-trained deep model, namely the pre-trained VGG-16 models. Sensitivity (Scn), specificity (Spe), and harmonic mean Sen and Spe (HM) were used for the evaluation. These parameters can be defined as follows;

$$Sensitivity\left(Sen\right) = TP / \left(TP + FN\right)$$

$$Specificity\left(Spe\right) = TN / \left(TN + FP\right)$$

Harmonic Mean (HM) = (2 × (Sen × Spe)/Sen + Spe) here, TP is the numbers of true positive samples, TN is true negative samples, FP is positive samples, and FN is false negative samples, respectively.

The goal of our method is to generate realistic synthetic images as is shown in Fig. 35.2. It is assumed that the synthesis anonymized data will be as successful as the real classification problems data. We demonstrated the efficiency of the MR image classification using synthetic data as a quantitative evaluation. The results of the GAN-train classification are given in Table 35.1. From the results, we can understand that PGGAN outperformed the relative conventional method in MR image classification efficiency. One fact to be noted that, performance does not outperform when actual images were used as training data. Also, the model trained at

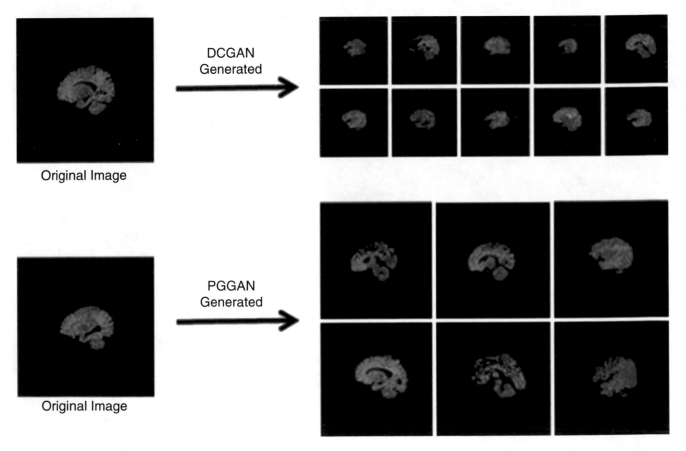

Fig. 35.2 Synthetic brain generation results. It shows that DCGAN can generate brain images with low-image resolution and quality while our proposed image generation method PGGAN can generate high-resolution synthetic images from original MR images

Table 35.1 Classification performance of PGGAN using generated images

Training data	Sen	Spe	HM
PGGAN	**0.762**	**0.712**	**0.736**
DCGAN	0.584	0.613	0.598
Real Data	0.883	0.885	0.883

DCGAN generated images cannot correctly identify the actual data. As a Whole, we have shown that the progressively growing network architecture is successful in detecting real data distribution.

35.5 Conclusion

It is known that, if a set of images can be generated by an optimized GAN model that captures the target distribution perfectly, they are indistinguishable from the original training set. As anonymized generated images were castoff for classification in our analysis, we used GAN-train as our evaluation index.

In this chapter, a synthetic brain MR image generation approach with gradually increasing adversarial learning PGGAN has been explained. It is a high-quality image generation system for easier understanding, sharing, and upgrading of clinical data based on deep learning techniques. Besides the fact that our anonymized generated images were useful for the classification of MR images, it sheds light on diagnostic and prognostic medical applications.

There are some limitations of this research. The classification performance of MR images in this study is not sufficient for clinical applications. In the experiment, instead of using deep neural networks that involve complicated parameter tuning processes, we used the simplest SVM models as our estimator since we focused on assessing the quality of the images produced. Our future work will address these issues.

References

1. Tanenbaum LN, et al. Synthetic MRI for clinical neuroimaging: results of the Magnetic Resonance Image Compilation (MAGiC) prospective, multicenter, multireader trial. Am J Neuroradiol. 2017;38(6):1103–10.
2. Hagiwara A, et al. Symri of the brain: rapid quantification of relaxation rates and proton density, with synthetic mri, automatic brain segmentation, and myelin measurement. Invest Radiol. 2017;52(10):647–57.
3. Krizhevsky A, Sutskever I, Hinton GE. Imagenet classification with deep convolutional neural networks. Advances in Neural Information Processing Systems; 2012.
4. Zhang K, Zuo W, Chen Y, Meng D, Zhang L. Beyond a Gaussian denoiser: residual learning of deep cnn for image denoising. IEEE Transactions on Image Processing; 2017.
5. Dong C, Loy CC, He K, Tang X. Image super-resolution using deep convolutional networks. IEEE transactions on pattern analysis and machine intelligence; 2016.
6. Zhu J-Y, Park T, Isola P, Efros AA. Unpaired image-to-image translation using cycleconsistent adversarial networks. arXiv preprint. 2017.
7. Choi Y, et al. StarGAN: unified generative adversarial networks for multi-domain image-to-image translation. arXiv preprint 1711; 2017.
8. Goodfellow I, et al. Generative adversarial nets. Advances in neural information processing systems; 2014.
9. Wolterink JM, et al. Deep MR to CT synthesis using unpaired data. International Workshop on Simulation and Synthesis in Medical Imaging; 2017.
10. Liu M-Y, Breuel T, Kautz J. Unsupervised image-to-image translation networks. In: Proceedings of the International Conference on Neural Information Processing Systems (NeurIPS); 2017.
11. Liu M-Y, Tuzel O. Coupled generative adversarial networks. In Proceedings of the International Conference on Neural Information Processing Systems (NeurIPS); 2016.
12. Zhu J-Y, Park T, Isola P, Efros AA. Unpaired image-to-image translation using cycle-consistent adversarial networks. In: Proceedings of the IEEE International Conference on Computer Vision (ICCV); 2017.
13. Kim T, Cha M, Kim H, Lee JK, Kim J. Learning to discover cross-domain relations with generative adversarial networks. In: Proceedings of the International Conference on Machine Learning (ICML); 2017.
14. Gulrajani I, Ahmed F, Arjovsky M, Dumoulin V, Courville AC. Improved training of Wasserstein GANs. In: Advances in Neural Information Processing Systems; 2017.
15. Radford A, Metz L, Chintala S. Unsupervised representation learning with deep convolutional generative adversarial networks. In: Proceedings of the International Conference on Learning Representations (ICLR). arXiv preprint arXiv:1511.06434; 2016.
16. Kwak H, Zhang B. Generating images part by part with composite generative adversarial networks. arXiv preprint; 2016.
17. Karras T, Aila T, Laine S, Lehtinen J. Progressive growing of GANs for improved quality, stability, and variation. In: Proceedings of the International Conference on Learning Representations (ICLR), arXiv preprint; 2018.
18. Salimans T, Goodfellow I, Zaremba W, Cheung V, Radford A, Chen X. Improved techniques for training GANs. In: Proceedings of the International Conference on Advances in Neural Information Processing Systems (NeurIPS); 2016.
19. Heusel M, Ramsauer H, Unterthiner T, Nessler B, Hochreiter S. GANs trained by a two time-scale update rule converge to a local nash equilibrium. In: Proceedings of the International Conference on Advances in Neural Information Processing Systems (NeurIPS); 2017.
20. Karras T, Aila T, Laine S, Lehtinen J. Progressive growing of GANs for improved quality, stability, and variation (ICLR); 2017.

Prediction of Personalized Postoperative Implanted Knee Kinematics with Statistical Temporal Modeling

Belayat Hossain and Syoji Kobashi

Abstract

One of the extension axes of multidisciplinary computational anatomy (MCA) is the temporal axis. This chapter introduces a temporal prediction model of MCA approaches to predict postoperative implanted knee kinematics of a patient before total knee arthroplasty (TKA) since it is necessary for surgery plans and for patients to more readily understand the TKA result. The primary challenge includes quantifying individual varieties of the kinematic patterns. This study proposes a statistical feature extraction method from a set of paired pre- and postoperative kinematics data and then derive a mapping function from preoperative to postoperative feature space using the machine learning technique. The statistical feature extraction method is a temporal version of the statistical shape model (SSM). We employed a CT-free navigation system to measure the kinematics because of its popularity in the TKA. The method is applied to two types of kinematics patterns and tested by cross-validation procedure. The experiment result shows that it is possible to predict the postoperative kinematic patterns by measuring the preoperative pattern using CT-free navigation system and predictive analytics.

Keywords

Total knee arthroplasty · Implanted knee · Knee kinematics · Machine learning

B. Hossain (✉) · S. Kobashi
Advanced Medical Engineering Center, University of Hyogo, Himeji, Japan
e-mail: belayat@ieee.org

36.1 Introduction

Total knee arthroplasty (TKA) is a common practice in knee surgery for the treatment of knee osteoarthritis (OA), rheumatoid arthritis (RA), and other complications. Because it resurfaces an injured entire knee joint with artificial one to regain normal functionality of patient's knee within a few weeks after surgery, and also to diminish severe knee joint pain [1]. The artificial knee joint (TKA prosthesis) has three main components—femoral, tibial, and tibial insert. The TKA prosthesis can be divided into three major types—such as cruciate-retaining (CR), posterior stabilized (PS), and cruciate substituting (CS). There are some functional differences among prostheses and also among providers. Anatomical morphology and function of the knee joint vary from patient to patient, therefore, an appropriate TKA implant should be chosen for each individual for personalized treatment. However, currently, the surgeon has to select a TKA surgery method among many types and a prosthesis product model without any quantitative analysis of its postoperative performance [2, 3].

Existing studies on investigating implanted knee kinematics (i.e., postoperative knee function) can be divided into two major categories—(a) invasive [4] and (b) non-invasive technique, which is the most popular due to patient safety, and its mainly based on image matching and/or sensor-based tracker technique [5]. In 2-D/3-D image registration technique, a 3-D computer-aided implant design is matched to 2-D X-ray digital radiograph image or movie [6–8]. Such in-vivo 3-D kinematics quantification of the implanted knee is essential for evaluating 3-D design of TKA prosthesis and surgical technique of a TKA patient [9], because the good fitting of prosthesis implies less-postoperative complication [10, 11]. Onsem et al. [12] proposed a prediction model for predicting patient's satisfaction before the surgery because some patients are dissatisfied with the operation outcome. The prediction of postoperative knee kinematics before the TKA was studied in [13–16].

Multidisciplinary computation anatomy (MCA) has been integrated with the information related to spatial axis, time series axis, functional axis, and pathological axis [17]. It has been applied to organ segmentation, disease detection, outcome prediction, etc. In this chapter, we discuss a statistical temporal model, which is a temporal version of statistical shape model (SSM) [17], and machine learning approach for predicting postoperative personalized implanted knee kinematics before TKA operation to evaluate the knee functional mobility with minimally acquired data, i.e., preoperative knee function using a navigation system. We employ the statistical features extraction method to quantifying individual varieties of the knee kinematics using high-dimensional clinical data, and then predictive models are constructed using a machine learning (ML) algorithm. We also compare their performance to find the best ML method because the characteristics features of the kinematic pattern differ from each type.

36.2 Kinematics Measurement Using CT-Free Navigation System

Computer-assisted OrthoPilot CT-free navigation system (B. Braun Aesculap, Tuttlingen, Germany) supports clinicians in achieving optimal implant alignment and enhanced surgical workflow [18] (Fig. 36.1a). It represents gold standard that allows for intraoperative visualization of the leg axis, real time intraoperative measurement of 3-D kinematics inspection aided by special software

for 3-D kinematic analysis before and after surgery, both at the tibiofemoral and patellofemoral joints [19]. Knee kinematics are measured in 3D space in navigation, following Grood's coordinate system (Fig. 36.1b). Three orientation angles are—flexion-extension (f-e), valgus-varus (v-v), and internal-external (i-e), and respective three position coordinates along the X, Y, and Z axes are called medial-lateral (M-L), anterior-posterior (A-P), and superior-inferior (S-I).

Firstly, patient's anatomical reference planes are registered (referencing) by the surgeon into the machine using trackers and pointers. Then the A-P translations and the i-e rotations were measured for every 10° f-e angle (i.e., 10°, 20°, ... 100°) by passively flexing the knee joint from zero degrees of extension to maximum flexion with the patient in a supine position under a non-load-bearing condition and the implants are attached using cement, if required, known as A-P and i-e patterns, respectively, because those two patterns are typically measured for better implant fitting during TKA. Intraoperative passive flexion 3D-kinematics were measured using the dedicated software (OrthoPilot TKA Version 4.2 Kobe version). During the kinematic measurement, the assistant surgeon held the thigh to align it perpendicularly while the operating surgeon gently held the heel and passively moved the knee from full extension to full flexion by inducing unconstrained motion. The kinematics measurements were taken both before (preoperative) and after the TKA surgery (postoperative) of every patient in the operating room using the OrthoPilot system (Fig. 36.1c).

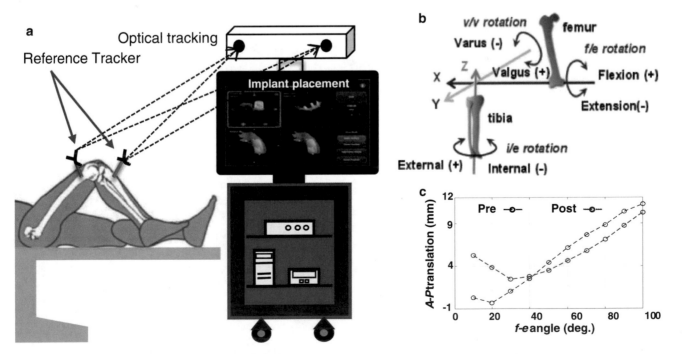

Fig. 36.1 (**a**) Main elements of a CT Image-free knee navigation system. Two references are attached to tibia and femur to measure their position and orientation by an optial tracking system. Computer controls the sys-

tem, run algorithm, dispaly meaured data and visualize anatoical information such as orientation of bone cut-plane and limb alignment, etc. (**b**) Knee coordinate system. (**c**) Kinematic pattern

36.3 Method

The method constructs predictive models using a set of pairs of clinical preoperative kinematics and postoperative implanted knee kinematics data. In the first stage, the variability of knee kinematics pattern among individuals in the training dataset is quantized using statistical techniques such as principal component analysis (PCA) [20], and then supervised ML algorithm is trained on the extracted features of pre- and postoperative data. Multiple models are trained to forecast the entire postoperative knee functions with optimized performance.

36.3.1 Feature Extraction

Let f^{pr} be training data of preoperative kinematic of size, $N_s \times N_m$, (Eq. 36.1),

$$f^{pr} = \begin{pmatrix} f_1^{pr}(1) & f_1^{pr}(2) & \cdots & f_1^{pr}(N_m) \\ f_2^{pr}(1) & f_2^{pr}(2) & \cdots & f_2^{pr}(N_m) \\ \vdots & \vdots & \ddots & \vdots \\ f_{N_s}^{pr}(1) & f_{N_s}^{pr}(2) & \cdots & f_{N_s}^{pr}(N_m) \end{pmatrix} \quad (36.1)$$

where N_s, total subjects, N_m, total measurement points, and $f_i^{pr}(j)$ represents reading (A-P translation, i-e rotation) of subject i at the measurement point, j, i.e., each row represents a preoperative kinematic of one subject. Likewise, training data of postoperative kinematic (f^{po}) of the paired subjects is also structured.

PCA is a linear dimensionality reduction technique for extracting features from a high-dimensional space by projecting them into a lower dimensional sub-space. It preserves significant features of maximum variation of the data and removes less-essential features with the fewer variant. We use PCA as feature extraction method for removing data redundancy in kinematics pattern (of $N_s \times N_m$ dimensions) in order to increase ML model performance.

At first, training data of preoperative (f^{pr}) and postoperative (f^{po}) kinematics are projected individually into a lower dimensional sub-space using PCA (with scaling). Later on, let us say principal components (PCs), principal axes (PAs), and mean kinematic pattern to be λ^{pr} of size $N_s \times N_m$, V^{pr} of size $N_m \times N_m$, and μ^{pr} of size $N_m \times 1$ for preoperative training data, and let λ^{po}, V^{po}, and μ^{po} are same parameters and size represented above but relating to postoperative training data.

36.3.2 Machine Learning Algorithms

A supervised ML method is applied for predictive model construction by deriving a mapping function to quantify individual variability between paired pre- and postoperative kinematics features in the PCA space. Features from pre- and postoperative kinematic patterns are defined as predictive variable and response variable, respectively.

36.3.2.1 Generalized Linear Regression

A predictive model based on a generalized linear model (GLM) [21] is constructed from PCs features of first D_{pr}-PAs of preoperative and PCs features of first PAs of postoperative kinematic. The regression model is given by Eq. (36.3),

$$Y = X\beta \quad (36.2)$$

here predictor variables,

$$X = \begin{pmatrix} x_{1,1} & x_{1,2} & \cdots & x_{1,D_{pr}} \\ x_{2,1} & x_{2,2} & \cdots & x_{2,D_{pr}} \\ \vdots & \vdots & \ddots & \vdots \\ x_{N_s,1} & x_{N_s,2} & \cdots & x_{N_s,D_{pr}} \end{pmatrix} \quad (36.3)$$

are defined from first D_{pr}-PCs taken from λ^{pr}. $x_{i,j}$ is the jth PCs of subject i; where $i = 1, 2, \ldots N_s$, $j = 1, 2, \ldots D_{pr}$. Response variable, Y is given by Eq. (36.4), obtained from PCs of first PAs taken from λ^{po}.

$$Y = \begin{pmatrix} y_1 \\ y_2 \\ \vdots \\ y_{N_s} \end{pmatrix} \quad (36.4)$$

Regression coefficients vector, $\beta = \left(\beta_1, \beta_2, \cdots, \beta_{D_{pr}} \right)^T$ is estimated from the training data. Akaike's Information Criterion (AIC) [22] is used to test statistically significant predictor variables during learning for optimizing the model.

36.3.2.2 Artificial Neural Network (ANN)

In ANN for regression, neurons at the output layers produce values that represent target values of continuous real numbers. A target function is learned during the training stage by adjusting the weights of the NN, and the tuning of the weight is performed by minimizing an optimization function (LSE, SGD, etc.) through back-propagation algorithm. For our kinematic problem, we exploit a three-layer feedforward ANN— input, hidden, and output layer. There are five input neurons to match 5-dimensional input PCs features, 30 hidden neurons and one output neurons to produce predicted results of the kinematics features in our NN for regression analysis. Back-propagation method with stochastic gradient descent (SGD) optimization is used by training the network with a set of preoperative PCs features of sample pairs for learning the network weights. A well-trained ANN can be thought of as an "expert" to grasp the entire kinematics of the test patient.

36.3.2.3 Support Vector Regression

Support vector regression (SVR), a supervised ML technique to perform regression analysis [23], includes all main features of SVM that characterize the maximum margin algorithm. In epsilon-SVM regression, let training dataset includes predictor variables (x_n) and response values (y_n) of pre-and postoperative PCs features. The objective in SVR is to derive a function, $f(x)$, that deviates from y_n by a value not larger than ε (deviation bound) for each training point x, and also as flat as possible. It does not consider errors as long as they are less than ε. For a particular training set (xi, yi), $i = 1, 2, \ldots n$, it minimizes the following function,

$$\frac{1}{2}w^2 + C\sum_{i=1}^{n}\left(\xi_i - \xi_i^*\right)$$

under the following constraints:

$$\begin{cases} w.x_i + b - y_i \leq \varepsilon + \xi_i \\ w.x_i + b - y_i \leq \varepsilon + \xi_i^* \\ \xi_i, \xi_i^* \geq 0, i = 1, 2, \ldots, n \end{cases}$$

In non-linear SVR, data are transformed into higher dimensional feature space by kernel function (i.e., polynomial, radial basis function) to make it feasible for linear separation. Thus SVR is controlled by the parameters instead of the dimensionality of the feature space.

36.3.3 Postoperative Kinematics Prediction

Using PCA, the first predictive model (m_1) is made between first D_{pr} preoperative PCs as predictive variables with first postoperative PCs as response variable. Likewise, D_{po} predictive models ($m_2, \ldots m_{D_{po}}$) are constructed to predict the whole postoperative knee functions with optimized performance by considering subsequent 3rd, $\ldots$, and D_{po}th postoperative PCs as response and the first D_{pr} predictor variables. Algorithm 36.1 is used to derive postoperative kinematic patterns from the preoperative PCs of maximum D_{pr} dimensions and postoperative PCs of D_{po} dimensions.

Algorithm 36.1 Prediction of Postoperative Implanted Kinematics of a Test Patient

Input: Preoperative kinematic (f_{new}^{pr})

   Output: Predicted postoperative kinematic (f_{po}^{pred})
   $i = 1, 2, \ldots, N_m$ (total measurement points)
   $j = 1, 2, \ldots, D_{pr}$ (total PAs at preoperative)

1. Project f_{new}^{pr} onto a lower dimension of size, D_{pr}

   using PC; $\lambda_{new}^{pr}(j) = \sum_{1}^{i=N_m}\left(\left(f_{new}^{pr} - \mu^{pr}\right).V^{pr}(i,j).\right)$

2. Postoperative PCs (λ_{new}^{po}) are predicted using λ_{new}^{pr} from the trained models, ($k = m_1, m_2, \ldots m_{D_{po}}$);
   $\lambda_{new}^{po}(k) = \left(\lambda_{new}^{po}(1), \lambda_{new}^{po}(2), \ldots \lambda_{new}^{po}(D_{po})\right)$

3. Postoperative kinematic (f_{po}^{pred}) is estimated from λ_{new}^{po}; $f_{po}^{pred}(i) = \sum_{k=1}^{D_{po}}\left(\lambda_{new}^{po}(k).V^{po}(i,k)\right) + \mu^{po}$

36.4 Experimental Study

36.4.1 Subjects and Dataset

We employed knee joint kinematics of 35 OA patients (24 females, 11 males; age 74.08 ± 6.90, mean ± standard deviation (SD)). Two experienced surgeons carried out the TKA surgery of the subjects with posterior stabilized (PS) type implant (Vega, Aesculap, B/Braun, Germany). The local Ethics Committee has approved our study, and each subject provided informed consent.

Each type of pattern (A-P and i-e pattern) has 35 pairs of samples. Each pair consists of one preoperative and one postoperative kinematic of the same patient. Kinematic patterns of each sample include non-linearity, and discrete (A-P/i-e values) features found among preoperative and postoperative kinematic. No direct relationships among the patterns are found to be represented by any general algebra equation.

36.4.2 Experiments

Firstly, the A-P or i-e kinematics of preoperative and postoperative data were organized as shown in (36.5). Any missing value was estimated by mean calculated from consecutive elements.

$$f_{ap/ie}^{pr} = f^{pr}(i,j)$$

$$f_{ap/ie}^{po} = f^{po}(i,j) \tag{36.5}$$

Here, subject's number, $i = 1, 2, \ldots 35$; measurement points, $j = 1, 2, \ldots 10$. $f_{ap/ie}^{pr}$ and $f_{ap/ie}^{po}$ were defined as the preoperative and postoperative A-P (or i-e) training data matrices, respectively. Then following experiments were performed individually for both patterns (A-P or i-e).

Experiment 36.1 Training Without PCA (nPCA)
Regression models were trained without extracting features by PCA-10 models (1 for each variable) were constructed using predictive and response variables from the pre-and postoperative kinematic data (Eq. 36.5), respectively.

Experiment 36.2 Training with Features Extracted from PCA (wPCA)
We used PCA to project pre- and postoperative kinematics (Eq. 36.5) data individually into a lower dimensional space. Predictor variables in each predictive model were defined from the first 5-PCs feature of preoperative ($D_{pr} = 5$), and first PCs feature of postoperative as the response variable. Therefore, three prediction models ($D_{po} = 3$) were constructed to consider the entire kinematic pattern.

36.5 Results and Discussion

We used statistical R Package v.3.2.2 [24] to implement our method. In Exp. 36.1, we did not apply PCA and trained 10 models (each for every f/e angles of 10° interval) from the paired pre-and post-op kinematics data. In Exp. 36.2, firstly, we applied PCA for feature reduction, and the ML models were trained with the extracted features (i.e., PCs). We varied PCA dimensions (1–10) to optimize prediction performance with high mean Pearson's correlation coefficient (cc), and low mean root mean squared error (RMSE) over all subjects for regression analysis. In this work, the best results were obtained with PCA dimension of 5 for preoperative and 3 for postoperative, retaining >95% of the total variance. In Exp. 36.1, a new patient's kinematic is directly (no need of features (PCs) vector to kinematic conversion) predicted from the trained models; however, for Exp. 36.2, we firstly predicted postoperative PCs of the test patient and then converted the PCs to the kinematic pattern (Algorithm 36.1).

Here we trained three popular ML predictive models (GLM, NN, SVR) separately for constructing to test their performance in predicting the most likely postoperative kinematics before the surgery. A test patient's predicted kinematics curves are shown in Fig. 36.2, and it confirms that the prediction performance differs from ML models. We found that PCA-based feature extraction (Exp. 36.2) outperforms over the without feature extraction (Exp. 36.1) (Table 36.1). We observed from Exp. 36.2, the predicted postoperative pattern follows most like the ground truth postoperative pattern, and its shape varies from subject to subject, which implies models's good capability of grasping pattern variability during the learning phase due to extracted good features.

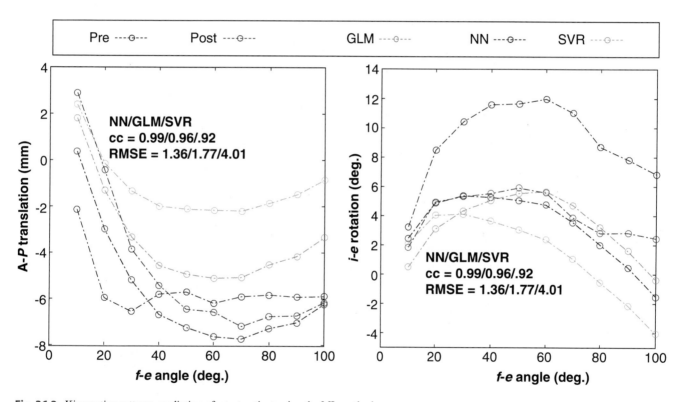

Fig. 36.2 Kinematics patterns prediction of a test patient using the ML method

Table 36.1 Prediction performance comparison among ML methods

ML model	A-P pattern cc/ RMSE		i-e pattern cc/RMSE	
	Exp. 36.1 (nPCA)	Exp. 36.2 (wPCA)	Exp. 36.1 (nPCA)	Exp. 36.2 (wPCA)
GLM	0.71 ± 0.26/3.99 ± 1.91	0.84 ± 0.15/3.55 ± 1.83	0.86 ± 0.17/4.55 ± 2.32	0.88 ± 0.12/4.43 ± 1.98
NN	0.72 ± 0.39/5.59 ± 2.95	0.84 ± 0.17/3.45 ± 1.79	0.85 ± 0.16/3.27 ± 1.42	0.87 ± 0.20/4.71 ± 2.36
SVR (radial)	0.70 ± 0.34/5.19 ± 2.82	0.84 ± 0.13/4.22 ± 2.75	0.82 ± 0.20/3.30 ± 1.38	0.84 ± 0.23/4.94 ± 2.89

In SVR, among the four kernel tricks [16], the radial basis was chosen because it offers high prediction performance in terms of pattern shape (high cc value) for both patterns. For A-P pattern, all methods in Exp. 36.2 had comparable performance (nearly same cc value); however, NN provided the smallest prediction error (3.45 ± 1.79 mm). In contrast, GLM was superior (prediction error, 4.34 ± 1.98°) to others for predicting i-e pattern and also offered optimized prediction performance in both types of pattern. Therefore, we observed prediction performance of the predictive models somewhat varies among three algorithms (SVR, GLM, NN). One probable cause could be quantifying the individual varieties of the kinematic patterns, which differ from each type. The kinematics pattern predicted by the methods (GLM and NN) has a good correlation (high CC value) which implies the shape of the predicted pattern has good agreement with that of the actual pattern. Therefore, the best method can predict the postoperative outcome of a new patient with a Pearson's CC of 0.84 ± 0.17 and RMSE of 3.45 ± 1.79 mm (NN) for A-P pattern, and a CC of 0.88 ± 0.12 and RMSE of 4.34 ± 1.98° (GLM) for i-e pattern.

This study could help the surgeon to examine a patient's postoperative outcome before surgery using a CT-free navigation system, and the prediction result of personalized kinematics could also serve as an inspiration to patients to carry on with TKA surgery. The method was validated here in PS type implant, which could be appropriate for other types as well, because knee kinematic definition is the same for all types of the implant in a CT-free navigation system. Additionally, our method can be extended in the future to select appropriate TKA implant in personalized knee surgery in TKA planning, if the model is trained by the kinematics of other prosthesess and by surgery code. Our study is valid for both load- and non-load bearing conditions, because basic kinematic patterns are similar for both conditions [25].

We assumed the same calibration of the anatomical coordinate of the patient's knee in the CT-free navigation during both pre-and postoperative pattern measurement; however, in some cases, surgeon has to revise the coordinate intentionally with some rotation angle to ensure good knee joint alignment and it could affect the zero-position and orientation of the coordinate system. This limitation of calibration in those cases could be improved by measuring kinematics after few months of the surgery using 2D/3D image registration technique [6, 7].

36.6 Concluding Remarks

This study introduced PCA-based statistical feature extraction and ML-based approach to predict personalized postoperative kinematics in TKA before the surgery by measuring preoperative knee functions using CT-free knee navigation. We achieved this by quantifying individual varieties of the kinematic patterns using the statistical features extraction method from a set of paired pre-and postoperative kinematics followed by ML algorithm and validated the method for two types of kinematic patterns, and satisfactory performance was accomplished. This study suggests the best ML method (NN for A-P pattern and GLM for i-e pattern) with high prediction performance for predicting postoperative kinematics of a patient. Finally, this work could help clinicians to choose the optimal treatment and to patients better understand TKA outcomes. By integrating additional training data, calibration error-free data, and investigating preoperative kinematics, the model's performance could be improved in the future.

Acknowledgments Authors would like to acknowledge Prof. Shinichi Yoshiya and his team, Nishinomiya Kaisei Hospital, Hyogo, Japan for their supports in data acquisition and their thoughtful discussion in the research meetings.

References

1. Murray DW, MacLennan GS, Breeman S, Dakin HA, Johnston L, Campbell MK, et al. A randomised controlled trial of the clinical effectiveness and cost-effectiveness of different knee prostheses: the Knee Arthroplasty Trial (KAT). Health Technol Assess. 2014;18(19):1.
2. Berend KR, Lombardi AV Jr, Adams JB. Which total knee replacement implant should I pick? Bone Joint J. 2013;95-B(11_Supple_A):129–32.
3. Victor J, Mueller JKP, Komistek RD, Sharma A, Nadaud MC, Bellemans J. In vivo kinematics after a cruciate-substituting TKA. Clin Orthopaed Relat Res®. 2009;468(3):807–14.
4. Horiuchi H, Akizuki S, Tomita T, Sugamoto K, Yamazaki T, Shimizu N. In vivo kinematic analysis of cruciate-retaining total knee arthroplasty during weight-bearing and non–weight-bearing deep knee bending. J Arthroplast. 2012;27(6):1196–202.
5. Katsumasa T, Nao S, Seiji K, Tomoyuki M, Akio M, Hiroomi T, et al. Kinematic analysis of mobile-bearing total knee arthroplasty using image matching technique. Orthopaed Proc. 2012;94-B(Suppl XXV):242.
6. Yamazaki T, Watanabe T, Nakajima Y, Sugamoto K, Tomita T, Yoshikawa H, et al. Improvement of depth position in 2-D/3-D registration of knee implants using single-plane fluoroscopy. IEEE Trans Med Imaging. 2004;23(5):602–12.
7. Kobashi S, Tomosada T, Shibanuma N, Yamaguchi M, Muratsu H, Kondo K, et al. Fuzzy image matching for pose recognition of occluded knee implants using fluoroscopy images. J Adv Comput Intell Intell Inform. 2005;9(2):181–95.
8. Tomaru A, Kobashi S, Tsumori Y, Yoshiya S, Kuramoto K, Imawaki S, et al. A 3-DOF knee joint angle measurement system with inertial and magnetic sensors. In: 2010 IEEE international conference on systems, man and cybernetics; 2010. p. 1261–6.
9. Morita K, Nii M, Ikoma N, Morooka T, Yoshiya S, Kobashi S. Implanted knee joint kinematics recognition in digital radiograph images using particle filter. J Adv Comput Intell Intell Inform. 2018;22(1):113–20.

10. Choi Y-J, Ra HJ. Patient satisfaction after total knee arthroplasty. Knee Surg Relat Res. 2016;28(1):1–15.

11. Seon JK, Park JK, Jeong MS, Jung WB, Park KS, Yoon TR, et al. Correlation between preoperative and postoperative knee kinematics in total knee arthroplasty using cruciate retaining designs. Int Orthop. 2010;35(4):515–20.

12. Van Onsem S, Van Der Straeten C, Arnout N, Deprez P, Van Damme G, Victor J. A new prediction model for patient satisfaction after total knee arthroplasty. J Arthroplast. 2016;31(12):2660–7.e1.

13. Kobashi S, Hossain B, Nii M, Kambara S, Morooka T, Okuno M, et al. Prediction of post-operative implanted knee function using machine learning in clinical big data. In: 2016 International conference on machine learning and cybernetics (ICMLC)2016. p. 195–200.

14. Hossain BM, Nii M, Morooka T, Okuno M, Yoshiya S, Kobashi S. Post-operative implanted knee kinematics prediction in total knee arthroscopy using clinical big data. In: Intelligent robotics and applications. Lecture notes in computer science. Cham: Springer; 2016. p. 405–12.

15. Hossain B, Morooka T, Okuno M, Nii M, Yoshiya S, Kobashi S. Surgical outcome prediction in total knee arthroplasty using machine learning. Intell Automat Soft Comput. 2018;25:1–17.

16. Hossain B, Morooka T, Okuno M, Nii M, Yoshiya S, Kobashi S. Implanted knee kinematics prediction: comparative performance analysis of machine learning techniques. In: 2018 Joint 7th international conference on informatics, electronics & vision (ICIEV) and 2018 2nd international conference on imaging, vision & pattern recognition (icIVPR); 2018. p. 544–9.

17. Hashizume M. Perspective for future medicine: multidisciplinary computational anatomy-based medicine with artificial intelligence. Cyborg Bionic Syst. 2021;2021:1–3. 9160478. https://doi.org/10.34133/2021/9160478.

18. Bae DK, Song SJ. Computer assisted navigation in knee arthroplasty. Clin Orthoped Surg. 2011;3(4):259–67.

19. Catani F, Zaffagnini S. Knee surgery using computer assisted surgery and robotics; 2013.

20. Jolliffe IT. Principal component analysis; 2002.

21. McCullagh P, Nelder JA. Generalized linear models. 2nd ed. London: Chapman and Hall; 1989.

22. Akaike H. A new look at the statistical model identification. In: IEEE transactions on automatic control, vol. 19, no. 6. 1974. p. 716–723. http://dx.doi.org/10.1109/TAC.1974.1100705.

23. Vapnik V. Statistical learning theory. New York: Wiley; 1998.

24. R Core Team. R: A language and environment for statistical computing. R Foundation for Statistical Computing, Vienna; 2016. http://www.Rproject.org/.

25. Yoshiya S, Matsui N, Komistek RD, Dennis DA, Mahfouz M, Kurosaka M. In vivo kinematic comparison of posterior cruciate-retaining and posterior stabilized Total knee arthroplasties under passive and weight-bearing conditions. J Arthroplast. 2005;20(6):777–83.

Jain Wang, Yutaro Iwamoto, Xian-Hua Han, Lanfen Lin, Hongjie Hu, and Yen-Wei Chen

Abstract

Multi-phase CT images, which is also known as dynamic CT images, are widely used for the diagnosis of focal liver lesions. In addition to 3-dimensional spatial information, the multi-phase CT image also has temporal information. In this chapter, we propose a tensor-based sparse coding model for the efficient representation of multi-phase CT images and apply it to classification and content-based medical image retrieval of focal liver lesions. The effectiveness of the proposed method has been validated by experiments.

Keywords

Sparse coding · Tensor · Multi-phase CT image · Focal liver lesion · Content-based medical image retrieval

37.1 Introduction

In recent years, with the remarkable progress of medical imaging devices and computer technologies, various high definition medical images can be obtained, and image-based computer-aided diagnosis or multidisciplinary computational anatomy (MCA) plays an important role in medicine and healthcare. Unlike 2-dimensional natural images, medical images (i.e., CT images) are volumetric images. In addition to 3-dimensional spatial information, some medical images, such as multi-phase CT images, have temporal information. Efficient representation of such multi-dimensional medical images is a key issue for MCA-based computer-aided diagnosis and multidisciplinary computational anatomy. In this chapter, we focus our study on the efficient representation of focal liver lesions (FLLs) in multi-phase CT images and the classification of FLLs.

Liver cancer is one of the leading causes of death worldwide [1]. Early detection of liver cancers by analysis of medical images is a helpful way to reduce death due to liver cancer. Multi-phase contrast-enhanced computer-tomography (CT) images are widely used for the diagnosis of focal liver lesions (FLLs). In the multi-phase contrast-enhanced CT scan procedure, non-contrast-enhanced (NC) phase images are obtained from scans before contrast injection. Three additional phase images are obtained after contrast injection, i.e., arterial (ART) phase in 25–40 s, portal venous (PV) phase in 60–75 s, and delayed (DL) phase in 3–5 min after contrast injection, respectively. Typical examples of five types of FLLs (CYST, Focal Nodular Hyperplasia (FNH), Hepatic Cell Carcinoma (HCC), Hemangioma (HEM), metastatis (METS)) on three-phases are shown in Fig. 37.1. As we can see that different types of FLLs exhibit different enhancement patterns after intravenous contrast injection.

Extracting effective features and incorporate multi-phase information into feature descriptors is a fundamental issue for computer-aided diagnosis of FLLs. One of the popular methods or characterization of FLLs is the bag of visual words (BoVW) method [2–7]. The BoVW method represents an image by using a normalized histogram of visual words, which is based on codebook learning using training images. In conventional BoVW, the K-means method is the most widely used for codebook learning, which can be considered a hard assignment. In this chapter, we present a sparse modeling method for soft assignment of BoVW to improve the representation of multi-phase CT images of FLLs. Furthermore, we present a tensor-based

J. Wang · Y. Iwamoto · X.-H. Han · Y.-W. Chen (✉)
College of Information Science and Engineering, Ritsumeikan University, Kusatsu, Shiga, Japan
e-mail: chen@is.ritsumei.ac.jp

L. Lin
College of Computer Science and Technology, Zhejiang University, Hangzhou, Zhejiang, China

H. Hu
Department of Radiology, Sir Run Run Shaw Hospital, Zhejiang University, Hangzhou, Zhejiang, China

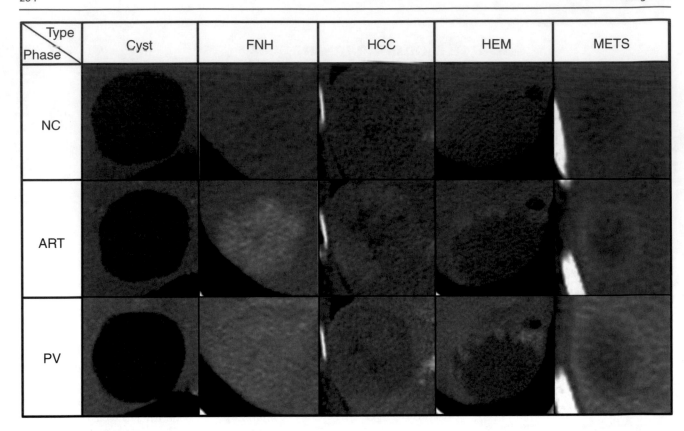

Fig. 37.1 Typical enhancement patterns of five types of FLLs (CYST, FNH, HCC, HEM, METS) on three-phases

sparse modeling method to extract spatial-temporal features, in which the multi-phase CT image is treated as a tensor.

37.2 Sparse Modeling-Based BoVW

As a generalization of the *K-means* algorithm, the sparse coding technique employs a linear combination of codewords for the representation of each signal, which means more than one non-zero entry in coding, and the weights can be calculated to be arbitrary values but not limited to 1. The intuitive way for the sparse coding problem can be formulated to optimize the following objective function:

$$\arg\min_{\mathbf{D},\mathbf{X}} \left\| \mathbf{Y} - \mathbf{D}\mathbf{X} \right\|_2^2, \quad s.t. \left\| \mathbf{X} \right\|_0 \leq \alpha K$$

where $\mathbf{X}$ is the sparse approximation of signals, $\mathbf{Y}$ on codebook $\mathbf{D}$. α is a sparsity measure, which is a ratio between the number of non-zero entries in x_i and the total number of codewords in $\mathbf{D}$. $\alpha*K$ controls the maximum number of codewords that can be used for approximation of the input signal y_i.

Similar to realization of *K-means* algorithm, there are two stages: sparse coding stage and codebook update stages, in the codebook learning of sparse coding. In consideration of simplicity and efficiency, we employ the Orthogonal Matching Pursuit (OMP) algorithm for coefficient calculation and K-SVD method for codebook updating in the two stages, respectively.

With the fixed codebook, OMP algorithm is a simple and efficient way to solve the sparse approximation problem, which is NP-hard because of the overcomplete codebook $\mathbf{D}$. K-SVD was proposed for generating a dictionary of spare representation via singular value decomposition (SVD) [8, 9]. It is a generalization of the *K-means* clustering method. K-SVD works by iteratively alternating between the coefficient calculation of the input data based on the current dictionary and updating the atoms in the dictionary to fit the data better.

We validated the representation accuracy of sparse coding comparing to *K-means* method. In this evaluation, we used 5000 local patches from training ART images and approximated them with *K-means* and sparse coding methods with codebook size 100 for both. Then the reconstruction errors (RE) can be calculated for all the selected patches, and the distributions of the RE of the samples using both methods are plotted in Fig. 37.2 [10]. As illustrated in Fig. 37.2, the sparse coding method achieves smaller reconstruction errors than using *K-means* method, which means more accurate approximations, and thus more accurate diagnosis performance can be expected.

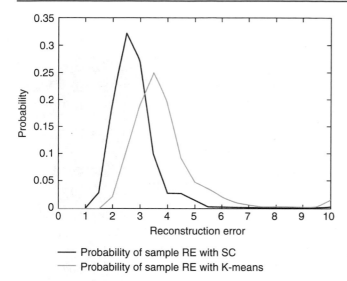

Fig. 37.2 Distribution of sample RE using the sparse modeling and *K-means* method

37.3 Tensor Sparse Modeling-Based BoVW

37.3.1 Tensor Codebook Learning

Given a set of tensor training samples $\mathcal{Y}$, we proposed a K-CP method to learn tensor codebook $\mathcal{D}$. Implementation of the proposed K-CP method also comprises two iterated stages: calculation of sparse coefficients, assuming that the codebook is fixed, and codeword update based on the calculated sparse coefficients. The first stage can be solved easily by using the tensor generalization of OMP algorithm. In tensor OMP, given a collection of samples $\mathcal{Y} = [\mathcal{Y}_1, \mathcal{Y}_2, ..., \mathcal{Y}_N]$, where $\mathcal{Y}_i$ is an Mth-order tensor, and $\mathcal{Y}$ is an $(M + 1)^{th}$-order tensor. Suppose a codebook $\mathcal{D}$ comprises of K M^{th}-order tensor codewords $\mathcal{D}_k$. Then, $\mathcal{D}$ is a $(M + 1)^{th}$-order tensor. The tensor OMP can be formulated as follows:

$$i = 1,2,...,N \quad \min_{x_i} \left\| \mathcal{Y}_i - \mathcal{D} \bar{\times}_{(M+1)} x_i \right\|_2^2, \quad s.t. \ \|x_i\|_0 \leq T, \forall i$$

In the codeword update stage, each tensor codeword is updated individually. To update codeword $\mathcal{D}_k$, we first find the row vector x_k^T in $\mathcal{X}$, in which each entry corresponds to the coefficient of a sample in $\mathcal{Y}$ to $\mathcal{D}_k$. Then, we define the approximation error without using codeword $\mathcal{D}_k$ as follows:

$$\mathcal{E}_k = \mathcal{Y} - \sum_{j \neq k}^{K} \mathcal{D}_j \circ x_j^T$$

where $\circ$ denotes the outer product. The total reconstruction error can be written as follows:

$$\left\| \mathcal{Y} - \mathcal{D} \times_{(M+1)} \mathcal{X} \right\|^2 = \left\| \mathcal{E}_k - \mathcal{D}_k \circ x_k^T \right\|$$

Our aim is to find the optimal $\mathcal{D}_k$ that well approximates the reconstruction error $\mathcal{E}_k$, which can be solved easily by applying CP decomposition on $\mathcal{E}_k$. CP (CANDECOMP/ PARAFAC decomposition) decomposes a Pth-order tensor $\mathcal{D}$ into a sum of rank-one tensors [11].

$$D \approx \sum_{r=1}^{R} \lambda_r \left(\mathbf{d}_r^{1\circ} \mathbf{d}_r^{2\circ} ... \circ \mathbf{d}_r^P \right)$$

We suppose the vector $\mathbf{d}^p_r$ is normalized to unit length, and the weight of each rank-one tensor is λ_r.

The process of applying the CP decomposition to the residual reconstruction tensor is executed K times to update each of the K tensor codewords in each iteration. Thus this method is called K-CP method. The above two stages are iterated until a pre-specified reconstruction error is achieved or the maximum iteration number is reached.

37.3.2 FLL Spatiotemporal Feature Extraction Using Tensor Sparse Modeling

To capture the temporal feature of multi-phase CT images, corresponding slices from triple-phase CT images were center-aligned according to the tumor masks and stacked to form three-layer volumes. By this operation, the temporal co-occurrence information is transformed into spatial information in the third dimension of the constructed volumes. A spatiotemporal codebook can be learned by applying our proposed method on the tensor training samples, which are local descriptors extracted from three-layer volumes. The spatiotemporal feature of each medical case can be then calculated by summarizing the representations of local descriptors using the mean pooling method.

37.3.3 Results

We compared the classification performance of the tensor-based sparse modeling method with the conventional sparse representation method over multi-phase medical images, single-phase medical images with 2-dimensional patches, and single-phase medical images with 3-dimensional patches, as shown in Fig. 37.3. It is interesting that both two methods got exactly the same results using single-phase images. The accuracy is more significantly improved, however, by the tensor-based sparse modeling method than the conventional one when using multi-phase images, which emphasizes that the tensor-based method is more effective in capturing the temporal information from multi-phase images [12].

Fig. 37.3 Comparing classification performance by the tensor sparse modeling and the conventional sparse modeling using single-/ multi-phase CT images

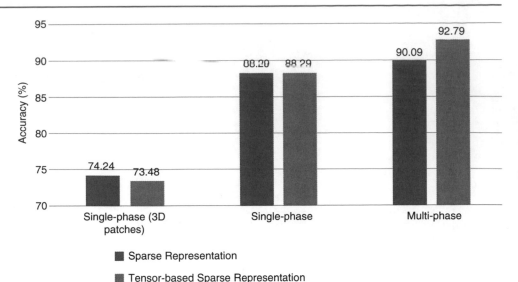

■ Sparse Representation

■ Tensor-based Sparse Representation

37.4 Conclusion

Multi-phase medical images contain both spatial and temporal information. In this chapter, we propose a tensor sparse modeling method for the efficient representation of multiphase CT images and apply it to FLL classification and retrieval. Experimental results show that the tensor-based method achieved much more significant improvement from single-phase to multi-phase images than the conventional sparse representation method, which illustrated the effectiveness of the tensor sparse modeling method on capturing spatial and temporal features.

References

1. Bray F, Ferlay J, Soerjomataram I, Siegel RL, Torre LA, Jemal A. Global cancer statistics 2018: GLOBOCAN estimates of incidence and mortality worldwide for 36 cancers in 185 countries. CA Cancer J Clin. 2018;68(6):394–424.
2. Roy S, Chi Y, Liu J, Venkatesh SK, Brown MS. Three-dimensional spatiotem-poral features for fast content-based retrieval of focal liver lesions. IEEE Trans Biomed Eng. 2014;61(11):2768–78.
3. Diamant I, Hoogi A, Beaulieu CF, Safdari M, Klang E, Amitai M, Greenspan H, Rubin DL. Improved patch based automated liver lesion clas- sification by separate analysis of the interior and boundary regions. IEEE J Biomed Health Inform. 2016;20(6):1585–94.
4. Diamant I, Klang E, Amitai M, Konen E, Goldberger J, Greenspan H. Task–driven dictionary learning based on mutual information for medical image classification. IEEE Trans Biomed Eng. 2017;64(6):1380–92.
5. Xu Y, Lin L, Hu H, Wang D, Chen Y-W. Bag of temporal co-occurrence words for retrieval of focal liver lesions using 3D multiphase contrast-enhanced CT images. In: 2016 23rd International conference on pattern recognition (ICPR 2016); 2016.
6. Xu Y, Lin L, Hu H, Wang D, Zhu W, Wang J, Han X, Chen Y-W. Texture-specific bag of visual words model and spatial cone matching based method for the retrieval of focal liver lesions using multiphase contrast-enhanced CT images. Int J Comput Assist Radiol Surg vol. 2018;13(1):151–64.
7. Xu J, Napel S, Greenspan H, Beaulieu CF, Agrawal N, Rubin D. Quantifying the margin sharpness of lesions on radiological images for conten-based image retrieval. Med Phys vol. 2012;39(9):5405.
8. Aharon M, Elad M, Bruckstein AM. K-SVD: an algorithm for designing overcomplete dictionaries for sparse representation. IEEE Trans Signal Process. 2006;54(11):4311–22.
9. Rubinstein R, Zibulevsky M, Elad M. Efficient implementation of the K-SVD algorithm using batch orthogonal matching pursuit. CS Technion. 2008.
10. Wang J, Han X-H, Xu Y, Lin L, Hu H, Jin C, Chen Y-W. Sparse codebook model of local structures for retrieval of focal liver lesions using multi-phase medical images. Int J Biomed Imaging. 2017;2017:ID1413297.
11. Kolda TG, Bader BW. Tensor decompositions and applications. SIAM Rev. 2009;51(3):455–500.
12. Wang J, Li J, Han X-H, Lin L, Hu H, Xu Y, Jin C, Chen Q, Chen Y-W. Tensor-based sparse representations of multi-phase medical images for classification of focal liver lesions. Pattern Recogn Lett. 2020;130:207–15.

Super Computing: Creation of Large-Scale and High-Performance Technology for Processes of MCA by Utilizing Supercomputers

38

Takahiro Katagiri and Daichi Nakajima

Abstract

In this study, we develop the parallelization and high-performance implementation of an application program for multidisciplinary computational anatomy (MCA) with a message passing interface (MPI). We also adapt the program to a supercomputer and evaluate its performance using advanced technologies for code optimization and parallel processing. To establish high-performance computations, we adapt loop collapse techniques for the target loops of a typical MCA program, called large deformation diffeomorphic metric mapping (LDDMM). The loop collapse enables us to obtain a high thread execution using OpenMP. In addition, we adapt an advanced parallel execution style, called hybrid MPI/OpenMP execution, to maximize the execution performance in a supercomputing environment. Hybrid MPI/OpenMP execution enables us to reduce the MPI communication time. This is crucial for massively parallel execution in a current supercomputer environment. The results of our implementation indicate that 18-day sequential processing can be reduced to a 43 h using eight nodes of a Fujitsu PRIMEHPC FX100, which is installed in the Information Technology Center, Nagoya University Japan.

Keywords

LDDMM · Hybrid MPI/OpenMP execution
Loop collapse · Supercomputing

T. Katagiri (✉)
Information Technology Center, Nagoya University,
Nagoya, Aichi, Japan
e-mail: katagiri@cc.nagoya-u.ac.jp

D. Nakajima
Graduate School of Informatics, Nagoya University,
Nagoya, Aichi, Japan

38.1 Introduction

A model of multidisciplinary computational anatomy (MCA) has been constructed on computers and applied to medical images from four different axes, namely, space, time, function, and pathology. With respect to the space axis, huge amounts of data are processed to treat three-dimensionalized data from a micromodel, such as a microscope, to a micromodel, such as Magnetic Resonance (MR). For example, three-dimensionalizing 2000 images of 2D data having a resolution of 20,000 pixels × 10,000 pixels requires several TBs for each 3D image. According to this increase in data, the computation time is also dramatically increased. Thus, such a task cannot be processed with a personal computer (PC).

By contrast, the progress made in supercomputing technologies has been remarkable, with the K-computer illustrating such progress. The supercomputer "Fugaku", which is predicted to achieve an EXA-FLOPS capability by the 2021 Financial Year, has been planned for installation in the RIKEN Center for Computational Science (R-CCS), Japan. High-performance computing (HPC) technologies for system software for I/O and parallelization have also progressed. Hence, a breakthrough in the processes for MCA is expected to be achieved by applying advanced HPC technologies.

In this study, we aim to make a breakthrough in the execution time by collaborating with researchers in the fields of HPC and MCA to develop novel algorithms and implementations. The representative of the project, Professor T. Katagiri, has been studying HPC technologies and auto-tuning (AT) [1] for supercomputers for 15 years, which is one of the major topics in the HPC field. In addition, Professor Hontani has provided major contributions toward the construction of a pancreas model for the imaging of a KPC mouse model [2]. By collaborating with both researchers from these different fields, we aim to establish a breakthrough in the execution time for MCA applications and create large-scale and high-performance parallel algorithms and implementations by applying advanced supercomputing technologies.

The present paper provides the following:

1. A discussion of non-rigid image registration, large deformation metric mappings, and tensor completion from the viewpoint of HPC to clarify the effective parallel algorithms and HPC implementations.
2. A clarification of the performance of supercomputers for achieving their implementation based on the results of the discussion in item 1.
3. A clarification of the effectiveness of the performance using real problems incurred through the application of MCA.

38.2 LDDMM Code

The target code is a registration processing code developed by Hontani Group, Nagoya Institute of Technology, Japan. The algorithm of the code for registration is based on large deformation diffeomorphic metric mapping (LDDMM) [3].

The Hontani Group developed a new method, i.e., partially rigid large deformation diffeomorphic metric mapping (PR-LDDMM). Given two images, this method can compute large deformation diffeomorphic metric mapping, in which pre-specified regions (tumor regions) in a source image are rigidly mapped. A source image is non-rigidly deformed, and the peripheral regions are matched with the corresponding regions in the target image, whereas the region maintains its shape and rigid alignment with the corresponding region.

The PR-LDDMM method requires iterative computations of LDDMM, which necessitates an extremely large computational complexity compared with the memory space complexity. Hence, our group is collaborating with Professor Hontani in parallelizing the process of LDDMM and has realized efficient computations of PR-LDDMM using supercomputers.

By contrast, the following four STEPs are major aspects of the computational complexity of the LDDMM code.

1. STEP 1—Calculate the Jacobians
2. STEP 2—Apply backward integration
3. STEP 3—Update the velocities; and
4. STEP 4—Update the paths and landmarks

We focus on STEP 1 through STEP 3 for adaptation of the parallelization because these steps require more than 90% of the total execution time of the LDDMM code.

38.3 Parallelization Method

38.3.1 MPI Process Parallelization

Because the program requires a high computational complexity rather than a large memory space, as mentioned earlier, we apply the following methodology to the LDDMM code for a 3D registration problem for MPI process parallelization.

1. Allocate memory space in a manner comparable to that of a sequential program.
2. Separate computations into each message passing interface (MPI) process by changing the loop length.
3. Store partially computed results in step 2 to sending buffer.
4. Gather the results in step 3 using MPI_Allgather.

The following code shows an overview of the above methodology.

```
// STEP 1 - Calculate Jacobians.
for (unsigned int t=0; t<M; t++) {
  for (int n=nhead; n<=ntail; n++) {
    float w1x = LINT::linterp3(L[t][n].x[0]+1,L[t]
[n].x[1],L[t][n].x[2],B[t].v[0],x,y,z);
    float w1y = LINT::linterp3(L[t][n].x[0]+1,L[t]
[n].x[1],L[t][n].x[2],B[t].v[1],x,y,z);
    float w1z =LINT::linterp3(L[t][n].x[0]+1,L[t]
[n].x[1],L[t][n].x[2],B[t].v[2],x,y,z);

        ...

  Some Computations;
      MPI Communication (1);
  }
MPI Communication (2);
}
```

The details of the MPI Communication (1) part are as follows:

```
nsend[c++] = (w1x-w4x)/2;
nsend[c++] = (w2x-w5x)/2;
nsend[c++] = (w3x-w6x)/2;

  ...
```

In addition, the following are the details of the MPI Communication (2) part.

```
MPI_Allgather(nsend, itmp[2], MPI::FLOAT,
nrecv, itmp[2], MPI::FLOAT, MPI_COMM_WORLD);
  c = 0; cc = 0;
  for (int i=0; i<pe*2; i=i+2) {
   c = itmp[2]*cc;
   for (unsigned int t=0; t<M; t++){
    for (int n=norder[i]; n<=norder[i+1]; n++) {
    L[t][n].Dv[0][0] = nrecv[c++];
    L[t][n].Dv[0, 1] = nrecv[c++];
    L[t][n].Dv[0, 2] = nrecv[c++];
      ...
    }
   }
   cc++;
  }
```

The main loop divides each MPI process using *n=nhead; n<=ntail;*, where *nhead* indicates a starting index, and *ntail* indicates an ending index for each MPI process. Let *p* be the number of MPI processes, *x* be the loop length, and *i* be the MPI rank number. The values of *nhead* and *ntail* can be calculated as follows:

```
nhead = floor (x / p) * ( i - 1 );
if ( i != p - 1 )
 ntail = floor (x / p) * i;
else
 ntail = x;
```

38.4 OpenMP Thread Parallelization and Loop Collapses for Code Optimization

In STEP 3, shown in Sect. 38.2, approximately 90% of the execution time is taken up by the LDDMM code. Hence, the code of STEP 3 is the heaviest among the computational kernels. We focus on the kernel to optimize the code.

We optimize the original LDDMM code using a code optimization technique, which is called a loop collapse. The following code shows the loop collapse for the code of STEP 3 in PR-LDDMM.

```
for (unsigned int t=0;  t<M;  t++) {
...
#pragma omp parallel for private(...)
for (i=0; i<x; i++)  {
  for (j=0; j<y; j++)  {
   for (k=0; k<z; k++)  {
     for (unsigned int n=0; n<N; n++)  {
```

```
      float f1 = L[t][n].x[0] - (float)i;  float
      f2 = L[t][n].x[1] - (float)j;
      float f3 = L[t][n].x[2] - (float)k;  float
      f4 = (f1*f1 + f2*f2 + f3*f3)/Sigma;
      if (f4 > minExp) {
        float f0 = fexp(f4);  Kx += f0*K[n][0];
        Ky += f0*K[n][1];  Kz += f0*K[n][2];
      }
     }
     B[t].v[0][i][j][k] -= Epsilon*(2*B[t].v[0]
     [i][j][k] - Rho*Kx);
     B[t].v[1][i][j][k] -= Epsilon*(2*B[t].v[1]
     [i][j][k] - Rho*Ky);
     B[t].v[2][i][j][k] -= Epsilon*(2*B[t].v[2]
     [i][j][k] - Rho*Kz);
   }
  }
```

For OpenMP thread parallelization, we can specify the OpenMP directive to the outer loop because there is natural parallelism in the code. See the directive of *#pragma omp parallel for* in the code above.

With the above code, a loop collapse for a twofolded loop can be written as follows:

```
...
#pragma omp parallel for private(...)
for (int ij=0; ij<x*y; ij++){
 unsigned int i = ij / y;  unsigned int j = ij
 % y;
 for (int k=0; k<z; k++) {
   ...
 }
}
```

The threefolded loop can be described as follows:

```
...
#pragma omp parallel for private (...)
for (int ijk=0; ijk<x*y*z; ijk++) {
 unsigned int i = ijk / ( y*z );
 unsigned int j = (ijk / z) % y;
 unsigned int k = ijk % z;
 ...
}
```

According to the above codes, the loop length to an OpenMP parallelization when adapting a loop collapse becomes longer than that of original loops (e.g., *x* versus *x*y* or *x*y*z*). Hence, the parallelism of the thread execution in the collapse codes increases to the parallelism of the original loop. Thus, there is adequate room to obtain a speedup for the collapsed loops when the loop length is small.

38.5   Performance Evaluation

38.5.1 Computer Environments

We use the Fujitsu PRIMEHPC FX100 (hereafter, denoted as FX100) at the Information Technology Center, Nagoya University. In the FX100, the SPARC64 XIfx processor or a single compute node is composed of two core memory groups. The specifications of the SPARC64 XIfx are as follows: 1.1264 TFLOPS, a 2.2-Ghz clock frequency, 32 cores, 2 assistant cores, a memory capacity of 32 GB per node, and a theoretical memory bandwidth of 480 GB/s. For the networking of the FX100, the Tofu Interconnect2 (six-dimensional Torus) with a theoretical networking bandwidth of 12.5 GB/s (in both directions) is used. The total system of the FX100 in Nagoya University includes 3.2 PFLOPS and 2880 nodes (92,160 cores).

The SPARC64 XIfx configuration uses "Non-Uniform Memory Access (NUMA) nodes" or "sockets." Each socket consists of 16 cores, an assistant core, and 12 MB of shared L2 cache and is equipped with 16 GB of its own local memory.

38.5.2 Problem Size and Estimated Execution Time for Final Target

The problem size comes from the target 3D medical image sizes of the pixels. In this performance evaluation, we apply $x * y * z$ and $100 \times 100 \times 100$.

The target loop of the MPI parallelization is an x loop. In the problem, x is set to 100. Note that in this problem, 100 is the maximum parallelism if we do not apply a loop collapse.

We check the execution time of the original code using the FX100. The size is set to $100 \times 100 \times 100$, and the time step is set to 50. Under this condition, it takes approximately 26 min for a sequential execution. To process the actual problems, a problem size of $100 \times 100 \times 100$ and a time step of 1000 are required. In addition, at least 49 input data are needed. Hence, we need 26 min $\times$ (1000/50) time steps $\times$ 49 input data = 18 days to process the actual problem.

38.5.3 Results and Discussion

In this session, a parallel LDDMM code with a pure MPI execution and a hybrid MPI/OpenMP execution is evaluated.

First, Fig. 38.1 shows the speedup ratios of pure MPI executions with one node on the FX100. Because the FX100 has 32 cores, the number of MPI processes can reach up to 32.

According to Fig. 38.1, the two- and threefolded loops have several better performance aspects than an execution without a loop collapse. The maximum speedup ratios of two- and threefolded loops as compared to without a loop collapse are 1.09 and 1.07, respectively. One of the reasons for the decreased performance is an increase in the cache miss hit ratios owing to non-continuous access by a loop col-

Fig. 38.1 Speedup ratios of loop collapse for pure MPI executions on the FX100 for 1–32 MPI processes on a single node. The ratios are normalized based on the execution time without a loop collapse during each MPI process. The time step is set to 50

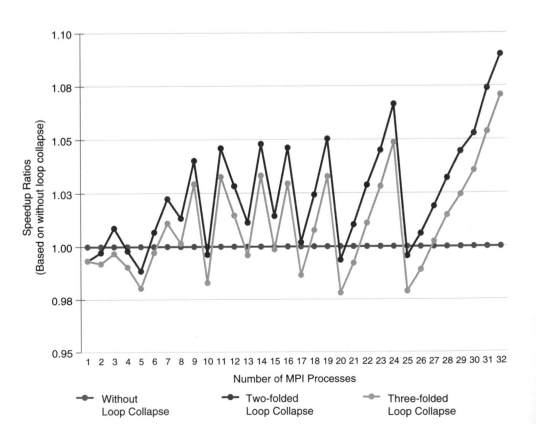

Fig. 38.2 Speedup ratios of loop collapse on several hybrid MPI executions in the FX100 with eight nodes. The ratios are normalized by the execution time without a loop collapse in 256Px1T. For aP × bT, aP indicates the a MPI processes and bT indicates the b OpenMP threads. The time step is set to 200

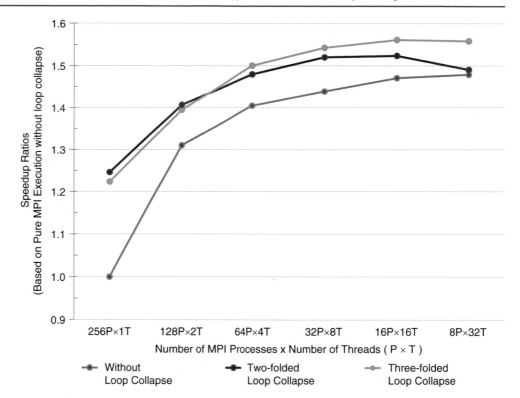

lapse from the viewpoint of each thread. However, we found that a twofolded loop collapse gives us an advantage in performance during a pure MPI execution.

Second, Fig. 38.2 shows the speedup ratios of hybrid MPI/OpenMP executions with eight nodes of the FX100. By adapting a hybrid MPI/OpenMP execution, we can take several combinations of numbers between the MPI processes and OpenMP threads. The notation aP × bT represents MPI processes and b OpenMP thread executions per MPI process.

Figure 38.2 shows that all hybrid MPI/OpenMP executions can obtain speedups to a pure MPI execution (256Px1T) with a loop collapse. The maximum speedup factors for a pure MPI execution without a loop collapse among the no-collapse and two-folded collapse and three-folded collapse conditions are 1.48x (8Px32T), 1.52x (16Px16T), and 1.56x (16Px16T), respectively. One of the contributions toward a speedup is reducing the MPI processes while increasing the OpenMP threads to reduce the communication time of such processes.

In addition, a gain in speedup for a loop collapse is high, as shown in Fig. 38.2. As one of the reasons for this, the speedup achieved by increasing the number of OpenMP threads is higher than the speedup achieved by increasing the number of MPI processes because of the shorter loop length compared to the loop lengths from two- and threefolded collapses.

Through the above two performance evaluations, we demonstrate that a loop collapse is an effective technique for

optimizing the LDDMM code. Moreover, a hybrid MPI/OpenMP execution is also a crucial way to speed up the LDDMM code.

38.6 Conclusion

In this chapter, we describe the development of a parallel and high-performance implementation for the application of MCA with an MPI. In addition, we adapted the program for use on a supercomputer and evaluated its performance.

The results of the performance evaluation indicate that an 18-day sequential processing, as shown in Sect. 38.5.2, can be reduced to (1) 64 h with a 0.87x speedup in pure MPI execution (256 MPI processes × 1 OpenMP thread) and (2) 43 h with an 11.52x speedup in hybrid MPI/OpenMP execution (16 MPI processes × 16 OpenMP threads), when using eight nodes of the Fujitsu PRIMEHPC FX100, which is installed in the Information Technology Center, Nagoya University, Japan.

The above results show an example of small-scale parallelism, because the eight nodes in the FX100 apply 256 parallelism. However, the increase in speed is significant. With respect to the full nodes used in the FX100 at Nagoya University, 2880/8 nodes = 360 sets can be processed within 43 hour. This has led to enormous computing power in the medical image processing field.

This is only one example showing a breakthrough in the execution time for applications of MCA achieved through collaborations with computer scientists and researchers in this field.

For future studies, we need to adapt other high-performance techniques, such as a modification of the data structures. An adaptation of auto-tuning technology [4] for selecting the best implementation of the computational kernels of the LDDMM code is also an important area of future study.

Acknowledgments The authors would like to thank Professor Hidekata Hontani at Nagoya Institute of Technology for providing the information regarding the LDDMM code.

This study was supported by JSPS KAKENHI, Grant Number JP17H05290 and JP18H03262. In addition, this work is supported by "Joint Usage/Research Center for Interdisciplinary Large-scale Information Infrastructures" in Japan (Project ID: jh180027-DAJ).

References

1. Katagiri T, Matsumoto M, Ohshima S. Auto-tuning of hybrid MPI/OpenMP execution with code selection by ppOpen-AT. In: Proceedings of IEEE IPDPSW2017. 2017. https://doi.org/10.1109/IPDPSW.2016.49
2. Goto H, Naito T, Hontani H. A new parametric description for line structures in 3D medical images by means of a weighted integral method. VISIGRAPP. 2016:208–17.
3. Beg MF, Miller M, Trouve A, Younes L. Computing large deformation metric mappings via geodesic flows of diffeomorphisms. Int J Comput Vis. 2015;61(2):139–57.
4. Katagiri T, Takahashi D. Japanese auto-tuning research: auto-tuning languages and FFT. Proc IEEE. 2018;106(11):2056–67.

MRI: Quantitative Evaluation of Diseased Tissue by Viscoelastic Imaging Systems

39

Mikio Suga

Abstract

Magnetic resonance elastography (MRE) and ultrasound elastography (USE) are imaging techniques that non-invasively quantify the mechanical properties of tissue by using magnetic resonance imaging and ultrasound imaging systems. In this study, we aim to develop a system that can quantitatively obtain details such as the viscoelasticity of small organs and tissues in the body by using clinical magnetic resonance imaging (MRI) and MR microscopy. By evaluating MRE and USE using a soft tissue-equivalent gel phantom with a known viscoelastic coefficient, we aim to investigate the characteristics of both devices and promote their standardization. The purpose of this study is to confirm the frequency characteristics of the developed phantom and optimize the scatterer material for ultrasound measurement. We confirmed that the phantoms are in good agreement with a physical model of the liver, and the developed phantoms are considered effective for the quantitative assessment of the MRE and USE systems.

Keywords

Magnetic resonance imaging · Magnetic resonance elastography · Ultrasound elastography · Phantom Quantitative assessment

39.1 Introduction

The mechanical property of a tissue is related to physiological and pathological states. Magnetic resonance elastography (MRE) and ultrasound elastography (USE) are imaging techniques that non-invasively quantify the mechanical properties

M. Suga (✉)
Center for Frontier Medical Engineering, Chiba University, Chiba, Japan
e-mail: mikio.suga@faculty.chiba-u.jp

of tissue by using magnetic resonance imaging and ultrasound imaging systems [1, 2]. It is expected that measuring the mechanical properties of tissues will be useful in the diagnosis of diseases such as hepatic fibrosis and cancer. MRE visualizes shear-wave patterns within a tissue using a modified phase-contrast MR sequence. In order to generate shear waves within the tissue, external vibration systems are used (Fig. 39.1). The local quantitative values of tissue viscoelasticity (stiffness) are calculated from the shear-wave pattern by using an inversion algorithm [3–5]. We have been developing an MRE system using clinical MRI and MR microscope for measuring viscoelasticity at multi-scale and multi-frequency (Fig. 39.2) [6, 7].

The measured viscoelasticity can be used as an imaging biomarker [8]. A quantitative phantom is required to assess the accuracy and repeatability of elastography systems. We have been developing tough and stable polyacrylamide (PAAm) gel phantoms for this purpose [9–12]. We developed a phantom with viscoelasticity close to that of living tissue by using glycerin as a solvent. For ultrasonic measurements, a scatterer is necessary for the phantom. The material and concentration of the scatterer are related to the stability of ultrasonic measurement and uniformity of the MRI image.

In this study, we compared the mechanical properties of our phantoms with those of living tissue and optimized the scatterer material and concentration for ultrasound measurement with the developed MRE system and commercial USE system.

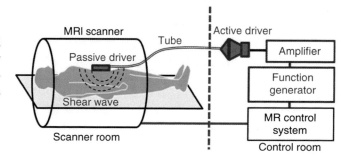

Fig. 39.1 Schematic of the MR elastography system

Fig. 39.2 Multi-modality equipment for measuring viscoelasticity at multi-scale and multi-frequency

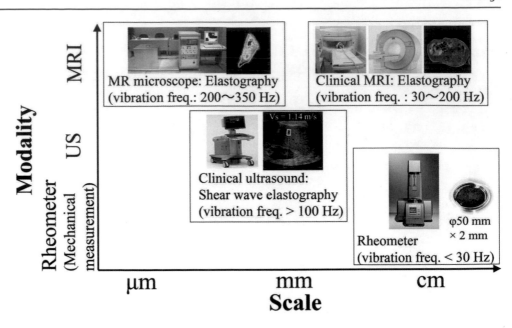

Table 39.1 Rheometer measurement parameters

Frequency	Hz	1–30
Strain amplitude	%	1
Normal force	N	0.7
Diameter	mm	50
Thickness	mm	1.5–2.0 (phantom) 2.0–2.7 (liver)
Temperature	°C	23

Table 39.2 Multi-frequency MRE imaging parameters by 3-T MRI

Sequence		Spin-echo-EPI-MRE (work in progress)			
Vibrational frequency	Hz	30	40	50	60
Repetition time (TR)	s	4.5	4.5	3.0	3.0
Echo time (TE)	ms	97			
FOV	mm²	384 × 384			
Matrix size	pixel	128 × 128			
Pixel size	mm	3			
Number of slices		15			
Slice thickness	mm	3			
Temperature	°C	23			

39.1.1 Development of a Tissue-Mimicking Viscoelastic Phantom for Quantitative Assessment of MRE

We fabricated gel phantom sheets with high loss moduli (G''). The phantom sheets were designed to have a storage modulus (G') of 1.5 kPa and a loss tangent (tan $\delta = G''/G'$) greater than 0.2 at 60 Hz [13]. The storage and loss moduli of the phantom and a fresh in vitro bovine liver were measured using a parallel-disc rheometer (MCR302, Anton-Parr). We repeated the measurements on these three samples with the strain amplitude set to 1% (Table 39.1).

The phantom (diameter: 120 mm, height: 150 mm) was measured with MRE using 3-T MRI (MAGNETOM Skyra, Siemens) at 30, 40, 50, and 60 Hz (Table 39.2) by spin-echo-EPI-MRE pulse sequence (work in progress) and the vibration from a pneumatic driver system with active driver to a passive driver to create the shear wave in the phantom. The passive driver with a diameter of 18 cm was placed on the center of the phantom. The fresh in vitro bovine liver (width: 135 mm, height: 70 mm, length: 250 mm) was measured with MRE using 0.3-T MRI (Hitachi) at 31.25, 62.5, 100, and 200 Hz (Table 39.3) by spin-echo-EPI-MRE pulse sequence (a motion encoding gradient (MEG) was added to

Table 39.3 Multi-frequency MRE imaging parameters by 0.3-T MRI

Sequence		Spin-echo-EPI-MRE			
Vibrational frequency	Hz	31.25	62.5	100	200
Repetition time (TR)	s	4.8			
Echo time (TE)	ms	67			
FOV	mm²	348 × 348			
Matrix size	pixel	116 × 116			
Pixel size	mm	3			
Number of slices		15			
Slice thickness	mm	3			
Temperature	°C	23			

SE-EPI using the sequence development environment of Hitachi) and ultrasound system (ACUSON S3000, Siemens) with virtual touch IQ (VTIQ). At high-frequency measurement of the fresh in vitro bovine liver (width: 15 mm, depth: 15 mm, height: 50 mm), MRE using 1-T MR microscope (MR-MICRO, MRTecnology) was used (Table 39.4).

To evaluate the temporal changes, the phantom designed to have a storage modulus of 3 kPa was examined by MRE for the duration of 1 year.

Table 39.4 MRE imaging parameters by 1-T MR microscope

Sequence		Spin-echo-MRE
Vibrational frequency	Hz	200
Repetition time (TR)	s	0.5
Echo time (TE)	ms	21
FOV	mm^2	25 × 25
Matrix size	pixel	128 × 128
Pixel size	mm	0.2
Number of slices		1
Slice thickness	mm	1.8
Average		3
Temperature	°C	23

Table 39.5 SWE measurement parameters

Probe		9 MHz linear (9 L4)	
Depth range	cm	0.3–4.0	
Detect pulse	MHz	6.0	
Method		VTIQ	VTQ
ROI size	mm^2	1.5 × 1.5	5 × 5
Push pulse	MHz	4.4, 5.7	4.0
Number of measurement		4 × 3	5
Depth	cm	2	
Temperature	°C	22	

39.1.2 Ultrasound-Based Shear-Wave Speed Measurement on a Highly Viscous Embedded Phantom

The PAAm gel is composed of a three-dimensional network polymer and a large amount of liquid. The storage modulus (stiffness) of the PAAm gel depends mainly on the quantity of acrylamide. Additionally, the density of the three-dimensional network polymer depends mainly on the quantity of cross-linker. The loss modulus (viscosity) depends mainly on the ratio of water and glycerin. To make compatible phantoms for MRE and USE, the aluminum oxide powder was added to the PAAm gel for the scatterer.

A highly viscoelastic embedded phantom was measured with US-based shear-wave elastography (SWE). The phantom composed square soft part (background part; width: 130 mm, depth: 130 mm, height: 160 mm) and embed two cylindrical hard parts (embedded part; length: 130 mm, diameter: 10 mm and 20 mm). The weight percent of acrylamide in the embedded part is 1.5 times higher than the background part. In addition, we have created a homogeneous phantom that is content the same as the embedded part.

The SWS was measured with virtual touch quantification (VTQ) and VTIQ (Table 39.5). VTQ provides only single-point SWS measurement, and VTIQ provides two-dimensional color-coded SWE (2D SWE), which displays 2D color velocity maps and allows for multiple measurements to be obtained. The stiffness is proportional to squaring of the SWS.

We measured the embedded parts, and the background parts were 5 mm apart from the outline of the embedded ones. VTIQ measurements were repeated 4 points on the same depth and three times at each part, VTQ measurements were repeated five times at each part, and the mean value and SD of the SWS were calculated. Reference values of the embedded part were measured in a homogenous phantom made with the same material component. The reference value of the background part was measured at a deeper area of the embedded phantom.

39.2 Results

39.2.1 Development of a Tissue-Mimicking Viscoelastic Phantom for Quantitative Assessment of MRE

Figure 39.3 shows the stiffness (square root of the sum of squares of the storage and loss modulus) obtained with MRE, the rheometer, and USE. The stiffness of the phantom and bovine liver increased with frequency. Figure 39.4 shows the change in the mechanical properties of the phantom over a year. The change in the storage and loss modulus during the 1-year period was within ±3%.

39.2.2 Ultrasound-Based Shear-Wave Speed Measurement on a Highly Viscous Embedded Phantom

Table 39.6 shows the SWS in the embedded part and background part. In the background part, the SWS was equivalent to the reference value. The SWS of the embedded part with a diameter of 20 mm in a highly viscous phantom was measured accurately with the SWE; however, with a diameter of 10 mm was lower than the reference value. Figure 39.5 shows the B-mode image and the VTIQ image around the embedded part with a diameter of 10 mm. On the VTIQ image, the embedded part was demarcated from the background; however, the border was not sharp. In addition, the embedded part on the VTIQ was visualized to be larger than the part on the B-mode image. This phantom has the potential to be used as a quality control phantom to mimic living tissues.

Fig. 39.3 Frequency
characteristics of gel
phantoms and in-vitro bovine
liver

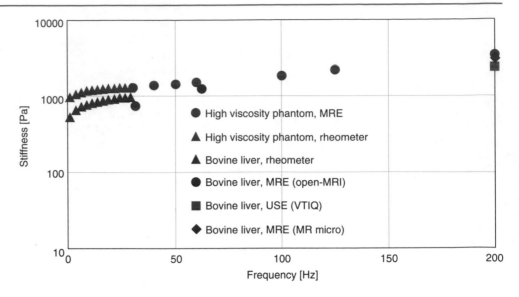

- ● High viscosity phantom, MRE
- ▲ High viscosity phantom, rheometer
- ▲ Bovine liver, rheometer
- ● Bovine liver, MRE (open-MRI)
- ■ Bovine liver, USE (VTIQ)
- ◆ Bovine liver, MRE (MR micro)

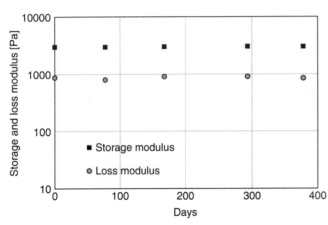

- ■ Storage modulus
- ● Loss modulus

Table 39.6 Results of SWS measurement

	Embedded part		Background part	
	VTIQ	VTQ	VTIQ	VTQ
10 mm	3.0 ± 0.2	2.5 ± 0.0	2.3 ± 0.1	2.0 ± 0.2
20 mm	3.3 ± 0.1	2.8 ± 0.0	2.3 ± 0.1	2.0 ± 0.2
Reference values	3.4 ± 0.2	2.8 ± 0.2	2.3 ± 0.1	2.0 ± 0.2

Fig. 39.4 Changes in the mechanical properties of high-viscosity phantom over time

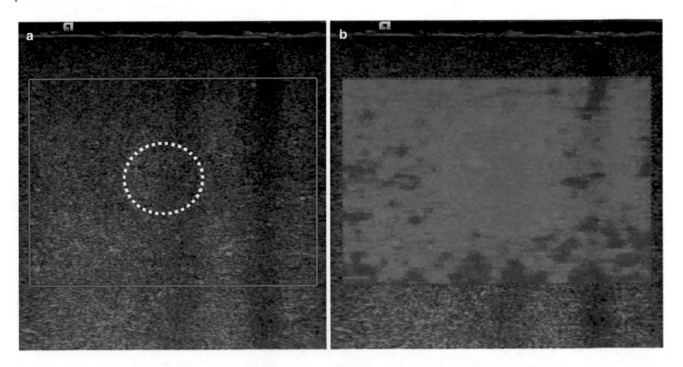

Fig. 39.5 Image of the phantom containing an embedded part with a diameter of 10 mm. (**a**) B mode image. (**b**) VTIQ image

39.3   Conclusion

The purpose of this study was to confirm the frequency characteristics of the developed phantom with commercial USE system and the developed MRE system at multi-scale and multi-frequency. We confirmed that the phantoms are in good agreement with a physical model of the liver, and the developed phantoms are considered effective for the quantitative assessment of the MRE and USE system.

Furthermore, an MRE system using an MR microscope and clinical MRI was developed. We succeeded in performing measurements at multiple frequencies. These developed systems allow quantitative multi-scale and multi-parameter imaging of diseased tissues.

Acknowledgments The viscoelastic evaluation of the phantom was performed under the collaboration of Dr. Takayuki Obata and Dr. Riwa Kishimoto of the National Institute of Radiological Sciences, QST, and Prof. Richard L. Ehman and Prof. Jun Chen of Mayo Clinic in the USA. The ultrasonic characteristics evaluation of the phantom was performed under the collaboration of Prof. Tsuyoshi Shiina and Prof. Makoto Yamakawa of Kyoto University and Prof. Tadashi Yamaguchi of Chiba University. This study was partly supported by JSPS, Grant No. JP17H05279, JP17H02115.

References

1. Muthupillai R, Lomas DJ, Rossman PJ, Greenleaf JF, Manduca A, Ehman RL. Magnetic resonance elastography by direct visualization of propagating acoustic strain waves. Science. 1995;269(5232):1854–7. https://doi.org/10.1126/science.7569924.
2. Parker KJ, Doyley MM, Rubens DJ. Imaging the elastic properties of tissue: the 20 year perspective. Phys Med Biol. 2011;56(1):R1–R29. https://doi.org/10.1088/0031-9155/56/1/R01.
3. Manduca A, Oliphant TE, Dresner MA, Mahowald JL, Kruse SA, Amromin E, Felmlee JP, Greenleaf JF, Ehman RL. Magnetic resonance elastography: non-invasive mapping of tissue elastic-

ity. Med Image Anal. 2001;5(4):237–54. https://doi.org/10.1016/s1361-8415(00)00039-6.
4. Suga M, Matsuda T, Minato K, Oshiro O, Chihara K, Okamoto J, Takizawa O, Komori M, Takahashi T. Measurement of in vivo local shear modulus using MR elastography multiple-phase patchwork offsets. IEEE Trans Biomed Eng. 2003;50(7):908–15. https://doi.org/10.1109/TBME.2003.813540.
5. Suga M, Miura H, Fujiwara T, Tanaka T, Yu Q, Arai K, et al., editors. Inversion algorithm by integral type reconstruction formula for magnetic resonance elastography. Proc Int Soc Magn Reson Med. 2009;2506.
6. Tomita S, Suzuki H, Kajiwara I, Nakamura G, Jiang Y, Suga M, Obata T, Tadano SA. Numerical simulations of magnetic resonance elastography using finite element analysis with a linear heterogeneous viscoelastic model. J Vis. 2017. https://doi.org/10.1007/s12650-017-0436-4
7. Haneishi H, Yamaguchi T, Suga M. Multimodal and multiscale medhical engineering. J Jpn Soc Appl Phys. 2018;87(5):350–6.
8. Quantitative Imaging Biomarkers Alliance. https://www.rsna.org/qiba/
9. Suga M, Mori T, Kishimoto R, Kurokawa T, Abe T, Tsuji H, Obata T. Development of a tissue-simulating viscoelastic gel phantom for MR elastography. Proc Eur Cong Radiol. 2015; https://doi.org/10.1594/ecr2015/C-0757.
10. Ishii K, Suga M, Kishimoto R, Hotta E, Obata T. Development of a tissue-mimicking Visco-elastic phantom for quantitative assessment of MRE. In: First international MRE workshop, vol 49; 2017.
11. Usumura M, Suga M, Kishimoto R, Obata T. Ultrasound-based shear-wave speed measurement on a highly viscous embedded phantom. In: Proceedings of international workshop on advanced imaging technology & international forum on medical imaging in Asia, vol 48; 2019.
12. Kishimoto R, Suga M, Koyama A, Omatsu T, Tachibana Y, Ebner DK, Obata T. Measuring shear-wave speed with point shear-wave elastography and MR elastography: a phantom study. BMJ Open. 2017;7(1):e013925. https://doi.org/10.1136/bmjopen-2016-013925en.
13. Asbach P, Klatt D, Schlosser B, Biermer M, Muche M, Rieger A, Loddenkemper C, Somasundaram R, Berg T, Hamm B, Braun J, Sack I. Viscoelasticity-based staging of hepatic fibrosis staging of hepatic fibrosis elastography. Radiology. 2010;257(1):80–6. https://doi.org/10.1148/radiol.10092489.

US: Development of General Biophysical Model for Realization of Ultrasonic Qualitative Real-Time Pathological Diagnosis

Tadashi Yamaguchi

Abstract

In order to quantitatively evaluate the properties of biological tissues from the body surface using ultrasound, it is necessary to comprehensively understand the acoustic properties of both microscopic acoustic properties at the cell level and tissues with macroscopic structure. This chapter introduces the results of evaluating the frequency dependence of the speed of sound of multiple organs using ultrasound in an extremely wide frequency band.

Keywords

Ultrasound · Tissue characterization · Quantitative diagnosis · Acoustic property · Speed of sound · Acoustic microscopy · Frequency dependency · Tissue structure

40.1 Introduction

Medical ultrasound is widely used as a diagnostic tool in clinical applications because it is non-invasive and provides images in real time. Research on medical ultrasound conducted around the world has led to the development of quantitative diagnostic methods for various diseases of biological tissue such as the liver, mammary glands, and lymph nodes [1–4]. For this reason, many studies have focused on quantitative ultrasound (QUS) diagnosis rather than qualitative. One example of such a QUS method is shear-wave elastography, which involves the generation of shear waves in the measurement object and the quantitative evaluation of the stiffness of the tissue based on the propagation of the waves [5, 6]. Furthermore, understanding the relationship between acoustic characteristics and tissue structure, which are quantitative evaluation metrics obtained from the echo signal, is being promoted more fundamentally. Several research groups have measured the speed of sound (SoS), attenuation, and backscatter coefficients in tissue using ultrasound with center frequencies ranging from 1 MHz to approximately 50 MHz [7–9].

In addition, many studies have been carried out at a cellular or subcellular scale that are finer than the connecting tissue structures. These studies used scanning acoustic microscopy (SAM) systems with an ultrasound band of several hundred megahertz, which enables the discrimination of different cell organelles to measure the SoS and acoustic impedance of biological tissue [10–14]. The spatial resolution of SAM is close to that of optical microscopy, and the relationship between acoustic and optical characteristics has been researched [15]. Therefore, it is possible to seamlessly integrate QUS metrics into diagnosis by understanding how the cell-level acoustic characteristics of a tissue relate to their optical characteristics and considering them in the whole organ. These correspond to the research area called tissue characterization, which uses ultrasound to understand the properties of tissues.

However, conventional SAM systems have a limited measurement area (the upper limit is only several square millimeters) because they have typically been used to observe targets that are, at most, the same size as optical microscopy slides. This measurable size is extremely small compared to clinical echo information. In other words, to apply SAM to the required organ level observations in QUS, it would be necessary to build a novel SAM system that can measure the whole organ without sacrificing the cell-level information. In our study, a new SAM system that can observe 100 mm^2 area was developed to enable this type of multiscale measurement. In addition, the conventional analysis method was modified to realize SoS analysis over a wide area.

In this chapter, the effectiveness of acoustic characteristics analysis in a wide area is introduced by exemplifying the results of SoS evaluation by observing rat kidneys with the

T. Yamaguchi (✉)
Center for Frontier Medical Engineering, Chiba University, Chiba, Japan
e-mail: yamaguchi@faculty.chiba-u.jp

newly developed SAM. Additionally, the SoS evaluation result from microscale resolution to macro-scale resolution by several kinds of frequencies in rat livers is also introduced. By comparing the SoS in each organ evaluated obtained at wideband frequencies, the spatial resolution of each transducer and the relationship between the center frequency and the SoS of the liver were investigated.

40.2   SoS Evaluation in High Resolution and High Region Using New SAM

In the developed SAM system, the moving pitch range of the XY linear stage controlled by a computer is 0.1 μm to 100 mm for each direction. The center frequency of transmission and reception can be set in the range of 1–500 MHz by changing the transducers. For acquiring the RF echo signal from an excised rat kidney, a ZnO membrane focuses transducer that has a 250-MHz center frequency was installed to SAM. The spatial resolution of the 250 MHz transducer is 7 μm (lateral) and 50 μm (depth). The spatial resolution defined for the transducer was within −6 dB from the maximum value of the sound pressure, and it is good enough to visualize the details of the tissue structure or cell-level information. To acquire RF echo signals from a sliced specimen on the glass plate, a two-dimensional (2D) scan was obtained by the transducer on the moving stage. In this observation, after scanning in the x-direction along each scan line, the RF echo signal was first passed through an attenuator, a high-pass filter (41–800 MHz), and a low-pass filter (DC-720 MHz) and then transferred from a digitizer to a computer. The bandwidth of each filter was determined so that noise could be effectively removed in consideration of the frequency characteristics of the transducer. The amplified RF echo data from each scan line were acquired with a sampling frequency of 2.5 GHz and digitized with 12 bits.

The surface and bottom echoes of the tissue sample were detected from the observed RF echo signal using a fifth-order AR model to estimate the SoS. The components of the echo signal from the tissue sample area were roughly divided into three. The first was the echo component from the surface of the tissue sample, the second was the echo component from the bottom of the tissue sample (sample–glass interface), and the third was the echo component of multiple reflections from the bottom of the sample surface. The RF echo signals were separated by including two random noise components generated during electrical and digitizing in addition to the three echo components above [12, 16].

Figure 40.1 shows the evaluated SoS image and the histological image of the excised healthy kidney of a 17-week-

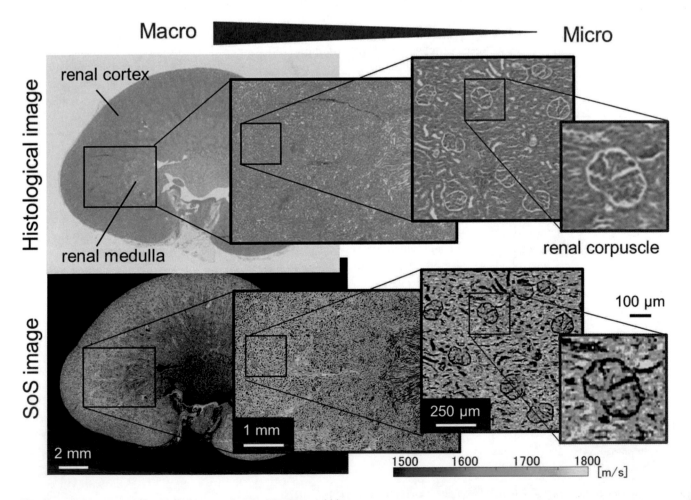

Fig. 40.1 SoS image and histological image of excised healthy rat kidney

old rat (SLC:SD, male). It was fixed with formalin, infiltrated with paraffin, and sliced to a thickness of 7 μm. After the paraffin was removed, the sliced specimen was placed on a glass plate and submerged in a water tank filled with degassed water. The scanning area was 21 mm × 16 mm, which encompasses the whole kidney, and the scanning intervals in both the x- and y-directions were 10 μm. That is, three-dimensional (3D) RF data associated with scanning line information on the 2100 × 1600 sampling points was acquired. After scanning, the sliced specimen of kidney was stained using the hematoxylin–eosin (HE) method, and a digital histological image was observed using a virtual slide scanner.

The difference of texture (difference of tissue density) and boundaries of the renal cortex and renal medulla are clearly shown with the SoS image of the whole kidney in the wide-field image. The structures inside the renal corpuscles can be observed in the zoomed micro image. The structures in the renal corpuscle are assumed to be glomeruli. Because it is difficult to recognize this structure in the corresponding histological images, the SoS map can be said to reveal physical properties that are not observable by staining. The average and standard deviation of the SoS obtained at 250 MHz was 1591.7 ± 106.8 m/s in the renal medulla, and 1678.1 ± 131.6 m/s in the renal cortex. These average values are close to the conventional evaluation result of the biological SoS in a normal kidney fixed with formalin and infiltrated with paraffin [17]. This result shows that the SoS of biological tissue can be evaluated with high accuracy over the macro-region while ensuring cell-level resolution.

40.3 Evaluation of Frequency Dependency and Tissue Structure Dependency of SoS

The realization of a SAM capable of observing a wide area has made it possible to evaluate the acoustic characteristics of biological tissues on a multiscale. In this section, the results of evaluating the frequency dependence of the SoS in rat liver using ultrasound in an extremely wide frequency band are introduced. As mentioned earlier, our SAM allows observation of samples (that is, acquisition of RF echo signals) with ultrasound of various frequencies by exchanging ultrasonic transducers. Of course, the setup of various amplifiers and digitizers will change depending on the type of transducer. Live observation for in vivo tissue is also possible at frequencies below 25 MHz, which are used clinically.

Immediately after being removed, the whole rat livers (normal and fibrosis) were first observed by PZT 5 MHz (most frequently used frequency in clinic) unfocused transducer, and then at a high frequency of 15 MHz. The defini-

tion of spatial resolution does not apply because they transmit and receive plane waves without setting the focus. The scanning interval of the transducer for the XY-plane was 30 μm for each transducer. The amplified RF echo data from each scan line were acquired with a sampling frequency of 100 MHz and digitized with 12 bits. After observation by 15 MHz, each liver was fixed with formalin, infiltrated with paraffin, and sliced to a thickness of 7 μm as same as the case of the kidney. The sliced specimens were observed by PVDF-TrFE 60 and 80 MHz focused transducers, and by ZnO 250 MHz focused transducer. The spatial resolution of the 60, 80, 250 MHz transducers are 28, 20, 7 μm (lateral) and 300, 200, 50 μm (depth), respectively. The scanning interval was set to 2 μm considering the spatial resolution of the 250 MHz transducer, which has the highest spatial resolution. The amplified RF echo data from each scan line were acquired with a sampling frequency of 2.5 GHz and digitized with 12 bits. From the three-dimensional RF echo data acquired by five transducers, the speed of sound of the target liver was evaluated.

Figure 40.2 shows the evaluated SoS image and the histological image of the excised rat livers of a 10-week-old healthy rat (SLC:SD, male) and 10-week-old fibrosis rat which was injected with a mixture of carbon tetrachloride twice a week for 4 weeks. In both livers, the texture of the speed of the sound image is blurred at 60 MHz; however, even the difference between the cell nucleus and the cytoplasm can be confirmed at 250 MHz. The difference in the contrast of the texture in the liver depending on the frequency is clearly greater in fibrosis than in the normal liver. As can be seen from the fact that the structure of the fibers, which is difficult to recognize at 60 MHz, can be clearly confirmed at 80 MHz or more, the frequency (spatial resolution) also strongly influences the physical property evaluation in microscopic ultrasound observation.

In Fig. 40.3, the results of the macroscopic evaluation of the liver as in the clinical diagnosis are compared with the results of the microscopic evaluation shown in Fig. 40.2. As the first feature, it can be confirmed that there is a small difference in speed of sound between clinically equivalent frequencies of 5 and 15 MHz, as is known in the past, whereas a strong frequency dependence is exhibited at frequencies above 60 MHz. It has been confirmed that there is no difference in speed of sound at a general frequency of 5 MHz, a high frequency of 15 MHz, and ultrahigh frequencies of over 60 MHz in a homogeneous medium such as an agar phantom in our pilot study. In other words, it can be said that Fig. 40.3 shows the characteristics peculiar to liver tissue. However, it should be noted that the difference in speed of sound between 15 and 60 MHz includes the effect of formalin. Another feature is that there is a difference in the frequency dependence of the speed of sound between normal liver and cirrhotic liver at the ultrahigh frequency. If only the spatial resolution

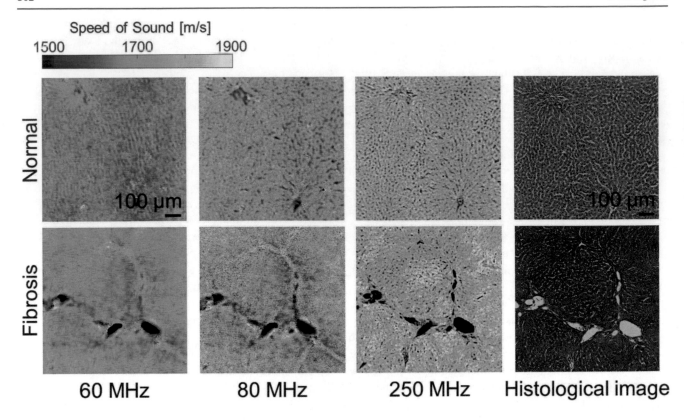

Fig. 40.2 SoS image and histological image of excised frat livers

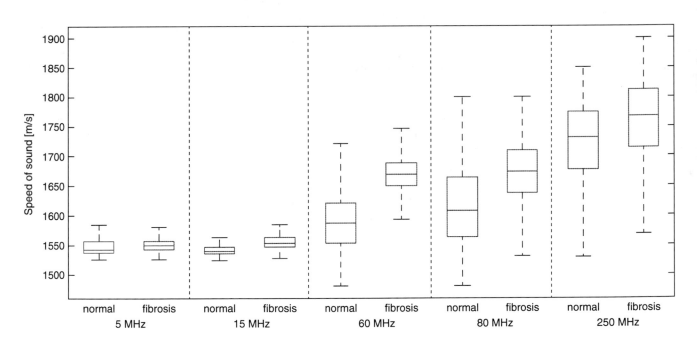

Fig. 40.3 Frequency dependency of SoS in rat livers

is different, the average value of the speed of sound will be similar in each liver. However, the results in Fig. 40.3 do not represent a simple average value because the mixing ratios and distributions of cytoplasm, cell nucleus, and fibers with their respective value of the speed of sound differ. The characteristic speed of sound of each tissue can be obtained by comparing the SoS image and the histological image by one-by-one matching [12].

40.4 Conclusion

An SAM system was constructed that enables the multiscale evaluation of the acoustic characteristics of biological tissue using ultra-wideband ultrasound. The system is also allowed to function like a clinical ultrasound equipment. Although the results of speed of sound evaluation were introduced in this chapter, the acoustic impedance, scatterer density, backscattering coefficient, etc. can also be evaluated on a multiscale by applying other signal acquisition and analysis methods. It is also possible to construct 3D acoustic characteristic maps by observing multiple tissue sections and use it as a simulation model for sound wave propagation analysis.

References

1. Omura M, Yoshida K, Akita S, Yamaguchi T. Frequency dependence of attenuation and backscatter coefficient of ex vivo human lymphedema dermis. J Med Ultras. 2020;47:25–34.
2. Tamura K, Yoshida K, Maruyama H, Hachiya H, Yamaguchi T. Proposal of compound amplitude envelope statistical analysis model considering low scatterer concentration. Jpn J Appl Phys. 2018;57:07LD19.
3. Mori S, Hirata S, Yamaguchi T, Hachiya H. Quantitative evaluation method for liver fibrosis based on multi-Rayleigh model with estimation of number of tissue components in ultrasound B-mode image. Jpn J Appl Phys. 2018;57:07LF17.
4. Saegusa-Beecroft E, Machi J, Mamou J, Hata M, Coron A, Yanagihara ET, Yamaguchi T, Oelze ML, Laugier P, Feleppa EJ. Three-dimensional high-frequency quantitative ultrasound for detecting lymph-node metastases. J Surg Res. 2013;183:258–69.
5. Hayashi A, Mori S, Arakawa M, Kanai H. Local two-dimensional distribution of propagation speed of myocardial contraction for ultrasonic visualization of contraction propagation. Jpn J Appl Phys. 2019;58:SGGE05.
6. Goss SA, Johnston RL, Dunn F. Comprehensive compilation of empirical ultrasonic properties of mammalian tissues. J Acoust Soc Am. 1978;64:423.
7. Hasegawa H, Nagaoka R. Initial phantom study on estimation of speed of sound in medium using coherence among received echo signals. J Med Ultras. 2019;46:297–307.
8. Brewin MP, Srodon PD, Greenwald SE, Birch MJ. Carotid atherosclerotic plaque characterisation by measurement of ultrasound sound speed in vitro at high frequency 20 MHz. Ultrasonics. 2014;54:428–41.
9. Agemura DH, O'Brien WD Jr, Olerud JE, Chun LE, Eyre DE. Ultrasonic propagation properties of articular cartilage at 100 MHz. J Acoust Soc Am. 1990;87:1786–91.
10. Ito K, Yoshida K, Maruyama H, Mamou J, Yamaguchi T. Acoustic impedance analysis with high-frequency ultrasound for identification of fatty acid species in the liver. Ultras Med Biol. 2017;43:700–11.
11. Rohrbach D, Jakob A, Lloyd H, Tretbar S, Silverman RH, Mamou J. A novel quantitative 500-MHz acoustic microscopy system for ophthalmologic tissues. J IEEE Trans Biomed Eng. 2016;64:715–24.
12. Irie S, Inoue K, Yoshida K, Mamou J, Kobayashi K, Maruyama H, Yamaguchi T. Speed of sound in diseased liver observed by scanning acoustic microscopy with 80 MHz and 250 MHz. J Acoust Soc Am. 2016;139:512–9.
13. Hattori K, Sano H, Saijo Y, Kita A, Hatori M, Kokubun S, Itoi E. Measurement of soft tissue elasticity in the congenital clubfoot using scanning acoustic microscope. J Pediatr Orthop B. 2007;16:357–62.
14. Raum K. Microelastic imaging of bone. IEEE Trans Ultrason Ferroelectr Freq Control. 2008;55:1417–31.
15. Takanashi K, Washiya M, Ota K, Yoshida S, Huzumi N, Kobayashi K. Quantitative evaluation method for differentiation of C2C12 myoblasts by ultrasonic microscopy. Jpn J Appl Phys. 2017;56:07JF11.
16. Ogawa T, Yoshida K, Yamaguchi T. Speed of sound evaluation considering spatial resolution in a scanning acoustic microscopy system capable of observing wide spatial area. Jpn J Appl Phys. 2020;59:SKKE13.
17. Sasaki H, Saijo Y, Tanaka M, Okawai H, Terasawa Y, Yambe T, Nitta S. Influence of tissue preparation on the high-frequency acoustic properties of normal kidney tissue. Ultras Med Biol. 1996;22:1261–5.

OCT: Ultrahigh Resolution Optical Coherence Tomography at Visible to Near-Infrared Wavelength Region

41

Norihiko Nishizawa and Masahito Yamanaka

Abstract

Optical coherence tomography (OCT) is an imaging modality using a low coherence interferometer with a spectrally broadband light source, which allows us to perform non-invasive observations of internal structures of living samples with a micrometer-scale resolution. Since recent advancements of OCT technologies have offered significant improvement in image acquisition speed, OCT is now used in a wide variety of medical fields, especially in ophthalmology. In our group, we have been working on the development of ultra-broadband light sources, so-called supercontinuum (SC) light sources, and ultrahigh-resolution OCT systems with the SC light sources. While the 0.8 μm wavelength window is typically used for OCT systems and 0.8 μm, OCTs offer high axial resolution in tissue imaging, and the 1.7 μm wavelength window has begun to be recognized as an excellent choice to achieve high penetration depth due to the lower scattering coefficient and the existence of a local minimum of light absorption by water in the wavelength band. In this chapter, we give a brief overview of the basic principle of OCT imaging and our recent works in the development of ultrahigh-resolution OCT imaging techniques.

Keywords

Optical coherence tomography (OCT) · Non-invasive Supercontinuum · Optical coherence microscopy (OCM)

N. Nishizawa (✉) · M. Yamanaka
Department of Electronics, Nagoya University, Nagoya, Aichi, Japan
e-mail: nishizawa@nuee.nagoya-u.ac.jp;
yamanaka.masahito@h.mbox.nagoya-u.ac.jp

41.1 Introduction

Optical coherence tomography (OCT) is an optical imaging technique that allows us to observe internal structures of biomedical specimens, such as the human eye, skin, and living small animals, with sub-μm to 10-μm axial resolution in non-contact and non-invasive manners [1]. Since OCT is a significantly useful technique to visualize retinal structures in patients, this technique is currently widely used for clinical diagnosis in ophthalmology.

Figure 41.1 shows a comparison of the various major non-invasive imaging techniques to observe the internal structures of biomedical specimens. Ultrasonography is one of the representatives and familiar clinical imaging modalities. In ultrasonography, although a sensor to detect ultrasound waves is required to be placed in contact with a surface of specimens, the penetration depth usually reaches several centimeters, and the millimeter-scale spatial resolution is achieved. Currently, Ultrasonography is used for a medical health checkup, fetal diagnosis, and so on. X-ray CT and MRI provide a larger penetration depth, but they require large-scale equipment. In addition, there are safety issues, and the spatial resolution is typically not so high in standard systems. Optical microscopy is known as a high-resolution imaging technique to provide several 100 nm to a few micrometer resolutions. However, generally, it is difficult to achieve a penetration depth of more than 1 mm in in-vivo imaging.

As mentioned previously, OCT techniques provide non-contact, non-invasive, and in-vivo imaging capabilities of internal micrometer-scale structures of biomedical specimens. Generally, sub-μm to 10-μm axial resolution and few-millimeter penetration depth was achieved in OCT. In addition, recent improvement of OCT image acquisition speed has made it possible to perform real-time OCT cross-sectional imaging and rapid volumetric imaging with an acquisition time of a few seconds to few tens seconds.

M. Hashizume (ed.), *Multidisciplinary Computational Anatomy*, https://doi.org/10.1007/978-981-16-4325-5_41

The basic optical setup for OCT is shown in Fig. 41.2 light beam output from a spectrally broadband light source is divided into a sample and reference arms by a beam splitter. Backscattered and reflected light from a sample is interfered with light returned from the reference arm. The interfered light is detected with a photodetector. The example of an interference signal is also shown in Fig. 41.2. A spectrally broadband light consists of a lot of waves at different wavelengths. Therefore, only when the optical path length difference between the sample and reference arm is zero, the phases of all wavelength components are matched, meaning that all wavelength components interfere constructively and the highest intensity of an interference signal is achieved. Then, with the increase of the optical path length difference between the sample and the reference arm, the intensity of the interference signal is reduced due to the phase mismatch

of wavelength components. Assuming that the spectral shape of a broadband light source is Gaussian, a full-width at half maximum (FWHM) of an interference signal (Δz) is,

$$\Delta z = \frac{2\ln 2}{n\pi} \cdot \frac{\lambda^2}{\Delta\lambda} \qquad (41.1)$$

where λ is a center wavelength, $\Delta\lambda$ is a spectral width of a light source, and n is a refractive index. Δz is corresponding to a coherence length determined by the spectral width of a light source [1]. As shown in Eq. (41.1), Δz is proportional to the square of the center wavelength and inversely proportional to the spectral width. Therefore, by using a light source with broader spectral width, it is possible to achieve a higher axial resolution. In addition, although the intensity of signal light (a backscattered and reflected light) from a sample is usually significantly weak, high signal detection sensitivity, such as 100 dB, is achieved in OCT. This is because an interference signal is achieved as the result of the product of signal light and strong reference light, meaning that the signal intensity is enhanced by the strong reference light. The sensitivity of 100 dB means that it allows us to detect pW level signal light under 1 mW light illumination on a sample.

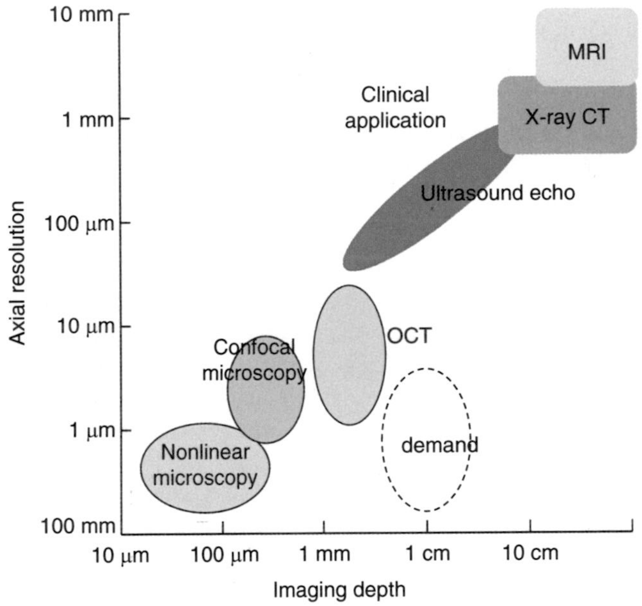

Fig. 41.1 The relationship between the axial resolution and penetration depth in various non-invasive imaging techniques

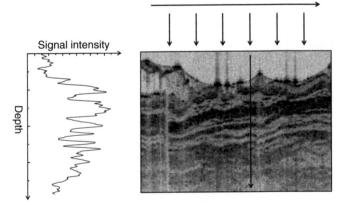

Fig. 41.3 Cross-sectional imaging with OCT

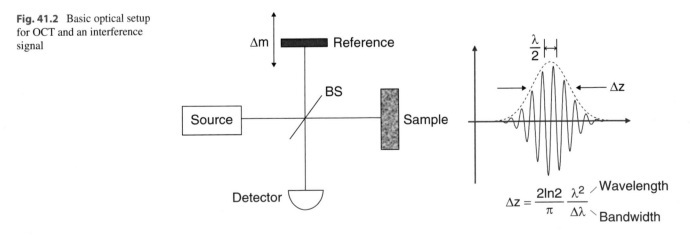

Fig. 41.2 Basic optical setup for OCT and an interference signal

Figure 41.3 shows the principle of cross-sectional imaging by OCT. The spectrally broadband light is focused on a sample with a low numerical aperture (NA) objective lens equipped in a sample arm. A backscattered and reflected light from a sample is collected with the same objective lens, and the collected light interferes with the reference light. The signal intensity distribution along the depth direction is achieved by scanning the optical path length of the reference arm because an interference signal appears when the optical path length between the sample and the reference arm is matched. Then, by scanning a sample along the horizontal direction and achieving the depth intensity distribution at each horizontal scanning position, a cross-sectional image is obtained.

41.2   Time and Fourier-Domain OCT

As shown in Fig. 41.4, there are several types of OCT systems, which are mainly classified into OCT systems with Time-domain (TD) and Fourier-domain (FD) detection schemes. In TD-OCT, depth information is achieved as the function of time. On the other hand, in FD-OCT, depth information is extracted from spectrally resolved interference fringes in recorded signals.

The example of an experimental setup for TD-OCT is illustrated in Fig. 41.5. An interferometer consists of optical fiber devices (optical fiber couplers) operating in a wide wavelength range. In the fiber couplers, the light output from a broadband light source is separated into the sample and reference arms. In the sample arm, the light beam is focused on a sample with a low NA objective lens (achromatic lens) and X–Y scanning is performed with a two-axis galvanometer scanner. Backscattered and reflected light from a sample is collected with the same objective lens and is coupled into the fiber couplers. Then, light from the sample and reference arms is made to interfere in the fiber coupler and then detected with a balanced detector. Depth information of a sample is achieved by scanning the optical path length of the reference arm with a corner cube prism mirror mounted on a single-axis galvanometer scanner.

Because a spectrally broadband light is used for OCT, the difference of the wavelength dependence on the refractive index between a sample and reference arms affects the FWHM and intensity of an interference signal. To cancel out the difference of the wavelength dependence, additional glass plates are usually utilized. In Fig. 41.5, glass plates are placed in the reference arm. This is called chromatic dispersion compensation. By performing this chromatic dispersion compensation and matching polarization states in two arms with polarization controllers, it is possible to achieve a strong interference signal with a narrower FWHM. Interference signals are recorded with a PC and are processed to construct an OCT image. In TD-OCT, the signal detection sensitivity of around 100 dB is achieved.

In TD-OCT, the image acquisition speed is restricted by a relatively low-speed mechanical scanner in the reference arm. To overcome this limitation, FD-OCT was developed. In FD-OCT, interference signals are measured with a spectrometer and so on. Because FD-OCT does not require a mechanical scanner to achieve interference signals, image

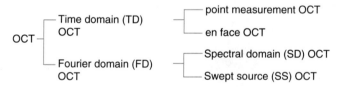

Fig. 41.4 Classification of OCT

Fig. 41.5 Experimental setup of TD-OCT

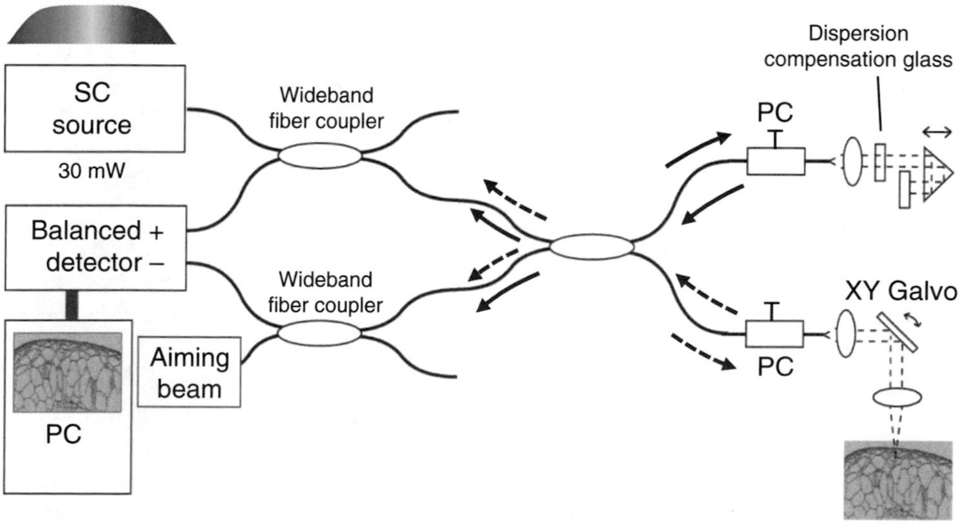

Fig. 41.6 Experimental setup of SD-OCT

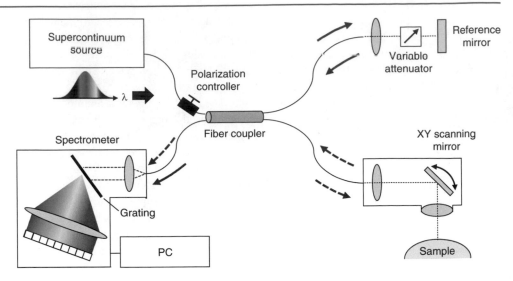

acquisition speed has been significantly improved. There are mainly two types of FD-OCT, which are (1) spectral-domain (SD) OCT using a broadband light source and spectrometer equipped with a high-speed line camera and (2) swept-source (SS) OCT using a high-speed wavelength-tunable light source, respectively.

The example of an experimental setup for SD-OCT is illustrated in Fig. 41.6. A spectrometer consists of a diffraction grating, lens, and a high-speed single line scan camera. In SD-OCT, spectral interference fringes in a detected spectrum are recorded. By performing a Fourier transform of the detected spectrum, an OCT signal intensity profile along the depth direction is achieved. This depth information is achieved by a single spectral measurement. Typically, a high-speed single line scan camera offers a line rate of 40 kHz. Therefore, when a cross-sectional image is constructed by 512 spectral measurements, it is possible to achieve about 80 cross-sectional images per second. A three-dimensional volumetric image is also achieved within a few seconds to few tens seconds. Even though signal averaging is applied to improve the signal-to-noise ratio in imaging, the image acquisition speed is still high enough to perform in-vivo observations.

In SS-OCT, instead of a broadband light source and spectrometer, a high-speed wavelength-tunable light source with a narrow spectral linewidth is used. In this case, signals are detected with a balanced detector, and then interference fringes on the optical spectrum are recorded as the function of time. Because optical power is concentrated at a specific wavelength and a certain time, it is known that the highest signal detection sensitivity among the various OCT methods is achieved. However, it is still challenging to achieve a high axial resolution in SS-OCT because there are technical issues in the development of spectrally broadband and high-speed wavelength-tunable light source with a narrow spectral linewidth.

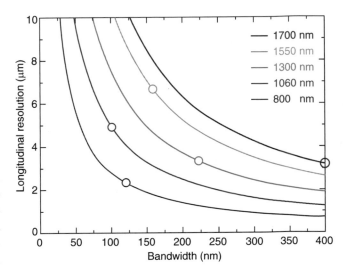

Fig. 41.7 The relationship between the axial (longitudinal) resolution and spectral bandwidth. The hollow circles indicate the axial resolution and bandwidth obtained with SC light sources developed by our group

41.3 Ultrahigh-Resolution OCT in the 0.8 μm Wavelength Window and Retinal Imaging

In OCT imaging, the axial resolution (longitudinal resolution) depends on the wavelength of a light source. Figure 41.7 shows the relationship between the spectral bandwidth of the light source and axial resolution at each wavelength window. As indicated in the Eq. (41.1), under the condition where the spectral bandwidths of light sources at different wavelengths are the same, a light source at a shorter wavelength window provides a higher axial resolution. On the other hand, when the wavelength of light sources is the same, a light source with a broader spectral bandwidth offers a higher axial resolution.

The use of the 0.8 µm wavelength window for OCT provides high axial resolution imaging capability and is widely utilized for retinal diagnosis due to the low light absorption coefficient of water. Figure 41.8a shows the optical spectrum of our high-power spectrally broadband light source, so-called supercontinuum (SC) light source, in the 0.8 µm wavelength window [2]. Figure 41.8b shows the representative interference signal obtained by using this SC light source light for the OCT system shown in Fig. 41.5. This SC light is generated by utilizing optical nonlinear effects in optical fibers. By coupling ultrashort pulsed light with the pulse width of about 100 fs into optical fibers, the spectrum of the input pulsed laser is broadened due to optical nonlinear effects, and then SC light with a single-peaked smooth spectral shape is achieved as shown in Fig. 41.8a. The shape of interference signals depends on the shape of the spectrum of the SC light source. Therefore, the use of the better spectral shape of the SC light for OCT enables us to suppress the sidelobes in interference signals and artifacts, such as ghost images. Here, the spectral bandwidth of the 0.8 µm SC light was 140 nm, and the theoretical limit of the axial resolution is 3 µm in air. Since the coherence length is inversely proportional to the refractive index, as shown in Eq. (41.1), the axial resolution becomes higher when the refractive index becomes higher. When using the 0.8 µm SC light for the TD-OCT system shown in Fig. 41.5, the axial resolution of 2.9 µm in the air (a refractive index $n = 1$) was achieved, which corresponds to 2.0 µm in the tissue under the assumption of $n = 1.38$.

Figures 41.9 and 41.10 show the cross-sectional images of the human retina around the fovea, rat airway and cartilage, and brain of swimming young medaka fish obtained with the

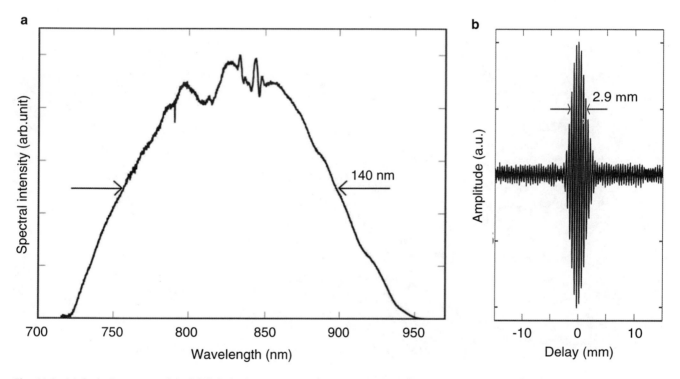

Fig. 41.8 (**a**) Optical spectrum of the SC light in the 0.8 µm wavelength window and (**b**) the interference signal obtained by using this SC light for OCT system shown in Fig. 41.5

Fig. 41.9 OCT image of human retina around fovea

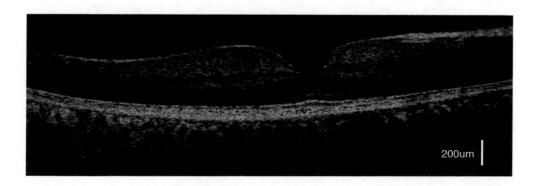

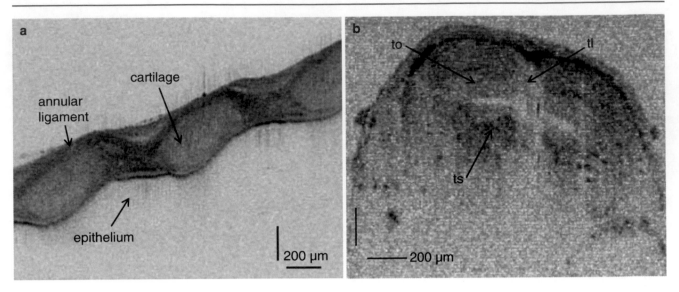

Fig. 41.10 OCT images of (**a**) rat airway and cartilage and (**b**) brain of swimming young medaka fish

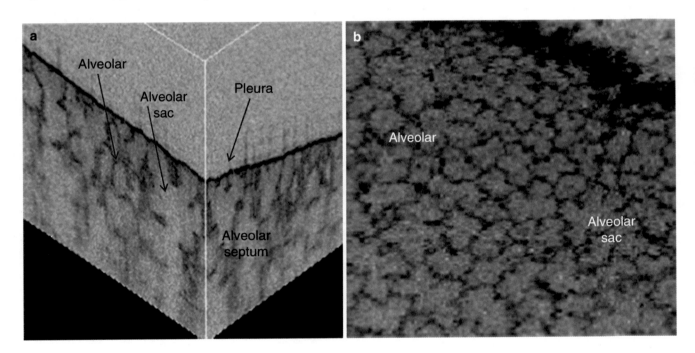

Fig. 41.11 OCT images of rt. lung tissue. (**a**) 3D image and (**b**) *en-face* image

0.8 μm ultrahigh-resolution OCT system. As shown in the OCT images, our ultrahigh-resolution OCT systems clearly revealed the layer structures of the human retina and cross-sectional structures of rat airway and cartilage. We also applied the 0.8 μm SC light for the SD-OCT system shown in Fig. 41.6 and confirmed that the 0.8 μm ultrahigh-resolution SD-OCT system allows us to perform real-time observations of the brain structures of swimming medaka fish.

Figure 41.11 shows the OCT images of rat lung tissue. In this measurement, because the difference of the refractive indices between air and the alveolar membrane is large and significantly reduced the penetration depth of OCT imaging,

phosphate-buffered saline was instilled into the lung to reduce the difference of the refractive indices. In the result, as shown in Fig. 41.11, alveoli, alveolar sac, alveolar membrane, etc., in the deep part of rat lung tissue were successfully visualized with the high axial resolution.

In Fig. 41.12, OCT images of normal rat lung and pulmonary disease (COPD) rat lung are shown. The 3D images were constructed by using the 3D display/analysis software NewVES, which is developed in Mori's group in Nagoya University. As shown in the 3D images, the volume of alveoli became larger in rat lungs with COPD than those in normal rat lungs. From these 3D data, the volume of alveoli was also successfully estimated.

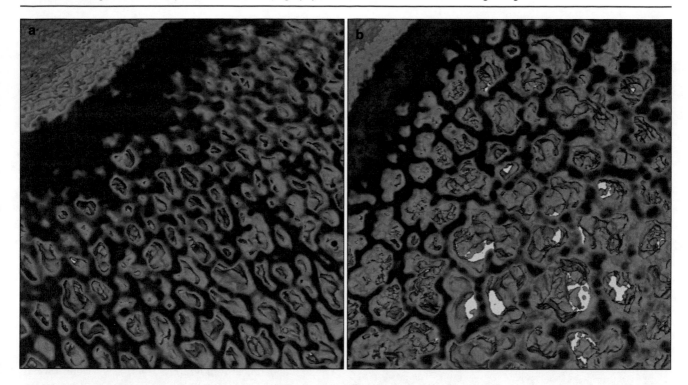

Fig. 41.12 3D OCT images of (**a**) normal rat lung and (**b**) pulmonary disease (COPD) rat lung

41.4   Ultrahigh-Resolution OCT in the 1.7 μm Wavelength Window

In tissue samples, light absorption and scattering coefficients strongly depend on an incident light wavelength because there are various substances, such as hemoglobin, lipid, water, etc. As shown in Fig. 41.13, hemoglobin strongly absorbs light in the shorter wavelength region, and water strongly absorbs light in the longer wavelength region. So far, for bioimaging, light sources in the 0.8–1.3 μm wavelength window have been widely utilized. For retinal imaging, because an eyeball in front of a retina is so thick and contains a huge amount of water, the wavelength window of 0.8–1.06 μm is often used. For imaging of turbid scattering tissues, such as skin and brain, 1.3 μm wavelength window is commonly chosen to reduce the attenuation effect by light scattering.

One of the challenges in OCT developments is the improvement of the penetration depth. Currently, the 1.7 μm wavelength window is considered as a promising choice to enhance the penetration depth because the light scattering coefficient becomes smaller, and there is a local minimum of light absorption coefficient by water in the wavelength window. This wavelength window is now called the third optical tissue window (first: 0.65 to 0.95 μm, second: 1.0 to 1.35 μm). In our group, we developed a high-power SC light source in the 1.7 μm wavelength window and realized an ultrahigh-

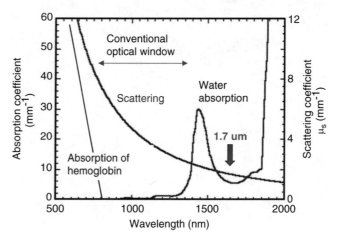

Fig. 41.13 The wavelength dependence of light absorption and scattering coefficients

resolution TD-OCT and SD-OCT system in the 1.7 μm wavelength window for the first time [3].

Figure 41.14 shows the cross-sectional and *en-face* images of mouse brains obtained with the 1.7 μm ultrahigh-resolution SD-OCT system. The studies of the attenuation length in brain samples reported that the largest penetration depth would be achieved in the 1.7 μm wavelength window. As shown in the OCT images, the hippocampus at the depth of 1.7 mm from the sample surface is clearly observed. By using this SD-OCT system, a volumetric image is achieved within a few seconds to several tens seconds.

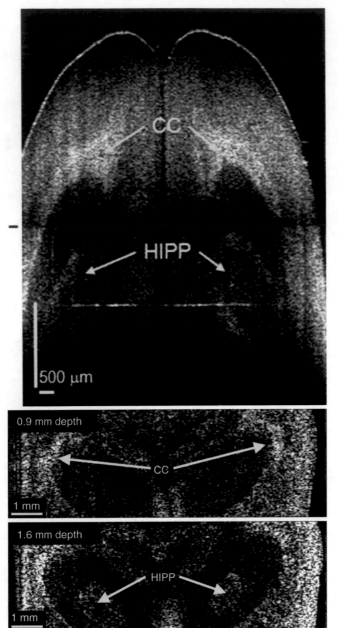

Fig. 41.14 OCT images of mouse brain obtained with the 1.7 μm ultrahigh-resolution SD-OCT system. (**a**) Cross-sectional and (**b**) *en-face* images

Recently, our group also developed optical coherence microscopy (OCM) in the 1.7 μm wavelength window. OCM is the 3D high-resolution imaging modality based on OCT and confocal microscopy with a high NA objective lens [4]. Although the depth of focus is dramatically reduced by using a high NA objective lens and confocal detection scheme, 3D high spatial resolution is achieved in OCM. While the lateral resolution of our 1.7 μm ultrahigh-resolution OCT system was around 50 μm, the lateral resolution of 3.4 μm was achieved in our 1.7 μm OCM system. The axial resolution of our 1.7 μm OCM system was 3.8 μm in tissue ($n = 1.38$). In OCM, although it is not possible to record a large 3D volume image with a single image acquisition like OCT imaging, a 3D high-resolution *en-face* image at a certain depth is achieved. As shown in Fig. 41.15, we confirmed that the 1.7 μm OCM system allows us to observe the distribution of myelin fibers (myelinated axons), which was not observable in OCT imaging.

41.5 Future Direction

Optical coherence tomography (OCT) is a non-invasive imaging technique of small internal structures in biomedical specimens and has been currently utilized for a wide variety of studies in biomedical fields. To apply OCT imaging to visualize various parts inside human bodies, fiber optic probes have been intensively developed.

Figure 41.16 shows a photograph of the fiber optic probe for intraoperative intraocular diagnosis, which was developed in collaboration with our group, a group in the medical school in Nagoya University, and a company. The diameter of the fiber optic probe is 0.6 mm. By rotating the lens equipped inside the fiber optic probe, surrounding structures around the fiber optic probe are observed. It is highly expected that further advancement of this kind of technology would open the door for the widespread applications of OCT in medical studies.

As mentioned in this chapter, OCT is useful to diagnose for non-invasive, real-time cross-sectional imaging. Our research of OCT contributes to the MCA as the highly functional future technology in medicine.

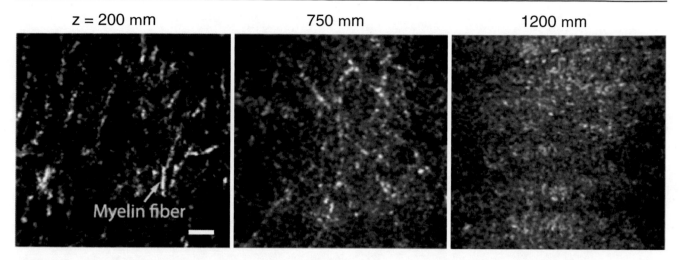

Fig. 41.15 *En-face* OCM images of mouse brain

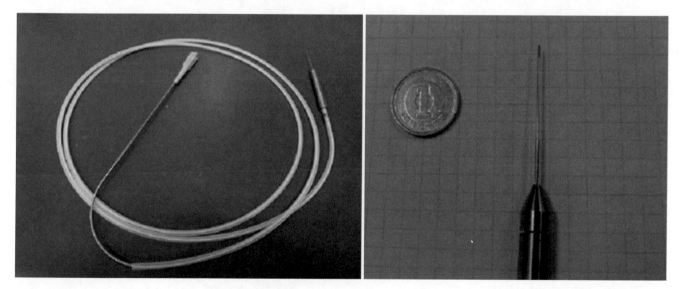

Fig. 41.16   OCT fiber optic probe for intraoperative intraocular diagnosis

Acknowledgments The experiments using rat lungs were carried out in collaboration with Professor Tsutomu Kawabe of the Graduate School of Medicine, Nagoya University, in accordance with the Animal Experimental Guides of Nagoya University Graduate School of Medicine.

References

1. Drexler W, Fujimoto JG. Optical coherence tomography, technology and applications, vol. 1–3. 2nd ed. Bern: Springer; 2015.

2. Nishizawa N. Generation and application of high quality super continuum sources. Opt Fiber Technol. 2012;18:394.
3. Kawagoe H, Ishida S, Nishizawa N, et al. Development of a high power supercontinuum source in the 1.7 um wavelength region for highly penetrative ultrahigh resolution optical coherence tomography. Biomed Opt Express. 2014;5:932–43.
4. Yamanaka M, Teranishi T, Kawagoe H, Nishizawa N. Optical coherence microscopy in 1700 nm spectral band for high-resolution label-free deep-tissue imaging. Sci Rep. 2016;6:31715.

MRI: Magnetic Resonance Q-Space Imaging Using Generating Function and Bayesian Inference

Eizou Umezawa, Yukiko Sonoda, and Itsuki Itoshiro

Abstract

Diffusional kurtosis imaging provides the kurtosis for the diffusion displacement of water molecules in vivo. Kurtosis-related metrics demonstrate high sensitivities in several disease diagnoses; however, they contain large systematic and statistical errors. The systematic error can be caused by the truncation of series expansions of the generating functions used for fitting to provide the diffusion parameters, and the cause of the statistical error is overfitting. If we increase the truncation order to reduce the systematic error, overfitting becomes increasingly severe. Hence, a Bayesian approach is developed, in which the arbitrariness regarding the determination of the posterior distributions or the regularization parameter is excluded as much as possible. The Bayesian approach effectively prevents overfitting and reduces imaging noise in kurtosis maps, thereby enabling the use of higher-order terms in the fitting. Our simulation shows that the use of higher-order terms reduced truncation-derived systematic errors in kurtosis estimation.

Keywords

Diffusion MRI · Kurtosis · Bayesian inference

E. Umezawa (✉)
School of Medical Sciences, Fujita Health University, Toyoake, Aichi, Japan
e-mail: umezawa@fujita-hu.ac.jp

Y. Sonoda
Department of Radiology, Tokai Memorial Hospital, Kasugai, Aichi, Japan

I. Itoshiro
Central Block of Radiology, Gifu Prefectural General Medical Center, Gifu, Japan

42.1 Introduction

Q-space imaging (QSI) is a diffusion magnetic resonance imaging (MRI) that uses the statistical properties of the thermal motion of water molecules [1]. Diffusional kurtosis imaging (DKI), which is a light version of QSI, is used to obtain the kurtosis of diffusion displacement and can be implemented using few MRI data [2–4]. Since the kurtosis varies depending on the microstructure of tissues, DKI is a promising diagnostic and biofunctional imaging tool [5–7]; however, the kurtosis map tends to be noisier than the conventional diffusion coefficient map. Furthermore, the estimated kurtosis contains considerable systematic errors.

The causes of these errors are as follows. The magnetic resonance (MR) signal of QSI corresponds to the characteristic function of the diffusion displacement, and its logarithm is the cumulant generating function, which are generating functions for moments [3] and cumulants [2], respectively. These functions are expressed as a series of b-values and used as fitting functions for the measured MR data. The simplest diffusion MRI analysis uses only the first-order terms of this series, whereas the conventional DKI uses up to the second-order terms. The statistical error is caused by overfitting owing to the increment of fitting parameters, resulting in a noisy DKI map. Meanwhile, although the conventional DKI analysis is the next-to-leading-order approximation, truncation after the second order still causes systematic errors in kurtosis estimations.

A simple solution to reduce systematic errors is to use higher-order terms of the generating functions; however, the further increment in model parameters worsens the overfitting problem. Therefore, we developed a Bayesian approach

M. Hashizume (ed.), *Multidisciplinary Computational Anatomy*, https://doi.org/10.1007/978-981-16-4325-5_42

[8, 9] as an accurate and robust method to measure the descriptive statistics of the diffusion displacement.

In developing the Bayesian approach, we excluded the arbitrariness related to the determination of posterior distributions or the regularization parameter as much as possible. This approach can be widely applied to this arbitrariness problem that exists in optimization tasks in general, and reconciles the trade-off between fitting performance and generalization ability.

This study aims to construct a method for obtaining microanatomical information from MR signals and to understand the mathematical properties in the truncation approximation of the cumulant expansion for model fitting. These are related to one of the objectives of MCA, which is to realize a new imaging method and to establish a mathematical analysis basis for it.

42.2 Theory

42.2.1 DKI Algorithm

The logarithm of the normalized QSI signal intensity is expressed as [4]

$$\log \frac{S(b)}{S(0)} = -b n_i n_j D_{ij} + \frac{1}{6} b^2 n_i n_j n_k n_l \bar{D}^2 W_{ijkl} + o\left(b^3\right), \quad (42.1)$$

where b is the b-value, n_i is the i-th component of a unit direction vector $\mathbf{n}$ oriented in the diffusion gradient direction, and the Einstein notation is used: a summation over repeated indices is implied. The diffusion and kurtosis tensors are defined, respectively, as follows:

$$D_{ij} = \frac{E\left(R_i R_j\right)}{2 t_{\mathrm{d}}}, \quad (42.2)$$

$$W_{ijkl} = \frac{1}{\left(2 \bar{D} t_{\mathrm{d}}\right)^2} \left\{ E\left(R_i R_j R_k R_l\right) - E\left(R_i R_j\right) E\left(R_k R_l\right) - E\left(R_i R_k\right) E\left(R_j R_l\right) - E\left(R_i R_l\right) E\left(R_j R_k\right) \right\}, \quad (42.3)$$

where R_i is the i-th component of the diffusion displacement vector, $E[\cdot]$ is the expectation value, $\bar{D} = \frac{1}{3} D_{ii}$, and $t_{\mathrm{d}} = \Delta - \delta/3$, with Δ/δ being the separation/duration of the diffusion gradient pulses. For an isotropic diffusion, $W_{iiii} = \frac{E\left(R_i^4\right)}{E\left(R_i^2\right)} - 3$. The number of independent components of D_{ij} and W_{ijkl} was 6 and 15, respectively; hence, 21 measurement points in the q-space other than the origin ($b = 0$) with 15 diffusion gradient directions were required at the least. For the full estimation of tensors appearing in higher order terms than the second, a larger number of diffusion gradient directions are required. To avoid this increment in the diffusion direction, we adapted the method proposed in [10, 11], in which the following rotational invariant, known as the mean kurtosis, was estimated:

$$\bar{W} \equiv \frac{1}{4\pi} \int_{S_2} dn\, W(\mathbf{n}) = \frac{1}{5} W_{iijj}, \quad (42.4)$$

where $W(\mathbf{n}) = n_i n_j n_k n_l W_{ijkl}$. From

$$\bar{W} = \frac{1}{15} \left\{ \sum_{i=1}^{3} W\left(n^{(i)}\right) + 2 \sum_{i=1}^{3} W\left(n^{(i+)}\right) + 2 \sum_{i=1}^{3} W\left(n^{(i-)}\right) \right\}, \quad (42.5)$$

we have

$$A(b) = -b\bar{D} + \frac{1}{6} b^2 \bar{D}^2 \bar{W} + o\left(b^3\right), \quad (42.6)$$

where

$$A(b) \equiv \frac{1}{15} \left\{ \sum_{i=1}^{3} \log \frac{S\left(b, n^{(i)}\right)}{S(0)} + 2 \sum_{i=1}^{3} \log \frac{S\left(b, n^{(i+)}\right)}{S(0)} + 2 \sum_{i=1}^{3} \log \frac{S\left(b, n^{(i-)}\right)}{S(0)} \right\} \quad (42.7)$$

is the quantity composed of the measured signals, and $n^{(1)} = (1,0,0)^T, n^{(2)} = (0,1,0)^T, n^{(3)} = (0,0,1)^T$, $n^{(1+)} = (0,1,1)^T/\sqrt{2}$, $n^{(2+)} = (1,0,1)^T/\sqrt{2}$, $n^{(3+)} = (1,1,0)^T/\sqrt{2}$, $n^{(1-)} = (0,1,1)^T/\sqrt{2}$, $n^{(2-)} = (1,0,-1)^T/\sqrt{2}$, and $n^{(3-)} = (1,-1,0)^T/\sqrt{2}$. Using Eq. (42.6), we can obtain $\bar{D}$ and $\bar{W}$ using the nine different diffusion gradient directions, even if higher-order b terms were used.

42.2.2 Maximum Likelihood (ML) Estimation

The Gaussian likelihood function is expressed as

$$p\left(\vec{A}|\vec{X}\right) \propto \exp\left\{-\frac{1}{2}\left(B\vec{X}-\vec{A}\right)^T \Sigma^{-1}\left(B\vec{X}-\vec{A}\right)\right\}, \quad (42.8)$$

where $B\vec{X}-\vec{A}$ is the residual vector composed of the statistical errors in the measured signals, and Σ is the variance–covariance matrix, which will be defined later. The vector $\vec{A}$ is

$$\vec{A} = \left[A\left(b_1\right), A\left(b_2\right), \cdots, A\left(b_M\right)\right]^T, \quad (42.9)$$

where M is the total number of non-zero b-values. The vector $\vec{X}$ is

$$\vec{X} = \left[\bar{D}, \bar{V}, c_3, c_4, \cdots, c_{N_t}\right]^T, \quad (42.10)$$

where $\bar{V} = \bar{D}^2 \bar{W}$, c_n is the coefficient parameter of term b^n on the right-hand side of Eq. (42.6), and N_t is the truncation order: higher-order terms than N_t in Eq. (42.1) and Eq. (42.6) are truncated. Matrix B is

$$B = \begin{bmatrix} -b_1, & \frac{1}{6}b_1^2, & b_1^3, & b_1^4, & \cdots, & b_1^{N_t} \\ -b_2, & \frac{1}{6}b_2^2, & b_2^3, & b_2^4, & \cdots, & b_2^{N_t} \\ \vdots & & & & \\ -b_M, & \frac{1}{6}b_M^2, & b_M^3, & b_m^4, & \cdots, & b_M^{N_t} \end{bmatrix}. \quad (42.11)$$

The variance–covariance matrix is the following diagonal matrix:

$$\Sigma = \mathrm{diag}\left[\sigma_1^2, \sigma_2^2, \cdots, \sigma_M^2\right], \quad (42.12)$$

where

$$\sigma_m^2 = \delta A\left(b_m\right)^2$$

$$= \frac{1}{9}\left[1 + \sum_{i=1}^{3}\left\{\frac{1}{5}\frac{S(0)}{S\left(b_m, n^{(i)}\right)}\right\}^2 + \sum_{i=1}^{3}\left\{\frac{2}{5}\frac{S(0)}{S(b_m, n^{(i+)})}\right\}^2 + \sum_{i=1}^{3}\left\{\frac{2}{5}\frac{S(0)}{S\left(b_m, n^{(i-)}\right)}\right\}^2\right]\left\{\frac{\delta S(0)}{S(0)}\right\}^2, \quad (42.13)$$

assuming $\delta S(b, n) = \delta S(0)$. For the signal values in Eq. (42.13), we used the measured values. Maximizing Eq. (42.8) is equivalent to minimizing

$$\left(B\vec{X}-\vec{A}\right)^T \Sigma^{-1}\left(B\vec{X}-\vec{A}\right). \quad (42.14)$$

The solution to the minimization problem is [9]

$$\hat{\vec{X}}_0 = \left(B^T \Sigma^{-1} B\right)^{-1} B^T \Sigma^{-1} \vec{A} = \left(B^T \tilde{\Sigma}^{-1} B\right)^{-1} B^T \tilde{\Sigma}^{-1} \vec{A}, \quad (42.15)$$

where

$$\tilde{\Sigma} = \left\{\frac{\delta S(0)}{S(0)}\right\}^{-2} \Sigma. \quad (42.16)$$

It is noteworthy that $\tilde{\Sigma}$ is independent of the factor $\delta S(0)/S(0)$, and an ML estimation is not required to obtain the value.

42.2.3 Bayesian Estimation

Bayes' theorem gives the posterior distribution as

$$p\left(\vec{X}|\vec{A}\right) \propto p\left(\vec{A}|\vec{X}\right)p\left(\vec{X}\right). \quad (42.17)$$

We assume that the prior distribution $p\left(\vec{X}\right)$ is Gaussian:

$$p\left(\vec{X}\right) \propto \exp\left\{-\frac{1}{2}\left(\vec{X}-\langle\vec{X}\rangle\right)^T \Sigma_p^{-1}\left(\vec{X}-\langle\vec{X}\rangle\right)\right\}. \quad (42.18)$$

The method used to determine $\langle\vec{X}\rangle$ and Σ_p will be explained in the next subsection. The posterior distribution becomes

$$p\left(\vec{X}|\vec{A}\right) \propto \exp\left\{-\frac{1}{2}f\left(\vec{X}\right)\right\}, \quad (42.19)$$

where

$$f\left(\vec{X}\right) = \left(B\vec{X}-\vec{A}\right)^T \Sigma^{-1}\left(B\vec{X}-\vec{A}\right) + \left(\vec{X}-\langle\vec{X}\rangle\right)^T \Sigma_p^{-1}\left(\vec{X}-\langle\vec{X}\rangle\right). \quad (42.20)$$

We used the expectation value of $\vec{X}$ as the estimate, which is the same as the solution of the minimization problem of $f\left(\vec{X}\right)$ expressed in Eq. (42.20). $f\left(\vec{X}\right)$ can be considered as the sum of the objective function defined in Eq. (42.14) and regularization terms. The solution to the minimization problem is [9]

$$\hat{\vec{X}} = \left(B^T \Sigma^{-1} B + \Sigma_p^{-1}\right)^{-1}\left(B^T \Sigma^{-1} \vec{A} - \Sigma_p^{-1}\langle\vec{X}\rangle\right)$$

$$= \left[B^T \tilde{\Sigma}^{-1} B + \left\{\frac{\delta S(0)}{S(0)}\right\}^2 \Sigma_p^{-1}\right]^{-1}\left[B^T \tilde{\Sigma}^{-1}\vec{A} - \left\{\frac{\delta(0)}{S(0)}\right\}^2 \Sigma_p^{-1}\langle\vec{X}\rangle\right]. \quad (42.21)$$

In contrast to ML estimation, we must determine the factor $\{\delta S(0)/S(0)\}^2$ to calculate $\widehat{\vec{X}}$. This factor corresponds to a quantity proportional to the hyperparameter of the regularized least squares method (LSM), which is a multiplication factor of the regularization term and modulates the degree of the effect. In fact, when we factor out $\{\delta S(0)/S(0)\}^{-2}$ on the right-hand side of Eq. (42.20), the first term becomes independent of the factor, and the second term is multiplied by $\{\delta S(0)/S(0)\}^2$. Whereas the hyperparameter is artificially selected in many cases of the regularized LSM, the parameter is determined by the signal-to-noise ratio (SNR) at $b = 0$ ($S(0)/\delta S(0)$) in the Bayesian approach. The method to obtain $\{\delta S(0)/S(0)\}^2$ is described in the next subsection.

42.2.4 Statistic Signal Errors and Prior Distribution

To obtain the values of $\langle \vec{X} \rangle$, Σ_p, and $\delta S(0)/S(0)$ in Eq. (42.21), we first implemented ML estimation, i.e., we obtained $\widehat{\vec{X}}_0$ using Eq. (42.15) for all voxels in a slice, except those indicating subnoise signals and have the set $\{\widehat{\vec{X}}_0\}$. We assign the median and the normalized interquartile range (NIQR) of $\{\widehat{\vec{X}}_0\}$ to $\langle \vec{X} \rangle$ and Σ_p, respectively.

The factor $\{\delta S(0)/S(0)\}^2$ is obtained by [9]

$$\left\{ \frac{\delta S(0)}{S(0)} \right\}^2 = \frac{E\left[\left(B\widehat{\vec{X}}_0 - \vec{A} \right)^T \dot{\Sigma}^{-1} \left(B\widehat{\vec{X}}_0 - \vec{A} \right) \right]}{M - N}, \quad (42.22)$$

where $\widehat{\vec{X}}_0$ is defined by Eq. (42.15). The expectation value in Eq. (42.22) was estimated as the median for a slice.

42.2.5 Updating of Prior Distribution and Successive Estimation

After one Bayesian estimation, we have a set of $\widehat{\vec{X}}$, from which a new prior distribution of $\vec{X}$ can be obtained. We successively estimated $\vec{X}$ using the same procedure as in Sects. 42.2.3 and 42.2.4 with the updated prior distribution. The Bayesian update was performed until all prior distributions remained unchanged.

42.3 Method

42.3.1 Human Data Study

The MRI data of a normal human brain available at https://doi.org/10.5061/dryad.9bc43 [12, 13] were used to test the proposed method. The data acquisition apparatus and the conditions were as follows [12]: A Siemens Trio 3 T equipped with a 32-channel head coil and a double-spin echo diffusion weighted (DW) echo planar imaging sequence. DW image data were recorded at $h = 0$ and $b = 200$–3000 mm/s² in steps of 200 mm/s² along 33 DW directions. Cerebrospinal fluid suppression (inversion recovery) was performed. $TR/TE/TI = 7200/116/2100$ ms, and 19 consecutive slices were acquired with isotropic resolution, 2.5 mm; matrix size, 96 × 96; phase encoding direction, A–P; SNR at $b = 0$, 39. We used the DW image data of b-values increasing from 200 to 2400 s/mm² in increments of 200 s/mm² in nine DW directions of $n^{(i)}$, $n^{(i+)}$, and $n^{(i-)}$ ($i = 1,2,3$) with one $b = 0$ image. The mean kurtosis maps were produced for truncation orders $N_t = 2$–6.

42.3.2 Numerical Phantom Study

We created several numerical phantoms corresponding to certain brain tissues by assuming multicomponent models comprising free and restricted diffusion signals [14–16]. Herein, we only present the results for white matter (WM). The simulated signal is expressed as

$$S_{WM}(q,n) = S_0 \left\{ f S_{cyl}(q,n) + (1-f) S_{free}(q) \right\}, \quad (42.23)$$

where

$$S_{free}(q) = \exp\left[-q^2 t_d D \right] \quad (42.24)$$

and

$$S_{cyl}(q,n) = S_{free}(q n_z) \Psi\left(q\sqrt{n_x^2 + n_y^2} \right) \quad (42.25)$$

with

$$\Psi(q) = \exp\left[-\frac{7R_1^4}{296 D_{in}} \left(\frac{q}{\delta} \right)^2 \left(TE - \frac{99 R_1^2}{112 D_{in}} \right) \right], \quad (42.26)$$

and $q = \sqrt{b/t_d}$. This is the signal model for a microstructure in which cylindrical fibers are arranged along the z-axis. f, R_1, D_{in}, and D are adjusted such that the mean kurtosis of the WM obtained from the human data study can be reproduced using natural values and S_0 is arbitrary. The region of interest (ROI) for the WM is shown in Fig. 42.2b. Rician random noise (SNR = 39) was added to the simulated signal. The sample size was 10^4.

42.4 Results

Figure 42.1 shows the mean kurtosis maps of the normal human brain. When Bayesian estimation was not applied, the maps obtained using higher-order terms ($N_t > 2$) became extremely noisy. When Bayesian estimation was applied, the noise has been improved on all maps, especially for $N_t > 2$.

Fig. 42.1 Mean kurtosis ($\bar{W}$) maps. N_t denotes the truncation order: the series of Eqs. (42.1) and (42.6) are truncated after the N_tth order. N_r denotes the number of Bayesian estimations performed. "$N_r = 0$" indicates the results of non-Bayesian ML estimation, and "final" indicates the result at the end of the Bayesian update iteration

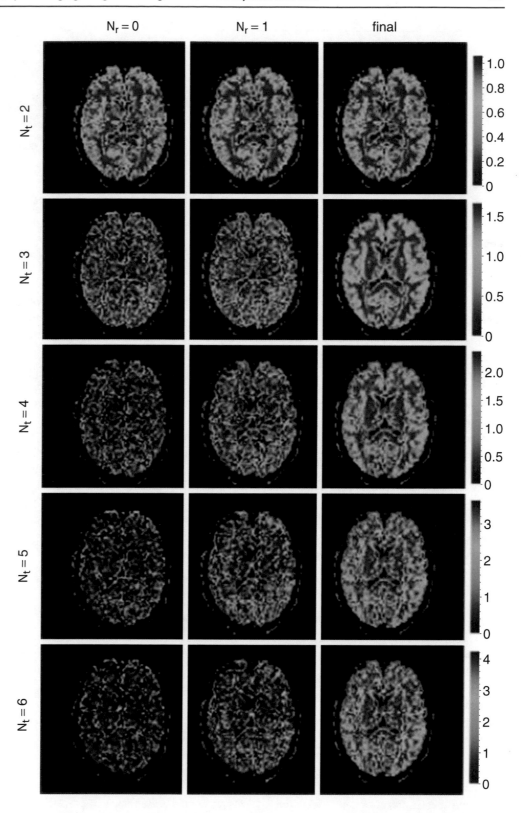

Furthermore, the figure shows that the estimated mean kurtosis increased with the truncation order (see the legend bar of each map).

Figure 42.2 shows the truncation order dependence of the mean kurtosis estimated for the numerical phantom and nor-mal brain WM. The plot and error bars denote the median and NIQR, respectively. The numerical phantom study results were obtained for the following model parameters: $f = 0.815$, $R_1 = 4.50$ µm, $D_{in} = 1.5 \times 10^{-3}$ mm²/s, and $D = 3.30 \times 10^{-3}$ mm²/s. These results approximate the

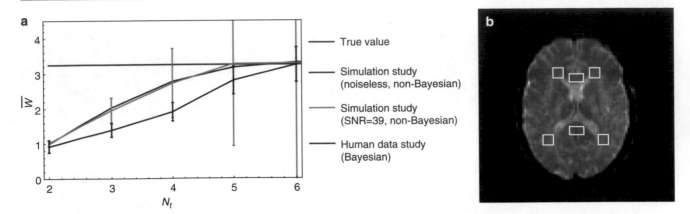

Fig. 42.2 Estimated mean kurtosis for numerical phantom and normal brain WM (**a**). Results of normal brain WM are the medians of ROI denoted in (**b**). The true value is theoretically calculated for the signal model of the numerical phantom. The plot and error bars denote the median and normalized interquartile range, respectively

results of the human data study, and the kurtoses approached the true value as the truncation order increased.

42.5 Discussion

When the truncation order was large, the mean kurtosis maps became extremely noisy unless Bayesian estimation was applied. The noise was generated from overfitting caused by the increase in the fitting parameters. The Bayesian approach effectively suppressed this overfitting and hence enabled the use of a large truncation order.

The medians of the Bayesian mean kurtosis maps were similar to those obtained by the non-Bayesian estimation (not shown). This implies that although the Bayesian approach improved the precision, it did not affect the accuracy.

Increasing the truncation order was expected to decrease the truncation error. We confirmed that the estimated kurtosis approached the true value as the truncation order increased by using a numerical phantom that reproduces the results of the human data study. Although the numerical phantom that reproduces the estimates of the human data study is not unique, this result suggests that the use of higher-order terms improved the accuracy.

In the proposed method, the repetition of the Bayesian update was terminated when all prior distributions remained unchanged. The procedure excluded the arbitrariness of the repetition number in Bayesian kurtosis estimation.

When the Bayesian update completed, the variances of the prior distributions for coefficients of higher order terms than the second become zero, whereas those of the first- and second-order term coefficients remained non-zero. This means that the higher-order term coefficients were fixed to certain constants in the fittings. In the conventional method ($N_t = 2$), the coefficients of all higher-order terms were fixed to zero. Thus, the proposed method can be said to generalize this zero fixing to non-zero fixing. Although the fixed coefficients for the higher-order terms may be unrealistic, the non-zero fixing was a better approximation than the conventional uniform zero fixing, and onsequently produced more accurate estimates.

An interesting phenomenon was obtained in the noiseless simulation of the higher-order coefficient estimations for $N_t = 3$–30: when $N_t > 6$, the estimated higher-order coefficients fluctuated with the truncation order, whereas the estimated first- and second-order term coefficients (i.e., the diffusivity and kurtosis) maintained their true values (not shown). A mathematical understanding of this phenomenon should be endeavored.

42.6 Conclusions

Bayesian estimation reduces the overfitting-derived statistical error in the mean kurtosis estimation, thereby enabling the use of large truncation order for the kurtosis estimation. The increasing truncation order decreases the truncation-derived systematic error. Hence, DKI analysis using higher-order cumulant terms with Bayesian updating provides a robust and accurate kurtosis estimation.

Acknowledgments We are grateful to Masutani Yoshitaka, Ph.D., of Hiroshima City University; Masayuki Yamada, Ph.D.; Kazuhiro Murayama, MD; Masato Abe, MD, of Fujita Health University; and Takayuki Enari, Ph.D., of Nihon University for their fruitful discussions and comments. This study was supported by JSPS KAKENHI Grant Number JP17H05307.

References

1. Callaghan PT. Principles of nuclear magnetic resonance microscopy. Oxford: Oxford University Press; 1993.
2. Jensen JH, Helpern JA, Ramani A, Lu H, Kaczynski K. Diffusional kurtosis imaging: the quantification of non-Gaussian water diffusion by means of magnetic resonance imaging. Magn Reson Med. 2005;53(6):1432–40.
3. Umezawa E, Yoshikawa M, Yamaguchi K, Ueoku S, Tanaka E. q-Space imaging using small magnetic field gradient. Magn Reson Med Sci. 2006;5(4):179–89.
4. Tabesh A, Jensen JH, Ardekani BA, Helpern JA. Estimation of tensors and tensor-derived measures in diffusional kurtosis imaging. Magn Reson Med. 2011;65(3):823–36.
5. Cauter V, Veraart J, Sijbers J, et al. Gliomas: diffusion kurtosis MR imaging in grading. Radiology. 2012;263(2):492–501.
6. Maier SE, Sun Y, Mulkern RV. Diffusion imaging of brain tumors. NMR Biomed. 2010;23(7):849–64.
7. Umezawa E, Kawasaki M, Sonoda Y, Fukuba T, Murayama K, Takano K, Yamada M, Onodera T, Ida M. A method to estimate the product of perfusion fraction f and pseudodiffusion coefficient Dp of IVIM without estimating f and Dp. Proceedings of 26th International Society for Magnetic Resonance in Medicine 2018;26:1581.
8. Orton MR, Collins DJ, Koh DM. Improved intravoxel incoherent motion analysis of diffusion weighted imaging by data driven Bayesian modeling. Magn Reson Med. 2014;71(1):411–20.
9. Umezawa E, Ishihara D, Kato R. A Bayesian approach to diffusional kurtosis imaging. Magn Reson Med. 2021;86(2):1110–24.
10. Hansen B, Lund TE, Sangill R, Jespersen SN. Experimentally and computationally fast method for estimation of a mean kurtosis. Magn Reson Med. 2013;69(6):1754–60.
11. Hansen B, Lund TE, Sangill R, Stubbe E, Finsterbusch J, Jespersen SN. Experimental considerations for fast kurtosis imaging. Magn Reson Med. 2016;76(5):1455–68.
12. Hansen B, Jespersen SN. Data for evaluation of fast kurtosis strategies, b-value optimization and exploration of diffusion MRI contrast. Sci Data. 2016;3:160072. https://doi.org/10.1038/sdata.2016.72.
13. Hansen B, Jespersen SN. Data from: Data for evaluation of fast kurtosis imaging, b-value optimization and exploration of diffusion MRI contrast. Dryad Digital Repository. 2016; https://doi.org/10.5061/dryad.9bc43
14. Neuman CH. Spin echo of spins diffusing in a bounded medium. J Chem Phys. 1974;60:4508–11.
15. Advances in magnetic resonance, Vol. 12. Edited by Waugh JS. Academic Press, London, 1988.
16. Assaf Y, Freidlin RZ, Rohde GK, Basser PJ. New modeling and experimental framework to characterize hindered and restricted water diffusion in brain white matter. Magn Reson Med. 2004;52:965–78.

Micro-CT and Lungs

43

Shota Nakamura

Abstract

Micro-computed tomography (μCT) provides extremely high-resolution images of samples as a non-destructive inspection tool, and then we can obtain images comparable with microscopic images. We have attempted to take high-resolution images of the human lung using μCT. In future, if histopathological diagnoses of pulmonary nodules can be obtained by μCT with the living body, there will be no need for lung cancer patients to be examined for bronchoscopic or surgical lung biopsy preoperatively. We compared μCT images with microscopic images. Resected human lungs were fixed by the Heitzman methods, and after then were taken by μCT. Those images gained by conventional HRCT and μCT were compared in details. Alveolar ducts and pulmonary alveoli could be identified on the μCT images. The resolution of these images was comparable with that of images gained by × 40 magnification on HE-stained samples. On μCT images, ground glass opacity areas on HRCT were seen as thick walls of alveoli, and those areas were corresponded with lepidic growth pattern microscopically. μCT images well divided the resected lung into normal lung area and tumor area. This article provides an overview of our study findings regarding the μCT and lungs.

Keywords

Micro-CT · Pulmonary resection · Histopathological diagnosis · Lung cancer · High-resolution CT (HRCT) Lepidic growth pattern · Ground Glass Opacity (GGO)

S. Nakamura (✉)
Department of Thoracic Surgery, Nagoya University Graduate School of Medicine, Nagoya, Japan
e-mail: shota197065@med.nagoya-u.ac.jp

43.1 Micro-CT and Lungs

In recent years, Micro-computed tomography (μCT) has been developed, and it provides high-resolution images in μ-meter order of small samples and can be employed as a non-destructive inspection tool. It also enables those μ-scale images in sectional or three dimensional. In the past, prepared slides for histological diagnosis were made by laboratory technicians spent many hours and put much effort. We believe it is possible to make a more accurate, fast, and three-dimensional histological diagnosis of lung diseases using this new imaging modality. In other words, establishing this histological diagnostic technique allows the preparation process of making tissue specimens to be substantially omitted. Although pleural or vascular invasion could only be observed in fragments up until now, this technology enables us to evaluate them throughout the entire tumor in addition to potentially understanding invasion of the tumor and measuring the tumor diameter in three dimensions. If the μCT can show images of a living body in the future, conventional invasive examination may be omitted such as a bronchoscopic biopsy and open lung biopsy under general anesthesia for histopathological diagnoses prior to treatments. To achieve these goals, we have studied to gain clear images by μCT and analyzed images of excised lungs. This article provides an overview of our study findings regarding the μCT and lungs.

43.2 Motivation and Findings of Study

We have investigated the relationship between CT imaging and histological findings in lung cancer with great interest [1]. As our study collaborator, the group of Prof. Kensaku Mori et al. at the Faculty of Information of Nagoya University has been working on studies involving Multidisciplinary Computational Anatomy [2]. One of these includes Micro-CT. If we can image the excised lungs in our routine practice with Micro-CT and make a histological diagnosis from these images, we will be able to reduce the time spent on the preparation of histopathological specimens requiring a lot of time and processing. Based on this idea, we conducted a study called "Establishing Histological Diagnostic Technology using Micro-CT Imaging." One achievement so far is that we were able to obtain highly precise Micro-CT images of the excised lungs and discovered the fixation conditions and imaging conditions of the lungs. We also found that regions exhibiting a lepidic growth pattern, as a subcategory of lung adenocarcinoma, could be identified using Micro-CT imaging. In addition, we were able to three-dimensionally match the positions of high-resolution CT (HRCT) images used in clinical practice, Micro-CT images, and histopathological images [3]. Where do regions in the HRCT images match in the histological images? Regarding this question, the existence of Micro-CT between the two images can deepen the understanding of the relationship between the two. This demonstrates the current clinical importance of Micro-CT images. We will use actual images to explain this.

43.3   Precision of Images

In order to obtain precise Micro-CT images of excised lung specimens obtained from surgery, we have repeatedly improved the method of extending and fixing the specimens along with the conditions of Micro-CT imaging. As a result, we have obtained the most precise Micro-CT images ever among those which have been reported for the lungs. Method of collection of specimens, method of extension/fixation of specimens, and imaging conditions are shown below. These study activities (specimen collection/extension and fixation/ Micro-CT imaging/analysis and publication) are conducted upon receiving approval from the bioethics review committee at our facility as well as the informed consent of patients.

Collection of specimens: Among the excised specimens, the parts required for the diagnosis and treatment of the patients are used to prepare histopathological samples for histopathological diagnosis after extension/fixation using formalin. Among the other parts not associated with histopathological diagnosis, a part sized 2 cm × 2 cm × 1 cm in which the border between tumor margin and the normal lung can be observed in the center is excised and collected.

Method of extension/fixation: Based on the Heitzman method as a classical lung extension/fixation method [4], we improved the drug ratio by raising the viscosity slightly for extension/fixation. The mixing ratio of the solution for extension/fixation was 11:5:2:2 (polyethylene glycol 400: 95% ethyl alcohol: 40% formalin: water). After injecting the extension/fixation solution with a syringe into the specimen by avoiding the area most wanted to observe, pressure was applied to 30 cm H_2O and the specimen was dipped in the solution. The specimen was removed from the extension/

fixation solution after three days, wrapped gently with soft water-absorbing paper, in practice, wrapping with commercial tissue paper in multiple layers, then wrapping with KimWipes in multiple layers and sealing in a plastic container. After 3 days of dehydration, the creation of a specimen fixed for imaging was completed (Fig. 43.1).

Imaging device and conditions: The Micro-CT used for imaging is the "inspeXio SMX-90CT Plus," (Fig. 43.2) a desktop microfocus X-ray CT system made by Shimadzu Corporation. Conventional CT devices take images using the difference in the "degree of absorption" of X rays transmitting through the subject, then non-destructively imaging the internal structure of a subject after reconstruction processing. On the other hand, Micro-CT uses a microfocus X-ray device capable of focusing on an extremely small subject at the X-ray source; therefore, it is able to obtain a high-resolution image by geometrical enlargement. This allows images of the inside of the subject with a high resolution to

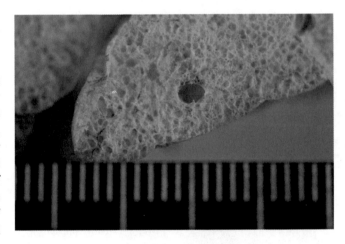

Fig. 43.1 Specimen of the resected lung after fixation

Fig. 43.2 We used this desktop microfocus X-ray CT system for taking images produced by Shimadzu CO.LTD. "inspeXio SMX-90CT Plus" http://www.an.shimadzu.co.jp/ndi/products/x_rylk/inspexio1.htm

be non-destructively obtained. The device used in this study realizes spatial resolution up to 6 μm and with high contrast, allowing it to be used in R&D, quality control, and accident analysis in a variety of fields. For medical purposes, it is mainly used for research purposes, focusing on bones. For the lungs, being an organ with a high contrast under pneumatic conditions, imaging studies using this device are conducted around the world. Regarding the imaging conditions, the noise is lowest and the resolution highest at an X-ray tube voltage of 90 kV and X-ray tube current of 110 μA. Imaging specimens are shot in a stick bin. The vibration of device may be passed to the specimens in order to blur the image or dislocate the specimens. As a measure, X-ray transmitting Styrofoam is stuffed as a fixative in the stick bin so that the specimens are not dislocated. We used a 3D printer to create a fixture having the shape of the cap of the stick bin containing the specimens. When taking pictures, the specimens were fixed with the fixture to prevent blurred images.

Micro-CT image: Terminal bronchioles and alveolar walls, as the terminal units of the airway, could be observed (Fig. 43.3). We used this image data for 3D construction and could observe the inside of the air space as desired (Fig. 43.4).

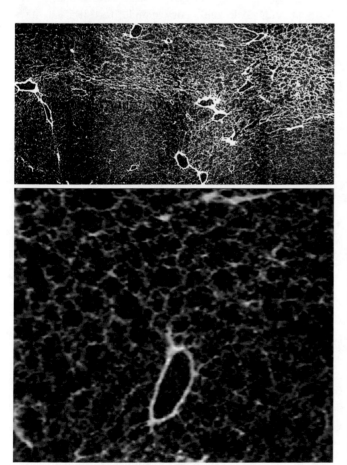

Fig. 43.3 Terminal bronchioles and alveolar walls, as the terminal units of the airway, could be observed on micro-CT images

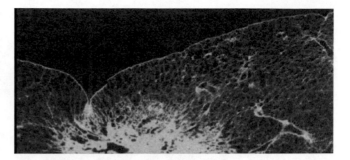

Fig. 43.4 Normal lung around the lung cancer. Pleural indentation was seen

43.4 Histological Diagnosis of Lung Cancer Using Micro-CT Imaging

The images taken by Micro-CT of lung adenocarcinoma showing Ground Glass Opacity (GGO) in HRCT images, made it visually clear those of differences between the normal lungs and the lung cancer region in thickness of alveolar wall (Fig. 43.5). Lung adenocarcinoma visualized as a GGO lesion on HRCT images is a proliferation form characteristic of a subtype of lung adenocarcinoma, histopathologically called a lepidic growth pattern. It is equivalent to the part in which the cancer cells are growing along the normal alveolar wall. With the alveolar space maintaining aeration, the alveolar wall became thicker than the normal alveolar wall. This is visualized as GGO lesions in HRCT images. The HRCT images cannot visualize the thickness of the alveolar wall with any degree of certainty. We are only able to recognize regions with slightly reduced permeability called Ground Glass Opacity. The HRCT image can visualize the normal alveolar wall thinly and the alveolar wall equivalent to the GGO lesions thickly, clearly differentiating the division of the two. In order to demonstrate this numerically, we compared the "thickness" of the normal alveolar wall and the thick alveolar wall equivalent to GGO lesions. For the alveolar wall thickness, we took measurements at 10 locations each for normal alveolar walls and the alveolar wall equivalent to GGO lesions. The average value was defined as the alveolar wall thickness. A total of 10 specimens were compared using the student t-test. The WHO classification of lung adenocarcinoma is shown as follows: three cases of non-invasive adenocarcinoma (one case of adenocarcinoma in situ, two cases of minimally invasive adenocarcinoma); and seven cases of invasive adenocarcinoma (three cases of papillary predominant, four cases of lepidic predominant). Upon measurement, the median of the normal alveolar wall thickness was found to be 0.039 (0.025–0.060) mm, while the median of the thickness of alveolar wall equivalent to GGO lesions was 0.088 (0.069–0.102) mm. The thickness of the alveolar wall was significantly different ($p < 0.001$). In

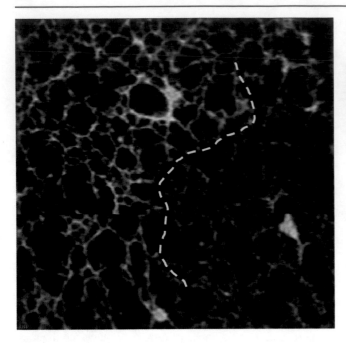

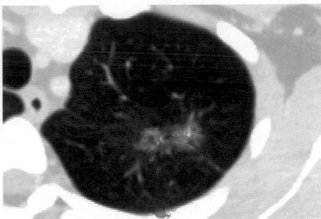

Fig. 43.6 We were able to match the anatomical positions in Micro-CT images, histopathological images, and the HRCT images used in clinical practice

Fig. 43.5 We can visually recognize the difference in thickness of the alveolar wall between the normal lungs and the lung cancer region

addition, matching of the positions in HRCT images/ Micro-CT images/histopathological images demonstrated that regions with thick alveolar walls in the Micro-CT images match those histopathologically exhibiting the lepidic growth pattern [5]. Based on these results, the regions visualized as GGO components in the HRCT images were visualized as alveolar walls with thickened stroma. This scientifically clearly differentiated between the division of the normal alveolar wall and the alveolar wall as the GGO component. In other words, the Micro-CT can differentiate the regions histopathologically demonstrated as cancer and the normal regions, as the first step in the histological diagnosis of cancer by Micro-CT.

43.5 Micro-CT Playing a Role in Filling the Gap Between HRCT Images and Histopathological Images

We were able to match the anatomical positions in Micro-CT images, histopathological images, and the HRCT images used in clinical practice [6] (Fig. 43.6). Based on this, the regions recognized as lung cancer in Micro-CT images can be histopathologically diagnosed as lung cancer and it can be confirmed whether the two diagnoses were consistent. In fact, Micro-CT imaging enabled the diagnosis of a lepidic growth pattern, as an advanced form of lung adenocarcinoma. "How regions of lesions recognized on histopathological images are visualized in HRCT images" could be more accurately grasped by matching the positions of the

three images using Micro-CT images as an intermediate between the two. Not only does this mean that if Micro-CT becomes capable of imaging living bodies in the future, it would be a very useful medical device, but also that Micro-CT plays a role in filling the gap between HRCT images and histopathological images. For example, for HRCT images and histopathological findings for diagnosis of interstitial pneumonia, the diagnosis method and sites are not related to each other. Using Micro-CT images as an intermediate between them may contribute to the pathological clarification and diagnosis of interstitial pneumonia.

43.5.1 Future of Micro-CT and the Lungs

The greatest advantage of Micro-CT is that it can nondestructively visualize the internal part of the subject in detail. When applied to lung imaging, the most important research enables imaging of living bodies and the realization of histological diagnoses without invading the human body. In other words, we believe that the major goal in the near future should be omitting conventional invasive tests, such as bronchoscopic biopsy and open lung biopsy under general anesthesia, performed for pre-treatment histological diagnosis for lung lesions. Moreover, by combining histopathological diagnosis using Micro-CT for living body imaging and operation navigation techniques, we may be able to histopathologically completely remove cancers without remnants during surgery.

Furthermore, Micro-CT images may facilitate our understanding of lung disease by filling the gap between HRCT images and histopathological images, as stated above. Among the previously reported studies of lung diseases using Micro-CT, some observed alveoli and terminal bronchioles, while others conducted image analyses on small airway diseases (cystic fibrosis and COPD) and reported them as useful in order to elucidate the pathology of these diseases [7, 8]. This may also be applied to the diagnosis of rejection responses in the early stages of interstitial pneumonia and after lung transplantation. Although the histological diagnosis of excised lungs using Micro-CT has nearly been achieved, there are some challenges. To realize the histological diagnosis of lung lesions by Micro-CT, detailed images of the inside part of the alveolar stroma and the tumor must be obtained. This requires the evolution of Micro-CT itself, improvement of the method of extension/fixation and imaging conditions, and adding contrast inside the specimen. We must also prepare for the imaging of living bodies using Micro-CT. It is necessary to establish the grading system in the order of features that are most likely to be cancer on μCT images. This requires the collection and analysis of a large amount of specimens and image data. Above all, the evolution of the Micro-CT device itself is awaited, particularly improvement of the resolution, miniaturization, and improved technology for living body imaging. We believe these will be accomplished in the near future, so we need to prepare for it under the given conditions.

43.6 Conclusion

Based on previous studies of the Micro-CT on pulmonary images, Micro-CT must contribute to the elucidation of lung diseases. Using advanced image processing techniques from the information engineering field, a large amount of data obtained from images of excised lungs by Micro-CT can directly lead to histological diagnosis and be reflected and applied to clinical practices. This area may also contribute to the accelerated development of Multidisciplinary Computational Anatomy defined by space axis, function axis, and pathology axis.

References

1. Nakamura S, Fukui T, Taniguchi T, et al. Prognostic impact of tumor size eliminating the ground glass opacity component: modified clinical T descriptors of the tumor, node, metastasis classification of lung cancer. J Thorac Oncol. 2013;8:1551–7.
2. Mori K. From macro-scale to micro-scale computational anatomy: perspective of the next 20 years. Med Image Anatomy. 2016;11:837–45.
3. Nakamura S, Mori K, Iwano S, et al. Micro-computed tomography images of lung adenocarcinoma: detection of lepidic growth patterns. Nagoya J Med Sci. 2020;82(1):25–31.
4. Itoh H, Tokunaga S, Asamoto H, et al. Radiologic-pathologic correlations of small lung nodules with special reference to peribronchiolar nodules. Am J Roentgenol. 1978;130:223–31.
5. Nakamura S, Mori K, Okasaka T, et al. Micro-computed tomography of the lung: imaging of alveolar duct and alveolus in human lung. Am J Respir Crit Care Med. 2016;193:A7411.
6. Nagara K, Oda H, Nakamura S, et al. Cascade registration of micro CT volumes taken in multiple resolutions. International conference on medical imaging and virtual reality 2016;269–280.
7. Watz H, Breithecker A, Rau WS, et al. Micro-CT of the human lung: imaging of alveoli and virtual endoscopy of an alveolar duct in a normal lung and in a lung with centrilobular emphysema-initial observations. Radiology. 2005;236:1053–8.
8. McDonough JE, Yuan R, Suzuki M, et al. Small-airway obstruction and emphysema in chronic obstructive pulmonary disease. N Engl J Med. 2011;365:1567–75.

Real-Time Endoscopic Computer Vision Technologies and Their Applications That Help Improve the Level of Autonomy of Surgical Assistant Robots

Atsushi Nishikawa and Noriyasu Iwamoto

Abstract

As AI technologies develop at a rapid rate, studies about the autonomy of surgical assistant robots are drawing increasing attention. In this chapter, we briefly introduce new endoscopic computer vision technologies (a stereo matching engine and a surgical instrument tracking system) and their applications (calculation of the distance between the tip of the instrument and the organ surface, and estimation of the load of the tip of the instrument during counter-traction against the organ) for the autonomous control of surgical assistant robots.

Keywords

Surgical assistant robots · Levels of autonomy
Stereoscopic endoscope · Instrument tracking
Counter-traction

44.1 Introduction

As AI technologies develop rapidly, studies about the autonomy of surgical assistant robots are drawing increasing attention [1]. Chinzei et al. discussed the importance of risk analysis for auto-diagnosing devices with AI technologies and AI-based surgical robots from the viewpoint of regulatory science [2]. Based on the six levels of auto-driving cars [3] defined by the Society of Automotive Engineers

A. Nishikawa (✉)
Department of Mechanical Science and Bioengineering, Graduate School of Engineering Science, Osaka University, Osaka, Japan
e-mail: atsushi@me.es.osaka-u.ac.jp

N. Iwamoto
Department of Mechanical Engineering and Robotics, Faculty of Textile Science and Technology, Shinshu University, Nagano, Japan
e-mail: iwamoto@shinshu-u.ac.jp

International, Yang et al. [4] proposed six levels of autonomy (LoAs) for surgical robots and Haidegger [1] revised them: LoA 0 (no autonomy), LoA 1 (robot assistance), LoA 2 (task-level autonomy), LoA 3 (supervised autonomy), LoA 4 (high-level autonomy), and LoA 5 (full autonomy). According to a recent study that gives an overview of currently available autonomous functions in surgical robots [1], even the da Vinci (Intuitive Surgical Inc.) [5], the most widely used surgical assistant robot, belongs to LoA 1. Therefore, LoA improvements for surgical assistant robots are becoming important areas. In this chapter, we first introduce the two fundamental real-time computer vision systems: (1) a high-performance stereo matching engine that calculates depths from any stereoscopic endoscope output and (2) a vison-based maker-less surgical instrument tracking system applicable to different endoscopic surgeries. Then, we describe their applications (calculation of the distance between the tip of the instrument and the organ surface, and estimation of the load of the tip of the instrument during counter-traction against the organ). These applications demonstrate that our approach may help improve the LoAs of surgical assistant robots.

44.2 Technology I: Calculation of Depths

Suppose that a stereoscopic endoscope observes a three-dimensional (3D) object point during surgery. Without loss of generality, we can assume a standard (parallel) stereo endoscope realized through image rectification (Fig. 44.1a). In this case, we can denote the projected positions of the 3D point onto the right and left images, and the corresponding binocular disparities can be denoted by (i,h), (j,h), and $d = j - i$ (Fig. 44.1b). Detecting the disparities d by stereo matching, we can easily calculate the depths (the distance between the endoscopic camera and the object) as bf/d, where b and f indicate the baseline and focal lengths, respectively. As has been known for more than 35 years, dynamic programming (DP) can be used to match each pixel on the

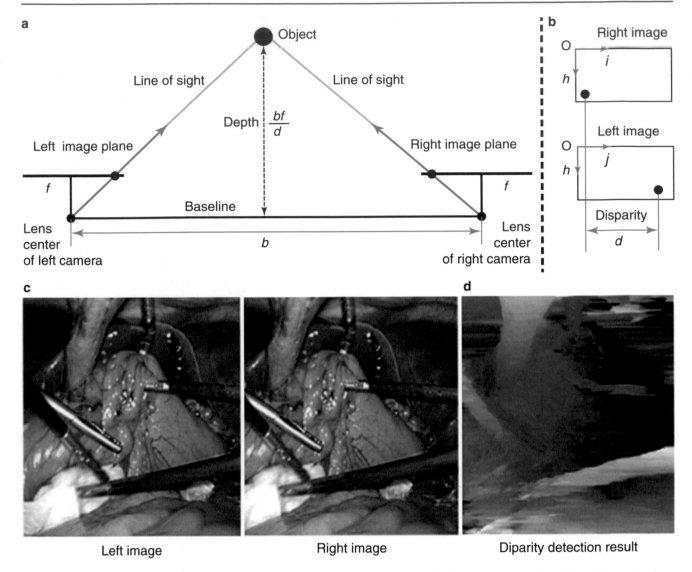

Fig. 44.1 Calculation of depths from a stereoscopic endoscope output. The standard stereo camera geometry (**a**), projections of a 3D point and disparities (**b**), example of a stereo–image pair (**c**), and the disparity detection result (**d**) (Images courtesy of Prof. Hisashi Suzuki and Dr. Hitoshi Katai)

left image with the corresponding pixel on the right image [6]. Suzuki et al. [7] developed high-performance DP stereo matching engine that controlled the trade-off between the correct-match percentage and the processing speed. This engine achieved correct-matching rates of 94.4% and 92.1% on typical stereo images in the Middlebury Stereo Datasets [8] at processing speeds of 21.0 MDE/s and 198.3 MDE/s, respectively (MDE/s denotes one million disparity estimations per second). Suzuki's group then showed that this engine was also applicable for calculating the depths of organ surfaces and surgical instruments simultaneously, from any stereoscopic endoscope output. Figure 44.1c shows a stereo–image pair during laparoscopic surgery, and Fig. 44.1d shows the disparity detection result (disparity map) of the proposed system.

44.3 Technology II: Instrument Tracking

A new visual tracking method for the tips of surgical instruments is developed in this research as an important reference for the autonomous control of endoscope-holding robots. The position of the tip of surgical instruments in endoscopy is important. In this technology, only RGB images are used; no external sensor or instrument-mounted marker is needed. Unlike the conventional vison-based, marker-less surgical instrument tracking methods [9] and recent deep-learning-based methods [10], our method is applicable to different endoscopic images, such as cholecystectomy and sigmoidectomy images, and the simultaneous detection and tracking of more than one surgical instrument are feasible. First, binarization is applied to detect the instrument region. The proposed binarization algorithm focuses on the relative relationships among the R, G, and B values of the organ surface and detects the non-organ region as the instrument region. The binarization processes for endoscopic images in cholecystectomy and sigmoidectomy are formulated as Eqs. (44.1) and (44.2), respectively.

$$p_{jh} = \begin{cases} 0 & \text{if}\left(\left(r_{jh} > g_{jh}\right) \cap \left(r_{jh} > b_{jh}\right)\right) \cup \left(\left(b_{jh} > r_{jh}\right) \cap \left(b_{jh} > g_{jh}\right)\right) \\ 1 & \text{otherwise} \left(\text{instrument region in case of cholecystectomy}\right) \end{cases}, \tag{44.1}$$

$$p_{jh} = \begin{cases} 0 & \text{if}\left(\left(r_{jh} > g_{jh}\right) \cap \left(r_{jh} > b_{jh}\right)\right) \cap \left(r_{jh} - g_{jh} \geq threshold\right) \\ 1 & \text{otherwise} \left(\text{instrument region in case of sigmoidectomy}\right) \end{cases}, \tag{44.2}$$

In Eqs. (44.1) and (44.2), r_{jh}, g_{jh}, and b_{jh} (in our case, r_{jh}, g_{jh}, $b_{jh} \in \{0, 1, 2, \cdots, 255\}$) indicate the R, G, and B values of the two-dimensional (2D) endoscopic image coordinates (j,h) (Fig. 44.1b). p_{jh} is the binarization result, and in our case, *threshold* is set to 10. After the binarization, the conventional morphological operations (dilation followed by erosion) and labeling are conducted. Then, the instrument region is narrowed down to the region that satisfies the following conditions: (i) the area of the region is more than a threshold (in our case, 300 pixels), (ii) the region contacts the image boundary, and (iii) $(n_p - n_b)/n_b > 1$, where n_p indicates the perimeter of the region and n_b indicates the number of pixels that contact the image boundary. After the relabeling of the remaining instrument region, the centroid, principal axis, and tip position are calculated per instrument region, and region matching between adjacent frames is conducted based on the Euclidean distance between the centroids. Figure 44.2 shows an example of the use of four kinds of image sequences: #1, #2, #3, and #4 (100 frames each). In this technology, the mean pixel error (mean Euclidean distance between the tip position of the surgical instrument and its actual value) is below 35 pixels (the image sizes are either 300×225 pixels or 400×225 pixels). The average image processing time is 3.6 ms/frame. As for the detection rate of instruments, three out of four image sequences show 100%, and the overall detection rate is 94%.

44.4 Application I: Controlling a Surgical Instrument-Holding Robot

With the integration of a stereo camera with the depth calculation function described in Sect. 44.2 and an optical 3D measurement device, two fundamental technologies for controlling surgical instrument-holding robots were developed in this research. Their functions are as follows: (1) estimation of the distance between the tip of the surgical instrument and the organ surface (the insertion direction distance), and (2) directional control of the instrument. The proposed system uses the commercial optical tracking system Polaris (Northern Digital Inc.) [13] to measure the 3D position/pose of both the instrument and the endoscope in real time and virtually project any point on the longitudinal axis of the instrument onto the stereoscopic endoscopic images without any image processing techniques under the assumption that the instrument is a rigid body (deformation is zero) [14]. To do this, precise camera calibration (calculation of the 3D–2D projection matrix) is required in advance. We performed this using Zhang's method [15]. The most beneficial feature of this technology is that we can obtain two pieces of disparity information (real and virtual) simultaneously using the stereo camera and 3D measurement device, respectively. As shown in Fig. 44.3a, we can estimate the intersection of the longitudinal axis of the instrument and the organ surface as the point at which the two disparities (real and virtual) are the same. Thus, we can calculate the distance between the tip of the surgical instrument and the organ surface (the insertion direction distance), which is essential for controlling surgical instrument-holding robots. Based on the above technology, two functions were installed in ZEUS (Computer Motion Inc.) [16], an instrument-holding surgical robot normally categorized as LoA 1 (Fig. 44.3b): (1) autonomous control for adjusting any out-of-sight instrument to the center of the image and (2) calculation and display of the insertion direction distance such that a target point on the image, given by one mouse click, can be tracked by the robot. These new functions have received positive comments from surgeons, which indicates that they are definitely needed in practice and successfully promote a robot to LoA 2 (task-level autonomy).

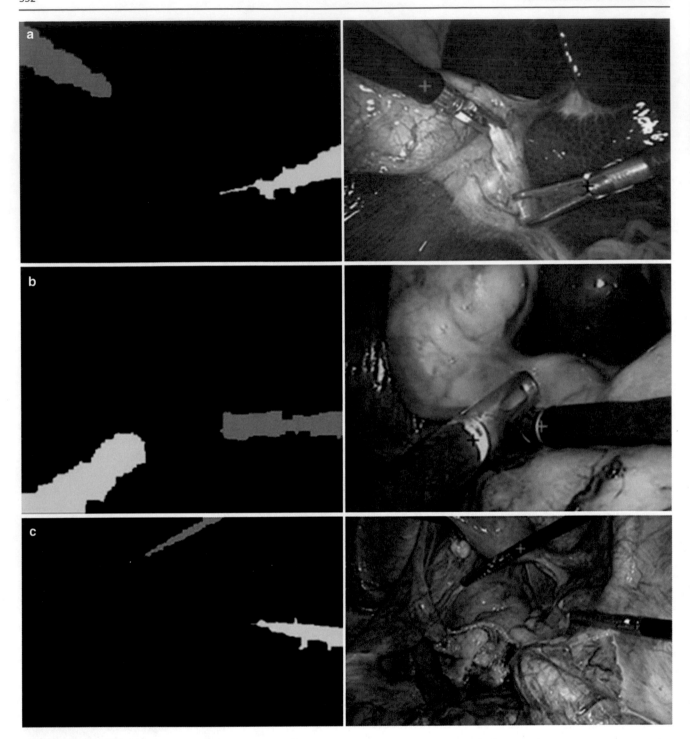

Fig. 44.2 Visual tracking of the tips of surgical instruments. The left figures show the detected instrument regions, and the right figures show the estimated instrument tip positions. Example from porcine cholecystectomy sequence #1 (**a**), example from human cholecystectomy sequence #2 (**b**), example from human sigmoidectomy sequence #3 (**c**), and example from human sigmoidectomy sequence #4 (**d**) (Images courtesy of Mr. Wataru Endo). Original image sequences #2 and #4 are available in DVD-ROM from published books [11, 12], respectively

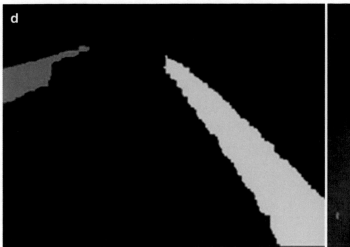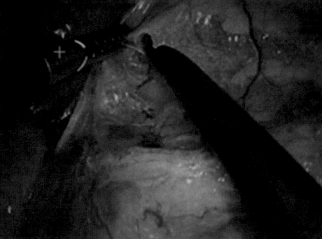

Fig. 44.2 (continued)

44.5 Application II: A Tip Load Calculation System for Surgical Instruments

A tip load calculation system for surgical instruments that uses an optical 3D measurement device and image processing was developed with consideration for the autonomous control of instrument-holding robots. As described in Sect. 44.4, with the use of the commercial optical tracking system and the conventional camera calibration technique, the position of the tip of the instrument projected onto endoscopic images can be precisely predicted under the assumption that the deformation of the instrument is zero. In our system, the real 2D and 3D positions of the tip of the instrument can also be extracted from the real images using the techniques described in Sects. 44.2–44.3. When the instrument is not contacting the organ (such as in Sect. 44.4), the predicted tip position based on the 3D optical tracker will coincide with the real tip position (Fig. 44.4a). When the instrument provides counter-traction against the organ, however, gaps arise between the virtual and real positions (Fig. 44.4b) because the instrument would bend (Fig. 44.4c, d). We can calculate the 3D gaps, called "deflection (say v)," if we have the stereoscopic endoscope described in Sect. 44.2. In this case, the load on the tip of the instrument can be estimated as follows. As the surgeon always holds the handle part of the instrument during surgery, the instrument can be basically modeled as a cantilever under concentrated load (say P_{free}) at the tip end. In practice, we can directly obtain the relationship between P_{free} and v by fitting a regression line in advance. That is, once the deflection v is measured, the load $P_{\text{free}}(v)$ is estimated. For real endoscopic surgery, the instrument should be modeled as a cantilever with support at the trocar port under concentrated load (say P_{support}) at the tip end part. In this case, we have

$$P_{\text{support}}(v,a) = \frac{4L^3}{a^2(3L+a)}P_{\text{free}}(v), \qquad (44.3)$$

In Eq. (44.3), L means the length of the instrument and a indicates the insertion length from the trocar port, which can be easily estimated by measuring the 3D positions of the port (fixed) and the tip of the instrument simultaneously with the optical tracking system during surgery. Since the instrument length L is known in advance, once the deflection v and the insertion length a are measured, the load P_{support} is estimated. Experiments are conducted to show the difference between an expert surgeon and novices, with a focus on the difference in their counter-traction (Fig. 44.4e). The counter-traction load P_{support} of the expert is more stable, and his task completion time is the shortest; the relationship between the counter-traction and dissection quality is thus confirmed. Therefore, the system is applicable for surgical education and training in addition to surgical robots. Kawai et al. developed LoA 1 forceps-holding robots for grasping and pulling organs [17, 18]. The counter-traction force estimation function is useful, especially for keeping the pulling state after the organ is grasped, which may help improve the LoA.

44.6 Conclusion

This chapter briefly reports the research progress about the improvement of the LoAs of surgical assistant robots. The stereo endoscope system developed through this study is an intraoperative 3D reconstruction system that can probably be a new modality in the construction of a multidisciplinary computational anatomy (MCA) model. The integration of other MCA modalities is promising for the further improvement of LoAs in robotic surgery.

a

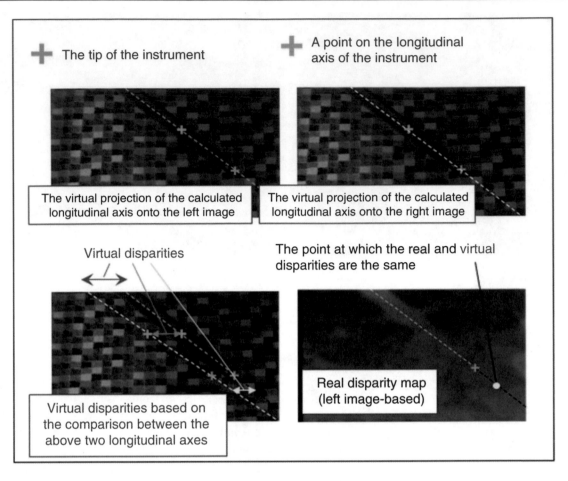

b

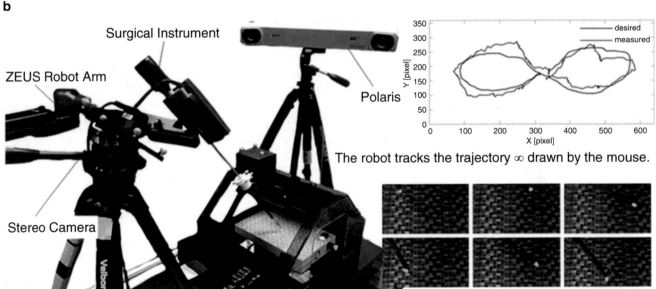

Fig. 44.3 Calculation of the distance between the tip of the surgical instrument and the organ surface. Calculation principle (**a**) and example for controlling a surgical instrument-holding robot (**b**) (Photos courtesy of Mr. Shinsuke Katsumata)

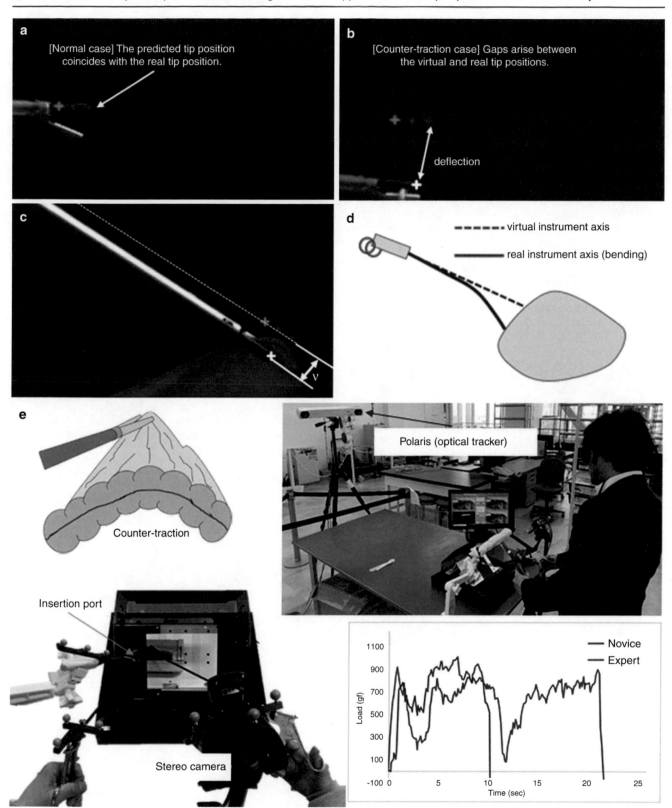

Fig. 44.4 Estimation of the load of the tip of the surgical instrument. Comparison of the integration of 3D measurement and image processing when the instrument is not tugging the organ (**a**) and when it is tugging the organ (**b**), example of deflection on the image (**c**), deflection diagram of the instrument (**d**), and the counter-traction force estimation results (**e**) (Photos courtesy of Mr. Shougo Kumaou, Dr. Yuji Nishizawa and Prof. Toshikazu Kawai)

References

1. Haidegger T. Autonomy for surgical robots: concepts and paradigms. IEEE Trans Med Robot Bionics. 2019,1.65–76. https://doi.org/10.1109/TMRB.2019.2913282.

2. Chinzei K, Shimizu A, Mori K, Harada K, Takeda H, Hashizume M, Ishizuka M, Kato N, Kawamori R, Kyo S, Nagata K, Yamane T, Sakuma I, Ohe K, Mitsuishi M. Regulatory science on AI-based medical devices and systems. Adv Biomed Eng. 2018;7:118–23. https://doi.org/10.14326/abe.7.118.

3. Takács A, Rudas I, Bösl D, Haidegger T. Highly automated vehicles and self-driving cars [Industry Tutorial]. IEEE Robot Automat Mag. 2018;25:106–12. https://doi.org/10.1109/MRA.2018.2874301.

4. Yang GZ, Cambias J, Cleary K, Daimler E, Drake J, Dupont PE, Hata N, Kazanzides P, Martel S, Patel RV, Santos VJ, Taylor RH. Medical robotics-regulatory, ethical, and legal considerations for increasing levels of autonomy. Sci Robot. 2017;2:eaam8638. https://doi.org/10.1126/scirobotics.aam8638.

5. Intuitive Surgical Inc. Da Vinci, robotic surgical systems. https://www.intuitive.com/en-us/products-and-services/da-vinci. Accessed 11 August 2020.

6. Ohta Y, Kanade T. Stereo by intra- and inter-scanline search using dynamic programming. IEEE Trans Pattern Anal Machine Intell. 1985;7:139–54. https://doi.org/10.1109/tpami.1985.4767639.

7. Lim JZ, Suzuki H, Utsugi S, Katai H. Experimental development of a multi-view stereo endoscope system, 2017 Second Russia and Pacific Conference on Computer Technology and Applications (RPC). Vladivostok. 2017:14–8. https://doi.org/10.1109/RPC.2017.8168058.

8. The Middlebury computer vision pages, Middlebury stereo datasets. http://vision.middlebury.edu/stereo/data/. Accessed 11 August 2020.

9. Bouget D, Allan M, Stoyanov D, Jannin P. Vision-based and markerless surgical tool detection and tracking: a review of the literature. Med Image Anal. 2017;35:633–54. https://doi.org/10.1016/j.media.2016.09.003.

10. Zhang J, Gao X. Object extraction via deep learning-based marker-free tracking framework of surgical instruments for laparoscope-holder robots. Int J Comput Assist Radiol Surg. 2020;15:1335–45. https://doi.org/10.1007/s11548-020-02214-y.

11. Matsumoto S, Kimura T, Yamashita Y, Mori T, Tokumura H. Laparoscopic cholecystectomy in videos, from basic to technical certification. Nakayama Shoten, 2008. ISBN: 978-4-521-73046-2.

12. Ito M. Thorough lecture on laparoscopic sigmoidectomy for certification. Kanehara Shuppan, 2015. ISBN: 978-4-307-20341-8.

13. Northern Digital Inc. Medical Polaris Spectra and Vicra. https://www.ndigital.com/medical/products/polaris-family/ Accessed 11 August 2020.

14. Nishikawa A, Ito K, Nakagoe H, Taniguchi K, Sekimoto M, Takiguchi S, Seki Y, Yasui M, Okada K, Monden M, Miyazaki F. Automatic positioning of a laparoscope by preoperative workspace planning and intraoperative 3D instrument tracking Proceedings, Workshop on Medical Robotics: Systems and Technology towards Open Architecture, 2006. MICCAI 2006 Workshop: 82–91.

15. Zhang Z. A flexible new technique for camera calibration. IEEE Trans Pattern Anal Machine Intell. 2000;22:1330–4. https://doi.org/10.1109/34.888718.

16. Iwamoto N, Nishikawa A, Kawai T, Horise Y, Masamune K. A novel medical robot architecture with ORiN for efficient development of telesurgical robots. Int J Comput Assist Radiol Surg 2018;13;S40.

17. Kawai T, Shin M, Nishizawa Y, Horise Y, Nishikawa A, Nakamura T. Mobile locally operated detachable end-effector manipulator for endoscopic surgery. Int J Comput Assist Radiol Surg. 2015;10:161–9. https://doi.org/10.1007/s11548-014-1062-4.

18. Kawai T, Hayashi H, Nishizawa Y, Nishikawa A, Nakamura R, Kawahira H, Ito M, Nakamura T. Compact forceps manipulator with a spherical-coordinate linear and circular telescopic rail mechanism for endoscopic surgery. Int J Comput Assist Radiol Surg. 2017;12:1345–53. https://doi.org/10.1007/s11548-017-1595-4.

Endoscopy: Computer-Aided Diagnostic System Based on Deep Learning Which Supports Endoscopists' Decision-Making on the Treatment of Colorectal Polyps

45

Yuichi Mori and Kensaku Mori

Abstract

The quality and quantity of endoscopic images that physicians obtain are dramatically increasing along with the recent advance in imaging technologies. However, benefits coming from these rich data may not be effectively utilized by all the endoscopists due to the lack of shared knowledge and experience. The use of artificial intelligence (AI) as a decision support during endoscopy is catching great attention as a measure to overcome this issue. In the colonoscopy field, AI is expected to facilitate polyp detection, prediction of polyp pathology, and prediction of invasion depth of colorectal cancer. With the use of AI technologies, the macroscopic anatomy (i.e., endoscopic view) is matched or fused with microscopic findings (i.e., pathological finding) in real time during the endoscopic examination. This new methodology allows clinical doctors to make a decision much easier than the conventional method. These research concepts and results well fit in the anatomy-pathology axis of the multidisciplinary computational anatomy (MCA) model.

Keywords

Artificial intelligence · Machine learning · Cancer Colonoscopy · Detection

Y. Mori (✉)
Clinical Effectiveness Research Group, Institute of Health and Society, University of Oslo, Oslo, Japan

Digestive Disease Center, Showa University Northern Yokohama Hospital, Yokohama, Japan

K. Mori
Graduate School of Informatics, Nagoya University, Nagoya, Japan
e-mail: kensaku@is.nagoya-u.ac.jp

45.1 Interpretation of Endoscopic Images: Challenging!

The quality and quantity of endoscopic images that physicians obtain are dramatically increasing along with the recent advance in imaging technologies available in endoscopy units. However, benefits coming from these rich data may be effectively utilized only by expert endoscopists because considerable expertise in image interpretation is required to effectively utilize the value of the data. It is broadly known that endoscopists miss around 20–40% polyps during colonoscopy [1] and many endoscopists cannot identify neoplastic change of colorectal polyps with the accuracy of over 90% [2]. AI tools are catching attention as an attractive measure to overcome these limitations which inherently exist in endoscopy practice. They are expected to bridge the gap between advanced technology and physician interpretation skill. Theoretically, excellent AI tools are expected to contribute to excellent standards of diagnosis irrespective of endoscopists' skill; however, there are a lot of hurdles for us to address before enjoying valuable benefits from the use of AI in real colonoscopy practice.

In this chapter, we would like to go over the current status of research and development in AI for colonoscopy and share some of the challenges we are facing in this academic field.

45.2 AI Tools in Colonoscopy

AI tools designed to help colonoscopy practice can be largely classified into three categories: (1) polyp detection (Fig. 45.1), (2) prediction of poly pathology (Fig. 45.2), (3) prediction of invasion depth of colorecta cancer (Fig. 45.3). Colonoscopy is the hottest area of research in the field of AI in endoscopy; multiple prospective studies have already been reported and multiple AI tools have already been on the market. Thus, understanding the situation in AI tools for colonoscopy can help have a future perspective on AI-assisted endoscopy in general.

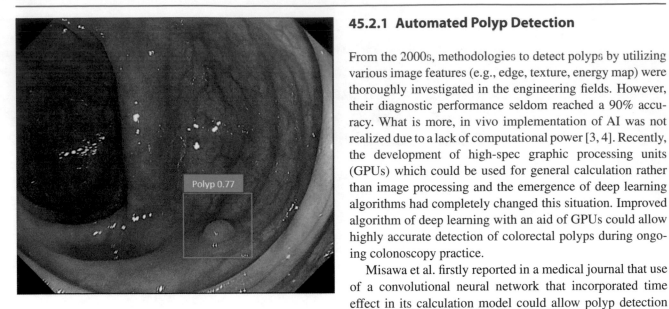

Fig. 45.1 Artificial intelligence tool designed to detect a colorectal polyp. The location of the polyp is identified with a bounding box with a probability of confidence

45.2.1 Automated Polyp Detection

From the 2000s, methodologies to detect polyps by utilizing various image features (e.g., edge, texture, energy map) were thoroughly investigated in the engineering fields. However, their diagnostic performance seldom reached a 90% accuracy. What is more, in vivo implementation of AI was not realized due to a lack of computational power [3, 4]. Recently, the development of high-spec graphic processing units (GPUs) which could be used for general calculation rather than image processing and the emergence of deep learning algorithms had completely changed this situation. Improved algorithm of deep learning with an aid of GPUs could allow highly accurate detection of colorectal polyps during ongoing colonoscopy practice.

Misawa et al. firstly reported in a medical journal that use of a convolutional neural network that incorporated time effect in its calculation model could allow polyp detection with a 90% sensitivity and a 63% specificity according to frame-based analysis [5]. Subsequently, Urban et al. showed

Fig. 45.2 Artificial intelligence tool that can identify neoplastic change. This algorithm was developed for images obtained with endocytoscopy

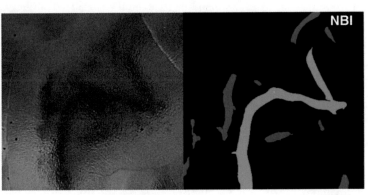

Neoplastic: 99 %
Non-neoplastic: 0.0 %

Fig. 45.3 Artificial intelligence tool that can predict cancer invasion. This algorithm was developed for images obtained with endocytoscopy

Non-neoplastic: 5 %
Adenoma: 11 %
Invasive Cancer: 85 %

much higher accuracy (i.e., a 93% sensitivity and a 93% specificity for polyp detection) by using highly variable endoscopic images as learning material [6]. However, these studies were ex vivo researches that used retrospectively composed test data; the study results were likely to be in favor of the use of AI. Actually, the first prospective study in which endoscopists used the AI in a real-time fashion during colonoscopy showed controversial results for the use of AI; in this study, Klare et al. evaluated the performance of AI in 55 patients who underwent colonoscopy [7]. The adenoma detection rate (ADR, a rate of colonoscopies in which one or more adenomas are detected. ADR is considered one of the most important quality measures in colonoscopy) was 29.1%, not a bad value; however, a quarter of the polyps that endoscopists could identify were actually missed by the AI tool.

However, six randomized controlled trials conducted between 2019 and 2020 have turned over the previous study results dramatically and provided a strong and positive impression for the use of AI for polyp detection [8–13]. All the RCTs were conducted to compare the ADR between colonoscopy with and without AI and provided favorable results for the use of AI during colonoscopy. Wang et al. randomized 1058 patients which were allocated into the two groups and found colonoscopy with AI yielded a 29% ADR while the control group showed a 20% ADR [9]. The results of the remaining five RCTs were basically in line with this study. Among them, only one study was conducted in a multicenter fashion while the others were single-center trials; Repici et al. conducted the multicenter trial in collaboration with three Italian hospitals including a total of 685 patients [11]. The participants of the trial underwent colonoscopies for various indications such as screening colonoscopy, surveillance colonoscopy after polypectomy, diagnostic colonoscopy as a follow-up exam to positive fecal immunochemical testing, which means the participants had a relatively high risk for the presence of adenomas. This RCT revealed a 55% ADR in the colonoscopies with AI while a 40% ADR in the standard colonoscopies. The use of AI contributed to the statistically significant increment of ADR by approximately 1.3 times.

On the other side, we may find some limitations of the AI tools in polyp detection. According to Wang et al., increment of ADR was largely attributed to the increased detection of tiny (<= 5 mm) adenomas which are considered to have less malignant potential compared to bigger ones [9]. This tendency was also confirmed in the study by Repici et al.; the use of AI actually did not play any role in the detection of clinically more relevant lesions such as polyps larger than 9 mm in size. Thus, it is not currently expected for AI technology to find out polyps that are easy to be missed but biologically severe lesions such as non-granular type, laterally spreading tumors.

45.2.2 Prediction of Polyp Pathology

45.2.2.1 AI Designed for Magnified Endoscopy

Many researches had been conducted to evaluate the possibility of AI to predict polyp pathology based on magnified endoscopy images. Preliminary studies were done in the 2000s which were focused on the interpretation of pit pattern findings under magnified chromoendoscopy [14, 15]. Subsequently, magnified narrow-band imaging (NBI, images captured with narrow-band spectrum light, which enhances vessel appearance and glandular structure) became the next popular target of the research [16]. Some of the researchers in this area achieved real-time interpretation of the images with an aid of AI during ongoing colonoscopy [15, 17]. The central idea of the image-interpretation algorithm was to collect the image features as a "bag of features" concept and construct the prediction model by analyzing these features with a support vector machine. According to the prospective trial reported by Kominami et al., 118 polyps from 41 patients received a real-time assessment of the AI tool. Their pathologies were predicted with a 93% sensitivity and a 93% specificity for neoplasm [18].

In this area, use of deep learning is also enthusiastically investigated [19, 20]. Chen et al. and Byrne et al. reported that their developed software allowed tumor identification with over 90% sensitivity. Improvement of the diagnostic power and their implementation will be expected as the number of images used for machine learning increases.

The AI tools designed for magnified endoscopic images are considered to harbor two hurdles to be overcome before its implementation. First, it is required for endoscopists to indicate the area of interest manually for the AI tool to correctly analyze the relevant polyp. The second hurdle is the lack of evidence. Given, only one small prospective study exists in this field, it would be too early to conclude the benefits of the use of AI in the optical diagnosis of colorectal polyps. Once these barriers are cleared, AI tools for the optical diagnosis will be widely utilized in clinical practice.

45.2.2.2 AI Designed for Endocytoscopy

Endocytoscopy (520-fold contact-microscope, Olympus Corp.) may be one of the most suitable devices for AI application because of the following reasons. (1) The contact microscopic observation allows that the whole image is subject to analysis of AI, which is a simple principle. (2) Big difference between endomicrosopic imaging and conventional imaging makes it easy for the AI tool to identify images that should be analyzed.

The research consortium consisted of Showa University, Nagoya University, and Cybernet System Corporation developed the AI software by utilizing these features of AI tools and confirmed it can differentiate neoplastic changes with more than a 90% sensitivity in multiple benchmark tests

[21–26]. Following these preliminary studies, Mori et al. conducted a large-scale prospective study to evaluate the performance of the AI tools in a real time fashion [27]. The study included 791 patients who received colonoscopies with the use of endocytoscopy supported by the AI tool. A total of 466 diminutive polyps were assessed with a sensitivity of 93% and specificity of 90%, which is considered acceptable performance in clinical use. This software (EndoBRAIN®) was approved by the regulatory body in Japan as the first medical device using the AI concept. To obtain regulatory approval, the research team conducted multicenter benchmark testing [28].

Confocal laser endomicroscopy (Cellvizio, Mauna Kea Technologies) is another endomicroscopy on the market. There were researches to explore the application of AI for this endomicroscope; however, the quality and quantity of the researches were limited [29, 30].

45.2.2.3 AI for Autofluorescence Endoscopy

There are roughly two kinds of researches on the use of AI for autofluorescence endoscopy. One is for autofluorescence spectroscopy (WavSTAT4, Pentax, Japan), while the other is for autofluorescence imaging (Olympus Corp., Japan). Two small prospective trials were conducted to assess the first product with controversial results [31, 32]. However, it is not widely used so far although it has secured regulatory approval in the USA and Europe.

Various studies were actively conducted to develop and evaluate the AI tools designed to interpret the images taken by Olympus's autofluorescence endoscopy mainly in Japan from 2013. Two prospective studies have been conducted; Aihara et al. reported the system showed 94% sensitivity and 89% specificity in predicting neoplastic changes in the subject of 102 colorectal polyps [33]. Subsequently, Horinouchi reported 80% sensitivity and 95% specificity in predicting neoplastic changes in 258 lesions [34]. This favorable results in two prospective trials can support future implementation of this promising modality.

45.2.2.4 AI Designed for Conventional Endoscopy

Non-magnifying normal endoscopy is the basis of endoscopic observation and the easiest way to evaluate colorectal lesions. However, research and development of AI tools designed for conventional endoscopy have not been as advanced as those for the other endoscopic modalities listed above [35]. Komeda et al. developed the deep learning-based AI tools to interpret the conventional endoscopy images to enable accurate tumor differentiation; however, its performance was limited to 75% accuracy in the cross-validation method [36]. Renner et al. also tried to tackle this challenge, however could not achieve a diagnostic yield of over 80% in terms of sensitivity and specificity [37]. These unfavorable

results may come from the fact that the quantity of the data which conventional endoscopy image contains are less than those of magnifying endoscopy and endocytoscopy. On the other hand, some researchers showed non-deep learning algorithms (i.e., hand-crafted feature-based algorithms) provided excellent diagnostic discrimination capability with a sensitivity of 92% and a specificity of 89% [38]. Deep learning usually, but not always, outperforms the hand craft-based feature extraction, which should be considered according to the subjects one tackles [39].

A Korean research team recently conducted a first-in-world study to clarify the additional value of the use of AI during colonoscopy. [40] According to their study, which is a benchmark test using prospectively constructed test data, the AI tool improved the discrimination capability of neoplastic change from 74% to 86% in endoscopists who did not have enough competence in the optical diagnosis of polyps, a main target of the AI tool in practice.

45.2.3 Predicting Invasion Depth

Pre-treatment identification of submucosal invasion of colorectal neoplasia is considered important because treatment options differ according to the depth of its invasion. Usually, submucosally invasive colorectal cancers with minute invasion are subject to endoscopic treatment because of their minimum risks of metastasis, while deeply invasive submucosal cancers should be removed surgically together with surrounding lymph nodes. However, endoscopists can actually differentiate deeply invasive submucosal cancer with less than 80% accuracy even with highly advanced endoscopic imaging modalities [41], which should be addressed further. Therefore, AI is emerging modalities that people expect help precisely identify cancer invasion. Tamai et al. developed the AI tool to identify deeply invasive submucosal cancer with a sensitivity of 84% and a specificity of 83%, which was a clinically encouraging result [42].

Takeda et al. also developed the AI tool to identify invasive submucosal cancer. They utilized features of the endocytoscopic images of cancer cells that were exposed on the surface of the tumor for constructing the prediction model [43]. The AI model provided a sensitivity of 89% and a specificity of 99%; however, prospective evaluation will be needed.

Itoh et al. and Lui et al. explored the possibility of the AI tools in predicting the cancer invasion based on non-magnified images with the use of deep learning methodology. Ito et al. achieved a sensitivity of 68% and a specificity of 89% in discrimination cancer invasion [44], while Lui et al. developed software that enabled identification of endoscopically curable lesions (i.e., well-differentiated adenocarcinoma with no lymphatic and venous invasion whose invasion depth is less than 1000 μm) with a sensitivity of 88% and a specificity of 78% [45]. They were preliminary studies investigating limited numbers of samples; however, they may be a strong solution to overcome the challenges that current clinical practice harbors in pre-treatment assessment of invasive cancer in the future.

45.3 Regulatory Approval of AI Tools in Colonoscopy

As we go over in the above sections, many researches have been conducted in this field. However, there are big barriers to the use of these AI tools in clinical practice. They should secure official approval as medical devices from the responsible regulatory body in each country before clinical use. This is because the use of AI is considered to harbor various clinical risks. In 2018, a working group endorsed by the official regulatory body in Japan (PMDA) published a statement regarding the approval process of AI-related medical devices [46]. According to this paper, AI-related medical devices were classified into five categories according to their harboring risks. They defined the medical device with a risk level of 3 or more have severe clinical risks if they provide incorrect prediction during ongoing clinical practice. The American regulatory body (FAD) also announced a similar idea in the public statement. They categorized the AI-related medical device into four categories according to their clinical risks and the hurdles for the regulatory assessment differ according to their risks.

As of June 2020, there were two AI-related software for a colonoscopy that have secured regulatory approval (EndoBRAIN® and EndoBRAIN®-EYE; Cybernet System Corp.) in Japan. Outside Japan, there are four AI-related software for colonoscopy that have secured regulatory approval (WavSTAT4, Pentax Corp.; GI Genius, Medtronic Corp.; DISCOVERY, Pentax Corp.; CAD-EYE, Fujifilm Corp.). The number of approved AI-related devices is dramatically increasing in these couple of years, thus they are expected to be widely adopted spread in routine practice.

45.4 Role of the Research Topic in the Concept of Multidisciplinary Computational Anatomy (MCA)

By introducing AI, the macroscopic anatomy (i.e., endoscopic view in our project) is matched or fused with microscopic findings (i.e., pathological finding) in real time during the endoscopic examination. This new methodology allows clinical doctors to make the decision easier than the conventional method. This research concept and results well fit in the anatomy-pathology axis of the MCA model.

45.5 Summary

We share the latest information regarding the AI tools designed for colonoscopy in this chapter. Nowadays, the role of AI in routine practice is increasing as the number of approved devices increases; however, most of the products lack supporting evidence. Especially, we should recognize the number of prospective studies is not sufficient in the field of automated polyp characterization (i.e., prediction of polyp pathology). On the other hand, we have encouraging evidence for the use of automated polyp detection software.

We should use the AI-assisted software with well understanding of its advantage and limitation, which may contribute to enhancing the quality of patient care.

References

1. Kumar S, Thosani N, Ladabaum U, et al. Adenoma miss rates associated with a 3-minute versus 6-minute colonoscopy withdrawal time: a prospective, randomized trial. Gastrointest Endosc. 2017;85:1273–80.
2. Ladabaum U, Fioritto A, Mitani A, et al. Real-time optical biopsy of colon polyps with narrow band imaging in community practice does not yet meet key thresholds for clinical decisions. Gastroenterology. 2013;144:81–91.
3. Fernandez-Esparrach G, Bernal J, Lopez-Ceron M, et al. Exploring the clinical potential of an automatic colonic polyp detection method based on the creation of energy maps. Endoscopy. 2016;48:837–42.
4. Tajbakhsh N, Gurudu SR, Liang J. Automatic polyp detection using global geometric constraints and local intensity variation patterns. Med Image Comput Comput Assist Interv. 2014;17:179–87.
5. Misawa M, Kudo SE, Mori Y, et al. Artificial intelligence-assisted polyp detection for colonoscopy: initial experience. Gastroenterology. 2018;154:2027–2029.e3.
6. Urban G, Tripathi P, Alkayali T, et al. Deep learning localizes and identifies polyps in real time with 96% accuracy in screening colonoscopy. Gastroenterology. 2018;155:1069–1078.e8.
7. Klare P, Sander C, Prinzen M, et al. Automated polyp detection in the colorectum: a prospective study (with videos). Gastrointest Endosc. 2019;89(3):576–582.e1.
8. Wang P, Liu X, Berzin TM, et al. Effect of a deep-learning computer-aided detection system on adenoma detection during colonoscopy (CADe-DB trial): a double-blind randomised study. Lancet Gastroenterol Hepatol. 2020;
9. Wang P, Berzin TM, et al. Real-time automatic detection system increases colonoscopic polyp and adenoma detection rates: a prospective randomised controlled study. Gut. 2019;68(10):1813–9.
10. Gong D, Wu L, Zhang J, et al. Detection of colorectal adenomas with a real-time computer-aided system (ENDOANGEL): a randomised controlled study. Lancet Gastroenterol Hepatol. 2020;5(4):352–61.
11. Repici A, Badalamenti M, Maselli R, et al. Efficacy of real-time computer-aided detection of colorectal neoplasia in a randomized trial. Gastroenterology. 2020;159(2):512–520.e7.
12. Su JR, Li Z, Shao XJ, et al. Impact of a real-time automatic quality control system on colorectal polyp and adenoma detection: a prospective randomized controlled study (with videos). Gastrointest Endosc. 2020;91:415–424.e4.
13. Liu W, Zhang Y, Bian X, et al. Study on detection rate of polyps and adenomas in artificial-intelligence-aided colonoscopy. Saudi J Gastroenterol. 2020;26(1):13–9.

14. Hafner M, Liedlgruber M, Uhl A, et al. Delaunay triangulation-based pit density estimation for the classification of polyps in high-magnification chromo-colonoscopy. Comput Methods Prog Biomed. 2012;107:565–81.

15. Takemura Y, Yoshida S, Tanaka S, et al. Quantitative analysis and development of a computer-aided system for identification of regular pit patterns of colorectal lesions. Gastrointest Endosc. 2010;72:1047–51.

16. Tischendorf JJ, Gross S, Winograd R, et al. Computer-aided classification of colorectal polyps based on vascular patterns: a pilot study. Endoscopy. 2010;42:203–7.

17. Kominami Y, Yoshida S, Tanaka S, et al. Computer-aided diagnosis of colorectal polyp histology by using a real-time image recognition system and narrow-band imaging magnifying colonoscopy. Gastrointest Endosc. 2015 [Epub ahead of print].

18. Kominami Y, Yoshida S, Tanaka S, et al. Computer-aided diagnosis of colorectal polyp histology by using a real-time image recognition system and narrow-band imaging magnifying colonoscopy. Gastrointest Endosc. 2016;83:643–9.

19. Byrne MF, Chapados N, Soudan F, et al. Real-time differentiation of adenomatous and hyperplastic diminutive colorectal polyps during analysis of unaltered videos of standard colonoscopy using a deep learning model. Gut. 2017 [Epub ahead of print].

20. Chen PJ, Lin MC, Lai MJ, et al. Accurate classification of diminutive colorectal polyps using computer-aided analysis. Gastroenterology. 2018;154:568–75.

21. Mori Y, Kudo SE, Chiu PW, et al. Impact of an automated system for endocytoscopic diagnosis of small colorectal lesions: an international web-based study. Endoscopy. 2016;48:1110–8.

22. Mori Y, Kudo SE, Wakamura K, et al. Novel computer-aided diagnostic system for colorectal lesions by using endocytoscopy (with videos). Gastrointest Endosc. 2015;81:621–9.

23. Misawa M, Kudo SE, Mori Y, et al. Characterization of colorectal lesions using a computer-aided diagnostic system for narrow-band imaging endocytoscopy. Gastroenterology. 2016;150:1531–1532.e3.

24. Misawa M, Kudo SE, Mori Y, et al. Accuracy of computer-aided diagnosis based on narrow-band imaging endocytoscopy for diagnosing colorectal lesions: comparison with experts. Int J Comput Assist Radiol Surg. 2017;12:757–66.

25. Mori Y, Kudo S, Misawa M, et al. Simultaneous detection and characterization of diminutive polyps with the use of artificial intelligence during colonoscopy. Video GIE. 2019;4(1):7–10.

26. Mori Y, Kudo SE, Mori K. Potential of artificial intelligence-assisted colonoscopy using an endocytoscope (with video). Dig Endosc. 2018;30(Suppl 1):52–3.

27. Mori Y, Kudo SE, Misawa M, et al. Real-time use of artificial intelligence in identification of diminutive polyps during colonoscopy: a prospective study. Ann Intern Med. 2018;169:357–66.

28. Kudo SE, Misawa M, Mori Y, et al. Artificial intelligence-assisted system improves endoscopic identification of colorectal neoplasms. Clin Gastroenterol Hepatol. 2020;18(8):1874–1881.e2.

29. Stefanescu D, Streba C, Cartana ET, et al. Computer aided diagnosis for confocal laser endomicroscopy in advanced colorectal adenocarcinoma. PLoS One. 2016;11:e0154863.

30. Andre B, Vercauteren T, Buchner AM, et al. Software for automated classification of probe-based confocal laser endomicroscopy videos of colorectal polyps. World J Gastroenterol. 2012;18:5560–9.

31. Kuiper T, Alderlieste YA, Tytgat KM, et al. Automatic optical diagnosis of small colorectal lesions by laser-induced autofluorescence. Endoscopy. 2015;47:56–62.

32. Rath T, Tontini GE, Vieth M, et al. In vivo real-time assessment of colorectal polyp histology using an optical biopsy forceps system based on laser-induced fluorescence spectroscopy. Endoscopy. 2016;48:557–62.

33. Aihara H, Saito S, Inomata H, et al. Computer-aided diagnosis of neoplastic colorectal lesions using 'real-time' numerical color analysis during autofluorescence endoscopy. Eur J Gastroenterol Hepatol. 2013;25:488–94.

34. Horiuchi H, Tamai N, Kamba S, et al. Real-time computer-aided diagnosis of diminutive rectosigmoid polyps using an auto-fluorescence imaging system and novel color intensity analysis software. Scand J Gastroenterol. 2019:1–6.

35. Mori Y, Kudo SE, Berzin TM, et al. Computer-aided diagnosis for colonoscopy. Endoscopy. 2017;49:813–9.

36. Komeda Y, Handa H, Watanabe T, et al. Computer-aided diagnosis based on convolutional neural network system for colorectal polyp classification: preliminary experience. Oncology. 2017;93(Suppl 1):30–4.

37. Renner J, Phlipsen H, Haller B, et al. Optical classification of neoplastic colorectal polyps – a computer-assisted approach (the COACH study). Scand J Gastroenterol. 2018;53:1100–6.

38. Sanchez-Montes C, Sanchez FJ, Bernal J, et al. Computer-aided prediction of polyp histology on whilte-light colonoscopy using surface pattern analysis. Endoscopy. 2019;51(3):261–5.

39. Khan S, Yong S. A comparison of deep learning and hand crafted features in medical image modality classification. In 2016 3rd International Conference on Computer and Information Sciences (ICCOINS), 15–17 Aug. 2016, 2016.

40. Jin EH, Lee D, Bae JH, et al. Improved accuracy in optical diagnosis of colorectal polyps using convolutional neural networks with visual explanations. Gastroenterology. 2020;158(8):2169–2179.e8.

41. Shimura T, Ebi M, Yamada T, et al. Magnifying chromoendoscopy and endoscopic ultrasonography measure invasion depth of early stage colorectal cancer with equal accuracy on the basis of a prospective trial. Clin Gastroenterol Hepatol. 2014;12:662–8.e1–2.

42. Tamai N, Saito Y, Sakamoto T, et al. Effectiveness of computer-aided diagnosis of colorectal lesions using novel software for magnifying narrow-band imaging: a pilot study. Endosc Int Open. 2017;5:E690–4.

43. Takeda K, Kudo S, Mori Y, et al. Accuracy of diagnosing invasie colorectal cancer using computer-aided endocytoscopy. Endoscopy. 2017;49:798–802.

44. Ito N, Kawahira H, Nakashima H, et al. Endoscopic diagnostic support system for cT1b colorectal cancer using deep learning. Oncology. 2018:1–7.

45. Lui TKL, Wong KKY, Mak LLY, et al. Endoscopic prediction of deeply submucosal invasive carcinoma with use of artificial intelligence. Endosc Int Open. 2019;7:E514–e520.

46. Chinzei K, Shimizu A, Mori K, et al. Regulatory science on AI-based medical devices and systems. Adv Biomed Eng. 2018;7:118–23.

Endoscopy: Application of MCA Modeling to Abnormal Nerve Plexus in the GI Tract

46

Masakuni Kobayashi and Kazuki Sumiyama

Abstract

A series of GI disorders with abnormal gastrointestinal (GI) motility are associated with morphological changes of enteric nervous system (ENS). Pathological analysis that evaluates an isolated area harvested from the GI tract has been the standard examination for the morphological analysis of ENS. However, it is suboptimal to evaluate discrete changes in ENS morphology in motility disorder cases. Recently, our group reported a novel ENS observation method using a confocal laser endomicroscopy (CLE). In a previous *ex vivo* study using surgically resected sections of the gut wall considered as normally innervated, we demonstrated the topical use of a fluorescent agent, cresyl violet (CV), enabled ENS to be visualized as bright ladder-like structures within an CLE image. Nuclei of neuron-like cells located in the ganglion were also visualized as dark oval dots and easily identifiable.

Hirschsprung's disease is a pediatric digestive disease with a congenital abnormality of the ENS in the distal intestine with morphological abnormality or a lack of ENS in diseased colon segments. In CLE image acquired from the same technique as the above study, ENS in the normal segment of Hirschsprung's disease was also visualized as a ladder-like structure in common with normal ENS cases. Meanwhile, only the smooth muscle fibers were visualized in the dysfunctional segments. Evidence obtained from prior studies show the technical feasibility of *in vivo* endoscopic visualization of ENS. CLE is a less-invasive visualization approach that evaluates ENS over a wider area than conventional pathological analysis. The technology could be an innovative and provide new insights into neural GI disorders.

Keywords

Enteric nervous system · ENS · Confocal laser endomicroscopy · CLE · Cresyl violet · Hirschsprung's disease

46.1 Introduction

Gastrointestinal (GI) motility is regulated with the electrophysiologic function of the enteric nervous system (ENS). ENS is composed of neuronal cells, glial cells, submucosal plexus (Meissner's plexus), and myenteric plexus (Auerbach's plexus) [1]. Morphological changes of ENS accompany a series of GI disorders, such as achalasia, gastrointestinal motility disorders, gastroparesis, Hirschsprung's disease. Hirschsprung's disease is characterized as a congenital abnormality of ENS with a lack of ganglion cells in the distal colon and rectum [2–4]. Surgical resection of a difunctionally innervated portion of the colon is clinically required in most of cases of the disease. At present, pathological analysis is the sole technique available to evaluate the morphology of ENS. However, the ENS forms finely meshed structures along the entire GI tract and therefore it would be challenging to precisely investigate discrete ENS abnormalities with the pathological analysis even using surgically sampled specimens [2]. Obviously, sampling a wider area of the ENS, such as with a full-thickness biopsy sample, would provide more information than a single point area. For instance, in surgery for Hirschsprung's disease, the abnormal area of colon with a lack or denaturation of ganglion cells needs to be surgically resected according to an intraoperative pathological diagnosis using frozen sections of a surgically sampled full-thickness of the gut wall. However, the pathological information obtained from randomly selected areas is inherently restricted and requires multiple full-thickness biopsy during surgery. Therefore, a commonly practiced intraoperative pathological analysis is suboptimal as an on-site surgical navigation technique [5]. Our group previously reported a novel and less-invasive ENS observation method

M. Kobayashi · K. Sumiyama (✉)
Department of Endoscopy, The Jikei University School of Medicine, Tokyo, Japan
e-mail: masakuni@jikei.ac.jp; kaz_sum@jikei.ac.jp

using a confocal laser endomicroscopy (CLE) [2, 5]. CLE provides a real-time histological analysis of surface structures of GI mucosa during routine endoscopy and has been widely used to evaluate GI neoplasia as an alternative to conventional forceps biopsy [6–8]. In order visualize tissues with CLE, use of fluorescent dye is mandatory. Fluorescein sodium is a solely approved fluorescent dye for clinical use and most of available clinical evidences for GI observation with CLE are based on intravenous administration of fluorescent sodium. However, when we explored the technical feasibility of muscularis layer observation with fluorescein assisted CLE, the fluorescein sodium was not well-absorbed into the muscular tissues and ENS was not visualized, although the leakage of the stain into the connective tissues in the muscularis propria distinctly silhouetted the intricately running smooth muscle fibers [9]. In our preceding animal studies, various fluorescent dyes with known neuronal affinity such as NeuroTrace, FM 1–43, acriflavine, and cresyl violet (CV) were tested for ENS visualization and all of them

successfully visualize the ladder-like anatomy of ENS [2, 3, 5, 10–12]. The clinical feasibility of ENS visualization with the topical application of acriflavine onto post therapeutic ulcers after endoscopic resection of colonic mucosal lesions was confirmed in a case series under IRB approval with two different clinically available; CLE systems, a probe type CLE system (CellVizio Mauna Kea Technologies, Paris, France) and a scope-embedded type CLE (Optiscan, Pentax, Japan) [12]. However, the safety profile of acriflavine is still unclear especially for inherent mutagenic potential of the nuclear staining. Meanwhile, CV, one of the Nissl-stains, has been used for neuropathological analysis and also used in *in vivo* for magnifying chromoendoscopy for colorectal lesions since the 1980s in Japan [13, 14]. So far, clinically significant side effect and adverse event after the topical use of the stain have not been reported. In addition, CV is a fluorescent agent with an excitation peak at 585 nm and an emission peak at 630 nm [15], therefore CV was used as a fluorescent dye in inaugural clinical trials of CLE [16, 17] (Fig. 46.1).

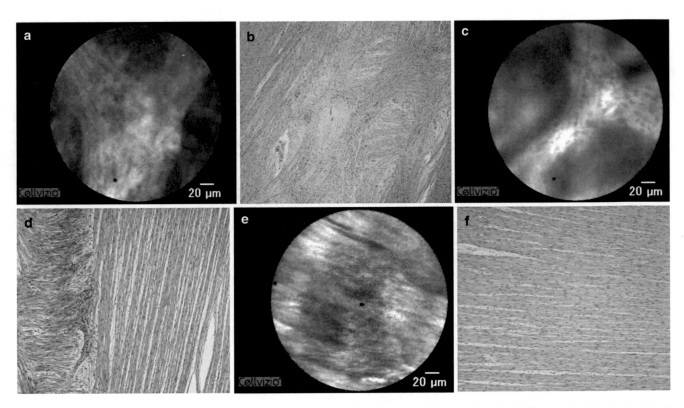

Fig. 46.1 Representative CLE and pathological images. (**a**) Representative CLE image in a human specimen considered as normally innervation. ENS was visualized as a white ladder-like structure in the image. Part of a forked or branching shape ENS was observed. (**b**) Horizontally sliced pathological image of a human specimen considered as normally innervation. ENS with neuron-like cells inside was observed as a meshed network pattern. (**c**) CLE image of a Hirschsprung's disease specimen in a normal segment. Although ENS was visualized as a white ladder-like structure as normally innervation

colon specimens, ENS was slender. (**d**) Horizontally sliced pathological image of a Hirschsprung's disease specimen in a normal segment. ENS and neuron cells appeared smaller than those of normally innervation colon specimens. (**e**) CLE image of a Hirschsprung's disease specimen in an abnormal segment. ENS was not identified and only smooth muscle fibers were observed. (**f**) Horizontally sliced pathological image of a Hirschsprung's disease specimen in an abnormal segment. ENS and neuron cells were not observed

46.2 ENS Visualization with CV-Assisted CLE

Our group further investigated details of CLE visualization of ENS using CV in an *in vivo* study using transgenic mouse model and an *ex vivo* study with surgically sampled human specimens [2]. A probe-based CLE system delivering excitation laser at 488 nm wavelength. The external diameter of the CLE probe used (GastroFlex-UHD: Mauna Kea Technologies, Paris, France) was 2.5 mm and the field-of-view of acquired image was 240 μm in diameter. The imaging rate and the image depth were fixed at 12 frames/s and 55–65 μm, respectively. In the *in vivo* animal study, a transgenic mouse in which neural crest-derived cells were labeled with green fluorescent protein (GFP) within various tissues, including the gut was used. The CLE observation without tissue staining was performed by gently applying onto the serosal side of the colon via a frontal abdominal incision in the transgenic mouse. The CLE was then repeated after trans-serosally applying 0.1% CV (Muto Pure Chemicals Co., Ltd., Tokyo, Japan). Consequently, the ladder like structures of ENS and spotty round defects presumed as nucleus of ganglionic or glial cells were observed in all models. The ENS structures acquired CV-assisted CLE image was morphologically reminiscent of a histological image of an endogenous GFP-positive network in the transgenic mouse. Thereby, we could confirm that CV could be well-absorbed into neural fibers and ganglionic cells at high contrast in the darker backgrounds of the smooth muscle layer.

In the *ex vivo* study using human specimens assumed with normal innervation, 11 patients (mean age was 5.7 years old) who underwent intestinal resection for treatment such as imperforate anus, Meckel's diverticulum, jejunal atresia were enrolled. Six small intestines and eight colon specimens in 11 patients were evaluated. The CLE probe was gently attached to the specimens and manually scanned from the serosa, dyed by topically applying 0.1% CV. Acquired CLE images were then assessed for the presence or absence of ENS. All examined specimens were pathologically evaluated as a gold standard. As a result, CV-assisted CLE visualized ENS as a white ladder-like structure. Clusters of spotty unstained areas were observed at transaction areas of the ladder-like structures of ENS. ENS was identified using CLE in 85.7% (12/14) with sensitivity and specificity of 92.3% and 100%, respectively. In the pathological analysis of horizontal slices of the muscularis propria, ENS appeared as a finely meshed network and its morphologically reminiscent of CLE images of ENS. We considered the morphology of normal ENS could be observed with CLE in this study, since the specimens were sampled from benign diseases without visible anatomical changes.

46.3 ENS Visualization of Hirschsprung's Disease

Previous studies have demonstrated that CV-assisted CLE visualized ENS as a white ladder-like structure. This procedure visualizes ENS in a less-invasive manner; therefore, CV-assisted CLE has the potential to replace neuropathological diagnosis. Although, neuropathological diagnosis analyzes ENS at a particular point in time, CV-assisted CLE has the potential to temporally evaluate ENS. Theoretically, CLE can visualize ENS repeatedly with less-invasiveness, thereby allowing the assessment of ENS changes over time. In our research we applied ENS visualization technology to diagnose Hirschsprung's disease because intraoperative pathological diagnosis to identify the abnormal ENS area in Hirschsprung's disease has a variety of technical challenges [5, 18]. The information obtained from an unselectively sampled specimen is too exiguous to evaluate widely spreading ENS structures. Tedious multiple tissue samplings as well as meticulous histopathological analysis for every specimen are inevitable to assure radical removal of an unfunctional portion of the colon. Therefore, if there is an endoscopic technique to visualize ENS in real-time, the full-thickness tissue sampling associated with surgical trauma would be unnecessary just for anatomical analysis of ENS within the muscularis propria. We surmised that morphological ENS analysis with CV-assisted CLE to identify the ENS abnormality would be greatly beneficial as a navigation method during surgeries for Hirschsprung's disease. As preclinical trial of CV-assisted CLE visualization of ENS in cases of Hirschsprung's disease we conducted an *ex vivo* study using surgical specimens [5]. Nine patients (mean age was 5.3-month-old) who underwent pull-through surgery for Hirschsprung's disease were enrolled. Each specimen contained both abnormally and normally innervated segments. As results, ENS was visualized as a white ladder-like structure in the normal segments of the colon of Hirschsprung's disease cases equivalent to ENS visualized in the normal colon. Although the ladder-like structures of ENS could be identified even in the transition zone between normal and aganglionic segments, the diameter of the plexus was slender compared with the normal ENS. Meanwhile, in aganglionic segments, the smooth muscle fibers were more brightly visualized and the ladder-like structures of ENS was not observed. For each observation, the accuracy of the ENS visualization was 88.4% (61/69), the sensitivity was 78.6% (22/28), and the specificity was 95.1% (39/45). We also compared the width of nerve strands located between ganglions between 9 Hirschsprung's disease cases (57 specimens) and 11 control cases (14 specimens). The maximum average width of nerve strands of ENS in Hirschsprung's disease cases was slender compared with those in normal cases (35.0 μm vs 70.5 μm, $p = 0.03$) [18].

46.4 Existing Technical Limitations and Future Perspectives for the ENS Observation with CLE

The probe type CLE system used for the ENS imaging was designed to be compliant to any type of flexible endoscope. Therefore, it is technically challenging for surgeons to manually scan tissues by stabilizing the tip of the flexible probe at a target on mobile serosal surface of the colon during operation. The focal plane of the probe-based CLE system is restricted to 10 μm. In order to visualize ENS, a thin layer of ENS needs to be adjusted into the range of focal plane depth. The scope-embedded type CLE has an adjustable scanning depth from 0 to 250 μm and would be more preferable for the ENS imaging. However, the system is not commercially available in current.

Recently, our research on the ENS visualization with CV-assisted CLE was promoted only for the serosal approach to ENS, because ENS exists shallower to the serosal surface than to the mucosal surface. However, we initially explored the endoluminal approach with a flexible endoscopy guidance, which allows access to ENS in a less-invasive fashion without need for laparotomy creation [9]. In order to directly apply the tip of the CLE probe onto the muscularis propria, we endoscopically created an artificial space within the submucosal layer or a mucosal defect. The endoluminal approach could be applicable to any GI diseases associated degeneration of ENS such as achalasia and gastroparesis and may provide their novel etiological information. However, we recognized the residual submucosal tissues could frequently hinder the CLE visualization of the muscularis propria and ENS. As a solution of the technical limitation with the access to the muscularis, Samarasena and colleagues used a needle-based CLE (nCLE) system, in which the imaging probe is passible through a 19-gauge needle. They inserted the nCLE probe into the muscularis propria of the esophagus under an ultrasonic guidance and succeeded to obtain CLE images of CNS with NeuroTrace tissue staining in porcine models [11].

We considered that CV would be one of most promising fluorescent stains for the ENS visualization and we have started clinical introduction in surgical volunteer cases, in whom a stained portion of the gut was eventually excised. However, the long-term biosafety profile should be carefully evaluated to expand indications to the endoluminal observation for benign diseases by tattooing CV into tissues.

The CLE system currently available for clinical use scans only one focal point of the target. While the current CV-assisted CLE evaluates ENS with only a two-dimensional (2D) image, it is possible to construct a three-dimensional (3D) image of ENS by scanning at multiple focal points with confocal imaging technology. The conventional pathological analysis only evaluates ENS running through the entire GI tract with a 2D image, however, we consider that the construction of 3D ENS images using the CLE system is theoretically feasible and urges the development of new technology. We believe that 3D images will provide a more accurate morphological ENS analysis.

If CV-assisted CLE can resolve the various technical challenges of morphological diagnosis of ENS, novel ENS diagnosis that considers the time axis may be possible. Accordingly, we consider that this novel diagnostic approach can contribute to the pathophysiology and/or therapeutic strategy of GI diseases. For instance, ENS analysis that considers the time axis has the potential to provide information on age-related and disease progression related changes to ENS in GI motility disorders. In addition, computer technology allows CLE images captured from ENS to be evaluated in detail. The *in vivo* CLE images are unaffected by factors associated with section preparation in a pathological analysis; therefore, morphological findings or width of nerve strands is accurately evaluable. For instance, in our study, we compared nerve strand widths between normal ENS and abnormal Hirschsprung's disease ENS by computer software.

46.5 Conclusion

CV-assisted CLE is an innovative diagnostic approach to gastrointestinal diseases. The technique evaluates a wider range of ENS than pathological analysis and has the potential to assess ENS changes over time in a less-invasive manner. Although the novel imaging has still some challenges to be resolved for broad clinical application, we believe that CV-assisted CLE would provide new insight into the gastroenterology field, especially for neuro-functional gastrointestinal disorders.

References

1. Goyal RK, Hirano I. The enteric nervous system. N Engl J Med. 1996;334(17):1106–15. https://doi.org/10.1056/NEJM199604253341707.
2. Kobayashi M, Sumiyama K, Shimojima N, et al. Technical feasibility of visualizing myenteric plexus using confocal laser endomicroscopy. J Gastroenterol Hepatol. 2017;32(9):1604–10. https://doi.org/10.1111/jgh.13754.
3. Ohya TR, Sumiyama K, Takahashi-Fujigasaki J, et al. In vivo histologic imaging of the muscularis propria and myenteric neurons with probe-based confocal laser endomicroscopy in porcine models (with videos). Gastrointest Endosc. 2012;75(2):405–10. https://doi.org/10.1016/j.gie.2011.09.045.
4. Rajan E, Gostout CJ, Lurken MS et al. Evaluation of endoscopic approaches for deep gastric-muscle-wall biopsies: what works?. Gastrointest Endosc. 2008;67(2):297–303. S0016-5107(07)02149–9 [pii].

5. Shimojima N, Kobayashi M, Kamba S, et al. Visualization of the human enteric nervous system by confocal laser endomicroscopy in Hirschsprung's disease: an alternative to intraoperative histopathological diagnosis? Neurogastroenterol Motil. 2020;32(5):e13805. https://doi.org/10.1111/nmo.13805.

6. Kobayashi M, Neumann H, Hino S, et al. Influence of reviewers' clinical backgrounds on interpretation of confocal laser endomicroscopy findings. Endoscopy. 2016;48(6):521–9. https://doi.org/10.1055/s-0042-101408.

7. Wallace M, Lauwers GY, Chen Y, et al. Miami classification for probe-based confocal laser endomicroscopy. Endoscopy. 2011;43(10):882–91. https://doi.org/10.1055/s-0030-1256632.

8. Kiesslich R, Burg J, Vieth M, et al. Confocal laser endoscopy for diagnosing intraepithelial neoplasias and colorectal cancer in vivo. Gastroenterology. 2004;127(3):706–13.

9. Kobayashi M, Sumiyama K, Matsui H et al (2014) Fluorescein assisted confocal LASER microscopy imaging of the muscularis propria in porcine models. Gastrointest Endosc. 79(5, Supplement):AB474.

10. Samarasena JB, Ahluwalia A, Shinoura S, et al. In vivo imaging of porcine gastric enteric nervous system using confocal laser endomicroscopy & molecular neuronal probe. J Gastroenterol Hepatol. 2016;31(4):802–7. https://doi.org/10.1111/jgh.13194.

11. Samarasena JB, Tarnawski AS, Ahluwalia A, et al. EUS-guided in vivo imaging of the porcine esophageal enteric nervous system by using needle-based confocal laser endomicroscopy. Gas-trointest Endosc. 2015;82(6):1116–20. https://doi.org/10.1016/j.gie.2015.06.048.

12. Sumiyama K, Kiesslich R, Ohya TR, et al. In vivo imaging of enteric neuronal networks in humans using confocal laser endomicroscopy. Gastroenterology. 2012;143(5):1152–3. https://doi.org/10.1053/j.gastro.2012.09.001.

13. Canto MI. Staining in gastrointestinal endoscopy: the basics. Endoscopy. 1999;31(6):479–86. https://doi.org/10.1055/s-1999-8041.

14. Furuta Y, Kobori O, Shimazu H, et al. A new in vivo staining method, cresyl violet staining, for fiberoptic magnified observation of carcinoma of the gastric mucosa. Gastroenterol Jpn. 1985;20(2):120–4.

15. Ostrowski PP, Fairn GD, Grinstein S, et al. Cresyl violet: a superior fluorescent lysosomal marker. Traffic. 2016;17(12):1313–21. https://doi.org/10.1111/tra.12447.

16. Goetz M, Toermer T, Vieth M, et al. Simultaneous confocal laser endomicroscopy and chromoendoscopy with topical cresyl violet. Gastrointest Endosc. 2009;70(5):959–68. https://doi.org/10.1016/j.gie.2009.04.016.

17. Goetz M, Fottner C, Schirrmacher E, et al. In-vivo confocal real-time mini-microscopy in animal models of human inflammatory and neoplastic diseases. Endoscopy. 2007;39(4):350–6. https://doi.org/10.1055/s-2007-966262.

18. Kobayashi M, Shimojima N, Takahashi-Fujigasaki J et al. Enteric nervous system visualization using confocal laser endomicroscopy. J Gastroenterol Hepatol. 2016;31, Supplement 3(53):212.

Optical Fluorescence: Application of Structured Light Illumination and Compressed Sensing to High-speed Laminar Optical Fluorescence Tomography

47

Ichiro Sakuma

Abstract

Spiral wave is known as a cause of arrhythmia. However, there is no report that a filament that is the center of the spiral wave (SW) in three dimensions was experimentally observed. Obtaining sufficient recording speed to observe cardiac excitation propagation remains a major challenge. We proposed an optical method called Compressed LOT (CLOT) which adapted compressed sensing to Laminar Optical Tomography (LOT) in order to estimate the three-dimensional membrane potential distribution inside the myocardial tissue. We examined the performance of the proposed method of identifying the location of SW three-dimensional membrane potential reconstructed with CLOT in simulation. We used phase variance analysis for the determination of SW filament. Obtained results show that the method can estimate the location of SW filaments up to 2.5 mm depth. We also constructed an experiment system to verify the principle of CLOT using optical phantoms containing fluorophores. We used a digital mirror device (DMD) with high-speed pattern switching capability (with an interval of 1 ms). Optical phantom mimicking biological tissue with fluorophore was prepared. Fluorescent dye contained in a capillary with an internal diameter of 1.4 mm was immersed in light scattering and absorbing medium contained intralipid. Reconstruction accuracy at the deep position (depth: 1.875 mm) was less than 04 mm. It is considered that accurate measurement of the optical constant is required for better reconstruction accuracy.

Keywords

Optical mapping · Laminar optical tomography · Compressed sensing

47.1 Introduction

In the field of Multidisciplinary Computational Anatomy researches, integration of physiological information with structural/anatomical information in three-dimensional space is important. For this purpose, a mapping method of physiological information is required. As an example of the technology, we have investigated three-dimensional mapping of electrophysiological information in Langendorff perfused rabbit hearts. The spatial organization and dynamics of reentrant activity in three-dimensional (3D) cardiac muscle is important to understanding arrhythmia such as ventricular tachycardia and fibrillation. Spiral reentry in three-dimensional space is called a scroll wave, and the scroll wave turns around a filament having a string shape connecting the centers of vortices of a two-dimensional space [1, 2]. Optical mapping using voltage-sensitive dye together with a high-speed video camera enables measurement with high spatial resolution and has been used in electrophysiological studies [3]. Optical mapping method can only acquire signals from the surface layer (~250 μm) of cardiac tissue. Several attempts to visualize scroll waves in actual myocardial tissues have been reported [4, 5]. Laminar optical tomography (LOT), which is a microscopic form of diffuse optical tomography, has been developed in the field of biomedical imaging [6, 7]. It can measure the 3D fluorescence distribution in tissue at the microscopic level. However, obtaining sufficient recording speed to determine rapidly changing dynamic fluorescence signals, such as fluorescence signals obtained in optical membrane potential mapping for cardiac excitation propagation measurements, remains a major challenge. The compressed sensing (CS) technique, which can reconstruct the original information from fewer measurements, is now

I. Sakuma (✉)
Medical Device Development and Regulation Research Center, School of Engineering, The University of Tokyo, Tokyo, Japan

Department of Precision Engineering, School of Engineering, The University of Tokyo, Tokyo, Japan

Department of Bioengineering, Graduate School of Engineering, The University of Tokyo, Tokyo, Japan
e-mail: sakuma@bmpe.t.u-tokyo.ac.jp

widely used for many imaging applications. In the chapter, a new LOT method using CS theory which is called CLOT is proposed for the measurement of three-dimensional membrane potential in a heart wall [8, 9].

47.2   Methods

47.2.1   Compressed Laminar Optical Tomography (CLOT)

Details of the method are available in the references [6–8]. The excitation light laser enters at the position r_s. The incident light diffuses inside the tissue and excites the dye located at position r where fluorescence is generated which also scatters inside the tissue and exits from the position r_d on the tissue surface as shown in Fig. 47.1a. Change in fluorescence $\Delta F_{x,m}(r_d, r_s, t)$ of voltage-sensitive dye due to change in membrane action potential $\Delta V(r, t)$ at position r in cardiac muscle as shown in Fig. 47.1a can be expressed as follows:

$$\Delta F_{x,m}(r_d, r_s, t) = \int w \Delta V(r,t) H_x(r,r_s) E_m(r,r_d) d^3 r \quad (47.1)$$

where, $H_x(r, r_s)$ represents the excitation light intensity at location originated from the incident excitation light with unit intensity at location r_s, and $E_m(r, r_d)$ represents the

intensity of fluorescence light scattered in the tissue detected at location r_d originated from fluorescence with unit intensity at location r. w is the scaling factor between quantum yield of the fluorophore and membrane action potential $V(r, t)$. $H_x(r, r_s)$ and $E_m(r, r_d)$ can be estimated by Monte Carlo simulations describing optical diffusion process in biological tissues. When the volume of tissue is discretized with the assumption of spatially homogeneous absorption and scattering properties, this Eq. (47.1) can be a linear equation [9]:

$$\Delta F_{x,m} = J \Delta V \quad (47.2)$$

Compressed sensing is to restore a high dimensional signal with sparsity (nature with many zero components) from a small observation value. By applying wavelet transform, we will be able to transfer 3D action potential distribution into sparse vector by applying wavelet transform since electrophysiological excitation wave is spatially localized having low spatial frequency components. By adopting a random excitation light pattern such as shown in Fig. 47.1c, J will have random property. We can transform the 3D action potential distribution into sparse vector by applying wavelet transform.

$$\Delta F_{x,m} = J \Delta V = J \Psi \alpha \quad (47.3)$$

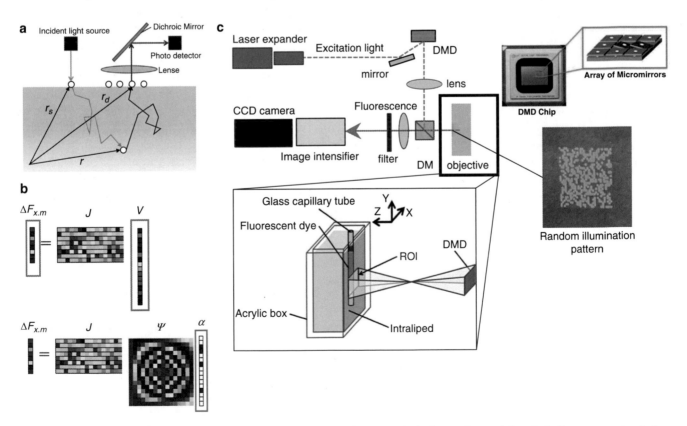

Fig. 47.1 Principle of CLOT and configuration of experimental CLOT system. (**a**) Optical scattering model to obtain fluorescence signals from fluorophore in a turbid medium. (**b**) Schematic representation of the equation of CLOT. (**c**) Experimental optical system of CLOT

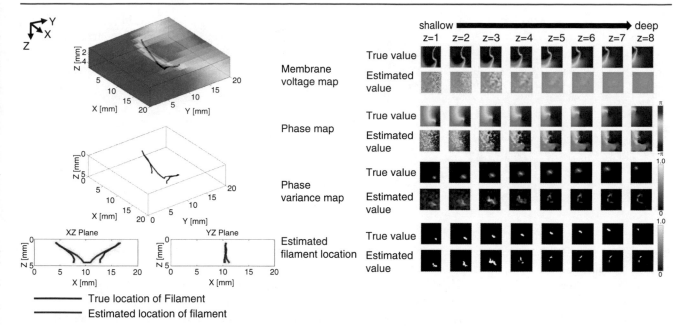

Fig. 47.2 Results of filament estimation in the simulation study. Upper left panel: Three-dimensional spiral reentry in a simulation model. Orange color and blue color stand for depolarized and repolarized states, respectively. Lower left panel: Estimated location of the fila-

ment. Right panel: Results of estimation in each layer along the depth direction. Membrane voltage map, phase map, phase variance map, and detected filament location are displayed at each layer along the depth direction

where Ψ is the discrete wavelet transformation matrix (Fig. 47.2b). We can get solution of equation by solving the following optimization problem by applying a compressed Sensing algorithm:

$$argmin\left\{\frac{1}{2}\left\|\Delta F_{x,m} - J\Psi\alpha\right\|_2^2 + \left\|\alpha\right\|_1\right\}$$
$$subject\ to\ \Delta F_{x,m} = J\Delta V = J\Psi\alpha \quad (47.4)$$

47.2.2 Simulation Study

We created a temporally continuous three-dimensional membrane potential distribution based on as a measurement object. We assumed amplitude of change in fluorescence signals due to membrane action potentials as large as 9% referring to typical experimental data found in optical mapping experiments using voltage-sensitive dye Di-4-ANEPPS (Ex. 532 nm, Em. 610 nm). We calculated 2D fluorescence image ΔF corresponding to the specific illumination pattern and three-dimensional membrane action potentials distribution based on optical diffusion model using Monte Carlo simulation. The size of the tissue model was 20 mm × 20 mm × 5 mm. We discretize the 5 mm thickness into 8 layers. Absorption and scattering coefficient of the tissue for the excitation light (532 nm) were 0.160 [mm⁻¹] and 8.820 [mm⁻¹] respectively. Those for fluorescence (610 nm) were 0.095 [mm⁻¹] and 8.830 [mm⁻¹], respectively.

Three-dimensional distributions of membrane action potentials were reconstructed from simulated ΔF using CLOT. Phase analysis was conducted on true value data and reconstructed three-dimensional membrane potential distribution to generate a phase map. We used phase variance analysis for the determination of SW filament described in [9]. An appropriate threshold is set, and a filament region was estimated. Finally, we calculated the average minimum distance error between the true value and the estimated filament.

47.2.3 Experimental System of CLOT for Optical Phantom Study [10]

The experimental optical measurement system as shown in Fig. 47.1c consisted of an excitation laser (SSL-532-1000-10TM-D-LED, Shanghai Sanctity Laser Technology, max output power: 1 W), a laser expander (BEHP-10-532, Optosigma), a dichroic mirror (DM) (Di 02-EM CCD camera (ADT-100, Flovel), a Digital Mirror Device (DMD) (DLP 6500, Texas Instruments), and lens system. The random pattern of 6 mm square generated by DMD was then enlarged to 10 mm square by an objective lens and was reflected by the dichroic mirror and irradiated onto the measurement object every 1 ms. Fluorescence light passed through a dichroic mirror and a long pass filter and was recorded by the EMCCD camera.

Table 47.1 Specification of the optical phantom used and its optical constants

Fluorescence dye	DY-521 XL (Excitation: 523 nm, Fluorescence: 668 nm)		
Capillary	Outer diameter:1.5 mm, Inner diameter 1.3 mm, Length: 75 mm		
Intralipd	0.5%		
Optical Properties			
Wavelength	Absorption Coefficient [mm^{-1}]	Absorption Coefficient [mm^{-1}]	Anisotropic parameter
532 nm	0.267	0.460	0.9
610 nm	0.110	0.424	0.9

Optical phantom mimicking biological tissue with fluorophore was prepared. Fluorescent dye contained in a capillary with an internal diameter of 1.4 mm was immersed in light scattering and absorbing medium contained intralipid. The specification of the phantom and its optical constant are shown in Table 47.1. The capillary was placed at depth of 0.625 mm and 1.875 mm from the surface.

47.3 Results and Discussion

47.3.1 Simulation Study

Figure 47.2a shows the result of estimating the filament and the average distance error for each layer along the depth direction. The absolute values of fluorescence signals were not accurate as shown in Fig. 47.2b. However, changes in spatio-temporal patterns of three-dimensional membrane potential distribution could be reconstructed to some extent. Thus, the phase map that can visualize the changes in temporal signal information regardless of its amplitude can be reconstructed until the fourth layer (2.5 mm). Consequently, the location of the filament of the spiral wave could be estimated with distance errors less than 0.625 mm that was the size of the voxel in the simulation. It was considered that the time change in each voxel of the three-dimensional membrane potential distribution could be reconstructed until the fourth layer (2.5 mm).

47.3.2 Optical Phantom Study

We have constructed an optical system for CLOT with random excitation illumination patterns generated by a digital mirror device (DMD) with high-speed pattern switching capability (every 1 ms). The CLOT was applied to identify the location of capillary obtaining fluorophore immersed in an optical scattering medium with intralipid. The reconstruction accuracy at the deep position (depth: 1.875 mm) was less than 0.4 mm. On the other hand, that at the shallow position was as large as 1 mm. Variation in fluorescence luminance value might be influenced by optical absorption properties. Thus, it is necessary to accurately measure the optical constant of the object to improve reconstruction accuracy.

47.4 Conclusion

Compressed sensing method was applied to high-speed laminar optical fluorescence tomography for identifying spatio-temporal patterns of optical mapping of cardiac action potential measurement. The basic concept of the proposed method was investigated with simulation studies and optical phantom studies. In the simulation study, we demonstrated that the filament can be estimated from the three-dimensional membrane potential reconstructed by CLOT down to 2.5 mm in the depth direction by using phase variance analysis. These results indicate the possibility of measurement with its speed sufficient for measuring the excitation propagation of the heart by applying compression sensing to LOT. We have constructed an optical system for CLOT where random excitation illumination patterns were generated using a digital mirror device (DMD) with high-speed pattern switching capability (with an interval of 1 ms). The CLOT was applied to identify the location of capillary obtaining fluorophore immersed in an optical scattering medium with intralipid. The reconstruction accuracy at the deep position (depth: 1.875 mm) was less than 0.4 mm. Variation in fluorescence luminance value might be influenced by optical absorption properties. Thus, it is necessary to accurately measure the optical constant of the object for increasing reconstruction accuracy in the actual experiment system.

References

1. Pertsov A, Vinson M. Dynamics of scroll waves in inhomogeneous excitable media. Phil Trans R Soc Lond A. 1994;347(1685):687–701.
2. Qu Z, Kil J, Xie F, Garfinkel A, Weiss JN. Scroll wave dynamics in a three-dimensional cardiac tissue model: roles of restitution, thickness, and fiber rotation. Biophys J. 2000;78(6):2761–75.
3. Efimov IR, Nikolski VP, Salama G. Optical imaging of the heart. Circ Res. 2004;95(1):21.
4. Bernus O, Mukund KS, Pertsov AM. Detection of intramyocardial scroll waves using absorptive transillumination imaging. J Biomed Opt. 2007;12:014035.
5. Hyatt CJ, Zemlin CW, Smith RM, Matiukas A, Pertsov AM, Bernus O. Reconstructing subsurface electrical wave orientation from cardiac epi-fluorescence recordings: Monte Carlo versus diffusion approximation. Opt Express. 2008;16(18):13758–5772.
6. Hillman E, Bernus O, Pease E, Bouchard MB, Pertsov A. Depth-resolved optical imaging of transmural electrical propagation in perfused heart. Opt Express. 2007;15(26):17827–41.
7. Hillman EM, Burgess SA. Sub-millimeter resolution 3D optical imaging of living tissue using laminar optical tomography. Laser Photonics Rev. 2009;3(1-2):159–79.

8. Harada T, Tomii N, Manago S, Kobayashi E, Sakuma I. Simulation study on compressive laminar optical tomography for cardiac action potential propagation. Biomed Opt Express. 2017;8(4):2339–58.

9. Tomii N, Yamazaki M, Arafune T, Honjo H, Shibata N, Sakuma I. Detection algorithm of phase singularity using phase variance analysis for epicardial optical mapping data. IEEE Trans Biomed Eng. 2016;63(9):1795–803.

10. Sakuma I, Kobayashi T, Seno H, Akagi Y, Nakagawa K, Yamazaki M et al., editors. Application of Structured Light Illumination and Compressed Sensing to High Speed Laminar Optical Fluorescence Tomography. Latin American Conference on Biomedical Engineering; 2019: Springer. pp. 1216–1219.

Magneto-stimulation System for Brain Based on Medical Images

48

Akimasa Hirata and Jose Gomez-Tames

Abstract

This chapter presents an individualized diagnosis system for non-invasive transcranial magnetic stimulation. The feature of this proposal is that highly-accurate cortical localization method in the magnetic stimulation is developed after validating the procedure by comparing with the clinical data obtained in neurosurgery. The method is based on the computation of individualized TMS-induced electric field on the brain based on physical head models developed from magnetic resonance images, and thus it can be applied to an individualized treatment system. This personalized method allows TMS localization with high precision, such as in preoperative mapping that can permit reduce time identifying the functional regions during intraoperative phase in brain surgery. Also, the proposed method allows the evaluation of TMS coil characteristics and coil positioning for daily clinical protocols.

Keywords

Transcranial magnetic stimulation · Computational model · Electric field · Personalized stimulation Magnetic resonance imaging

48.1 Introduction

Transcranial magnetic stimulation (TMS) is one type of non-invasive brain stimulation technologies aimed to generate or modify brain activity by placing a magnetic coil on the surface of the head [1]. The injected electric current through the coil windings produces time-varying magnetic fields that cause eddy currents in the brain that can effectively stimulate brain neurons. The generation of neuronal responses to TMS permits applications on therapy, diagnosis, and assessment of brain functions.

One application of TMS is brain function mapping in which the targeted cortical region is associated with the motor responses (measured as motor evoked potential: MEP) or speech (measured as linguistic arrest) produced by the stimulation. Preoperative mapping of brain functions is one application as a prescreening test to intraoperative mapping that uses direct electrical stimulation (DES). The preoperative mapping technique by TMS can help to delineate the tumor from cortical tissue in a non-invasive way instead of directly and invasively stimulating the brain cortex by electrodes as DES. If precise localization can be achieved by a preoperative procedure, time to identify the functional regions can be reduced in the intraoperative phase [2]. Also, it can impact the neurosurgical decision-making and lead to the modification of the initial treatment strategy [3]. It has been reported improved surgical and oncological outcomes in patients after the adoption of nTMS mapping [4, 5].

However, it is often reported that the estimated target area by TMS is somewhat different from the actual area confirmed by the direct electric stimulation (DES) in the neurosurgery. According to recent studies (e.g., [6]), this mislocalization of the target area is caused by the complicated anatomy of the brain, especially because of the existence of the cerebrospinal fluid whose conductivity is much higher than the remaining brain tissues. So far, the only method that permits in-vivo and real-time estimation of the target regions is via electromagnetic computational techniques. To estimate the exact stimulation location on the cortical region, it is essential to use medical images to construct a computational electromagnetic model of a human head (e.g., [7]).

This chapter presents an individualized diagnosis system for non-invasive transcranial magnetic stimulation. In the first part, we investigated the effects of coil design and positioning on targeting specific brain regions. In the second part, a highly-accurate localization method in the magnetic

A. Hirata (✉) · J. Gomez-Tames
Department of Electrical and Mechanical Engineering, Nagoya Institute of Technology, Nagoya, Japan

Center of Biomedical Physics and Information Technology, Nagoya Institute of Technology, Nagoya, Japan
e-mail: ahirata@nitech.ac.jp; jgomez@nitech.ac.jp

stimulation was developed after validating the procedure by comparing it with the clinical data obtained in neurosurgery using individualized patients model.

48.2 Modeling Methods

The modeling methods are based on the computation of the induced electric field as a metric of stimulation [8]. The following steps are considered to develop the computation model to determine the TMS-induced electric field in the brain. First, the development of a digital individualize human head model from MRI (magnetic resonance image) is presented. Second, electromagnetic computation of the induced electric field is described. Consideration of the magnetic coil modeling and location relative to the scalp is important to estimate the correct location.

48.2.1 Development of Human Head Models

Anatomically based human head models are developed from magnetic resonance images. The detailed procedure for generating head models is described in our previous study [9] and others [10–12]. Segmentation and surfaces of brain tissues are obtained via FreeSurfer image analysis software [13, 14] from T1-weighted MRIs [13] while a mesh of non-brain tissues is generated from T1- and T2-weighted MRIs via the FSL software library [15]. In our research group, non-brain tissues are obtained by using a semi-automatic procedure that includes region-growing and thresholding techniques. The head models are segmented into 14 anatomical tissues/fluids (skin, fat, muscle, outer skull, inner skull, gray matter, white matter, cerebellar gray matter, cerebellar white matter, brainstem, nuclei, ventricles, cerebrospinal fluid, and eye tissues) using T1 and T2 MRIs. The final volume conductor models were represented in a grid of cubical voxels with a resolution of 0.5 mm. The electrical conductivities of the tissues were determined at 10 kHz.

48.2.2 Computational Electromagnetic Method

The computational procedure is identical to that in our previous study [16]. The frequency component of typical pulses used in TMS is below 100 kHz or lower, and thus the magneto-quasi-static approximation was applicable. Under the approximation, the electric displacement current is ignored, and the induced currents in the human body are assumed not to perturb the external magnetic field.

The magnetic vector potential specific to the TMS coil under study in each voxel of the head model was computed by FEKO (EMSS-SA, Stellenbosch, South Africa), a commercial software package that is based on the method of moments. The electric fields induced in the human head by the magnetic vector potential were then determined using an in-house magneto-quasi-static solver by the scalar potential finite difference (SPFD) [16]. It was solved iteratively by the successive-over-relaxation method and a multigrid method [16]. The electric field along the edge of the voxel was obtained by dividing the difference in the potential between the nodes of the voxel by the distance across the nodes and adding the vector potential.

48.2.3 Experimental Protocol

Two datasets were used in the analysis. For the part of the coil evaluation, ten anatomical head models were constructed from T1- and T2-weighted images acquired from a magnetic resonance image scanner. For the evaluation of high-accuracy TMS localization for preoperative mapping, eight patients (29–64 years, four women) participated [17, 18]. The patients had intra-axial brain neoplasms located within or close to the motor eloquent area. Peritumoral mapping was conducted in the tumor-containing hemisphere (affected hemisphere). A TMS navigation system (nTMS, Brainsight, Rogue Resolutions, Canada) was used to record the orientation and position of the coil relative to the head position in real time during presurgical mapping. The sites in the brain cortex stimulated by DES during surgery were recorded using an optical tracking system as ground truth of the estimated values by computational model (Polaris, Northern Digital Instruments Co., Canada). The anatomical head model and coil position/ orientation were used to estimate the target cortical site.

48.3 Evaluation of TMS Coils

We discuss the variation of focality and targeting variability among different coils and subjects, as shown in Fig. 48.1a. Three different coils (figure-8 coil, figure-8 with isolation, and eccentric coils) were placed in the C3 position over the scalp according to International 10–20 EEG system among the ten different subjects. The C3 was chosen to target the hand motor area that is usually investigated for allowing measurements of physiological response to TMS. Localization on the same scalp landmark for all subjects is known as "one-for-all" placement approach, in which the coil localization is not based on the anatomical differences of each subject's head.

Figure 48.1b provides the computed group-level electric field strength of the 10 subjects mapped on a standardized brain model for coils with a radius of 20 mm and 60 mm [19]. Also, it presents the locations of the maximum electric field strength of each subject. It is confirmed that the electric field distributions are distributed around the target area and concentrated in particular when the coil radius is small. Similarly, the inter-subject variability of the peak electric field is smaller for a coil with a small radius as compared to a coil with a large radius. The probability that more than 90% of the times the hotspot is inside the target area is almost the same between the various coils with a minor higher focality when the radius of coils is smaller. Hence, we found that focality could be improved only to some extent when selecting the coil radius or type.

The individualized computation of the induced electric field indicates that one-for-all localization allows targeting with a certain dispersion of maximum values in the region of interest within subjects [20]. However, precise targeting is required in applications, such as preoperative mapping. The next section discusses high-accurate targeting in personalized stimulation.

48.4   High-Accurate TMS Localization

High-precision localization method is presented for the motor area using a post-processing method that combines computed electric fields for TMS that delivered high MEPs during peritumoral mapping. Peritumoral mapping by TMS was conducted on patients who had intra-axial brain neoplasms located within or close to the motor speech area. The method was compared to the stimulation site localized via intraoperative direct DES during awake craniotomy for the tumor-containing hemisphere navigated TMS. The localization method is described, and comparison with experiments is presented.

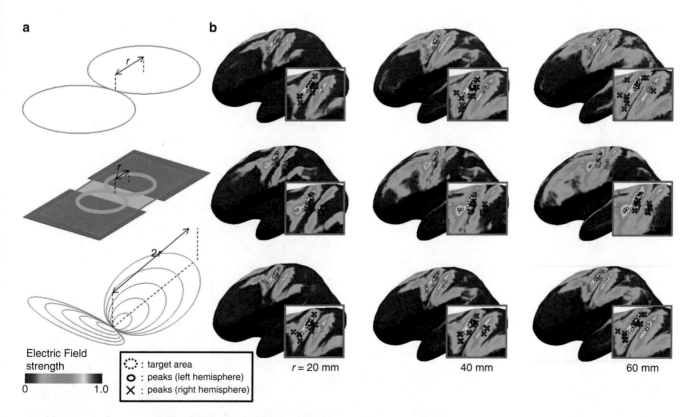

Fig. 48.1 TMS coils evaluation for one-for-all localization. (**a**) Commonly used figure-8 coil, figure-8 coil with isolation (covered by a magnetic sheet), and eccentric figure-8 coil with inclination (top-to-bottom order). (**b**) Induced electric field strength averaged over ten subjects on the standardized brain surfaces. The field strengths are normalized by the maximum value of each coil. Positions where peak induced field are indicated by dots (left hemisphere) and cross (right hemisphere) for each subject (10 subjects for each case). The target area (hand motor area) is indicated by a black circle

48.4.1 Localization Method

Multiple stimulation locations are considered using a commonly used figure-8 coil. MEPs with the highest peak were selected from more than thirty stimulations using different coil positioning and angles during peritumoral mapping by TMS (hereinafter, a sample refers to each peritumoral stimulation coil configuration). The induced electric field E_i for each sample, i, was multiplied to find the cortical regions where the electric fields were high for all samples (

$$E_{\text{focal}} = \Pi_i |E_i| \qquad (48.1)$$

The rationale for this approach was that, if all samples with the highest MEPs activate the same cortical region, the electric fields in the activated region should also be high for each sample. In this study, the hotspot area corresponds to the cortical surface area where E_{focal} is larger than the threshold ($0.7 \times \max(E_{\text{focal}})$). The estimated target area decreased with the number of samples and fell below 50 mm^2 from the fourth sample.

48.4.2 Localization in Healthy Hemisphere

Hotspot areas of eight non-affected hemispheres were determined, as shown in Fig. 48.2a [17]. The hotspots of each subject were registered on a standard brain space. The registered hotspots were summed and normalized to the maximum E_i. The resulting hotspot was compared with the location of the hand motor area in non-affected hemispheres. As seen in Fig. 48.2a, the hotspot can be detected in the crown of the precentral gyrus. The location is in agreement with data from the literature and agrees with the well-known position of the "hand-knob" as a landmark for neural elements involved in the motor hand function [21].

48.4.3 Localization in Hemisphere Containing Tumor

Peritumoral mapping was conducted in the tumor-containing hemisphere in four subjects [17]. Figure 48.2b shows the hotspot localization by DES, nTMS, and the proposed method. The hand motor areas estimated by the proposed method (22.1 ± 12.3 mm^2) were included in the hotspot area obtained by DES in the four cases. Hotspots predicted by nTMS were within DES area in two out of four cases. Euclidean distance between the proposed method at the center of gravity (CoG) and DES was 4.95 ± 0.72 (standard error [SE]). The distance between nTMS and DES was 7.78 ± 1.99 (SE) mm. We found that the proposed method was more accurate than nTMS. In the particular case of subject 3, there were two apparent hotspots. In this case, the criterion followed dictated that only hotspots located in the primary motor cortex were considered.

48.5 Conclusion

A high-precision localization method for TMS targeting a specific motor area was presented. The strategy was to estimate a stimulated area by synthesizing computed non-uniform current distributions in the brain during multiple stimulations considering the measurements of the MEP. This developed computational system was able to achieve a highly-accurate cortical localization method after validating the procedure with clinical data obtained in neurosurgery. The proposed method is much easier to use in clinical practice and may reduce the number of measurements required during preoperative mapping as compared to the original protocol. More importantly, a more accurate localization mapping method may benefit neurosurgical

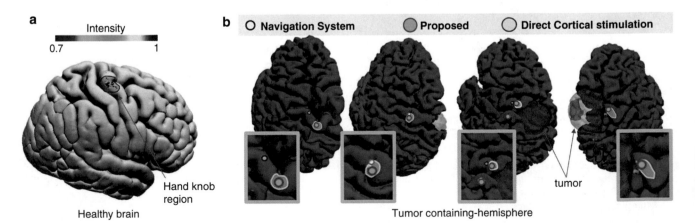

Fig. 48.2 High-accurate personalized localization for TMS. (**a**) Average mapping of the hand motor area in a healthy hemisphere. (**b**) Hotspot areas on the tumor-containing hemisphere for the proposed method, direct electrical stimulation (DES), and navigated TMS (nTMS)

decision-making in the surgery strategy. In contrast to high-accurate estimation method, we showed that one-for-all TMS localization could be used to generated high-electric fields distributed around the target region in which the concentration may be reduced by the coil design to a certain extend. Finally, the proposed computational method was able to detect the hotspot in the gyral crown of the motor area based on the electric field strength in agreement with the intraoperative DES experiment. Future work includes applying the proposed localization method to non-motor regions, such speech area.

References

1. Barker AT, Jalinous R, Freeston IL. Non-invasive magnetic stimulation of human motor cortex. Lancet (London, England). 1985;1:1106–7.
2. Takahashi S, Vajkoczy P, Picht T. Navigated transcranial magnetic stimulation for mapping the motor cortex in patients with rolandic brain tumors. Neurosurg Focus. 2013;34:E3.
3. Krieg SM, Shiban E, Buchmann N, Gempt J, Foerschler A, Meyer B, et al. Utility of presurgical navigated transcranial magnetic brain stimulation for the resection of tumors in eloquent motor areas. J Neurosurg. 2012;116:994–1001.
4. Picht T, Schulz J, Vajkoczy P. The preoperative use of navigated transcranial magnetic stimulation facilitates early resection of suspected low-grade gliomas in the motor cortex. Acta Neurochir. 2013;155:1813–21.
5. Krieg SM, Sabih J, Bulubasova L, Obermueller T, Negwer C, Janssen I, et al. Preoperative motor mapping by navigated transcranial magnetic brain stimulation improves outcome for motor eloquent lesions. Neuro Oncol. 2014;16:1274–82.
6. Laakso I, Hirata A, Ugawa Y. Effects of coil orientation on the electric field induced by TMS over the hand motor area. Phys Med Biol. 2014;59:203–18.
7. Wagner TA, Zahn M, Grodzinsky AJ, Pascual-Leone A. Three-dimensional head model simulation of transcranial magnetic stimulation. IEEE Trans Biomed Eng. 2004;51:1586–98.
8. Eaton H. Electric field induced in a spherical volume conductor from arbitrary coils: application to magnetic stimulation and MEG. Med Biol Eng Comput. 1992;30:433–40.
9. Laakso I, Tanaka S, Koyama S, De Santis V, Hirata A. Inter-subject variability in electric fields of motor cortical tDCS. Brain Stimul. 2015;8:906–13.
10. Huang Y, Datta A, Bikson M, Parra LC. Realistic volumetric-approach to simulate transcranial electric stimulation-ROAST-a fully automated open-source pipeline. J Neural Eng. 2019;16:056006.
11. Windhoff M, Opitz A, Thielscher A. Electric field calculations in brain stimulation based on finite elements: an optimized processing pipeline for the generation and usage of accurate individual head models. Hum Brain Mapp. 2013;34:923–35.
12. Rashed EA, Gomez-Tames J, Hirata A. Development of accurate human head models for personalized electromagnetic dosimetry using deep learning. NeuroImage. 2019;202:116132.
13. Dale AM, Fischl B, Sereno MI. Cortical surface-based analysis. NeuroImage. 1999;9:179–94.
14. Fischl B. FreeSurfer. NeuroImage. 2012;62:774–81.
15. Smith SM. Fast robust automated brain extraction. Hum Brain Mapp. 2002;17:143–55.
16. Laakso I, Hirata A. Fast multigrid-based computation of the induced electric field for transcranial magnetic stimulation. Phys Med Biol. 2012;57:7753–65.
17. Aonuma S, Gomez-Tames J, Laakso I, Hirata A, Takakura T, Tamura M, et al. A high-resolution computational localization method for transcranial magnetic stimulation mapping. NeuroImage. 2018;172:85–93.
18. Takakura T, Muragaki Y, Tamura M, Maruyama T, Nitta M, Niki C, et al. Navigated transcranial magnetic stimulation for glioma removal: prognostic value in motor function recovery from post-surgical neurological deficits. J Neurosurg Am Assoc Neurol Surgeons. 2017;127:1–15.
19. Iwahashi M, Gomez-Tames J, Laakso I, Hirata A. Evaluation method for *in situ* electric field in standardized human brain for different transcranial magnetic stimulation coils. Phys Med Biol. 2017;62:2224–38.
20. Gomez-Tames J, Hamasaka A, Laakso I, Hirata A, Ugawa Y. Atlas of optimal coil orientation and position for TMS: a computational study. Brain Stimul. 2018;11:839–48.
21. Yousry TA, Schmid UD, Alkadhi H, Schmidt D, Peraud A, Buettner A, et al. Localization of the motor hand area to a knob on the precentral gyrus. A new landmark. Brain. 1997;120:141–57.

AI: A Machine-Learning-Based Framework for Developing Various Computer-Aided Detection Systems with Generated Image Features

49

Mitsutaka Nemoto and Naoto Hayashi

Abstract

A computer-aided detection (CADe) system identifies features on a medical image and brings them to the radiologist's attention. However, the spread of CADe systems in clinical sites is limited. Our aim is to establish a generalized framework for developing various CADe systems. The framework would provide opportunities for many clinicians without the CADe system expertise to develop and use such systems that meet their specific needs. We proposed a pilot version of the framework, including four pretrained algorithms: preprocessing, extraction of a candidate area, candidate detection, and candidate classification. We experimentally confirmed that two different types of CADe system were developed successfully through the generalized framework using two different datasets. We also proposed two feature generation methods to improve the generalized framework. One is the use of multiple deep convolutional autoencoders (DCAEs) trained with a normal dataset. The other is by transfer using a deep convolutional neural network (DCNN) pretrained with an anatomical landmark dataset from whole-body CT. An evaluation of these methods using head MR angiography datasets shows that the DCAEs could extract useful features for lesion classification.

Keywords

Computer-aided detection (CADe) · Generalized framework for CADe development · Deep learning Transfer learning · Anatomical landmark

M. Nemoto (✉)
Faculty of Biology-Oriented Science and Technology, Kindai University, Wakayama, Japan
e-mail: nemoto@waka.kindai.ac.jp

N. Hayashi
Department of Computational Diagnostic Radiology and Preventive Medicine, 22nd Century Medical and Research Center, The University of Tokyo Hospital, Tokyo, Japan
e-mail: naoto-tky@umin.ac.jp

49.1 Introduction

Computer-aided detection (CADe) systems often speed up medical diagnosis, reduce diagnostic errors, and improve quantitative evaluation [1]. However, CADe systems are challenging to develop. The development requires specialized knowledge such as image processing, machine learning, and radiology. Generally, a CADe system consists of four components: preprocessing, candidate search area extraction, candidate detection, and candidate classification. Most of the algorithms in a CADe system must be manually designed according to the detection target, image modality, and so on. Machine learning methods are used to optimize some algorithms in a CADe system [2, 3]. It is difficult for clinicians to develop their own CADe system and use it. A new method of developing CADe systems easily is necessary for clinicians without expertise in medical image analysis.

We conceived an idea of a generalized framework to develop various CADe systems without any manual design of algorithms. When developing a CADe system using the framework, developers need to make a CADe training dataset, which is an annotated medical image dataset. The annotations in the CADe training dataset include the types of lesion and the areas of the lesion in medical images. The framework includes some pretrained algorithms. Those algorithms are designed generally and are prepared for each component of the CADe system. The pretrained algorithms can be optimized by machine learning methods with the CADe training dataset. With the help of the framework, users without technical knowledge can develop an arbitrary CADe system by only collecting the CADe training dataset.

In this study, we engaged in the establishment of the generalized framework for developing any CADe systems (Fig. 49.1). We developed a pilot version of the generalized framework and evaluated its feasibility [4]. We also developed a method of generating local image features by multiple deep convolutional autoencoders (DCAEs) for 2.5-dimensional images [5, 6]. The generated features could be used in some component algorithms of the CADe system.

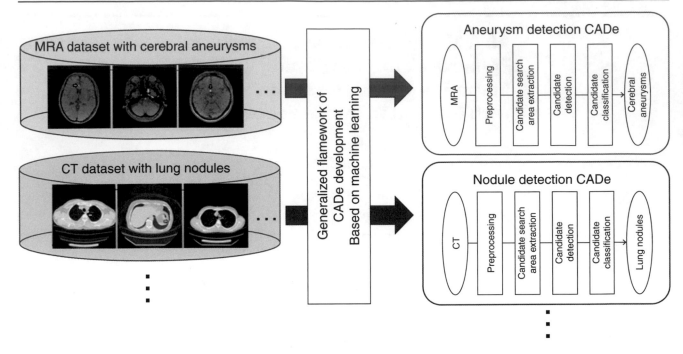

Fig. 49.1 Automated CADe system development by the generalized framework

Moreover, we studied another feature generation method, that is the transfer of a deep convolutional network pretrained with the local appearance of anatomical landmarks (LMs) to CADe tasks [7]. An LM is a unique local structure having anatomical meanings and plays a crucial role in analyzing medical images [8, 9]. LMs are the key aspects to understand patient anatomy in a medical image.

49.2 Generalized Framework for Developing CADe Systems

The proposed pilot version of the generalized framework for developing CADe systems (Fig. 49.2) [4] includes four pre-trained algorithms: preprocessing, candidate search area extraction, candidate detection, and candidate classification. The algorithms are designed to have a generality applicable to various diagnostic tasks. Most of the component algorithms are based on voxel classifications that could be optimized by machine learning and feature selection. In feature selection, a large-scale feature bank including a wide variety of voxel features is used.

1. Preprocessing: The first processing is for scaling input volume to isotropic volume by trilinear interpolation. The isotropic voxel size is automatically determined by the size of the lesions in the CADe training dataset.
2. Candidate search area extraction: The second processing is for extracting the search area for lesion candidates in which target lesions are likely to exist. The candidate

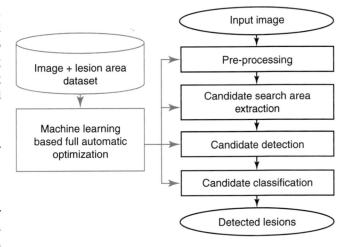

Fig. 49.2 Proposed prototype generalized CADe framework [4]

search area is extracted using a cascade [10] of voxel classifier ensembles. The ensembles consist of single or multiple weak classifiers, which are decision stumps [11] based on the voxel features from the feature bank. The ensembles are trained with the cost-sensitive [12] AdaBoost [4, 13] to classify the lesion-related voxels. The parameters of the classifier ensembles, such as the thresholds of the ensemble outputs, are optimized to achieve 99.95% voxel sensitivity. The numbers of weak classifiers included in the first nine ensembles are predetermined as 1, 10, 10, 25, 25, 50, 50, 100, and 100.

3. Candidate detection: The third processing is for detecting the lesion candidate points from the extracted candidate

search area. A candidate point is a local maximum of the lesion voxel likelihood. The lesion voxel likelihood, which is calculated at every voxel in the candidate area, is the output value of a voxel classifier ensemble consisting of 200 decision stumps. The voxel classifier ensemble is trained with the cost-sensitive AdaBoost. Each of the decision stumps is based on the voxel features from the feature bank or the outputs of the ensembles within the cascade for the candidate search area extraction.

4. Candidate classification: The fourth processing is for classifying the true positive lesion candidates and the false positives using a candidate classifier ensemble. The ensemble is constructed from 200 decision stumps. Each of the decision stumps is based on a pooling feature [14], which is calculated by applying a max-, min-, or average-pooling operator to a voxel feature. The voxel feature is a member of the feature bank or the output of the ensemble used in the candidate area extraction or the candidate point detection. The candidate classification ensemble is trained with the cost-sensitive AdaBoost.

To evaluate the feasible applicability of the generalized framework, we performed two experiments with two different CADe training datasets: (i) the head MR angiography dataset with cerebral aneurysms and (ii) the chest CT dataset with pulmonary nodules. The developments of the CADe systems using the framework and the performances of the developed systems were evaluated by threefold cross-validation. FROC curves and ANODE scores [15] were calculated to evaluate the lesion detection performances.

The CADe systems for detecting cerebral aneurysms in the brain MRA were successfully developed using the brain MRA volume dataset with cerebral aneurysms. The average time to develop the CADe systems in the cross-validation was 28.7 h with a standard deviation (SD) of 3.0 h. Since the dataset for developing CADe system changed with each validation, the development time varied. The cerebral aneurysm detection processes for all 300 sets were completed without any problems. The ANODE score had an average of 0.239 with an SD of 0.062. The average detection time per case was 106.2 s with an SD of 19.6 s.

The CADe systems for detecting lung nodules in chest CT were also successfully developed using the chest CT volume dataset with pulmonary nodules. The average time to develop the CADe systems in the cross-validation was 22.5 h with an SD of 0.4 h. The lung nodule detection process for all 129 sets was completed without any problems. The ANODE score had an average of 0.218 with an SD of 0.058. The average processing time per case was 200.2 s with an SD of 22.5 s.

The results of this study show promise of a new paradigm for CADe system development. This system would enable clinicians without expertise in medical image analysis to develop CADe systems for use in clinical settings. Anyone with access to an appropriate medical image dataset could develop various CADe systems required for their particular clinical situation. However, the CADe systems developed using the pilot version of the generalized framework are inferior to the state-of-the-art CADe system. One of the reasons is that the features were derived from a feature bank with a limited number of voxel features. Feature generation methods will be effective for obtaining the optimal features for the target lesion detection and for improving the performance of the CADe systems.

49.3 Feature Generation Using Multiple Deep Convolutional Autoencoders

The feature generation is essential for improving the generalized framework for developing CADe systems. Recently, deep learning has received worldwide attention. The use of deep learning algorithms provides not only the accurate pattern classifier but also the useful feature extractor. Various CADe studies using deep learning show accurate results [16]. However, deep learning often requires a large-scale training dataset. It is challenging to prepare a significant amount of lesion data, especially around the beginning of CADe system development. It is necessary to develop a feature generation method that could be trained with a small dataset, including limited data of lesions.

We have proposed a feature generation method using multiple DCAEs [5, 6]. The DCAEs are among the most commonly used deep neural networks and are trained with an unsupervised algorithm [17]. In the proposed method, the DCAEs are used to encode input volume patches as latent variable vectors. The main idea of the proposed method is based on anomaly detection. It is expected that the DCAEs trained with only the normal (not lesion) volume patches fail the encoding of the lesion volume patches. Moreover, the DCAEs are trained with the 2.5-dimensional (2.5D) volume patches [18], including three images: the axial, coronal, and sagittal slices passing across the center of the original volume patch. The 2.5D patch has a smaller number of voxels than the original 3D patch. It is expected that the application of the 2.5D patches reduces the difficulty in training the DCAEs.

To improve the convergence of DCAE learning, each DCAE should have a simple structure including three convolutional layers, two max-pooling layers, and a full-connection layer. To apply various types of lesion and image modality, four types of 2.5D projection technique are applied: the maximum intensity projection (MIP), the minimum intensity projection (Min-IP), the mean intensity projection (Mean-IP), and the extraction of center slices (Cent). These four types of 2.5D patch are encoded by a specific DCAE individually.

The number of latent variable dimensions is optimized experimentally by each DCAE. The features generated by a DCAE are described as follows:

1. The latent variables of the input image patch,
2. The mean squared error between pixel values of the input patch and the reproduced patch from the latent variables,
3. The Mahalanobis distance from the normal dataset in the latent variable space, and
4. Nine types of pixel value statistics of the difference image between the input patch and the reproduced patch.

To evaluate the DCAE feature generation performance, we use the generated features for classifying cerebral aneurysm patches and normal vessel patches extracted from the head MRA dataset including data on 150 cases. The 2.5D patch has an image size of 32 pixels × 32 pixels × 3 channels and an isotropic image resolution of 0.78 mm. In the experiments, feature generation was successfully performed regardless of the training dataset scale. The AdaBoosted ensemble with the DCAE features showed sufficient classification performance, equal to the performance of the aneurysm CADe system used in clinical settings. The experimental evaluation also showed that the proposed method could accurately generate image features from a small-scale training dataset including from 10 to 30 cases.

49.4 Feature Extraction Using the Deep Convolutional Neural Network Pretrained with Local Anatomical Structures in Medical Images

The deep transfer learning has recently attracted increasing attention from medical image analysis researchers and is applied to many clinical domains. Generally, general image datasets such as CIFER and ImageNet are usually used to pretrain deep convolutional neural networks (DCNNs). However, the difference between a pretraining domain and a target clinical domain often causes a negative transfer.

In this study, a DCNN pretrained with local anatomical structures on medical images is used to transfer to a computer-aided detection task. The volume patch dataset, including anatomical LM appearances on CT, is used as the pretraining dataset. Generally, collecting the LM appearance patch data is easier than collecting lesion appearance patch data, because the LM appearance patch data can be collected from a large amount of normal case data obtained by group examinations. In this pilot study, the pretrained DCNN is used as an image feature extractor for lesion classification. Its use does not require an adequate dataset to fine-tune the pretrained DCNN and easily provides image features for the target domain.

The deep CNN used in this study has a VGG-like structure [19], including thirteen convolutions layers, five max-pooling layers, and two full-connection layers. The input is a 2.5D patch; the size is 32 × 32 pixels. The batch normalization and the dropout are also appropriately inserted in the CNN layers. The output of the first full-connection layer, which consists of 512 units, is extracted as the image feature set. In the pretraining, the LM patch dataset, including twenty types of LM data and not-LM class data (21 classes in total), is used. The dataset is extracted from CT data on 80 cases and consists of about 1600 LM-related patches and about 24,000 not-LM patches. The isotropic resolution of every patch data is 2 mm.

The DCNN pretrained with the LM dataset (LM-DCNN) was applied to classify the aneurysm lesion patches and normal vessel patches extracted from the head MRA dataset with data on 150 cases. The patch classification was performed by an AdaBoosted classifier ensemble consisting of decision stumps with LM-DCNN based features. The area under the ROC curve reached 0.942 due to the classification with only the 20 decision stump classifiers. The decision stump is a thresholding function for one of the features calculated by the LM-DCNN. The experimental result showed that the obtained LM-DCNN features included a few effective features for the aneurysm classification.

49.5 Conclusion

We proposed a pilot version of the generalized framework to develop CADe systems without any manual design for various types of target detection. We experimentally confirmed that two different types of CADe system were developed successfully through the generalized framework using two different datasets. We also proposed a feature generation method using multiple DCAEs to improve the generalized framework. Additionally, we proposed a deep transfer using the LM-DCNN to obtain a useful feature set for analyzing medical image patterns. The evaluations using head MR angiography reveal that the DCAEs and the DCNN could extract useful features for lesion classification. The future works include the improvement of the generalized framework by the proposed feature generation and the LM appearance classification.

References

1. van Ginneken B, Schaefer-Prokop CM, Prokop M. Computer-aided diagnosis: how to move from the laboratory to the clinic. Radiology. 2011;261:719–32.
2. Chan HP, Lo SCB, Sahiner B, Lam KL, Helvie MA. Computer-aided detection of mammographic microcalcifications: pattern recognition with an artificial neural network. Med Phys. 1995;22:1555–67.

3. Gokturk SB, Tomasi C, Acar B, Beaulieu CF, Paik DS, Jeffrey RB Jr, Yee J, Napel S. A statistical 3-D pattern processing method for computer-aided detection of polyps in CT colonography. IEEE Trans Med Imaging. 2001;20:1251–60.

4. Nemoto M, Hayashi N, Hanaoka S, Nomura Y, Miki S, Yoshikawa T. Feasibility study of a generalized framework for developing computer-aided detection systems—a new paradigm. J Digit Imaging. 2017;30:629–39.

5. Nemoto M, Ushifusa K, Kimura Y, Hayashi N. Pilot study to generate image features by deep autoencoder for computer-aided detection systems. Int Forum on Medical Imaging in Asia (IFMIA), Proceedings SPIE. 2019;11050:10501J.

6. Ushifusa K, Nemoto M, Kimura Y, Nagaoka T, Yamada T, Tanaka A, Hayashi N. A generalized image feature generation based on unsupervised deep learning with small scale normal dataset. Int J Comput Assist Radiol Surg. 2020;15(Suppl 1):S210–2.

7. Nemoto M. A pilot study for transferring deep convolutional neural network pre-trained by local anatomical structures to computer-aided detection. Int J Comput Assist Radiol Surg. 2020;15(Suppl 1):S110–2.

8. Rohr K. Landmark-based image analysis: using geometric and intensity models. Amsterdam: Springer Netherlands; 2001.

9. Hanaoka S, Shimizu A, Nemoto M, et al. Automatic detection of over 100 anatomical landmarks in medical CT images – a framework with independent detectors and combinatorial optimization. Med Image Anal. 2016;35:192–214.

10. Viola P, Jones M. Rapid object detection using a boosted cascade of simple features. Proc IEEE Computer Society Conference on Computer Vision and Pattern Recognition (CVPR 2001). 2001;1:I511–I518.

11. Iba W, Langley P. Induction of one-level decision trees. Proc International Conference on Machine Learning (ICML 1992). 1992:233–240.

12. Sun Y, Kamel MS, Wong AK, Wang Y. Cost-sensitive boosting for classification of imbalanced data. Pattern Recogn. 2007;40:3358–78.

13. Tu Z, Zhou XS, Barbu A, Bogoni L, Comaniciu D. Probabilistic 3D polyp detection in CT images – The role of sample alignment. Proc IEEE Computer Society Conference on Computer Vision and Pattern Recognition (CVPR). 2006;

14. Boureau Y, Ponce J, LeCun Y. A theoretical analysis of feature pooling in visual recognition. Proc International Conference on Machine Learning (ICML). 2010:111–118.

15. van Ginneken B, Armato SG, de Hoop B, et al. Comparing and combining algorithms for computer-aided detection of pulmonary nodules in computed tomography scans – The ANODE09 study. Med Image Anal. 2010;14:707–22.

16. Litjens G, Kooi T, Bejnordi BE, et al. A survey on deep learning in medical image analysis. Med Image Anal. 2017;42:60–88.

17. Masci J, Meier U, Cireşan D, Schmidhuber J. Stacked convolutional auto-encoders for hierarchical feature extraction. Artificial Neural Networks and Machine Learning – ICANN. 2011;2011:52–9.

18. Roth HR, Lu L, Seff A, Cherry KM. A new 2.5D representation for lymph node detection using random sets of deep convolutional neural network observations. Medical Image Computing and Computer-Assisted Intervention – MICCAI. 2014;2014:520–7.

19. Simonyan K, Zisserman A. Very deep convolutional networks for large-scale image recognition. The International Conference on Learning Representations – ICLR, 2015.

Radiomics: Artificial Intelligence-Based Radiogenomic Diagnosis of Gliomas

50

Manabu Kinoshita

Abstract

Radiogenomics is a rapidly emerging research field. It is expected to aid molecular or prognostic diagnosis of cancer patients by radiological images. Gliomas, one kind of brain cancer, can be thought of as a "touchstone" for pursuing and expanding this research area. In this chapter, I would like to address the background, methods, and current challenges of radiogenomics in glioma.

Keywords

Glioma · IDH mutation · Radiogenomics · Convolutional neural network · Machine learning

50.1 The Necessity of Radiomics in Glioma Treatment

Recent research using large-scale cohorts has rapidly uncovered the impact of genetic signatures on treatment decisions and prognostic prediction in glioma. Clinicians have long recognized that some low-grade astrocytoma patients do better than others. It was the discovery of *IDH* mutation in WHO grade 2 and 3 astrocytomas that highlighted for the first time that gliomas should be diagnosed based on molecular signatures on top of pathological diagnosis [1]. The WHO 2016 CNS tumors classification system now divides WHO grade 2 and 3 gliomas, namely lower-grade gliomas (LrGG), into three major subgroups:

1. Diffuse or anaplastic astrocytoma *IDH*-mutant.
2. Diffuse or anaplastic astrocytoma *IDH*-wildtype.
3. Oligodendroglioma or anaplastic oligodendroglioma *IDH*-mutant and 1p/19q co-deleted.

M. Kinoshita (✉)
Department of Neurosurgery, Osaka University Graduate School of Medicine, Suita, Japan
e-mail: mail@manabukinoshita.com

Furthermore, a deeper understanding of this neoplasm from a molecular biology standpoint is rapidly adding new subcategories into the classification system, which movement is inevitable in the era of precision medicine [2]. As the prognosis of glioma patients dramatically differs according to the genetic signatures of the tumor [3–6], clinicians desire a secure method that can triage high-risk patients. Among various "pretreatment" information, radiological images have always played a pivotal role in glioma treatment. Magnetic resonance images (MRI) are the primary choice of modality as it provides a vast amount of information from the anatomical presentation of the lesion to the functional condition of the surrounding brain. In recent years, the neuro-oncology community has started to focus on developing technologies that enable the non-invasive molecular diagnosis of gliomas using pretreatment MRI [7–20]. "Radiomics" is a rapidly expanding research field in radiology, aiming to simultaneously and automatically analyze numerous imaging parameters within a single session. Combined with machine learning algorithms and artificial intelligence, "radiomics" is expected to provide a novel diagnostic strategy and clinical workflow in glioma treatment. This section aims to describe the current state of radiomics in the field of glioma.

50.2 *IDH* Mutation and 1p19q Co-deletion as Prognostic Biomarkers in Lower-Grade Gliomas

It has long been known that some portions of glioma patients present a favorable prognosis. Among lower-grade gliomas (LrGG), oligodendroglioma had been recognized as one of the most favorable glial tumors. Co-deletion of chromosome 1p and 19q was furthermore identified as a significant prognostic and predictive biomarker. Several retrospective and prospective studies showed that patients with 1p19q co-deleted tumors show more prolonged survival and benefit from chemoradiation. On the other hand, astrocytomas,

M. Hashizume (ed.), *Multidisciplinary Computational Anatomy*, https://doi.org/10.1007/978-981-16-4325-5_50

another subtype of LrGG, were known to contain tumors that present an abysmal and relatively favorable prognosis. The discovery of *IDH* mutation in astrocytomas was able to beautifully explain this phenomenon from a genetic point of view. Various cohorts have repeatedly confirmed the importance of 1p19q co-deletion and *IDH* mutation in LrGG, and the current most updated WHO classification for CNS tumors fully utilizes these genetic markers in their diagnosis workflow. It is important to note that patients harboring diffuse or anaplastic astrocytoma *IDH*-wildtype, present a dismal prognosis similar to that seen in glioblastoma. The treatment goal will be completely different between patients who cannot expect a survival time as short as 2 years and those who can expect longer than 5–10 years. Poor prognostic patients should be offered rapid and intensive treatment. On the other hand, favorable prognostic patients should be treated, taking both overall survival time and quality of daily life (QOL) into consideration. As presumed pretreatment diagnosis will significantly impact the treatment strategy of LrGG, a diagnostic method that enables to provide qualitative information of the tumor is desired in the neuro-oncology community.

50.3 Radiological Determination of *IDH* Mutation and 1p19q Co-deletion in LrGG

The unique feature of *IDH* mutation in oncology is that mutated *IDH* impacts the tumor's metabolic state rather than stimulating cell cycles such as common oncogenes. The mutated *IDH* disturbs the tricarboxylic acid (TCA) cycle, which leads to the malfunction of converting isocitrate to alpha-ketoglutarate (α-KG) and grains new function to convert α-KG to 2-hydroxyglutarate (2HG) [1]. As a result, the intracellular concentration of 2HG elevates within *IDH* mutated tumors, and direct detection of 2HG via MRI has been explored [21]. Several reports have provided evidence that 2HG is detectable on magnetic resonance spectroscopy (MRS) and that this is a promising technique for the non-invasive radiological diagnosis of *IDH* mutation [22]. However, this technique is still not widely distributed in clinical practice, and its real-world value is still under investigation. Another approach for detecting *IDH* mutation via MRI is by investigating the perfusion of the tumor. *IDH*-wildtype tumors tend to be more abundant in vascularity compared to *IDH*-mutant tumors. As perfusion MRI can directly assess tissue vascularity, several reports have proposed using perfusion MRI for discriminating *IDH*-wildtype from *IDH*-mutant tumors. Although this is a clinically applicable technique, diagnostic accuracy will significantly depend on cut-off values, which could differ in each institution. Radiomics, on the other hand, is thought to be, in a way, more robust as it analyzes images acquired in routine clinical practice, such as

contrast T1-weighted images (GdT1WI) and T2-weighted images (T2WI) or fluid-attenuated inversion recovery (FLAIR). However, the problem of radiomics lies in the fact that analytical pipelines and the analyzed cohort that each institution has significantly impact texture analysis methods and diagnostic modeling. These issues will be discussed in further detail in the "Radiomics for gliomas" section.

Moving on to 1p19q co-deletion, the most famous radiological feature of 1p19q co-deleted gliomas is calcification on computer tomography (CT). Although several other pathologies exhibit calcification on CT, this feature has long been used among neuroradiologists as a highly specific image marker for a 1p19 co-deleted oligodendroglioma. As the biological function of 1p19q co-deletion is not well described, direct or indirect detection of this biological feature using the radiological lens has not been easy, which situation is different from *IDH* mutation.

50.4 Radiomics for Gliomas; the Basic Methods

The image analysis workflow is presented in Fig. 50.1. The following four sequences are essential for glioma diagnosis and treatment; T1WI, T2WI, FLAIR, and GdT1WI. Diffusion-weighted images (DWI) and diffusion tensor images (DTI) are now routinely acquired in many neuro-oncological institutions worldwide. However, the detailed image acquisition and processing procedures differ among institutions, and images reproducibility and robustness are inferior to those mentioned above. Conventional radiomics in glioma aims to obtain multiparametric image texture features from the four images discussed above. Current MRIs, however, do not provide absolute values of T1 or T2 relaxation time on their images but provide only qualitative and relative metrics on the image. This issue must be solved before moving forward, as the variabilities of images obtained from different institutions or scanners are substantial. Various kinds of "intensity normalization" methods are proposed, and it is up to the researcher which method to choose. After image-intensity-normalization is completed, all four image sequences will be co-registered for further analysis.

The operator will then have to identify the lesions on the image and create regions-of-interest (ROI) that contains the lesion. ROI can be built either in two (2D) or three dimensions (3D). One should be aware that the annotating lesion is subjected to operator-dependent variability. Defining abnormal from healthy tissues within the MRI is a more challenging task than one might think. The diffusive nature of glioma renders it challenging to determine the tumor border as the signal change from tumor to healthy brain tissue occurs gradually. In glioma radiomics, two types of ROI are usually created. One is created on GdT1WI that mainly contains the

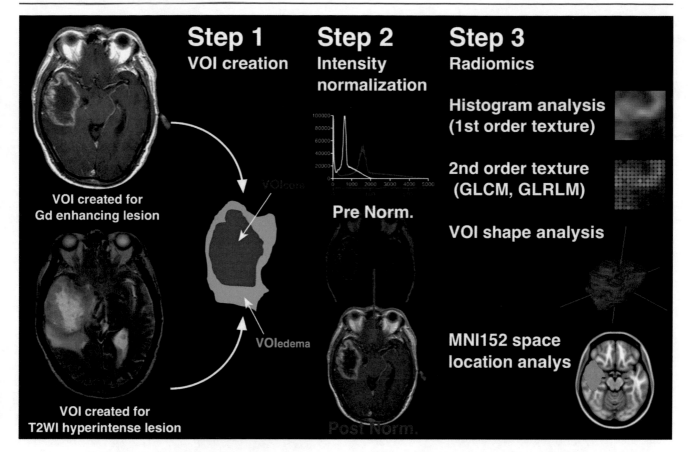

Fig. 50.1 Illustration showing the workflow for image analysis. Two types of VOIs were created based on Gd enhancement of the tumor and edema lesion identified on T2-weighted images. Both VOIs were co-registered and VOI$_{core}$ and VOI$_{edema}$ were generated. Subsequently, intensity normalization of all images was performed and first-order and second-order texture analysis, VOI shape analysis, and location analysis were performed

contrast-enhancing part of the lesion. The other is created on T2WI or FLAIR that represents a non-enhancing part of the lesion or brain edema due to compression of a healthy brain by the tumor. As LrGG tends to lack contrast enhancement, ROI concerning the contrast-enhancing lesion can be omitted specifically for LrGG.

Once the ROIs are defined for each case, image processing will then move onto calculating and retrieving radiomics parameters. Radiomic parameters will include, but not be limited to, features regarding shape, location, histogram, and texture analysis. The "histogram" analysis mainly consists of parameters related to the image's data distribution within the ROI. More specifically, histogram analysis includes parameters such as average, median, standard deviation, skewness, and entropy. Parameters concerning "shape" include volume, surface area, volume-to-surface ratio, and others. In addition to these basic image metrics, radiomics may analyze secondary texture features. Texture feature analysis investigates the "patterns" within images. Haralick texture features are one of the most commonly used methods for analysis. This analysis requires intensive image processing, as described in the following.

First, images usually in 256 grayscale must be "downgraded" into 16 or 8 gray scales to enable repetitive pattern analysis. This "down-grading" enables pattern analysis in cases where the texture of the target region consists of signals with substantial fluctuations. On the other hand, this procedure is again user-dependent, and the magnitude of degrading grayscale images is arbitrary. One should be fully aware that there is still no community standard that defines how radiomic parameters are calculated, and published data should be interpreted with great care in this regard.

Recent studies further utilize an "automated" approach to retrieve radiomic features via a convolutional neural network (CNN). CNN is classified among one of the deep-learning algorithms mainly designed for image analysis. This approach allows to analyze various features of images such as edges or textures without an a priori knowledge. Thus, this approach renders defining radiomics features before analysis obsolete and relies on the capability of the CNN to automatically search significant image features that represent abnormality of the region in interest.

Finally, the obtained radiomic features, whether by conventional texture feature analysis or by CNN, are subjected to building machine learning algorithms that aim to predict biological characteristics in interest such as specific genetic mutations or prognosis of the patient.

The current state of radiogenomics in LrGG.

The prediction accuracy of *IDH* mutation is currently the most often used benchmark addressing the built algorithm's performance. Most publications claim to achieve 80–90% accuracy. It is crucial to take caution in the type of cohort that each study uses to build its diagnostic algorithm. For example, if the analyzed cohort consists of both LrGG and glioblastoma (GBM), the diagnostic accuracy for detecting *IDH* mutation will be higher than in cases where the cohort was built upon solely on LrGG. The representation of GBM and LrGG on MRI is so different that the developed algorithm can easily distinguish *IDH*-wildtype GBM from *IDH*-mutant LrGG, resulting in a much higher diagnostic accuracy than real-world scenario. Another critical factor that should be taken into consideration is the clinical background of the cohort. *IDH*-mutant tumors arise more often in younger patients than in the elderly. Thus, simple clinical information such as age can significantly help to distinguish *IDH*-mutant from wildtype tumors. In a previous report, the authors were able to show that the diagnostic accuracy of *IDH*-mutation was as high as 80% by age alone. Adding radiomic analysis improved the diagnostic accuracy by less than 10% [12]. Thus the "real" positive effect of radiomics in predicting *IDH*-mutation was only 10% in this report, while our research showed that age was able to predict *IDH* mutation with an accuracy of as low as 60%, and *IDH* mutation was predicted with 80% accuracy with MRI alone. These research discrepancies emphasize the importance of looking into the characters of the cohort that is analyzed. Ideally, a more "difficult" cohort to solve is preferred to build a robust and universal prediction algorithm.

50.5 Current Limitation and Future Direction of Radiogenomics

When clinicians encounter LrGG patients, he or she will try their best to estimate the molecular characteristics of the tumor. Pretreatment knowledge of the tumor will triage high-risk patients and facilitate further intervention. During this process, clinicians are searching for clues in patients' neurological symptoms, present and past medical histories, clinical information such as age, and radiological images. Whether radiomics can outperform expert clinicians in diagnostic accuracy is an essential question that determines the future of this new research field. One research attempted to answer this question by comparing radiomics and expert neuro-radiologist in predicting 1p19q co-deletion status. While radiomics outperformed non-experts in diagnostic accuracy, expert neuro-radiologists were able to achieve as

precise a diagnosis as radiomics [23]. This result demonstrates that current radiomics can offer diagnostic accuracy of expert clinicians at their best and are not capable of outperforming them and provide something magical.

Another critical issue is the generalizability of the built model for prediction. Machine learning is known to depend on the cohort with which it was trained. One algorithm that exhibits superb performance may not be applicable for a different data set. This issue also relates to the problem that there is no community standard for radiomics analysis. It seems that it is high time that the research community defined technical standards for radiomics analysis.

References

1. Yan H, Parsons DW, Jin G, et al. IDH1 and IDH2 mutations in gliomas. New Engl J Med. 2009;360:765–73.
2. Louis DN, Wesseling P, Aldape K, et al. cIMPACT-NOW update 6: new entity and diagnostic principle recommendations of the cIMPACT-Utrecht meeting on future CNS tumor classification and grading. Brain Pathol. 2020;30(4):844–56. https://doi.org/10.1111/bpa.12832.
3. Shirahata M, Ono T, Stichel D, et al. Novel, improved grading system(s) for IDH-mutant astrocytic gliomas. Acta Neuropathol. 2018;20:1–14.
4. Suzuki H, Aoki K, Chiba K, et al. Mutational landscape and clonal architecture in grade II and III gliomas. Nat Genet. 2015;47:458–68.
5. Arita H, Yamasaki K, Matsushita Y, et al. A combination of TERT promoter mutation and MGMT methylation status predicts clinically relevant subgroups of newly diagnosed glioblastomas. Acta Neuropathol Commun. 2016;4:79.
6. Arita H, Narita Y, Fukushima S, et al. Upregulating mutations in the TERT promoter commonly occur in adult malignant gliomas and are strongly associated with total 1p19q loss. Acta Neuropathol. 2013;126:267–76.
7. Li L, Mu W, Wang Y, et al. A non-invasive radiomic method using 18F-FDG PET predicts isocitrate dehydrogenase genotype and prognosis in patients with glioma. Front Oncol. 2019;9:1183.
8. Liu X, Li Y, Qian Z, et al. A radiomic signature as a non-invasive predictor of progression-free survival in patients with lower-grade gliomas. Neuroimage Clin. 2018;20:1070–7.
9. Li Z, Wang Y, Yu J, Guo Y, Cao W. Deep Learning based Radiomics (DLR) and its usage in noninvasive IDH1 prediction for low grade glioma. Sci Rep-uk. 2017;7:5467.
10. Arita H, Kinoshita M, Kawaguchi A, et al. Lesion location implemented magnetic resonance imaging radiomics for predicting IDH and TERT promoter mutations in grade II/III gliomas. Sci Rep-uk. 2018;8:11773.
11. Rudie JD, Rauschecker AM, Bryan RN, Davatzikos C, Mohan S. Emerging applications of artificial intelligence in neuro-oncology. Radiology. 2019;290:607–18.
12. Lu C-F, Hsu F-T, Hsieh KL-C, et al. Machine learning-based radiomics for molecular subtyping of gliomas. Clin Cancer Res. 2018;24(18):4429–36.
13. Zhou H, Vallières M, Bai HX, et al. MRI features predict survival and molecular markers in diffuse lower-grade gliomas. Neuro-Oncology. 2017;19:862–70.
14. Lohmann P, Lerche C, Bauer EK, et al. Predicting IDH genotype in gliomas using FET PET radiomics. Sci Rep-uk. 2018;8:13328.
15. Fukuma R, Yanagisawa T, Kinoshita M, et al. Prediction of IDH and TERT promoter mutations in low-grade glioma from magnetic res-

onance images using a convolutional neural network. Sci Rep-uk. 2019;9:20311.

16. Kickingereder P, Neuberger U, Bonekamp D, et al. Radiomic subtyping improves disease stratification beyond key molecular, clinical, and standard imaging characteristics in patients with glioblastoma. Neuro-Oncology. 2018;20:848–57.

17. Kickingereder P, Burth S, Wick A, et al. Radiomic profiling of glioblastoma: identifying an imaging predictor of patient survival with improved performance over established clinical and radiologic risk models. Radiology. 2016;280:880–9.

18. Sasaki T, Kinoshita M, Fujita K, et al. Radiomics and MGMT promoter methylation for prognostication of newly diagnosed glioblastoma. Sci Rep-uk. 2019;9:14435.

19. Zhang B, Tian J, Dong D, et al. Radiomics features of multiparametric MRI as novel prognostic factors in advanced nasopharyngeal carcinoma. Clin Cancer Res. 2017;23:4259–69.

20. Mouraviev A, Detsky J, Sahgal A, Ruschin M, Lee YK, Karam I, Heyn C, Stanisz GJ, Martel AL. Use of radiomics for the prediction of local control of brain metastases after stereotactic radiosurgery. Neuro-Oncology. 2020;22(6):797–805. https://doi.org/10.1093/neuonc/noaa007.

21. Choi C, Ganji SK, DeBerardinis RJ, et al. 2-hydroxyglutarate detection by magnetic resonance spectroscopy in IDH-mutated patients with gliomas. Nat Med. 2012;18:624.

22. Kickingereder P, Sahm F, Radbruch A, Wick W, Heiland S, von Deimling A, Bendszus M, Wiestler B. IDH mutation status is associated with a distinct hypoxia/angiogenesis transcriptome signature which is non-invasively predictable with rCBV imaging in human glioma. Sci Rep-uk. 2015;5:16238.

23. van der Voort SR, Incekara F, Wijnenga MMJ, et al. Predicting the 1p/19q codeletion status of presumed low-grade glioma with an externally validated machine learning algorithm. Clin Cancer Res. 2019;25:7455–62.

4D—Four-Dimensional Dynamic Images: Principle and Future Application

51

Naoki Suzuki and Asaki Hattori

Abstract

This chapter touches on the acquisition of four-dimensional (4D) phenomena in living organisms and their clinical application. Originally, the human body is a subject that undergoes 4D changes in every part, that is, undergoes spatial and temporal changes. We think that measuring this 4D change of the human body quantitatively and making it possible to analyze the data in spatiotemporal space will open up a new frontier in medicine. Therefore, we have been developing measurement devices and analysis methods for those purposes. First, we show the measurement result of the cardiac dynamics by the development of the 4D X-ray CT device for measuring the 4D phenomenon of the living body, and the MRI measurement result of the dynamics of the thigh muscle in the repetitive movement of the lower limbs with supportive equipment. Then, in order to visualize the whole-body motion in virtual space, we describe how we developed a whole-body model with anatomical features for each patient and how we developed a Dynamic Spatial Video Camera (DSVC) that can measure the surface shape of the whole-body motion without restraint. The spatiotemporal changes of the living body that occur within a short time such as heartbeat and movement of limbs are not the only 4D phenomena. Slow changes of the living body that occur over several months and years are also 4D phenomena. Therefore, at the end of this chapter, we will also discuss an example of the development of a method for visualizing changes in human growth, changes in the treatment process of the liver and lungs, and the prediction of pathological conditions in the near future using this method.

Keywords

Times · Four-dimensional · 4D measurement
4D analysis

51.1 4D imaging

Humans, like any other living things, exhibit four-dimensional (4D) phenomena in every part of their bodies. A typical example of this is the heart, which is a structure made of muscle tissue with a unique shape that has four internal lumens. The appearance of blood flow with different pressures in each blood vessel is a 4D phenomenon with a spatiotemporal spread. Furthermore, even the brain, which is often thought to be the most static organ, actually exhibits a four-dimensional phenomenon in which it slightly expands and contracts with the pulsation of the heart. Likewise, it is a 4D phenomenon, with voluntary and involuntary movements generated by muscles. From a spatiotemporal perspective, any small part of a living body can be regarded as displaying a 4D phenomenon.

However, modern science has not advanced enough to quantitatively measure all 4D phenomena. Therefore, we thought to combine several scientific methods to clinically acquire 4D phenomenon data of the living body. The virtual space in the computer plays an important role as a space for displaying a 4D image of the living body and allowing the user to recognize it.

The 4D image referred to here can be defined as images that have the function that can measure the temporal changes of a three-dimensional (3D) subject by manipulating the spatial axis and the temporal axis to understand, and measure

N. Suzuki (✉)
Global Information and Telecommunication Institute, Waseda University, Tokyo, Japan

Institute for High Dimensional Medical Imaging, The Jikei University School of Medicine, Tokyo, Japan
e-mail: nsuzuki@jikei.ac.jp

A. Hattori
Institute for High Dimensional Medical Imaging, The Jikei University School of Medicine, Tokyo, Japan
e-mail: hat@jikei.ac.jp

the state, if necessary, and generally analyze the temporal and spatial changes of the subject.

At present, the acquired quantitative numerical data of 4D phenomena is expressed as a group of 3D images arranged in time series. So, we took the changes in the anatomical structure derived from the patient and the structure of the lesion of the individual patient on the time axis to try to construct 4D imaging technology and to apply it to clinical trials to utilize the 4D dynamics of the patient for diagnosis and treatment.

51.2   Trial for Measurement of 4D Phenomenon

Recently, with the progress of biometrics technology and faster processing, it has become possible to directly measure the 4D phenomenon of live body parts. We can collect 4D data of body parts by temporal resolution, by spatial

resolution, and although in different measure ranges, by MDCT, 4D ultrasound tomography, and MRI using a high-speed tomographic sequence. We also started the development in 1998 of a high-speed cone-beam X-ray CT device that can obtain 4D information of cardiac dynamics or 4D X-ray CT and completed the prototype of the device in March 2002. With this device, using the projection data obtained by 10 s of imaging, we reconstructed 180 sets of volume data (512 × 512 × 216) in time series and evaluated its clinical usefulness, and especially evaluated the observation and analysis of the dynamics of the human heart [1–3]. Figure 51.1a shows 3D images of 4D data in time series obtained from transvenous imaging of normal human volunteers.

Next, we show an example of our development of a measurement method to visualize in 4D, muscle deformation using MRI. With the current MRI capabilities, it is still difficult to acquire 4D data with anatomically sufficient spatio-

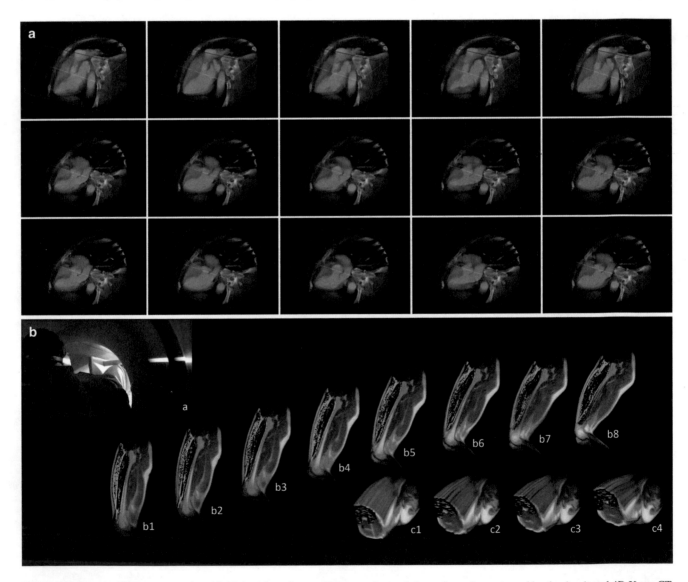

Fig. 51.1 Dynamic 4D image examples. (**a**) 4D imaging of normal human volunteer's heart dynamics measured by the developed 4D X-ray CT. (**b**) 4D imaging of deformation of major thigh muscles during flexion and extension of the lower limbs measured by MRI

temporal resolution for one motion in the space covering the thigh. Therefore, we had the thigh repeatedly flex and extend using a supportive device and took images of it moving the tomographic plane. In this way, we obtained 25 time-series volume data and acquired the dynamics of the thigh muscle group as 4D data (Fig. 51.1b).

51.3 Development of a 4D Human Body Model with Dynamic Deformation

We think that the human body can be comprehensively understood by grasping and analyzing the dynamics of the human body in 4D. In other words, if such a 4D model can be constructed for each individual patient, it will be possible to grasp the state of the illness in more diverse ways than before, and it will be possible to obtain more information for a detailed diagnosis of the illness and better determine how to cure it.

Therefore, we tried to construct a 4D human body model in which it includes not only the skeletal structure but also soft internal structure such as major organs, blood vessels, and skeletal muscles. We attempted to drive this model in virtual space and manipulate it in real time so that we could observe and analyze its dynamics. We used MRI considering the invasiveness to the subject's body and constructed a whole-body model that has each individual's anatomical shape. It has a function that drives each model using each subject's motion data using motion capture. We used the linear blend skinning method for the deformation of the whole-body model other than the skeleton model [4].

The advantage of this method is that for a living body subject with a layered structure such as skin, skeletal muscle under it, and skeletal system under that, the inversion of layers is maintained and the protrusion of the internal structure does not occur and the deformation similar to the dynamics of the living body can be generated while maintaining high-speed calculation and volume of the target object.

We used VICON 612 (Vicon Motion Systems, UK) to measure the whole-body motion of the subject. We attached 42 optical markers to the whole body of the subject to measure the motion of the subject and drove the whole-body model constructed from the measurement.

Figure 51.2a shows a video image of the subject's movements, as well as a display image of the whole-body model driven by the motion-captured data during walking movements. The whole-body model displays the body surface, skeleton, blood vessels, and skeletal muscles of the lower limbs.

In Fig. 51.2b, we show the whole-body model from a different viewpoint. We show how the internal structure such as the main organs of the thoracoabdominal region is deformed by the movement, and show how the skin deforms in its natural state as it houses those internal structures.

The primary purpose of this method is to visualize the mutual changes of skeletal structure and skeletal muscles and use it for motion analysis, but we believe there is a possibility of wider medical application.

51.4 4D Observation of Human Body Motion by DSVC

A new image method has been developed that enables the movement of an arbitrary viewpoint in the filming space by such methods as View-independent Scene Acquisition using multiple cameras. However, most of these studies are aimed at the acquisition of image information for communication methods such as telepresence, and they do not take into consideration the quantity of the subject. Therefore, we developed a Dynamic Spatial Video Camera (human body motion spatiotemporal imaging device, DSVC) for quantitative spatiotemporal analysis of the 4D dynamics of human body motion, with the whole-body motion as its maximum range [5]. The motion of the human body was captured by 65 synchronized digital video cameras, and a function was added to allow the user to select a viewpoint from any camera to observe the series of motions recorded. Among them, 60 cameras were fixed in a metal ring with a diameter of 4 m, and 5 cameras were installed above. This ring was hung by a mobile crane and was enabled to follow the movement of the subject to bring the cameras to their appropriate positions. Figure 51.3a shows the external view of the completed system. The results measured using this system are shown in Fig. 51.3b. This figure shows a state in which a motion at a certain moment at an arbitrary time is observed from around the subject. Furthermore, in Fig. 51.3c, the result of superimposing the human body model including the subject's skeleton and the simple muscle model on the image of the subject's walking DVSC is shown. It can be said that this 4D model makes it possible to simultaneously capture changes in the surface shape of each part and movements of the internal structure of the subject in a time series, enabling quantitative analysis of human movements.

51.5 4D Visualization of Body Surface and Skeleton Dynamics by DSVC

With the development of DSVC, we were able to observe human motion from the entire circumference in 4D from the viewpoints of arbitrary cameras. In this development, we

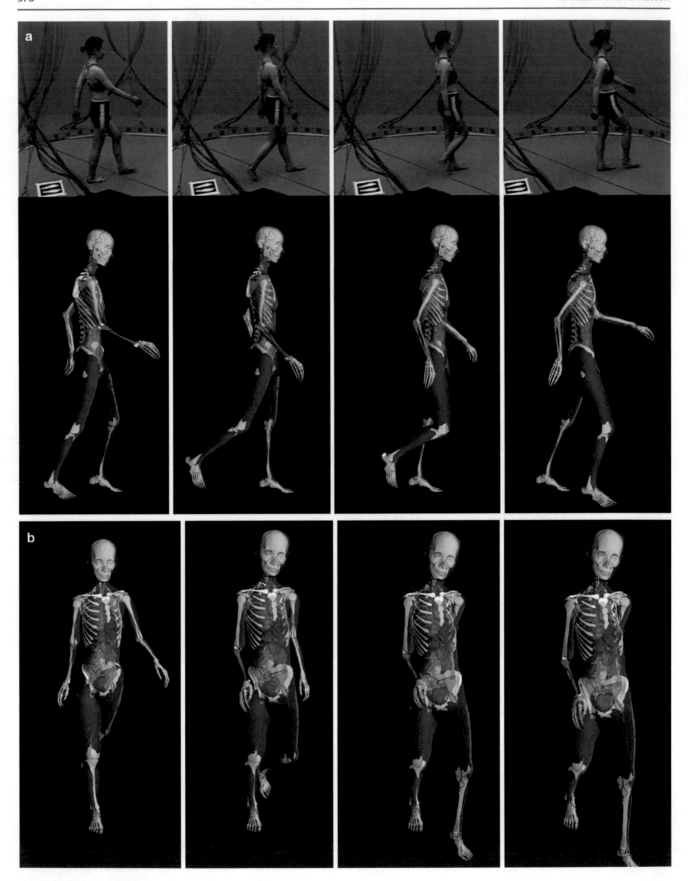

Fig. 51.2 Results of driving a 4D human body model using motion capture data during walking. (**a**) Video image of the subject (top) and human body model (bottom). (**b**) Viewpoint changed to the front

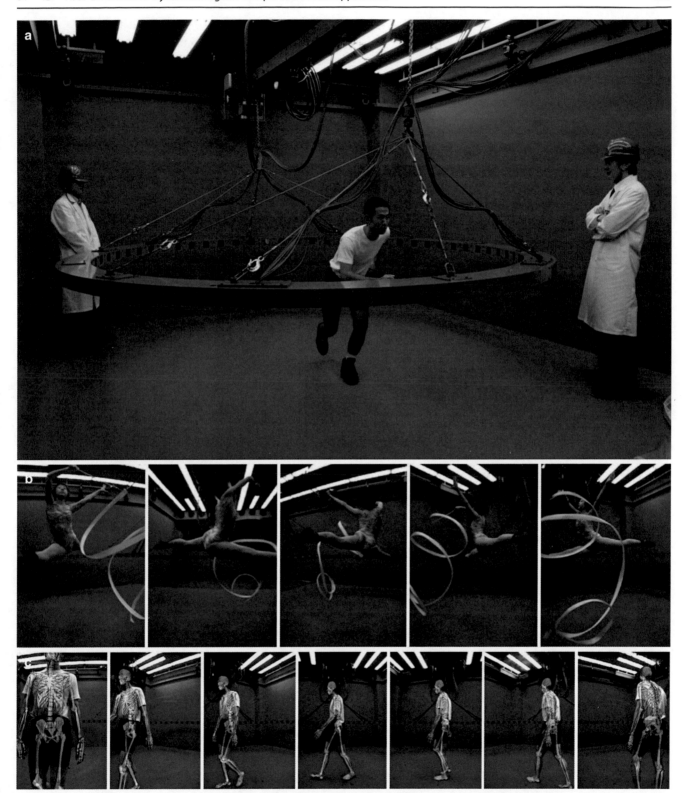

Fig. 51.3 Dynamic Spatial Video Camera (DSVC) and its applications. (**a**) The exterior of DSVC. (**b**) Measurement result by DSVC. Display of a certain moment of gymnastics ribbons performance action from various viewpoints. (**c**) Results of superimposing the skeleton and simple muscle model on the image of walking measured by DSVC

also aimed to record the movements of a human's whole-body movement in a completely unconstrained state such as without using optic markers, and to create a 4D body model from it. In addition, we aimed to visualize the changes in the internal skeletal muscles and skeletal structures in respect to the body surface during human motion [6, 7].

Therefore, we extracted the subject's area from the consecutive images of 65 cameras and modeled the shape of the subject's body surface at each instant based on the parameters such as the posture, position, and focal length of each camera. From this, we constructed a time-series model group of 3D models at 15 fps, i.e., a 4D model of the subject's body surface. This 4D model has color because it is constructed from images in the visible light region, and it includes facial expressions, clothes colors and patterns as well as body surface and hair colors. Figures show the estimated results of the body surface shape during swinging the right foot (Fig. 51.4a) and walking (Fig. 51.4b), and the driven internal skeleton. By using the subject's body surface shape and skeleton model acquired by DSVC, we think we were able to estimate and visualize the 4D dynamics of the subject's whole-body skeletal structure under unrestrained conditions.

51.6 Understanding the 4D Change that Occurs Over a Long Period of Time

We attempted to develop a method that visualizes phenomena that occurs over a long period of time such as human growth and changes in pathological conditions in 4D images

so that it could be used for diagnosis and treatment. We used multiple X-ray CT data sets that were measured discretely on a time axis of a patient, or common anatomical features from MRI data sets, and calculated the growth curve (variant amount) of each body part. From this, we developed a method to visualize as a 4D image of volume data that continuously changes on the time axis, and made it possible to predict the state of the future of the latest dataset from the amount of change.

We show the result of the visualized changes of the post-surgical state over several years using X-ray CT data of children after living donor liver transplantation (LDLT) and congenital diaphragmatic hernia (CDH) treatment. In the case of a child with LDLT (Fig. 51.5a), we displayed using X-ray CT data taken between the ages of 3 and 9 years. In the case of a child with CDH (Fig. 51.5b), segmentation of the lung region is performed and displayed for four X-ray CT data in the 3 years from one month to the age of 3 years. In this case, the affected (right) lung rapidly grew about 1 year after the operation, and the process in which the lung volume increased with the expansion of the rib cage could be visualized in detail. The growth curve obtained from past data is used to predict the state one year later, as shown in the rightmost image of Fig. 51.5b. We think that such near-future prediction is important for determining treatment policies such as whether or not the patient needs surgery in the future. Also, in Fig. 51.5c, we focused on the left lung and showed an example of how the upper lobe and lower lobe grow rapidly over the course of 3 years, along with structural changes inside the lung.

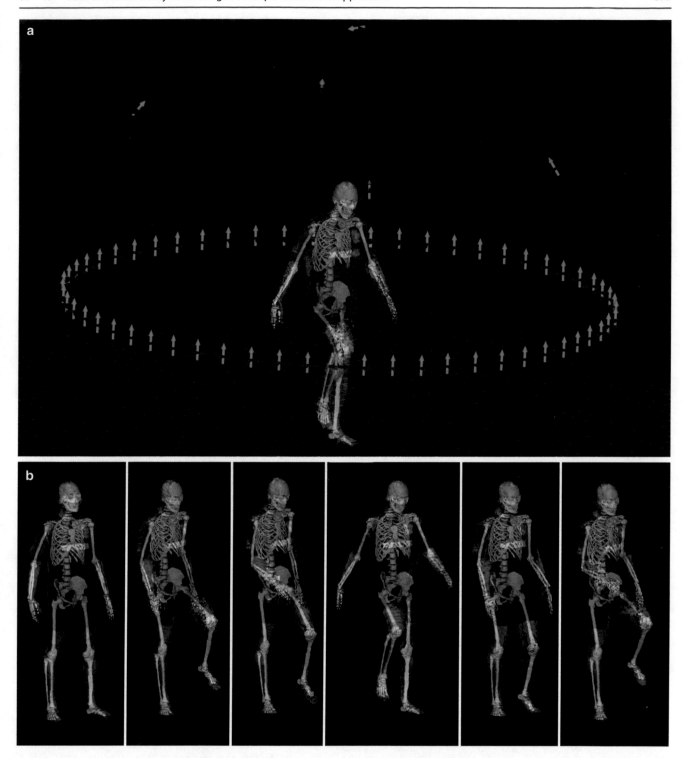

Fig. 51.4 4D estimation results. (**a**) Estimated body surface shape and internal skeleton drive status when swinging the right foot. (**b**) Results of estimated body surface shape and driving situation of the internal skeleton during the walking motion

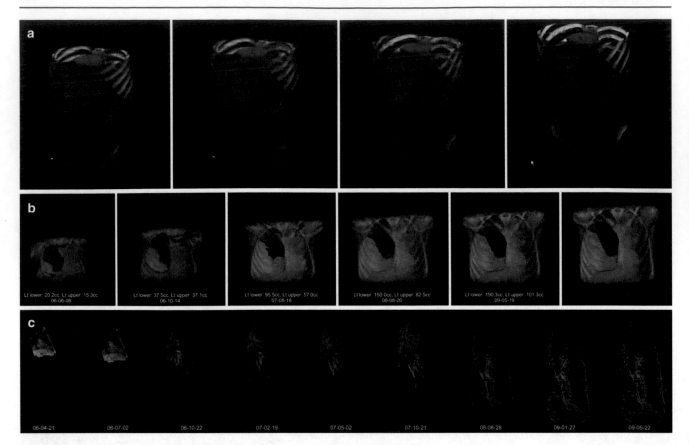

Fig. 51.5 4D changes over a long period of time. (**a**) Results of visualizing changes that occurred 6 years after living donor liver transplantation. (**b**) Results of visualizing changes that occurred 3 years after surgery for congenital diaphragmatic hernia. The right end is the result of predicting the state one month after the last X-ray CT scan. (**c**) Only the left lung is displayed, the upper lobe is rendered in blue using the segmented data, and the lower lobe is rendered with the original X-ray CT data

51.7 Conclusion

It has become gradually possible to collect 4D phenomena of a living body as 4D data and analyze them spatiotemporally by developing measurement methods and analysis devices. We believe that by visualizing and analyzing live body 4D motion not only in orthopedics and sports medicine but in any area of medicine, more accurate diagnosis and treatment can be realized in the field of cardiovascular dynamics, muscle dynamics, and digestive system dynamics.

References

1. Mori S, Kondo C, Suzuki N, Yamashita H, Hattori A, Kusakabe M, Endo M. Volumetric cine imaging for cardiovascular circulation using prototype 256-detector row computed tomography scanner (4-dimensional computed tomography): a preliminary study with a porcine model. J Comput Assist Tomogr. 2005;29(1):26–30. https://doi.org/10.1097/01.rct.0000151189.80473.2e.
2. Kondo C, Mori S, Endo M, Kusakabe K, Suzuki N, Hattori A, Kusakabe M. Real-time volumetric imaging of human heart without electrocardiographic gating by 256-detector row computed tomography: initial experience. J Comput Assist Tomogr. 2005;29(5):694–8. https://doi.org/10.1097/01.rct.0000173844.89988.37.
3. Mori S, Kondo C, Suzuki N, Hattori A, Kusakabe M, Endo M. Volumetric coronary angiography using the 256-detector row computed tomography scanner: comparison in vivo and in vitro with porcine models. Acta Radiol. 2006;47(2):186–91. https://doi.org/10.1080/02841850500479669.
4. Suzuki N, Hattori A, Hashizume M. Development of four dimensional human model that enables deformation of skin, organs and blood vessel system during body movement – visualizing movements of the musculoskeletal system. Stud Health Technol Inform. 2016;220:396–402. https://doi.org/10.3233/978-1-61499-625-5-396.
5. Suzuki N, Hattori A, Hayashibe M, Suzuki S, Otake Y. Development of Dynamic Spatial Video Camera (DSVC) for 4D observation, analysis and modeling of human body locomotion. Stud Health Technol Inform. 2003;94:346–8. https://doi.org/10.3233/978-1-60750-938-7-346.
6. Saito T, Suzuki N, Hattori A, Suzuki S, Hayashibe M, Otake Y. Estimation of skeletal movement of human locomotion from body surface shapes using Dynamic Spatial Video Camera (DSVC) and 4D human model. Stud Health Technol Inform. 2006;119:467–72.
7. Saito T, Suzuki N, Hattori A, Hayashibe M, Suzuki S, Otake Y. Marker-less whole body skeletal motion analysis based on a body surface model constructed from multi-camera images. Int J Comput Assist Radiol Surg. 2006; https://doi.org/10.1007/s11548-006-0010-3.

Cloud XR (Extended Reality: Virtual Reality, Augmented Reality, Mixed Reality) and 5G Mobile Communication System for Medical Image-Guided Holographic Surgery and Telemedicine

52

Maki Sugimoto

Abstract

In the field of medical image processing, three-dimensional image analysis has been further advanced and extended reality (XR) technologies such as virtual reality (VR), augmented reality (AR), and mixed reality (MR) have been applied to support 3D and spatial diagnosis and surgery by wearing smart XR goggles. We developed a cloud-based XR application "Holoeyes MD" for analyzing the patient individual medical images in 3D by CT, MRI, and US. The shape data of these organs is exported to polygons, the 3D shapes are digitized into coordinates. The 3D structures such as the depth and size of the organs usually used to be incompletely understood on a flat monitor screen. The introduction of the cloud-based XR and 5G has made it possible to understand more intuitively the 3D positional relationship of various anatomy and doctor's presence and existence in online telemedicine and remote space sharing.

Keywords

Extended reality · Virtual reality · Augmented reality
Mixed reality · Telemedicine

52.1 Introduction

Recently, the digital transformation of medical informatics including medical imaging has been rapidly progressing in the clinical setting. In the field of medical image processing, three-dimensional (3D) image analysis has been further

advanced and extended reality (XR) technologies such as virtual reality (VR), augmented reality (AR), and mixed reality (MR) have been applied to support 3D and spatial diagnosis and surgery [1, 2] (Fig. 52.1). They are widely used in clinical, academic, and educational fields because high-specification computers and versatile applications are now commercially available.

In the past, the only way to utilize such digital information was to browse the data by using the flat monitor or smart device. However, nowadays, the XR experience is being introduced into clinical practice by wearing smart devices such as a head-mounted display, goggles, and smart glasses and superimposing them on the real world. The digital transformation of medicine and healthcare is progressing mainly in the surgical field [3–5].

52.2 Methods

We have developed an application that digitally analyzes CT and MRI data of individual patients, automatically extracts feature points of the organ shape with artificial intelligence (AI), converts them into polygons, and views these organ shape coordinates as VR.

Then, using the mixed reality technology, the coordinates of the organ shape and the position of the real space calculated by the position sensor were integrated and displayed in the real world using a wearable holographic spatial computer. We have practically used these XR for simulation of surgery, support for treatment, surgical training, and medical education.

Based on patient-specific DICOM data from MDCT, after generating its surface polygons using OsiriX application, particularly abdominal organs can be segmented as polygonal selections from several 2D CT images. Polygon (.stl, . obj): 3D model format OBJ is a geometry definition file format, which is open and has been adopted by many 3D graph-

M. Sugimoto (✉)
Innovation Lab, Teikyo University Okinaga Research Institute, Tokyo, Japan
e-mail: sgmt@med.teikyo-u.ac.jp

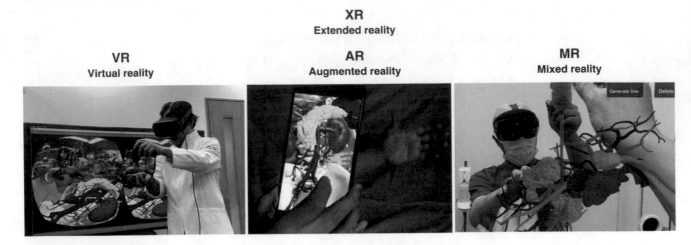

Fig. 52.1 Segmentation and exporting organ polygon files from DICOM sequence of the CT data

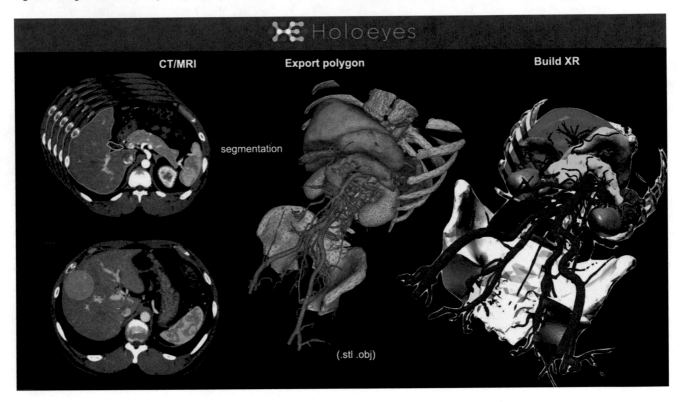

Fig. 52.2 Medical application of extended reality (XR)

ics application vendors. STL is also a file format native to the stereolithography CAD software created by 3D systems (Fig. 52.2).

Wearable devices such as a head-mounted display, goggles, and smart glasses that allow users to experience XR are already available in the market mainly for games and entertainment, and thus can be purchased at a low price.

At first, to use these devices for medical imaging, the individual medical images of patients are analyzed in 3D by CT, MRI, and ultrasonography (US). The shape data of these organs is exported to polygons, the 3D shapes are digi-

tized into coordinates. These are combined with the coordinates of the real space, the user's position, and motion. When the organs and lesions are then superimposed on the real space, they can be experienced as an immersive XR application.

We developed a cloud-based medical XR imaging service Holoeyes MD (Holoeyes Inc. Tokyo, JAPAN) [6]. We conducted this technology for more than 100 surgeries in brain, liver, biliary duct, pancreas, esophagus, stomach, colon, lung, heart, kidney, prostate, uterus, spine, bone, artery, vein, muscular, and sensory organs.

52.3 Results and Discussions

52.3.1 XR in Spatial Diagnostic Imaging and Image-Guided Surgery

Using our cloud service, the 3D structures such as the depth and size of the organs usually used to be incompletely understood on a flat monitor screen. The introduction of the XR technique has made it possible to understand more intuitively the 3D positional relationship of various organs, blood vessels, and the complex positional relationship of fat and musculature (Fig. 52.3).

In VR, the position of the organ and the user's gaze were corrected by three-dimensional stereoscopic vision using a position sensor, and displayed on the VR headset as if it were always in stereoscopic space. When this was linked to the movement of the surgeon, the immersive sense for understanding the patient's anatomy improved dramatically, and three factors were improved: three-dimensional spatiality, real-time interaction, and self-projection.

Particularly, VR for telemedicine enabled a more intuitive mode of interacting with information, and as a flexible environment that enhances the feeling of physical presence during the interaction. Users can feel as if they are floating in the air in front of their eyes, which is a 3D constructed image of CT, MRI, and ultrasonography (UC). Users could feel the presence and existence as if they were floating in the air.

Some XR wearable devices are capable of gesture-controlled user interfaces and can be operated while wearing a sterile glove during surgery. In particular, not only one user can view the device alone, but several people can share virtual and real space at the same time, leading to smooth telecommunication. Also, multiple users can share the experience of treatment planning and surgical simulation at the same time and achieve rich reproducibility.

In spatial holography by MR, 3D data was presented to a translucent wearable glass as a guide to support organs, surgical devices, and procedures. Using multiple sensors that measure the position and the movement inside during the surgery, using a transparent holographic wearable glasses with built-in IR position sensors (HoloLens and Magic Leap One), surgeons could watch the floating organ models beneath the surgical field.

This is effective in a sterilized environment, and even during surgery, the hologram of the organs could be confirmed from all directions in the air of the operative field. The location information of multiple MR devices was shared under the same Wi-Fi reception, multiple surgeons could view the same anatomy in the same air, improving their communication (Figs. 52.4 and 52.5).

The ability to spatial awareness for understanding the extent of resection, blood vessel processing, lymph node dissection, and sutures were improved before and after surgeries. We designed to guide surgeons through potentially complex procedures giving them step-by-step contextually accurate instructions. This can be used for guidance in complex surgeries for young surgeons and medical students. It could also be used for advanced training for skilled surgeons during actual surgery. In endoscopic surgery, this holography-guided spatial navigation system based on patient-specific CT/MRI improved surgeons' spatial recognition for overcoming limitations of visual field and forceps movement (Fig. 52.6).

Fig. 52.3 Virtual reality using a standalone headset

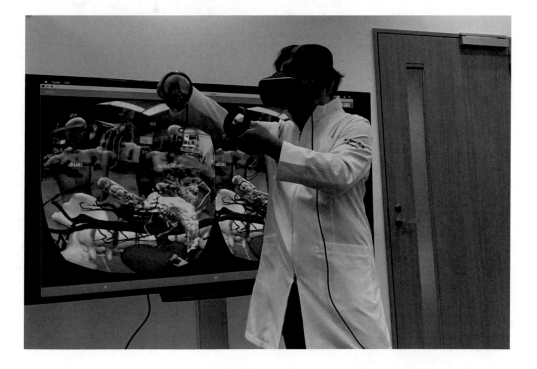

Fig. 52.4 Mixed reality in prostate cancer surgery using wearable holographic computer (HoloLens2)

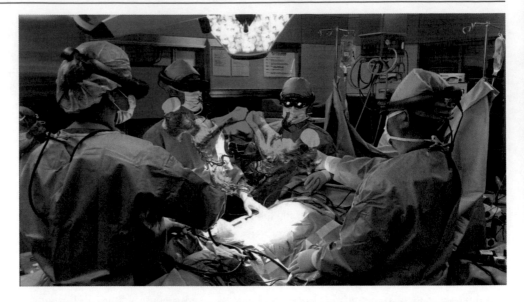

Fig. 52.5 Mixed reality for hepatic surgery using wearable holographic computer (Magic Leap 1)

52.3.2 Online Telemedicine and Remote Space Sharing Using Cloud XR and 5G

In recent years, online medicine has been attracting much attention in Japan because the revision of the national medical payment system was revised in 2020 and the scope of reimbursement has expanded. At the same time, the expansion of COVID-19 has become a social problem, and medical services have been converted to contactless and remote services. However, there are a lot of institutions that can apply only telephone correspondence without Internet or video chat. Under such a situation, voice-only information may lead to miscommunication in diagnosis that requires face-to-face or palpation.

The fifth generation of mobile technologies (5G) is expected to connect people, things, data, and applications in

smart networked communication environments. It should transport a huge amount of data much faster, reliably connect an extremely large number of devices, and process very high volumes of data with minimal delay. 5G is promising to be a ubiquitous force in our daily lives. One such area that has experienced several recent 5G developments in the medical field, where experts are changing the way doctors do their jobs and how patients receive care.

5G is 20 times faster than 4G. Its latency is 1/tenth of that of 4G, and it is said to be able to provide many simultaneous connections with ten times the number of connections at the same time. This is expected to advance online medical care and emergency services and reduce the disparity in the quality of care.

5G and cloud XR bring together cloud-based technologies and XR to deliver superior experiences that revolution-

Fig. 52.6 Mixed reality and surgical guide in spine surgery using wearable holographic computer (HoloLens2)

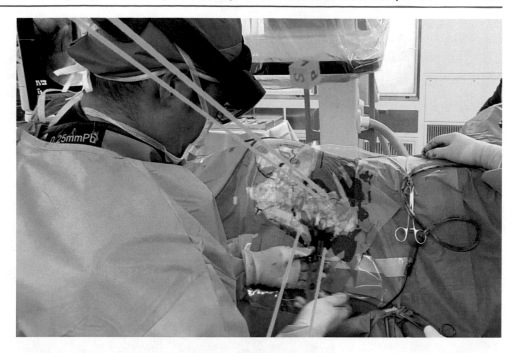

Fig. 52.7 Virtual session in real-time for telemedicine

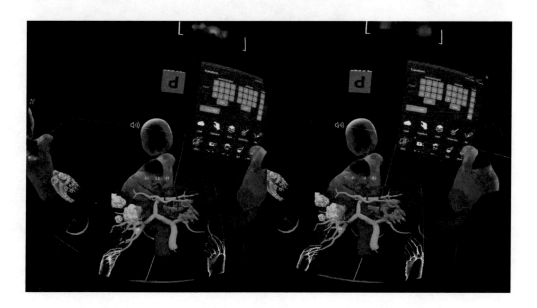

ize the use of content in both the doctors and patients. The system can not only transmit medical information of patients but also capture the motion of doctors and patients by sensing their movements and make them appear as avatars in the space.

Medical images and information on medical records are shared simultaneously with the movements of multiple people and spatial backgrounds, and a high degree of freedom and natural and intuitive communication, which is not possible with a flat-screen, can be realized in cloud XR (Fig. 52.7).

On the other hand, there are some problems such as cost, safety, and risk due to multiple simultaneous connections, and the need to learn new equipment and technology.

However, the expansion with 5G and XR using cloud technology will more quickly increase medical care coverage. The introduction of 5G with cloud XR is expected to improve the way of working of medical staff by solving the problems caused by the shortage of doctors and medical facilities, uneven distribution of doctors, medical disparity, and other issues (Fig. 52.8).

Fig. 52.8 The realistic avatar-based 3D visualization system for telemedicine

Fig. 52.9 VR education system using a smartphone and a 3D VR headset constructed of cardboard

52.3.3 Medical Education and Training Through XR

On-the-job training (on-JT) is widely used in medical education and training, which takes place in the actual clinical setting alongside their job. Depending on the opportunity for experience, the number of cases, time limitations, and facilities, there is a difference in the skills that can be learned, and the accuracy of those skills. In the clinical setting, patient safety and medical accuracy have the highest priority, on-JT has its limits in actual practice. There is a lot of tacit knowledge based on experience. Therefore, XR technology is used for off-the-job training (off-JT) to accurately reproduce the patient's condition and diagnostic and therapeutic techniques.

In the actual clinical situation, medical images of individual patients, movements of limbs, surgical procedures, and surgical plans can be reproduced and memorized along with visual information in virtual space through XR. In this way, the techniques of skilled doctors are quantified in terms of time and space. Medical technology can be formalized and handed down easily through XR (Fig. 52.9).

52.4 Prospects for Cloud XR Medicine

In the field of medical XR, there is a demand for enriched content, low-cost devices, high functionality, high variety, lightweight, small size, and simple operation. Furthermore, it is expected to be versatile enough to be used not only in medical facilities but also in the health care field. For this purpose, it is necessary to enhance the infrastructure, literacy, and environment of the equipment such as networks and devices. It is expected that the digital transformation will be promoted with an emphasis on corporate governance and compliance, while thoroughly protecting the security of the collection, management, and utilization of medical big data such as patient information (Fig. 52.9).

52.5 Conclusion

We developed a patient-specific cloud-based XR surgical navigation. This is highly effective in improving spatial recognition for surgeons in any kinds of surgery. These results are expected to share the technical gap between surgeons by formalizing the tacit knowledge of surgical techniques. This can further develop "Precision Medicine" into "Precision Surgery," which selects individual analysis and establishes the optimal treatment for each specific population.

References

1. Sugimoto M. Augmented tangibility surgical navigation using spatial interactive 3-D Hologram zSpace with OsiriX and bio-texture 3-D organ modeling. IEEE 2015 International Conference on Computer Application Technologies (CCATS 2015). 2015. p. 189–194. https://doi.org/10.1109/CCATS.2015.53
2. Iannessi A, Marcy PY, Clatz O, Bertrand AS, Sugimoto M. A review of existing and potential computer user interfaces for modern radiology. Insights Imaging. 2018;9:599–609. https://doi.org/10.1007/s13244-018-0620-7.
3. Saito Y, Sugimoto M, Imura S, Morine Y, Ikemoto T, Iwahashi S, et al. Intraoperative 3D Hologram support with mixed reality techniques in liver surgery. Ann Surg. 2020;271:e4–7. https://doi.org/10.1097/SLA.0000000000003552.
4. Sato Y, Sugimoto M, Tanaka Y, Suetsugu T, Imai T, Hatanaka Y, et al. Holographic image-guided thoracoscopic surgery: possibility of usefulness for esophageal cancer patients with abnormal artery. Esophagus. 2020; https://doi.org/10.1007/s10388-020-00737-6.
5. Yoshida S, Sugimoto M, Fukuda S, Taniguchi N, Saito K, Fuji Y. Mixed reality computed tomography-based surgical planning for partial nephrectomy using a head-mounted holographic computer. Int J Urol. 2019;26:681–2. https://doi.org/10.1111/iju.13954.
6. Holoeyes Inc. Web site. http://holoeyes.jp/en. Accessed 22 Nov 2020.

Smart Cyber Operating Theater (SCOT): Strategy for Future OR

53

Yoshihiro Muragaki, Jun Okamoto, Ken Masamune, and Hiroshi Iseki

Abstract

Digitization or digital transformation (DX) is finally progressing in the field of surgery. Historically, digitization of surgical record images from videotapes to ROMs and hard disks was regarded as DX. More recently, robots have begun to perform surgery via digital signals and artificial intelligence (AI) analysis of surgical procedures as part of DX. However, it is a system in which only the input (imaging diagnosis) and the output (surgical robot) are digital. The analog is mixed. Since mixed analog information poses a barrier, maximum efficiency cannot be expected. This type of DX is far from the connected machines featured in the internet of things (IoT).

We have made various developments in neurosurgery to improve the effectiveness of glioma resection and reduce its complications. The basis of this is DX. The process involves a surgical operation or surgical method that can make judgments based on objective and reproducible digital information, rather than analog judgments based on subjectivity and experiences. This journey started in 1995 with the development of navigation and has since led to the development of an intelligent operating room where intraoperative MRI can be performed (2000) and a smart treatment room in which medical equipment in the operating room is connected by a network (2016). Through this process, the IoT was realized. With the introduction of IoT, the smart treatment room has become a "medical device" that diagnoses and performs treatment as a single device, unlike the conventional operating room which only provides a sterilized space.

The smart treatment room digitizes almost all the information necessary for surgical decision-making and aims to analyze the currently accumulated digital data, make predictions using AI, and perform treatment using new robotic equipment. In this way, a digitized treatment system inclusive of all phases of input, analysis, and output is made possible. The room supports the decision-making of the surgeon by displaying the three-axis information of multi-dimensional computational anatomy in an integrated manner in time synchronization.

In this chapter, we report the development process of the smart treatment room SCOT (Smart Cyber Operating Theater: SCOT) and the actual conditions of the three types of SCOT. The future SCOT will also be described. The SCOT was developed with the support of the Japan Agency for Medical Research and Development (AMED).

Keywords

Intraoperative imaging · Internet of things · Digital transformation (DX) · Brain tumor · Artificial intelligence

53.1 Introduction

The operating room is a place that provides a space for performing surgical procedures that currently require sterilization. In addition to basic surgical equipment, necessary equipment is brought in according to the department and the type of surgery being performed. Moreover, even for the same function, different surgeons may use different models and have specific preferences. Therefore, a wide variety of surgical medical devices are stocked. In addition, various new intraoperative diagnostic and treatment devices have been introduced into modern surgery. Some surgeries require more equipment. The situation is further exacerbated by the potential risk of a mixture of old and new surgical equipment. At the time of the 2014 survey, 747 operating room medical devices were in stock at our hospital.

In 2013, Weerakkody et al. reviewed and analyzed a paper that quantitatively evaluated operating room safety [1], and reported that the average number of "errors" in one proce-

Y. Muragaki (✉) · J. Okamoto · K. Masamune · H. Iseki
Faculty of Advanced Techno-Surgery, Institute of Advanced Biomedical Engineering and Science, Tokyo Women's Medical University, Tokyo, Japan
e-mail: ymuragaki@twmu.ac.jp; okamoto.jun@twmu.ac.jp; masamune.ken@twmu.ac.jp

dure was as high as 15.51. Of these, 23.5% were due to malfunctions or failures of the equipment or technology, 37% did not have the necessary equipment or equipment, 43% had a combination of setting errors, and 34% were related to failures in the equipment itself.

We consider the potential risk of mixing old and new devices to be the main cause of errors in the operating room. To reduce this risk and improve the effectiveness of surgery and procedures, we developed the Smart Cyber Operating Theater (SCOT), with the support of the Japan Agency for Medical Research and Development (AMED).

Unlike the conventional operating room that only provides a sterilized space, the SCOT actually becomes the "medical device" that diagnoses and performs treatment as a single device. In this paper, we will explain and outline each element necessary for its realization, and describe the development of artificial intelligence (AI).

53.2 Smart Cyber Operating Theater (SCOT)

53.2.1 Overview and Development Requirements

The SCOT is different from the conventional operating room that provides a sterilized space because the entire SCOT becomes a "medical device" that performs diagnosis and treatment as a single device. Specifically, the necessary basic equipment is selected and packaged with the intraoperative diagnostic imaging device as the core packaging. The indoor medical equipment which is the component is connected to

the network by the industrial middleware open response interface for networking (ORiN) (networking). The visualized data is then integrated and displayed via the network, and the information necessary for intraoperative decision-making is presented (informationization). With the developed robot, we aim to realize ultra-minimally invasive and highly reproducible precision-guided therapy (robotization). By packaging, the above-mentioned surgical errors and risks will be reduced, and data will be integrated into an information database by the "internet of things (IoT)" that connects reality and cyberspace. Then, by robotizing the device, the overall surgical effect is improved, by integrating the entire process from diagnosis to treatment.

This was a 5-year AMED project which started in 2014. Different SCOTs have been installed and verified for each developmental element. In 2016, we introduced the basic SCOT, which was packaged as equipment, to Hiroshima University, and in 2018, we introduced the standard SCOT, in which all equipment, was networked to Shinshu University. In addition, the robotized hyper-SCOT was installed as a prototype in 2016 (Fig. 53.1), and as a clinical research version in 2019, at Tokyo Women's Medical University.

53.2.2 The Predecessor Intelligent Operating Room and Packaged Basic Smart Treatment Room (Basic SCOT)

The first step in establishing a basic SCOT is device packaging. We have experience in packaging an intelligent operating room that centered on intraoperative MRI [2, 3]. To improve the removal rate of malignant brain tumors, the

Fig. 53.1 High-performance smart treatment room prototype (hyper SCOT). With the intraoperative MRI as the core, all medical devices are connected by a network, and information is displayed on the navigation in time synchronization. It is also equipped with a robotized microscope, a robotized operating table, and a robot that supports the operator's hands. A clinical version was installed to start operation at Tokyo Women's Medical University in 2019

SCOT is used to determine the presence or absence of residual tumors by intraoperative MRI, and the MRI-compatible operating tables, anesthesia machines, surgical microscopes, monitoring devices, etc. are prepared for the actual surgery. All of these had to be integrated. Since 2000, 2023 cases of neurosurgery have been performed, mainly for glioma. In this "classic" SCOT, it is possible to perform surgery based on objectively visible information-information-guided surgery rather than rely on judgment based on conventional experience and intuition. For example, the mapping and motor evoked potential (MEP) by awake surgery is centered on anatomical information from the intraoperative MRI (AIRISII, 0.3 Tesla, Hitachi, Tokyo) and navigation devices updated with intraoperative images. Functional information is used for confirming the preservation of brain function, histological information is used for intraoperative rapid diagnosis, and intraoperative flow cytometry is used to determine whether the tissue is a tumor or its surroundings. Thus, SCOT is a treatment room where malignant brain tumors can be removed using information from 3 of the 4 axes of multi-dimensional computational anatomy below. The three axes are (1) the spatial axis from the cell level to the organ level-anatomical information, (2) the functional axis such as imaging modality, physiology, and metabolism-functional information, and (3) the pathological axis from normal to disease-histological information.

As a result of classic SCOT, the average removal rate of primary glioma was 89%, and the 5-year survival rates for WHO grades 2, 3, and 4, were 89%, 74%, and 18%, respectively.

The classic SCOT packaged intraoperative MRI and MRI-compatible equipment in a single product production. A basic version of SCOT (basic SCOT) was packaged as a system in 2016 for the Hiroshima University Neurosurgery department (Professor Kaoru Kurisu, part-time employee Dr. Taiichi Saito introduced SCOT to the doctors). To date, 23 patients have been operated upon and it has been used not only for brain tumors but also for epilepsy and bone tumors. Furthermore, lateral expansion has also begun. It was also been applied to external projects in several facilities, including private hospitals.

53.2.3 Networked Standard SCOT

In a conventional operating room, the devices are not independently connected by a network. The data remains inside the device, and the internal clock of the devices differ from each other making it extremely difficult to integrate the data. Networking is not possible in an environment where a wide variety of medical devices from many companies are introduced. In contrast, in the basic SCOT, the target model is selected, and information from a variety of sources, but it is

not networked. Therefore, we focused on the ORiN, which is an industrial middleware (managing software between the OS and apps) that connects a large number of robots to a network and controls them efficiently at the factory. If you create a software (provider) that corresponds to a device driver for personal computer peripherals, you can connect to the network without changing the inside of the device and control data input/output and robots. We have developed OPeLiNK (Denso, Aichi), which is a medical middleware for the SCOT project and have connected more than 30 devices using this software so far. This OPeLiNK aims for a global standard. It is a future goal to expand to include ICUs and wards, as well as operating rooms [4].

If each device can be networked, independent information can be integrated and time-synchronized. Moreover, if combined with navigation position information, spatial information can also be added. We have developed a strategy desk system that can display each piece of information alone and also integrate information together. From these, we created an application for removing malignant brain tumors. The MEP value was given as functional information, and the intraoperative flow cytometry value was given as histological information to the operation site on the navigation. As data between the devices are time-synchronized, the operation site where the risk of postoperative paralysis is high and the MEP value is low can be recorded on the navigation system. If the proportion of cells in the proliferative phase is high by flow cytometry, then, the malignancy is high. The identified part can be presented on navigation. The former is the integration of functional and anatomical information, and the latter is the integration of histological and anatomical information.

In 2018, standard SCOT, which is a network of almost all devices by OPeLiNK, was introduced at Shinshu University (Professor Kazuhiro Hongo, Lecturer Dr. Tsuya Goto). In the future, we will explore its effectiveness in clinical research.

53.2.4 Robotized High-performance Smart Treatment Room Hyper-SCOT

The packaged basic version and the networked standard version focus on acquiring and integrating information. In other words, it is the development of equipment and systems that should become the surgeon's new eyes and new brain, but in the future, surgery and procedures will be replaced by new robotized treatment methods as surgeons' new hands. In the hyper-SCOT, we introduced a robotized operating table and microscope and introduced a hand-held robot that supports the surgeon [5–7]. The robotized operating table automatically places the tumor in the center of the operating room. The robot moves the microscope so that the tip of the tool is at the center of the microscope field of view, and the handheld robot reduces the operator's tremors and fatigue. In 2016, we developed a prototype that embodied this idea (Fig. 53.1).

A clinically usable hyper-SCOT was installed at Tokyo Women's Medical University in 2019, and currently, a clinical trial to evaluate its efficacy has been started.

53.3 Internet of Things and Artificial Intelligence Utilization in SCOT

In the networked standard SCOT, a new mechanism called "IoT" will be launched. It is a system in which various "things" (medical devices) themselves are connected, and like the Internet, these "things" exchange information and control each other. Therefore, time-synchronized information is displayed on the map (location information), and congestion information and parking lot availability information are displayed on the navigation map. It becomes possible to support the decision-making of the operator (Fig. 53.2). More advanced surgical decisions require not only intraoperative information but also prognostic information [8]. For example, in the final phase of tumor removal, decision-making on whether or not to perform further removal to improve the survival rate requires historical data, which is the basis for predicting the extension of survival time due to the improvement in the removal rate. Currently, we are constructing a database (data warehouse) that makes it easier to analyze electronic medical record data. Risk maps may also be used to predict complications. It collects records of manipulation sites in the brain when the MEP drops [9], and displays the locations where there were many cases of statistically significant drops.

If prognosis prediction and risk map analysis progress and a large amount of structured data can be accumulated [10], it will be possible to support decision-making using AI in the future, such as machine learning and deep learning. We have also begun research on AI and have succeeded in predicting when the white blood cell count will drop the most in anti-cancer drugs [11]. We also created a risk map of areas that are likely to cause higher brain functions.

Furthermore, hyper-SCOT aims to use IoT to operate medical devices themselves via a network, similar to operating home appliances remotely from mobile phones. At first, the shadowless lights are turned on and off, and the operating bed is moved, but in the future, it will be possible to operate robotized treatment equipment.

In the operating room, the SCOT could realize the world that Industry 4.0 strives for, which is to operate the real world better by closely linking the real world with the network of various sensors and the high computer power of cyberspace. We believe that the SCOT will revolutionize the therapeutic world of Medicine 4.0.

53.4 Summary

We advocate precision-guided therapy as the goal of surgery in the twenty-first century. It integrates and analyzes various visualized information (surgeon's new eyes) as a strategy to support decision-making (surgeon's new brain) and performs ultra-minimally invasive treatment with new robotized treatment equipment (surgeon's new eyes and new hands). The SCOT is the place to perform this precision-guided treatment.

The SCOT not only performs surgery and treatment centered on intraoperative MRI for parenchymal organs such as

Fig. 53.2 Strategy desk consolidating time-synchronizing data tagged with location. Most medical devices are connected by network (OPeLiNK). Time-synchronized information is displayed on the map (location information). Values of motor evoked potentials (pink and red) and rapid flow cytometry (blue) were tagged with navigation. It becomes possible to support the decision-making of the operator

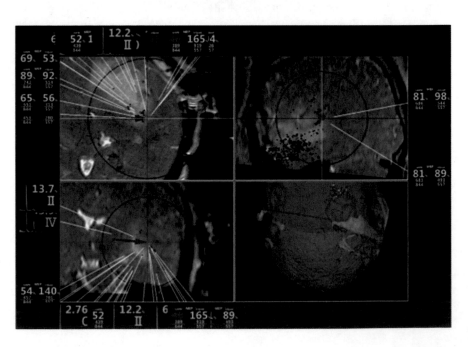

malignant brain tumors, but also intravascular surgery and treatment for vascular lesions, and luminal organs such as the stomach and intestines. It can be applied to targeted surgery and treatment. In addition, OPeLiNK, which can connect medical devices from different companies, has the potential to spread not only to operating rooms, but also to ICUs, wards, and hospitals as a whole. International standardization will be key in the future [12].

We chose to refer to "a smart treatment room" instead of "a smart operating room" because this treatment room should be a single treatment device where doctors perform all invasive procedures and treatments, not just surgery. We hope that the doctors from each specialty will be able to extract the diseases effectively in the SCOT and discuss with us what kind of information are necessary to optimize the treatment; and in so doing, expand various lateral developments.

Acknowledgments This SCOT project is supported by the medical device/system research and development project "Development of a smart treatment room that achieves both safety and improvement of medical efficiency" that realizes AMED future medical care. We would like to express our gratitude to Professor Masaki Kitajima, Program Supervisor of this project; Professor Kenjun Dohi, and Professor Masatake Kitano, Program Officers, for their guidance and guidance; and to the 133 registered researchers from 12 companies in five universities for their research execution. In addition, we would like to thank Dr. Kintomo Takakura, former president of Tokyo Women's Medical University, Dr. Tomokatsu Hori, former Professor of Neurosurgery, who had the opportunity to project the intelligent operating room, which is the predecessor of the smart treatment room. Moreover, the Ministry of Economy, Trade and Industry, and national research bodies must be thanked for supporting the project with research grants. Moreover, we would like to express our deep gratitude to the New Energy and Industrial Technology Development Organization (NEDO). In clinical surgery for malignant brain tumors, we would like to show our appreciation to Dr. Takashi Maruyama, Dr. Masayuki Nitta, and Dr. Taichi Saito, of Tokyo Women's Medical University Neurosurgery, and in the Smart Treatment Room Business, we are grateful to the Tokyo Women's Medical University Institute of Advanced Biomedical Sciences. Furthermore, we would like to express our deep gratitude to Dr. Yuki Horise, Dr. Kaori Kusuda, Dr. Mikio Izumi, Dr. Satoko Ikuta, and Dr. Michael Chernov, for their contributions.

References

1. Weerakkody RA, Cheshire NJ, Riga C, et al. Surgical technology and operating-room safety failures: a systematic review of quantitative studies. BMJ Qual Saf. 22:710–8. https://doi.org/10.1136/bmjqs-2012-001778. Epub 2013 Jul 25, 2013.
2. Muragaki Y, Iseki H, Maruyama T, et al. Information-guided surgical management of gliomas using low-field-strength intraoperative MRI. Acta Neurochir Suppl. 2011;109:67–72.
3. Muragaki Y, Iseki H, Maruyama T, et al. Usefulness of intraoperative magnetic resonance imaging for glioma surgery. Acta Neurochir Suppl. 2006;98:67–75.
4. Okamoto J, Masamune K, Iseki H, et al. Development concepts of a Smart Cyber Operating Theater (SCOT) using ORiN technology. Biomed Tech (Berl). 2018;63:31–7. https://doi.org/10.1515/bmt-2017-0006.
5. Goto T, Hongo K, Ogiwara T, et al. Intelligent Surgeon's arm supporting system iArmS in microscopic neurosurgery utilizing robotic technology. World Neurosurg. 119:e661–e665. https://doi.org/10.1016/j.wneu.2018.07.237. Epub 2018 Aug 6, 2018.
6. Okuda H, Okamoto J, Takumi Y, et al: The iArmS robotic armrest prolongs endoscope lens-wiping intervals in endoscopic sinus surgery. Surg Innov. 1553350620929864, 2020.
7. Ogiwara T, Goto T, Fujii Y, et al. Usefulness of a newly developed ultrasonic microdissector in neurosurgery: a preliminary experimental study. J Neurol Surg A Cent Eur Neurosurg. 2019;80:96–101.
8. Fukutomi Y, Yoshimitsu K, Tamura M, et al. Quantitative evaluation of efficacy of intraoperative examination monitor for awake surgery. World Neurosurg. 2019;126:e432–8.
9. Saito T, Muragaki Y, Tamura M, et al: Awake craniotomy with transcortical motor evoked potential monitoring for resection of gliomas in the precentral gyrus: utility for predicting motor function. J Neurosurg. 2019;1–11.
10. Tamura M, Sato I, Maruyama T, et al. Integrated datasets of normalized brain with functional localization using intra-operative electrical stimulation. Int J Comput Assist Radiol Surg. 2019;14:2109–22.
11. Shbahara T, Ikuta S, Muragaki Y' Machine-learning approach for modeling myelosuppression attributed to nimustine hydrochloride. JCO Clin Cancer Inform. 2018; https://doi.org/10.1200/CCI.17.00022.
12. Berger J, Rockstroh M, Schreiber E, et al. GATOR: connecting integrated operating room solutions based on the IEEE 11073 SDC and ORiN standards. Int J Comput Assist Radiol Surg. 2019;14:2233–43.

Supplement

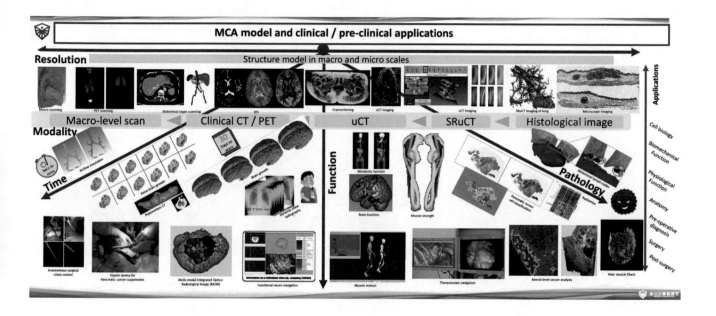

Mathematical Methods and Fundamental Technologies for MCA Modeling and Segmentation

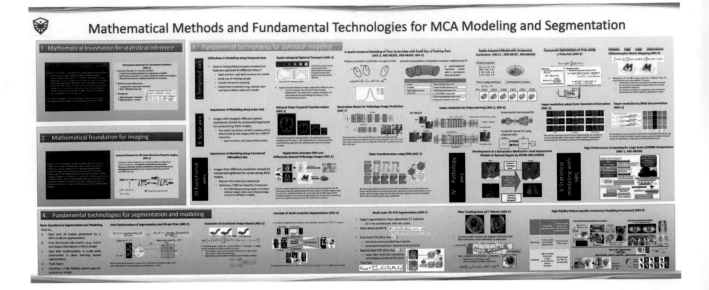

Lifetime MCA Modeling Scenario

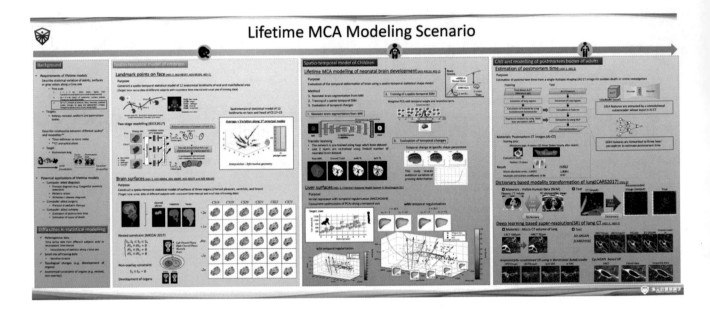

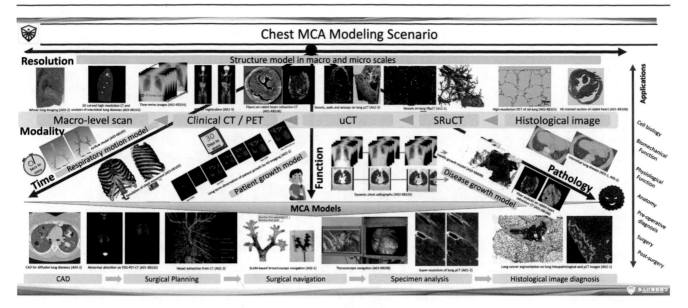

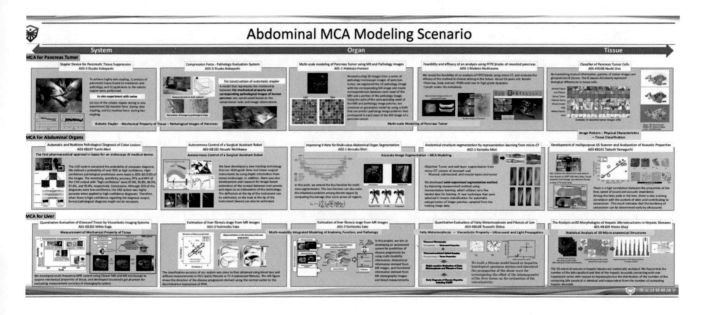

Brain MCA Scenario

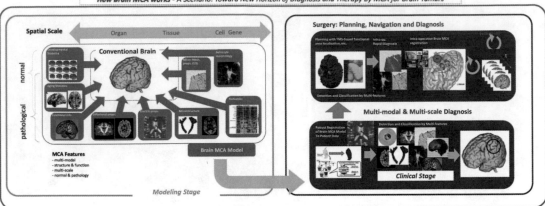

Musculo-skeletal MCA Scenario